Genome Editing and Engineering
From TALENs, ZFNs and CRISPRs to Molecular Surgery

Recent advances in genome editing techniques using endonucleases, such as TALENs and CRISPRs, combined with genomic engineering technologies have opened up a wide range of opportunities from applications in basic disease biology research to potential new diagnostic tools and clinical applications.

This complete guide to endonuclease-based genomic engineering gives readers a thorough understanding of this rapidly expanding field. Chapters cover the discovery, basic science, and application of these techniques, focusing particularly on their potential application to the treatment of cancer, and cardiovascular and immunological disease. The final section discusses the legal and ethical issues that accompany the technology. Providing authoritative coverage of the potential that genome editing and engineering carry, this is an ideal reference for researchers and graduate students and those working in the biotechnology and pharmaceutical industries, as well as in a clinical setting.

Krishnarao Appasani is the Founder and Chief Executive Officer of GeneExpression Systems. He is an award-winning scientist, and has edited several books, including: *Optogenetics: From Neuronal Function to Mapping and Disease Biology* (2017), *Genome-Wide Association Studies: From Polymorphism to Personalized Medicine* (2016), *Epigenomics: From Chromatin Biology to Therapeutics* (2012), *MicroRNAs: From Basic Science to Disease Biology* (2007) and *RNA Interference: From Basic Science to Drug Development* (2005).

Genome Editing and Engineering

From TALENs, ZFNs and CRISPRs to Molecular Surgery

Edited by

Krishnarao Appasani
GeneExpression Systems, Inc., Waltham, Massachusetts, USA

With a foreword by

George M. Church
Harvard University Medical School, Boston, Massachusetts, USA

CAMBRIDGE
UNIVERSITY PRESS

CAMBRIDGE
UNIVERSITY PRESS

University Printing House, Cambridge CB2 8BS, United Kingdom

One Liberty Plaza, 20th Floor, New York, NY 10006, USA

477 Williamstown Road, Port Melbourne, VIC 3207, Australia

314-321, 3rd Floor, Plot 3, Splendor Forum, Jasola District Centre, New Delhi - 110025, India

79 Anson Road, #06-04/06, Singapore 079906

Cambridge University Press is part of the University of Cambridge.

It furthers the University's mission by disseminating knowledge in the pursuit of education, learning and research at the highest international levels of excellence.

www.cambridge.org
Information on this title: www.cambridge.org/9781107170377
DOI: 10.1017/9781316756300

First published 2018

A catalogue record for this publication is available from the British Library

Library of Congress Cataloging in Publication data
Names: Appasani, Krishnarao, 1959– editor.
Title: Genome editing and engineering: from TALENs, ZFNs and CRISPRs to molecular surgery / edited by Krishnarao Appasani.
Description: Cambridge, United Kingdom; New York, NY: Cambridge University Press, 2018. | Includes bibliographical references.
Identifiers: LCCN 2017051165 | ISBN 9781107170377 (hardback)
Subjects: | MESH: Gene Editing | Transcription Activator-Like Effector Nucleases | Zinc Fingers | Clustered Regularly Interspaced Short Palindromic Repeats
Classification: LCC QH440 | NLM QU 550.5.G47 | DDC 576.5072–dc23
LC record available at https://lccn.loc.gov/2017051165

ISBN 978-1-107-17037-7 Hardback
ISBN 978-1-316-62146-2 Paperback

Dedicated to

Phillip A. Sharp (1944–) 1993 Nobel Laureate Medicine or Physiology and Institute Professor at the Koch Institute for Integrative Cancer Research, Massachusetts Institute of Technology, Cambridge, Massachusetts.

Phil is an American-born legendary molecular biologist, co-discoverer of split genes and RNA splicing, an entrepreneur and futurist, who foresaw the new field of RNA interference and genome editing.

Contents

The color plates are found between pages 194 and 195.

Contributors

Ramesh Akkina
Department of Microbiology, Immunology and Pathology
Colorado State University
Fort Collins, CO, USA

Krishnarao Appasani
GeneExpression Systems, Inc.
Waltham, MA, USA

Raghu Kiran Appasani
The MINDS Foundation,
Boston, MA, USA

Amanda Freire Assis
Department of Genetics
Ribeirão Preto Medical School
University of São Paulo (USP)
Ribeirão Preto, São Paulo, Brazil

Russell Bainer
Departments of Bioinformatics
Genentech Inc.,
South San Francisco, CA, USA

Rasmus Otkjaer Bak
Aarhus Institute of Advance Studies and Department of Biomedicine
Aarhus University
Aarhus, Denmark

Andrew R. Bassett
Cellular Operations
Wellcome Trust Sanger Institute
Hinxton, UK

Mark A. Behlke
Integrated DNA Technologies, Inc.
Coralville, IA, USA

Eric P. Bennett
Department of Odontology
Copenhagen Center for Glycomics
Copenhagen, Denmark

Martin F. Berthelsen
Department of Clinical Medicine
Faculty of Health
Aarhus University
Aarhus, Denmark

Andrew C. Birkeland
Department of Otolaryngology
University of Michigan School of Medicine
Ann Arbor, MI, USA

Ira L. Blitz
Department of Developmental and Cell Biology
University of California
Irvine, CA, USA

J. Chad Brenner
Department of Otolaryngology
University of Michigan School of Medicine
Ann Arbor, MI, USA

Alexander Brown
Department of Bioengineering
University of Illinois at Urbana-Champaign
Urbana, IL, USA

Marinella Callow
Departments of Discovery Oncology
Genentech Inc.,
South San Francisco, CA, USA

Jia Jia Chan
Cancer Science Institute of Singapore
National University of Singapore
Singapore

Sarah Chan
Usher Institute for Population Health Sciences and Informatics
University of Edinburgh
Edinburgh, UK

Paige Charlins
Department of Microbiology, Immunology and Pathology
Colorado State University
Fort Collins, CO, USA

Shreevathsa S. Chilkunda
Max Delbrück Center for Molecular Medicine in the Helmholtz Association
Berlin, Germany

Vincenzo Corbo
Department of Diagnostic and Public Health
University of Verona
Verona, Italy

Mike Costa
Departments of Discovery Oncology
Genentech Inc.
South San Francisco, CA, USA

Larissa Cotrim-Sousa
Department of Genetics
Ribeirão Preto Medical School
University of São Paulo (USP)
Ribeirão Preto, São Paulo, Brazil

Trinna L. Cuellar
Department of Molecular Biology
Genentech Inc.
South San Francisco, CA, USA

Ruby Emily Dawson
Department of Molecular and Cellular Biology
School of Biological Sciences
University of Adelaide
Adelaide, Australia

Samantha N. Devenport
Department of Otolaryngology
University of Michigan School of Medicine
Ann Arbor, MI, USA

Daniel Patrick Dever
Department of Pediatrics
Stanford University
Stanford, CA, USA

Behnom Farboud
Department of Molecular and Cell Biology
University of California
Berkeley, CA, USA

Mark W. Feinberg
Cardiology Division
Brigham and Women's Hospital
Harvard Medical School
Boston, MA, USA

Rafaela Martins Felicio
Department of Genetics
and Graduate Program in Basic and Applied Immunology
Ribeirão Preto Medical School
University of São Paulo (USP)
Ribeirão Preto, São Paulo, Brazil

Quentin R.V. Ferry
Weatherall Institute of Molecular Medicine
Radcliffe Department of Medicine
University of Oxford
Oxford, UK

Morten Frödin
Biotech Research & Innovation Center
University of Copenhagen
Copenhagen, Denmark

Hodaka Fujii
Department of Biochemistry and Genome Biology
Graduate School of Medicine
Hirosaki University
Hirosaki, Japan

Toshitsugu Fujita
Department of Biochemistry and Genome Biology
Graduate School of Medicine
Hirosaki University
Hirosaki, Japan

Tudor Fulga
Weatherall Institute of Molecular Medicine
Radcliffe Department of Medicine
University of Oxford
Oxford, UK

Michael Gapinske
Department of Bioengineering
University of Illinois at Urbana-Champaign
Urbana, IL, USA

Laurent Grant
Department of Bioengineering
University of Illinois at Urbana-Champaign
Urbana, IL, USA

Stefan Haemmig
Cardiology Division
Brigham and Women's Hospital
Harvard Medical School
Boston, MA, USA

Benjamin Haley
Department of Molecular Biology
Genentech Inc.
South San Francisco, CA, USA

Cesar Speck Hernandez
Department of Genetics
and Graduate Program in Basic and Applied Immunology
Ribeirão Preto Medical School
University of São Paulo (USP)
Ribeirão Preto, São Paulo, Brazil

Rebecca C. Hoesli
Department of Otolaryngology
University of Michigan School of Medicine
Ann Arbor, MI, USA

Wendy J.M. Huang
Department of Celluar and Molecular Medicine
University of California, San Diego
La Jolla, CA, USA

James Nicholas Hughes
Department of Molecular and Cellular Biology
School of Biological Sciences
University of Adelaide
Adelaide, SA, Australia

Tetsuya Ishii
Life Science Section
Office of Health and Safety
Hokkaido University
Sapporo, Japan

Zoltán Ivics
Division of Medical Biotechnology
Paul Ehrlich Institute
Langen, Germany

Zsuzsanna Izsvák
Max Delbrück Center for Molecular Medicine in the Helmholtz Association
Berlin, Germany

Ashley M. Jacobi
Integrated DNA Technologies, Inc.
Coralville, IA, USA

Vinod Jaskula-Ranga
Wilmer Eye Institute
Johns Hopkins University School of Medicine
Baltimore, MD, USA

Yasuhiko Kato
Department of Biotechnology, Graduate School of Engineering
Osaka University, Suita, Osaka, Japan and
Frontier Research Base for Global Young Researchers
Graduate School of Engineering, Osaka University
Suita, Japan

Justin B. Kinney
Cold Spring Harbor Laboratory
Cold Spring Harbor, NY, USA

Aditi S. Kulkarni
Department of Otolaryngology
University of Michigan School of Medicine
Ann Arbor, MI, USA

ZhiHao Kwok
Cancer Science Institute of Singapore
Centre for Translational Medicine
National University of Singapore
Singapore

Pranav V. Lalgudi
Department of Genetics
Stanford University School of Medicine
Stanford, CA, USA

Ted Lau
Departments of Discovery Oncology
Genentech Inc.
South San Francisco, CA, USA

Te-Wen Lo
Department of Biology
Ithaca College
Ithaca, NY, USA

Megan L. Ludwig
Department of Otolaryngology
University of Michigan School of Medicine
Ann Arbor, MI, USA

Yonglun Luo
Faculty of Health
Department of Biomedicine
Aarhus University
Aarhus, Denmark

Jaqueline E. Mann
Department of Otolaryngology
University of Michigan School of Medicine
Ann Arbor, MI, USA

Nicole M. Michmerhuizen
Department of Otolaryngology
University of Michigan School of Medicine
Ann Arbor, MI, USA

Madison Miller
Cold Spring Harbor Laboratory
Cold Spring Harbor, NY, USA

Takashi Nakanishi
Department of Biotechnology
Graduate School of Engineering
Osaka University
Suita, Osaka, Japan

Suneel A. Narayanavari
Mobile DNA Group
Max Delbrück Center for Molecular Medicine in the Helmholtz Association
Berlin, Germany

Francesco Niola
Biotech Research & Innovation Center
University of Copenhagen
Copenhagen, Denmark

Chad O'Gorman
Animal Science Research Center
Division of Animal Sciences
University of Missouri
Columbia, MO, USA

Ernna Hérida Oliveira
Department of Genetics
Ribeirão Preto Medical School
University of São Paulo (USP)
Ribeirão Preto, São Paulo, Brazil

Geraldo A. Passos
Department of Genetics
Ribeirão Preto Medical School
and Department of Morphology, Physiology and Basic Pathology
School of Dentistry of Ribeirão Preto
University of São Paulo (USP)
Ribeirão Preto, São Paulo, Brazil

Pablo Perez-Pinera
Department of Bioengineering
University of Illinois at Urbana-Champaign
Urbana, IL, USA

Matthew H. Porteus
Stanford University, School of Medicine
Department of Pediatrics
Stanford, CA, USA

Randall S. Prather
Animal Science Research Center
Division of Animal Sciences
University of Missouri
Columbia, MO, USA

Vaishnavi Raghunathan
Mobile DNA Group
Max Delbrück Center for Molecular Medicine in the Helmholtz Association
Berlin, Germany

Bethany K. Redel
Animal Science Research Center
Division of Animal Sciences
University of Missouri
Columbia, MO, USA

Garrett R. Rettig
Integrated DNA Technologies, Inc.
Coralville, IA, USA

Zhili Rong
Cancer Research Institute
Southern Medical University
Guangzhou, Guangdong, P.R.China

Nalani Sachan
Cold Spring Harbor Laboratory
Cold Spring Harbor, NY, USA

Serif Senturk
Dokuz Eylul University
Izmir International Biomedicine and Genome Institute
Izmir Province, Turkey

James L. Sherley
Director
Asymmetrex, LLC
Boston, MA, USA

Nitin H. Shirole
Cold Spring Harbor Laboratory
Cold Spring Harbor, NY, USA and
Graduate Program in Genetics
Stony Brook University
Stony Brook, NY, USA

Wilson Araújo Silva Jr.
Department of Genetics
Ribeirão Preto Medical School
University of São Paulo (USP)
Ribeirão Preto, São Paulo, Brazil

Raffaella Sordella
Cold Spring Harbor Laboratory
Cold Spring Harbor, NY, USA and
Graduate Program in Genetics
Stony Brook University
Stony Brook, NY, USA

Rodney L. Sparks
University of Virginia Licensing and Ventures Group
Charlottesville, VA, USA

Paul Quinton Thomas
Laboratory of Developmental Genetics
Department of Molecular and Cellular Biology
University of Adelaide
Adelaide, Australia

Martin K. Thomsen
Faculty of Health
Department of Clinical Medicine
Aarhus University
Aarhus, Denmark

Ella Paulina Thomson
Department of Molecular and Cellular Biology
School of Biological Sciences
University of Adelaide
Adelaide, Australia

Yvonne Tay
Department of Biochemistry
Yong Lo Lin School of Medicine
National University of Singapore
Cancer Science Institute of Singapore
Singapore

Lorena Alves Teixeira
Department of Genetics
Ribeirão Preto Medical School
University of São Paulo (USP)
Ribeirão Preto, Brazil

Valéria Valente
Department of Clinical Analysis
School of Pharmaceutical Sciences
UNESP State University
Araraquara, SP, Brazil

Chunxin Wang
Surgical Neurology Branch
National Institute of Neurological Disorders and Stroke
Bethesda, MD, USA

Colin Watanabe
Departments of Bioinformatics
Genentech Inc.
South San Francisco, CA, USA

Hajime Watanabe
Department of Biotechnology
Graduate School of Engineering
Osaka University
Suita, Japan

Philip Webber
Dehns Patent and Trademark Attorneys
Oxford, UK

Monte M. Winslow
Department of Genetics
Stanford University School of Medicine
Stanford, CA, USA

Ian P. Winters
Department of Genetics
Stanford University School of Medicine
Stanford, CA, USA

Kristin M. Whitworth
Animal Science Research Center
Division of Animal Sciences
University of Missouri
Columbia, MO, USA

Qianxin Wu
Weatherall Institute of Molecular Medicine
Radcliffe Department of Medicine
University of Oxford
Oxford, UK

Yingjie Wu
Institute of Genome Engineered
Animal Models for Human Diseases,
Dalian Medical University, Dalian
Liaoning, P.R. China

Youjun Wu
Department of Diabetes and Metabolic Diseases Research
Beckman Research Institute
City of Hope National Medical Center
Duarte, CA, USA

Yang Xu
Division of Biological Sciences
University of California
San Diego, CA, USA and
Cancer Research Institute
Southern Medical University
Guangzhou, Guangdong, P.R. China

Jiing-Kuan Yee
Department of Diabetes and Metabolic Diseases Research
Beckman Research Institute
City of Hope National Medical Center
Duarte, CA, USA

Foreword

I am thankful to Krishnarao Appasani for making the extraordinary effort to prepare this timely volume on the rapidly expanding field of genome editing and engineering, and for inviting me to offer a few of my own initial statements.

We live in exponentially exciting times. Times in which we can design and build new versions of industrial microbes, agricultural plants, animals and even ourselves. Our ability to sequence and synthesize DNA has improved over four-millionfold in just a few years, such that trillions of base pairs are now read and written by automated chip-based systems. The cost of editing a single base pair has improved about tenfold in that same time frame. It is important to keep in mind that the revolution is just as much about sequencing and synthesis as it is about editing. Without facile reading of whole genomes, we would lack key tools that permit avoidance of off-target editing. Without the synthesis of DNA on chips of large libraries of barcodes, guide-RNAs or sequence-specific DNA-binding protein domains, we would not be well positioned to conduct large screens for function.

When we speak of genome engineering, this is not just about one type of endonuclease, so it is refreshing to see representation in this book of meganucleases, zinc-finger nucleases, TALENs and CRISPRs. Moreover, genome engineering is not simply about nucleases, but other editing methods, especially those which intrinsically avoid the curse of non-homologous end-joining (NHEJ). These include phage-derived integrases/recombinases (such as lamba–int, cre, Bxb1), MAGE recombineering and adeno-associated virus (AAV) homologous recombination. In addition to increasing the on-target precision, these non-nuclease methods can reduce the toxicity of double-strand DNA breaks and hence increase the multiplicity. To put this in perspective, the record for multiplex edits per genome is 62 for mammalian cells, 25 for a whole living mammal (pig), and 321 for a bacterium (recoded *E. coli*). New efforts have begun, collectively called "Genome Project Write" (GPW), which involve 62 000 edits (*E. coli*) and 400 000 edits (human genome) to make cells resistant to all viruses. This indicates how crucial technology development will be in the near future for editing, screening, testing and debugging of complex functional interactions. So, it is noteworthy that this text contains over a half dozen chapters on these technology development topics.

This volume covers the full spectrum from model organisms to clinical applications. Organisms include representatives of crustaceans, nematodes, fish, frogs, mice, pigs and humans. Naturally, within experiments on human cells we have additional topics well resourced in this book, including pluripotent stem cells, organoids, transplantation, cancer and regenerative biology, modeling and treating disease biology. Thankfully, in this context the final four chapters cover the very pressing and timely issues of patents, safety and ethical concerns of genome editing.

This comprehensive book is welcome at this point of rapid growth in new technologies and discoveries, such that keeping up with applications of genome engineering is challenging. Here we have both a general introduction to the field and to deep mechanisms and glimpses of future applications and ethical considerations – with gems for newcomers and experts alike. It is a great pleasure to participate in this grand experiment of sharing protocols, reagents and viewpoints on how to proceed responsibly.

March 21, 2018

George M. Church, PhD
Professor of Genetics
Harvard University Medical School and
Massachusetts Institute of Technology
Boston, Massachusetts, USA

Preface

Science knows no country, because knowledge belongs to humanity, and is the torch which illuminates the world; and in the fields of observation chance favors only the prepared mind.

– *Louis Pasteur*, a French microbiologist (1822–1895) who discovered the principles of microbial fermentation and pasteurization

The famous Austrian British philosopher of science Karl R. Popper (1902–1994) believed that "before we can find the answers, we need the power to ask new questions, in other words, we need new technology." A key example of such a technological advance is the recent development of the new field of genome editing and/or genome engineering developed by several pioneers. Genome editing can be defined as "a powerful new tool for making precise additions, deletions, and alterations to the genome – an organism's complete set of genetic material." The development of efficient and reliable ways to make precise, targeted changes to the genome of living cells has been a long-standing goal for biomedical scientists. To reach this goal, researchers have utilized three different approaches in recent years: (1) zinc-finger nucleases (ZFNs), (2) transcription-activator like effector nucleases (TALENs) and (3) clustered regularly interspaced short palindromic repeats (CRISPRs).

Both ZFN and TALEN technology platforms have enabled researchers to generate permanent mutations by introducing double-stranded breaks to activate repair pathways. These approaches are relatively costly and time-consuming to engineer, limiting their widespread use for large-scale high-throughput studies. Recently, a new tool based on a bacterial CRISPR-associated protein-9 nuclease (Cas9) from *Streptococus pyogenes* has generated considerable excitement. This follows several attempts over the years to manipulate gene function, including homologous recombination and RNA interference, but they were hampered by providing only temporary inhibition of gene function and unpredictable off-target effects. The first section of this book details the biology of these endonucleases and their regulatory networks.

Genome Editing and Engineering: From Talens, ZFNs and CRISPRs to Molecular Surgery is intended for those in the genetic engineering, molecular agriculture, stem cell biology, biotechnology, genetics, genomics, pharmaco-genomics and molecular medicine fields. There are a few books already available covering genome editing and various endonucleases.[1–7] Most of these books primarily focus on methods and lab protocols, except the recent book by Doudna and Sternberg, which discusses details of the field of genome editing and its bioethical issues and evolutionary perspectives. This present book differs, in that it is the first text completely devoted to combining endonuclease biology and genome editing use in model organisms, technology platforms and applications in studying the pathology of many diseases, as well as stem cell biology. Special emphasis is given to highlight the studies of model organisms, novel screening platforms that have been applied to regenerative and disease biology. This authoritative volume brings together the views of global experts on the use of genome editing in disease research, including cancer, cardiovascular disease, infectious disease and immunity, as well as RNA biology, HIV, retinal diseases and the development of therapeutics. Importantly, this authoritative guide also covers the legal and bioethical issues related to this emerging field of modern biology.

Repeated sequences in *E. coli* were initially discovered by a Japanese group,[8] followed by Francisco Mojica from the University of Alicante, Spain in 1995.[9] In parallel, two French groups have also identified such sequences,[10,11] however its "adaptive immunity function" was demonstrated by Phillippe Horvath and his colleagues.[12] There are many more heroes who have played key roles in the identification of various nucleases, but their potential was realized in 2012 by Jennifer Doudna of the University of California, Berkeley, CA, USA[13] and Emmanuelle Charpentier of the Helmholtz Center for Infection Research, presently at the Max-Planck Institute of Infection Biology, Germany.[14] Later, Feng Zhang's group at MIT, USA also adopted this technique for precise engineering of genomes and developed several reagents, making them public to the research community worldwide.[15] Since then, targeted genome editing has brought about a new revolution in biological research.

Although genome editing arose from microbiology, it addresses a much broader unmet need in the study of biological systems beyond microbiology, in everything from agriculture to embryonic biology. The editing phenomenon has been observed in many model organisms, including bacteria, round worms, flies, fish, mice and humans. The second section of this volume covers genome editing processes in model organisms and the third section focuses on technology development and screening methodologies.

CRISPR/Cas-9 system-related genes are essential in "adaptive immunity" in selected bacteria, the archaea, enabling an organism to respond to and eliminate invading genetic material. The CRISPR/Cas-9 system originates from type II endonucleases, which provide bacteria with adaptive immunity to viruses and plasmids. This is a remarkable technology that can be used to precisely and efficiently target, edit, modify, regulate and mark genomic loci of a wide array of cells and organisms. Genome editing was hailed as the 2011 "Method of the Year" by *Nature Methods*, and in 2015 the CRISPR/Cas9 system was named the

"Breakthrough of the Year" by *Science* magazine. Additionally, a special issue of *Nature* was launched in 2016 with a cover headline reading "CRISPR Everywhere."

Genome engineering with the RNA-guided CRISPR/Cas-9 system in animals and plants is changing the face of biology. It is easier to use and more efficient than other genetic engineering tools, so it is already being applied in several thousand labs all over the world, just a few years after its discovery. The number of scientific publications on this technique also has been skyrocketing. The "CRISPR craze" has spread worldwide and researchers are adopting it for analysis of functional genomics, large-scale screening for drug targets and helping to engineer animal models that will benefit pharmacological studies. These endonucleases (especially CRISPR-Cas9) act like molecular scissors, cutting and replacing DNA letters in an organism's genome with exquisite precision and ease. The technique is revolutionizing the study of species from mice to potatoes, and is likely to open up powerful new avenues in gene therapy for the treatment of human diseases – "genome surgery" is becoming a reality.

Recently genome editing techniques have been used to alter embryos in monkeys and reduce unwanted mutations in the mitochondria in the eggs or embryos of mice. Such applications of genomic editing tools in stem cells and regenerative biology are covered in section four of this volume. Genome editing potentially offers a powerful approach to treat many human diseases, including HIV/AIDS, hemophilia, sickle-cell anemia and several forms of cancer, as we move closer to the clinic. The applications of genome editing and genome engineering techniques have been widely adopted to study disease biology, as detailed in the fifth section of this book. Nearly half a century ago, Bernard Davis, a Harvard microbiologist, envisioned the promise, the risks and the roadmaps for genetic research, including research on making heritable changes in the germline.[16] A note of caution is that genome editing in human embryos using current technologies could have unpredictable effects on future generations, and there are clearly important questions to be answered. The final section of this volume is devoted to cover such bioethical and legal (intellectual property) perspectives. Last year, the "Human Gene-Editing Initiative" was formed in collaboration with the US National Academy of Sciences and the US National Academy of Medicine, by inviting an international panel of scientists to set up a framework and guidelines for the use of genome editing tools in humans, especially relating to assisted reproductive technologies, stem cell therapies, gene transfer and mitochondrial replacement techniques (www.nap.edu/10766). As mentioned by Eric Lander of MIT and Harvard in his seminal review,[17] "the genome-editing discoveries are examples for both 'hypothesis-free' and 'hypothesis-driven' science and the partnership between these two approaches will be seen quite often in this 21st century."

The goal of this book is to serve as a reference for graduate students, postdoctoral researchers and primary investigators, as well as an explanatory analysis for executives and scientists in molecular medicine, molecular engineering, biotechnology and pharmaceutical companies. Our hope is that this volume will serve both as a prologue to the field for newcomers and as a reference for those already active in the field. Most importantly, this book serves as a bridge between the basic science of CRISPR and its diverse applications in areas such as agriculture and biomedicine.

We have carefully selected the chapters, written by experts in the field from both academia and industry, and have divided the chapters into appropriate sections to support the theme expressed in the subtitle of this book: From TALENs, ZFNs and CRISPRs to Molecular Surgery. Developing novel genome editing tools is likely to become a prerequisite for genome engineering studies used to alter the genetic makeup. The elegant and revolutionary genome editing approaches that are covered in this book will undoubtedly have great future commercial promise for the development of innovative molecular surgical procedures and for the study of biomedicine and human disease biology.

Many people have contributed to making our involvement in this project possible. We thank our teachers for their excellent teaching, guidance and mentorship, which have helped us to bring about this educational enterprise. We are extremely grateful to all of the contributors, without whose commitment this book would not have been possible. Many people have had a hand in the preparation of this book. Each chapter has been passed back and forth between the authors for criticism and revision; hence each chapter represents a joint contribution. We thank our reviewers, who have made the hours spent putting together this volume worthwhile. We are indebted to the staff of Cambridge University Press, and in particular to Katrina Halliday for her generosity and efficiency throughout the editing of this book; she truly understands the urgency and need of this volume. We also extend our appreciations to Noah Tate for their excellent cooperation during the development of this volume. We want to thank Professor George Church, an American-born geneticist from the Harvard University Medical School, USA and one of the pioneers in the field of modern genetics and genome engineering, for his kindness in writing a Foreword to this book. Last, but not least, we thank Shyamala Appasani for her understanding and support during the development of this volume.

This book is the ninth in the series *Gene Expression and Regulation* that we have worked on and the fourth joint project between father and son. A portion of the royalties will be contributed to the Dr. Appasani Foundation (a non-profit organization devoted to bringing social change through the education of youth in developing nations) and The MINDS Foundation (Mental Illness and Neurological Diseases), which is committed to taking a grassroots approach to providing high-quality mental healthcare in rural India.

Boston, Massachusetts, USA
February 2018

Krishnarao Appasani, PhD.
Raghu Kiran Appasani, MD.

REFERENCES

1. Witzany G. 2009. *Natural Genetic Engineering and Natural Genome Editing*. Hoboken: Wiley-Blackwell for the Annals of the New York Academy of Sciences.

2. Barrangou R, van der Oost J. 2013. *CRISPR-Cas9 System: RNA-mediated Adaptive Immunity in Bacteria and Archae*. New York: Springer Press.
3. Doudna JA, Sontheimer E. 2014. *The Use of CRISPR/Cas9, ZFNs, TALENs in Generating Site-specific Genome Alterations*. Amsterdam: Elsevier Publications.
4. Lundgren M, Charpentier E, Fineran P. 2015. *CRISPR*. New York: Springer Humana Press.
5. Yamamoto T. 2015. *Targeted Genome Editing Using Site-specific Nucleases*. Tokyo: Springer Verlag Tokyo Press.
6. Kozubek J. 2016. *Modern Prometheus: Editing the Human Genome with CRISPR-Cas9*. Cambridge University Press.
7. Doudna, JA, Sternberg, SH. 2017. *A Crack in Creation: Gene* Editing *and the Unthinkable Power to Control Evolution*. Boston: Houghton Mifflin Harcourt Press.
8. Ishino Y, Shinagawa H, Makino K, Amemura M, Nakata A. 1987. Nucleotide sequence of the iap gene, responsible for alkaline phosphatase isozyme conversion in *Escherichia coli*, and identification of the gene product. *J Bacteriol*. **169**: 5429–5433.
9. Mojica FJM, Díez-Villaseñor C, Soria E, Juez G. 1995. Biological significance of a family of regularly spaced repeats in the genomes of archaea, bacteria and mitochondria. *Mol Microbiol* **9**: 613–621.
10. Pourcel C, Salvignol G, Vergnaud G. 2005. CRISPR elements in *Yersinia pestis* acquire new repeats by preferential uptake of bacteriophage DNA, and provide additional tools for evolutionary studies. *Microbiology* **151**: 653–663.
11. Bolotin A, Quinquis B, Sorokin A, Ehrlich SD. 2005. Clustered regularly interspaced short palindrome repeats (CRISPRs) have spacers of extrachromosomal origin. *Microbiology* **151**: 2551–2561.
12. Barrangou R, Fremaux C, Deveau H. 2007. CRISPR provides acquired resistance against viruses in prokaryotes. *Science* **315**: 1709–1712.
13. Jinek M, Chylinski K, Fonfara I, *et al*. 2012. A programmable dual-RNA-guided DNA endonuclease in adaptive bacterial immunity. *Science* **337**: 816–821.
14. Deltcheva E, Chylinski K, Sharma CM, *et al*. 2011. CRISPR RNA maturation by trans-encoded small RNA and host factor RNase III. *Nature* **471**: 602–607.
15. Cong L, Ran FA, Cox D, *et al*. 2013. Multiplex genome engineering using CRISPR/Cas systems. *Science* **339**: 819–823.
16. Davis BD .1970. Prospects for genetic intervention in man. *Science* **170**: 1279–1283.
17. Lander ES .2016. The heroes of CRISPR. *Cell* **18**: 18–28.

Part I

Biology of Endonucleases (Zinc-Finger Nuclease, TALENs and CRISPRs) and Regulatory Networks

1 Targeted Genome Editing Techniques in *C. elegans* and Other Nematode Species

Behnom Farboud and Te-Wen Lo

1.1 Introduction

C. elegans is a genetically tractable model organism that has pioneered studies to understand cell fate decisions (Sulston *et al.*, 1983), apoptosis (Conradt and Xue, 2005; Ellis and Horvitz, 1986; Ellis *et al.*, 1991; Hengartner *et al.*, 1992), RNA interference (RNAi) (Fire *et al.*, 1998), micro RNAs (Lee *et al.*, 1993; Reinhart *et al.*, 2000) and cellular aging (Dorman *et al.*, 1995; Kenyon *et al.*, 1993; Kimura *et al.*, 1997; Lin *et al.*, 1997; Morris *et al.*, 1996; Ogg *et al.*, 1997), to mention a few. It is a 1 mm free-living, non-parasitic nematode that has two sexes: males and hermaphrodites. The ability to culture self-fertile hermaphrodites allows for easy maintenance of isogenic strains. Additionally, their transparent cuticle has allowed tracking of the fate of every cell, from fertilization to adulthood, to generate a complete cell lineage map that has been pivotal for cell fate studies (Sulston *et al.*, 1983). And in a seminal effort, *C. elegans* became the first multicellular eukaryotic organism to have its entire genome sequenced (*C. elegans* Sequencing Consortium, 1998), yielding a well-defined genetic map and opening the door for development of numerous molecular tools to dissect gene function.

To study gene function, *C. elegans* researchers have utilized both forward and reverse genetic strategies. A common forward genetic screen introduces genetic lesions using mutagens such as ethyl methane sulfonate (EMS) and screens for the desired phenotype (Brenner, 1974). In addition to forward genetic screens, two reverse genetic methods were developed. One powerful reverse genetic tool is RNA interference (RNAi) (Fire *et al.*, 1998). RNAi-mediated knockdown of known target genes results in loss-of- or reduction-of-function phenotypes. Due to the inability to guarantee complete reduction of gene function and the transient nature of RNAi, scientists developed additional methods to create and screen for gene deletions. Individuals could make their own frozen deletion library (Edgley *et al.*, 2002; Jansen *et al.*, 1997) or request pre-isolated mutant strains from two groups: the *C. elegans* Gene Knockout Consortium (1998), and the Million Mutation Project (Thompson *et al.*, 2013). Unfortunately, since these mutants are generated randomly, there are no guarantees that there is a mutation in your gene of interest or that the deletion occurs in a functional domain of the gene. Therefore, more precise methods for targeted gene manipulation were desired.

Introduction of custom transgenes into *C. elegans* allowed more direct dissection of genetic components (Figure 1.1). Microinjection of DNA plasmids encoding genes of interest results in the formation of heritable, multi-copy extra-chromosomal arrays (Mello *et al.*, 1991). These arrays, however, are not stable and are lost at an unpredictable rate during cell division. In order to create stable transgenic lines, extrachromosomal arrays can be integrated into the genome by UV irradiation (Mitani, 1995) or microparticle bombardment (Praitis *et al.*, 2001). These methods generate endogenous, low-copy transgenes, which are expressed at levels closer to that of native genes; however, single-copy integrates are rare. Also, using these methodologies, it is impossible to control for the site of integration. Utilization of existing Mos techniques (Boulin and Bessereau, 2007) overcame many of these challenges. MosSCI and miniMos techniques target transgenes for single-copy insertion at defined chromosomal loci (Frøkjær-Jensen *et al.*, 2012; 2014). Notably, single-copy transgenes can be expressed in the germline where extrachromosomal arrays are often silenced. One limitation of these methods, however, is that transgenes can only be inserted in locations where Mos sites already reside, thus limiting transgene placement.

The rapid development of next generation sequencing (NGS) technology has elevated the power of *C. elegans* as a model organism, by allowing molecular comparisons between *C. elegans* and closely related species for evolutionary

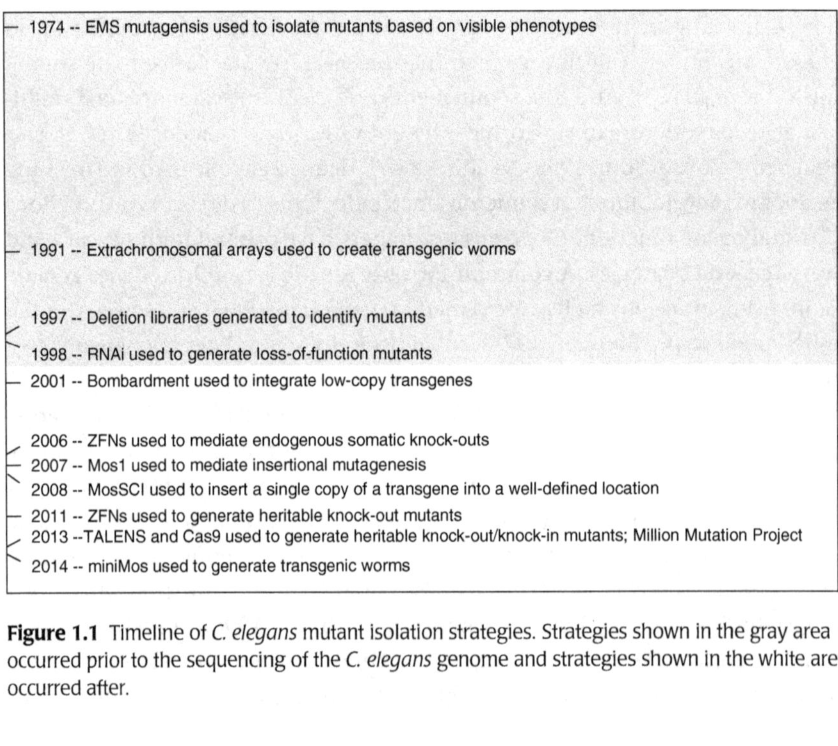

Figure 1.1 Timeline of *C. elegans* mutant isolation strategies. Strategies shown in the gray area occurred prior to the sequencing of the *C. elegans* genome and strategies shown in the white area occurred after.

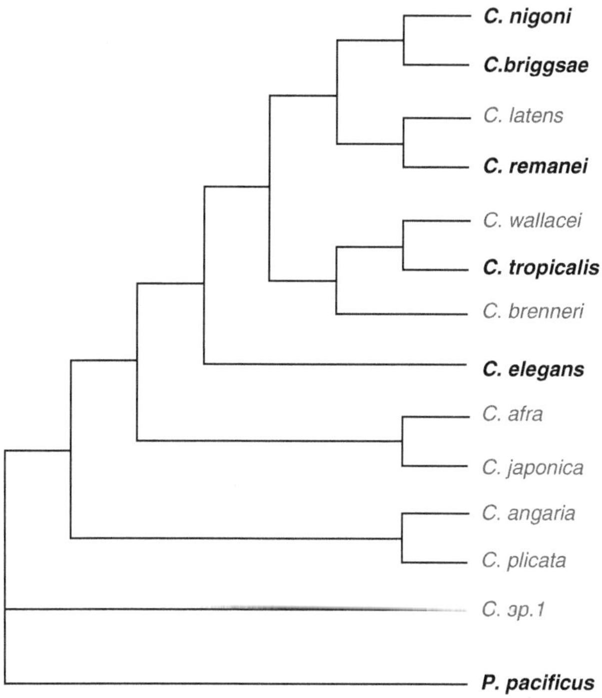

Figure 1.2 Partial *Caenorhabditis* phylogeny (using data from Félix *et al.*, 2014; Kiontke *et al.*, 2011) of sequenced or partially sequenced (gray) species. Genome-edited species are shown in bold. Sequencing status is according to 959 Nematode Genomes.

studies (Figures 1.1, 1.2). Just in the past decade, dozens of related nematodes have been identified, significantly broadening the nematode phylum. Improvements in NGS technology have reduced the time and cost of sequencing new species and the ease of evolutionary comparisons between species that can further our understanding of all aspects of biology that would not be possible by studying one species alone.

Nematodes, and particularly *C. elegans*, have benefited from numerous molecular tools available to study gene function. However, conspicuously absent was the ability to precisely edit the genome to introduce custom mutations. With ZFNs, TALENs and now CRISPR/Cas9, programmable nucleases have revolutionized the study of nematode biology. Below, we discuss how these nucleases have been used in nematodes and how they have the potential to transform any nematode species into a model organism amenable for biological research.

1.2 ZFN-mediated Genome Editing in Nematodes

Zinc-finger nucleases (ZFNs) are customizable DNA binding molecules fused to FokI (Figure 1.3A; Urnov *et al.*, 2010). FokI is a restriction endonuclease that functions as an obligate dimer. Once a FokI dimer is bound to duplex DNA, the DNA cleavage domains are activated and both DNA strands are cleaved. The DNA

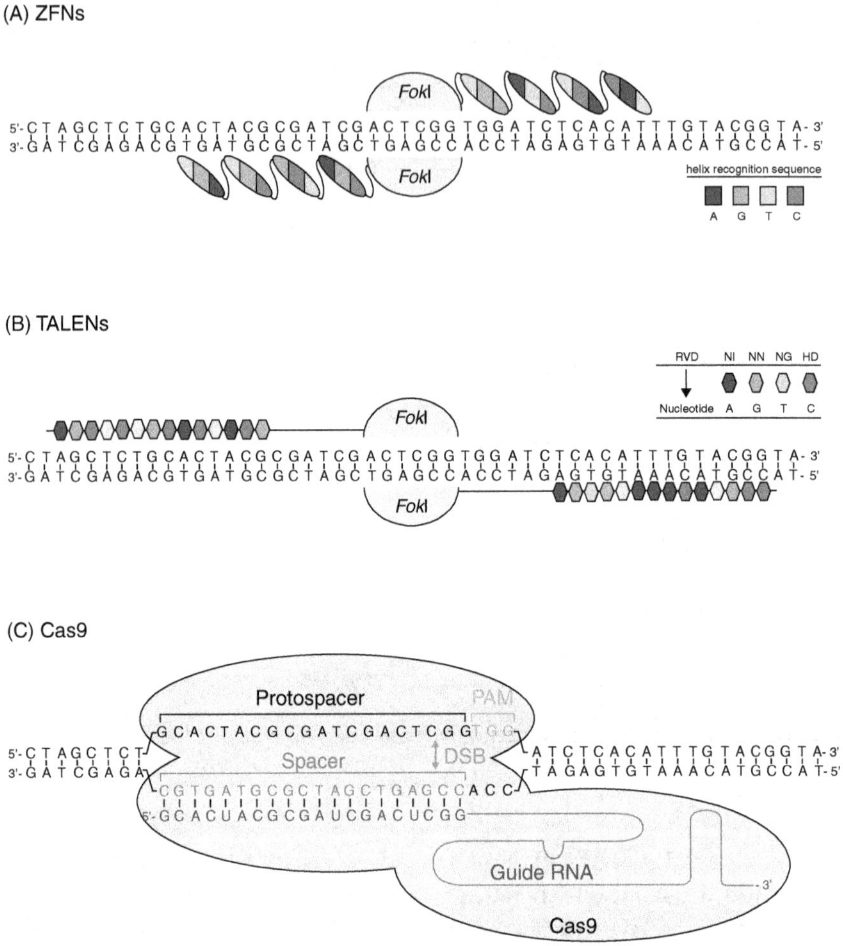

Figure 1.3 ZFN, TALEN and Cas9 DNA recognition. (A) Shown is a ZFN pair bound to its DNA target. Each monomer is composed of a DNA binding domain fused to a FokI cleavage domain. The DNA binding domain is assembled from several zinc-finger DNA recognition domains (four are shown, depicted as ovals) and each domain can recognize three unique nucleotides, noted in different colors. Strung together, the multiple domains can dictate DNA binding to unique genomic sequences with high specificity. When two monomers are brought together by binding adjacent DNA sequences, the FokI domains dimerize and the nuclease is active. (B) A TALEN pair bound to its DNA target. As with ZFNs, TALENs are composed of a DNA binding domain fused to the FokI cleavage domain. The binding specificity is dictated by the repeat variable diresidues (RVD) within each TALE repeat unit (shown are the different RVDs that bind each nucleotide). (C) Cas9 is targeted to DNA sequences by a bound sgRNA. If a protospacer adjacent motif (PAM) is present, the first 20 nucleotides of the guide RNA can form Watson–Crick base pairs with the target spacer sequence, and activate Cas9 nuclease activity. (A black-and-white version of this figure will appear in some formats. For the color version, please refer to the plate section.)

binding domains of ZFNs can be designed to target specific sequences in an organism's genome. Each zinc-finger DNA binding domain makes contact primarily with a separate DNA triplet (Pabo *et al.*, 2001; Pavletich and Pabo, 1991) and can be assembled in unique combinations to target specific DNA sequences within the genome. Increasing the number of zinc-finger domains in each zinc-finger protein increases specificity. Typically, 3–4 zinc-finger domains are assembled resulting in 9–12 bp specificity. Taken together with the need for FokI to form a dimer for cleavage, this results in 24 bp specificity for each ZFN pair.

In 2006, Morton *et al.* demonstrated that custom-engineered ZFNs can generate targeted double-strand breaks (DSBs) in somatic tissues of *C. elegans* by injecting plasmids encoding ZFNs (Morton *et al.*, 2006). Repair of these breaks by non-homologous end joining (NHEJ) results in small insertions and deletions at the desired locus. Unfortunately, these mutations are somatic and not heritable. The lack of heritable mutagenesis may result from the absence of germline nuclease expression. Morton *et al.* suggested that germline expression was undetectable because ZFN expression from transgene arrays was suppressed by RNAi. Building upon these early successes, Wood *et al.* demonstrated heritable germline mutagenesis with ZFNs in *C. elegans* and *C. briggsae* by injecting mRNA into the germline, in order to avoid silencing associated with transgene arrays (Wood *et al.*, 2011). Wood *et al.* chose to edit the endogenous *ben-1* locus due to its easily scored mutant phenotype; mutants are mobile on media containing benomyl. ZFN-encoding mRNAs with 5′ and 3′ untranslated regions (UTRs) permissive for germline expression were injected to obtain *ben-1/+* F1 progeny. Methods were also established to screen F1 progeny for mutations without a visible phenotype using the endonuclease CEL-1, which recognizes and cleaves heteroduplex DNA at sites of mismatches. PCR amplicons that contain the ZFN-targeted site were generated from potential F1 mutant heterozygotes. The amplicons from F1s were then pooled, denatured and slowly reannealed. Mutagenized amplicons will randomly anneal to wild-type amplicons and CEL-1 can recognize and cleave the mismatched bases. Mutant CEL-1-positive clones can be easily detected by resolving the cleaved, mismatched PCR product from the wild-type full-length PCR fragment by agarose gel electrophoresis.

To demonstrate that ZFN-directed mutagenesis could be ported to other nematode species, Wood *et al.* targeted the *sdc-2* gene in *C. briggsae*. Without any modifications to the protocols established in *C. elegans*, ZFNs were successfully utilized to isolate custom mutations in *C. briggsae* (Wood *et al.*, 2011), thus demonstrating that these genome editing methods could be applied across species to achieve targeted genome editing.

1.3 TALEN-mediated Genome Editing in Nematodes

Similar to ZFNs, transcription activator-like effector nucleases (TALENs) consist of a customizable DNA binding domain (TALE domain) fused to the FokI endonuclease (Figure 1.3B). The TALE domain consists of modular repeating units that are 33 or 34 amino acids long. Within each repeat unit, the 12th and 13th residues, termed the repeat variable diresidues (RVDs), are unique, and dictate which DNA base a particular repeat unit will bind. By stringing together repeat units with unique RVDs, TALENs could be easily designed to predictably target DNA (Boch *et al.*, 2009; Bogdanove *et al.*, 2010; Moscou and Bogdanove, 2009). As with ZFNs,

TALENs function as dimers, so two TALENs must be targeted to adjacent DNA sequences to reconstitute an active nuclease. Typical TALE domains contain 15–20 RVDs where the first RVD targets a "T" with 15–30 base pairs between TALEs. Similarly to ZFNs, there is a requirement of dimerization for FokI cleavage, so two TALEs are required for function, thus increasing specificity up to 40 base pairs.

Using protocols established for ZFN mutant isolation, Wood *et al.* engineered TALE nucleases to target genes in *C. elegans* and *C. briggsae* (Figure 1.2). The proof-of-principle experiment targeted TALENs to the same *ben-1* locus that was successfully targeted using ZFNs. Rates of TALEN-mediated genome editing of *ben-1* were comparable to rates of ZFN-mediated genome editing. To further demonstrate the portability of custom nucleases, TALENs were used to generate mutations in another nematode *C. nigoni* (previously named *C. species 9*) and *P. pacificus* (Figure 1.2). Wei *et al.* further adapted protocols to also utilize TALENs in both *C. briggsae* and *C. tropicalis* (Figure 1.2) (Wei *et al.*, 2014).

The editing discussed thus far is imprecise; after the double-strand break, DNA is repaired through NHEJ, resulting in small insertions and deletions. To demonstrate that specific site-directed changes could also be introduced at the DSB, Lo *et al.* used TALENs to insert three different restriction sites into the *ben-1* locus. The authors co-injected mRNA encoding TALENs directed at the *ben-1* locus with three single-stranded oligos. Each oligo possessed a unique restriction site flanked by sequence that was homologous to the TALEN's target site. Along with small insertion and deletion alleles, precise engineered knockins possessing the unique restriction site sequences were recovered (Lo *et al.*, 2013).

1.4 Cas9-mediated Genome Editing in Nematodes

Studies of the bacterial and Archael adaptive immune systems that utilize clustered regularly interspersed short palindromic repeats (CRISPRs) and CRISPR-associated (Cas) proteins have led to the identification of a family of RNA-guided endonucleases that can readily be repurposed to target and cleave double-stranded DNA sequences with very few constraints (reviewed in Carroll, 2014; Doudna and Charpentier, 2014; Hsu *et al.*, 2014; Mali *et al.*, 2013b). The most well-studied family member is Cas9, a component of the type IIa CRISPR system. Cas9 is targeted to genomic loci by a guide RNA that encodes a 20-nucleotide region of homology to the DNA target of choice (Figure 1.3C; Garneau *et al.*, 2010; Jinek *et al.*, 2012; Mojica *et al.*, 2009). Cas9-mediated DNA target recognition and cleavage has only one sequence requirement, a trinucleotide motif adjacent to the 20-nucleotide DNA target sequence, called the protospacer adjacent motif (PAM) (Anders *et al.*, 2014; Garneau *et al.*, 2010; Jinek *et al.*, 2012; Sapranauskas *et al.*, 2011; Sternberg *et al.*, 2014). For Cas9, the PAM is NGG. The RNA guide naturally consists of two separate RNAs: a CRISPR RNA (crRNA) that can form Watson–Crick base pairs with the DNA target (spacer) backbone and a transactivating RNA (trRNA) that binds the crRNA and forms a ternary complex with Cas9 (Figure 2.3C). This can be further simplified and a chimeric single-guide RNA (sgRNA) made from fusing the crRNA and trRNA (Jinek *et al.*, 2012).

The minimal requirement for target design and the straightforward construction of guide RNAs have made Cas9 a very attractive alternative to other site-specific nucleases, which have more stringent requirements for target selection and require more effort to design and construct. Ease of implementation has led to widespread use of CRISPR/Cas9, from bacteria and yeast to plants and metazoans, including nematodes.

In the following sections, we will discuss general considerations for performing Cas9-mediated genome editing in nematodes. While the majority of studies in nematodes have involved *C. elegans*, most principles can be applied to other nematode species and some of these principles have already been adapted to accomplish genome editing in other nematodes, including *C. briggsae, C. tropicalis, C. remanei* (Yin and Haag, unpublished data), *C. nigoni* and *P. pacificus* (Figure 1.2; Lo *et al.*, 2013; Markov *et al.*, 2016; Witte *et al.*, 2015). Finally, we discuss newer applications of CRISPR systems in nematodes.

1.4.1 Delivery of Cas9 and RNA Guides into Nematodes

Cas9 and guide RNAs can be introduced into nematodes using three different strategies, all relying on microinjection. With the first method, mRNA encoding Cas9 and RNA guides are delivered by co-injection into the gonad of young adult hermaphrodites (Chiu *et al.*, 2013; Katic and Großhans, 2013; Lo *et al.I*, 2013). Initial studies using this approach blindly scored offspring of injected animals for the presence of the desired mutation, assuming the mutation did not have a phenotype. While successful, this approach often required screening in excess of 1000 worms to identify the desired mutation. The second alternative approach of injecting DNA encoding Cas9 and the sgRNA along with transformation markers reduced the burden of screening (Figure 1.4A) (Dickinson *et al.*, 2013; Friedland *et al.*, 2013; Waaijers *et al.*, 2013). Only worms expressing transformation makers, and presumably Cas9 and sgRNAs, were examined and the odds of isolating the desired mutant were improved.

There are several parameters to consider when performing DNA-based introduction of Cas9 and guide RNAs into *C. elegans*. To ensure Cas9 is present early enough in development to generate heritable genomic modifications, constructs are codon-optimized, and germline-permissive 5′ and 3′ UTRs are utilized to maximize transcriptional and translational efficiency (Dickinson *et al.*, 2013; Friedland *et al.*, 2013; Waaijers *et al.*, 2013). Several constructs employ the *eft-3* promoter and the *tbb-2* 3′ UTR, while an alternate vector uses a heat shock promoter to drive expression of Cas9. Heat shock induction produces short bursts of Cas9 expression, potentially reducing embryonic lethality observed using the *eft-3* promoter and reducing the chances for off-target cleavage by the continual expression of Cas9 (Waaijers *et al.*, 2013).

Two of the most widely used vectors, *Peft-3::cas9-SV40-NLS::tbb-2* 3′UTR and pDD162, drive Cas9 expression with the *eft-3* promoter and *tbb-2* 3′ UTR (Dickinson *et al.*, 2013; Friedland *et al.*, 2013), but minimal differences between the two vectors have a dramatic impact on mutagenesis efficiency. A recent study found that the sequence of the flexible linker between Cas9 and the nuclear

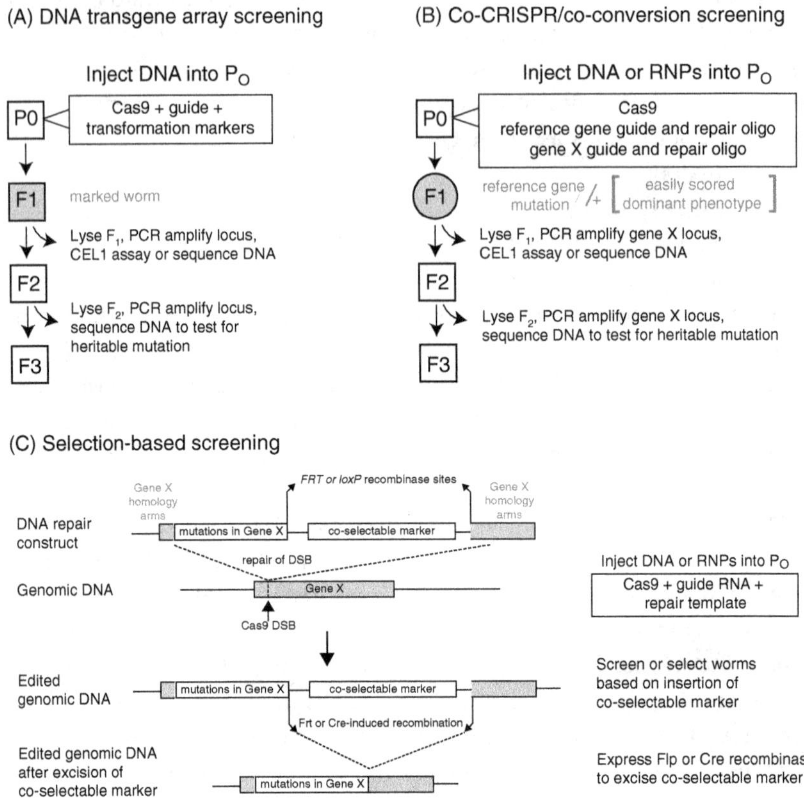

Figure 1.4 Three different strategies for isolating mutants generated by Cas9. (A) DNA transgene array screening: P0 hermaphrodites are microinjected with DNA encoding Cas9, a sgRNA and co-transformation markers. In some F1 offspring, the injected DNA forms transgene arrays and worms are selected based on the expression of the transformation markers from the arrays (e.g. expression of fluorescent makers). After F2 progeny are laid, worms are examined for the presence of the desired mutation. (B) Co-CRISPR/co-conversion screening: P0 animals are microinjected with DNA encoding Cas9 and sgRNAs or RNPs targeting a gene of interest and a co-conversion reference gene (Arribere *et al.*, 2014; Kim *et al.*, 2014; Ward, 2015). Appropriate editing of a reference gene using a co-injected repair oligonucleotide yields F1s with an easily scored dominant phenotype. F1 worms mutant for the reference gene are selected and the gene of interest is examined. (C) Selection-based screening: P0s are injected with DNA encoding Cas9, a sgRNA and a DNA repair construct. Repair using the template results in insertion of the desired mutation and a co-selectable marker flanked by loxP or Frt recombination sites (Chen *et al.*, 2013b; Dickinson *et al.*, 2013; 2015). After selection or screening of worms based on insertion of the co-selectable marker, expression of Flp or Cre recombinase leads to excision of the selectable marker, leaving behind the mutation of interest.

localization signal accounts for the impaired efficiency of *Peft-3::cas9-SV40-NLS::tbb-2* 3′UTR (Zhao *et al.*, 2016).

Guide RNA expression vectors utilize U6 small RNA promoters that are transcribed by RNA polymerase III (polIII) (Dickinson *et al.*, 2013; Friedland *et al.*, 2013; Waaijers *et al.*, 2013). The first 20 nucleotides of the guide are DNA-target-specific, so successful expression of guides depends on predictable transcriptional initiation and polIII transcriptional initiation from the U6 promoter is well characterized, as is the transcriptional termination signal. This, however, subjects guide design to the limitations of polIII transcription: polIII prefers to initiate from a guanine, so there is a preference for a 5′ G in the first position of the guide, a nucleotide that base pairs with the target DNA (Figure 1.3C; Fruscoloni *et al.*, 1995). The limitation of selecting targets that base pair with a G in the first position of the guide can be overcome by substituting a non-templated G as the 5′ base and only supplying 19 nucleotides with homology to the DNA target (Arribere *et al.*, 2014; Cong *et al.*, 2013; Gagnon *et al.*, 2014; Hwang *et al.*, 2013a; Ren *et al.*, 2014). Mismatches at the 5′ end of the guide are well tolerated and can still target Cas9 for efficient cleavage (Cencic *et al.*, 2014; Fu *et al.*, 2014; Hwang *et al.*, 2013b; Jinek *et al.*, 2012; Pattanayak *et al.*, 2013; Sternberg *et al.*, 2014; Tsai *et al.*, 2015). A second limitation imposed by polIII is the propensity to terminate transcription when encountering stretches of uracils, which would correspond to stretches of As in the spacer DNA (Wu *et al.*, 2014). This requires avoiding DNA targets that possess runs of As on the spacer strand.

The third method to introduce Cas9 is to preassemble recombinant Cas9 and RNAs *in vitro* prior to microinjection (Cho *et al.*, 2013; Paix *et al.*, 2015). This eliminates the need to identify species-appropriate germline promoters to drive expression of protein and guides, and also eliminates the need to codon-optimize vectors for each species injected, removing a roadblock for implementation of Cas9-mediated mutagenesis in a wide variety of nematode species. Additionally, since RNAs are synthesized *in vitro*, issues with expression of certain RNA sequences are eliminated. Both recombinant Cas9 and guide RNAs can be synthesized in the laboratory or purchased commercially.

1.4.2 Guide Design

The sequence of the 20-nucleotide target-specific region of the guide RNA is pivotal to successful Cas9-mediated genome editing (Figure 1.3C). Numerous studies in mammalian systems applied high-throughput library screens to identify effective guide sequences, and several patterns emerged (Doench *et al.*, 2014; 2016; Wang *et al.*, 2014; Wu *et al.*, 2014). The most consistent characteristic of a successful guide was the presence of a purine at the 3′ end of the target specific sequence with a notable preference for guanosine. Similar preferences have also been observed in *C. elegans, D. melanogaster* and *D. rerio* (Farboud and Meyer, 2015; Gagnon *et al.*, 2014; Ren *et al.*, 2014). Additionally, in *C. elegans* there is a preference for a guanosine as the penultimate base of the target-specific sequence (Farboud and Meyer, 2015). Choosing guides based solely on harboring a 3′ GG dinucleotide was highly predictive of successful editing. To aid in choosing efficient guides, there are several online tools available (Doench *et al.*, 2016; Liu *et al.*, 2015; Montague *et al.*, 2014; Moreno-Mateos *et al.*, 2015; Tsai *et al.*, 2015). These utilize sequence preferences collected from the library screening approaches above to assign predictive scores to guides.

While the guide's target-specific sequence is important, changes to the constant region of the guide can improve Cas9 targeting and cleavage efficiency. An A–U base pair flip was engineered to interrupt a stretch of four Us, a polIII transcriptional termination signal (Chen *et al.*, 2013a). This is particularly important if expressing the guide *in vivo*. Additionally, a hairpin loop was extended to help stabilize binding to Cas9. Both alterations improved Cas9 targeting in mammalian cells and improved editing efficiency in numerous systems, including *C. elegans* (Chen *et al.*, 2013a; Ward, 2015).

Cas9 guide selection is ultimately limited by the presence of NGG sequences near the DNA target. To overcome this limitation, several groups have identified other CRISPR class II endonucleases that utilize different PAM motifs (Shmakov *et al.*, 2015; Zetsche *et al.*, 2015). Zetsche *et al.* characterized Cpf1, a class II CRISPR endonuclease that utilizes a T-rich PAM motif, which should be particularly useful in targeting the A/T-rich intragenic regions in many nematodes. Additionally, Cas9 variants have been engineered to use a variety of different PAM motifs (Kleinstiver *et al.*, 2015). Already, two Cas9 variants that recognize NGA and NGCG PAMs have successfully been used to mutagenize *C. elegans* (Bell *et al.*, 2016).

The efficiency of the RNA guide can also depend on the chromatin landscape in which the target DNA resides. Nucleosomes have been shown to inhibit Cas9 binding and cleavage *in vitro* and *in vivo* (Hinz *et al.*, 2015; Horlbeck *et al.*, 2016; Isaac *et al.*, 2016). Therefore, consideration should be given to the chromatin state of target DNA, particularly if targeting Cas9 variants for fluorescent imaging or for transcriptional regulation of targeted genes, where Cas9 residency time is critical. If a custom genomic edit is the sole desire, then nucleosome breathing may allow for limited access to the DNA target and less frequent Cas9 cleavage and repair. In these cases, selection methods to isolate the more rare, edited worms then become more important.

1.4.3 Selecting and Screening for Mutagenized Nematodes

Not all embryos in Cas9-microinjected worms are targeted equally for mutagenesis. Differences may be due to the varying degrees of microinjection efficiency or inherent cleavage rates of different Cas9/RNA targets. Therefore, a method to enrich for mutagenized offspring was needed to reduce the burden of screening. Several strategies have been employed. Initial DNA microinjection experiments in *C. elegans* selected for embryos that possessed transgenic arrays (Figure 1.4A) (Friedland *et al.*, 2013; Waaijers *et al.*, 2013). Along with DNA encoding Cas9 and the sgRNA, vectors encoding for visible markers, such as fluorescent myosin II, were included. Worms expressing the marker and likely also Cas9 and the sgRNA were picked and the target loci examined. Using this approach, mutagenesis efficiencies of screened worms ranged widely from 0 to 80%, but the median efficiency was quite low (~4%).

A second method to enrich for mutagenized worms, termed co-CRISPR or co-conversion, soon followed (Figure 1.4B; Arribere *et al.*, 2014; Kim *et al.*, 2014; Ward, 2015). This approach simultaneously targets two unrelated loci with two different guides in the same worm. Repair at one loci results in a visible phenotype and selection of worms based on this phenotype significantly improves the odds of identifying a mutagenic event in the second. Rates of co-conversion ranged from 14 to 84%, demonstrating a marked improvement for mutant recovery from the previous screening method.

A third approach relies on inserting selectable makers in tandem with targeted mutations, similar to procedures used to target transgenes for insertion by the Mos1 transposase (Figure 1.4C; Chen *et al.*, 2013b; Dickinson *et al.*, 2013; 2015; Frøkjær-Jensen *et al.*, 2008; 2010; 2012; 2014). The benefit of inserting selectable markers is that efficiency of guides and the accessibility of the genomic target are less critical, as selection allows identification of low probability cleavage and repair events.

1.4.4 Repair of Cas9-mediated Double-strand Breaks

After the Cas9/RNP complex cleaves DNA, there are several repair pathways that can be used to mend the lesion. The predominant pathway is likely the polymerase-theta-mediated alternative end joining (AEJ) pathway that can use non-templated repair, a repair template in *cis* or a template in *trans* to repair the cut (Kent *et al.*, 2016; Thyme and Schier, 2016; van Schendel *et al.*, 2015). Without a repair template, imprecise repair prevails. This gives way to mostly small insertions made from copying short regions of microhomology in *cis* to the DSB, as well as small deletions.

If a repair template is provided in *trans* to the Cas9 cut, some of the DSBs are repaired using the donor template through homology-directed repair (HDR). Many factors can influence the efficiency of HDR. The choice of single-stranded vs. double-stranded DNA repair templates plays a role. If relatively small mutations are required, single-stranded repair templates are efficient substrates that are easily designed and commercially synthesized (Paix *et al.*, 2014; 2015; Zhao *et al.*, 2014). If large deletions are desired, two guides can be used to excise a large segment of DNA and a single-strand repair template, straddling the deletion, can be used for repair. Repair, however, is quite local, and may extend only 10 bases away from the Cas9 cut site (Arribere *et al.*, 2014; Paix *et al.*, 2015). Therefore, choosing a DNA target near the desired site of mutagenesis is critical for efficient template repair. When Cas9 cannot be targeted near the site of mutagenesis, the selection-based screening method described above can be utilized to isolate rare repair events occurring distal to the Cas9 cut site (Dickinson *et al.*, 2013).

To improve repair efficiency, care should be taken in selecting repair templates. When using single-stranded DNA repair templates, there is a potential repair bias for which DNA strand to choose as the donor repair template. One study in *C. elegans* found that using a repair template that matched the coding strand of the targeted gene is more efficient than one corresponding to the non-coding strand (Katic *et al.*, 2015). Additionally, strand repair bias may arise from Cas9's propensity to release the PAM proximal strand after cleavage, but continue to bind the other three DNA strands (Richardson *et al.*, 2016). The resulting asymmetric exposure of the non-targeted DNA strand biases repair from donor templates that can base pair with the non-targeted, protospacer DNA strand.

Targeting for large genomic insertions lends itself to using double-strand repair templates that can have as few as 35 base pairs to kilobases of homology on each side of the sequence to be inserted (Dickinson *et al.*, 2013; Paix *et al.*, 2014; 2015). Recent work has made construction of these larger templates simple. Paix *et al.* have demonstrated that *C. elegans* is capable of stitching together multiple DNA fragments with overlapping homology as short as 35 base pairs, to repair Cas9-generated DSBs (Paix *et al.*, 2015). Double-stranded DNA also works well as a donor template for small insertions and mutations, but requires more effort to construct than oligonucleotides.

1.4.5 Off-target Cleavage by Cas9

Base pairing of the full 20-nucleotide spacer to the guide RNA is not necessary to activate Cas9's nuclease activity. Several studies have examined Cas9's off-target cleavage rates and found that some base pair interactions are more critical for Cas9 activity than others (Cencic *et al.*, 2014; Fu *et al.*, 2014; Hwang *et al.*, 2013b; Jinek *et al.*, 2012; Pattanayak *et al.*, 2013; Sternberg *et al.*, 2014; Tsai *et al.*, 2015). For instance, base pairing of the RNA guide to nucleotides near the 5′ end of the spacer, termed the seed sequence, is essential to activate Cas9 while base pairing more 3′ is less critical. This necessarily means that guided Cas9 can engage and cut sequences that do not fully match the designated target and care should be taken to select a DNA target that is sufficiently unique to minimize off-target cleavage.

Nonetheless, in the few studies that have examined Cas9 off-target cleavage in *C. elegans*, no appreciable off-target effects have been observed (Chiu *et al.*, 2013; Lo *et al.*, 2013). However, this does not eliminate the potential for off-target cleavage in nematodes, and mutagenized strains should be outcrossed to reduce the unintended consequences of off-target mutations.

The use of alternative nucleases can also reduce off-target cleavage. Both TALENs and ZFNs have longer DNA binding sites, reducing the chances of multiple occurrences in the genome. Additionally, both TALENs and ZFNs function as obligate dimers and require the two independent binding sites of each half of the dimer pair be adjacent to one another. This limitation drastically reduces the chances of off-target cleavage and in cases where the potential for off-target activity is objectionable, TALENs and ZFNs should be considered.

1.4.6 Alternative Uses for CRISPR/Cas9 Technology

The basic CRISPR/Cas9 system has been repurposed to target Cas9 for objectives other than to simply cleave target DNA for heritable mutagenesis. Cas9 can be targeted to generate somatic lesions that can help elucidate the tissue-specific, developmental roles of target genes (Li *et al.*, 2015; Shen *et al.*, 2014; Tian *et al.*, 2015). Tissue-specific promoters and UTRs help ensure that Cas9 expression is restricted to certain cells. This is particularly useful when investigating the role of essential genes, where phenotypes of null alleles are not always informative. An alternative strategy to generating somatic mutations is to conditionally deplete gene products using the auxin-inducible degron system recently adapted for *C. elegans* (Zhang *et al.*, 2015). Zhang *et al.* employed Cas9 to heritably append the auxin degron coding sequence onto genes. In a strain expressing TIR1, the auxin-responsive F-box protein, rapid depletion of tagged proteins was observed. Unlike germline mutations that heritably compromise protein expression, auxin-induced protein depletion can be reversed. Removal of auxin from growth media returns steady state protein levels to normal.

Another approach to modulating gene expression relies on a catalytically dead Cas9 variant (dCas9) that can still target DNA and either directly or indirectly recruit transcriptional activators or repressors to the DNA target site (Cheng *et al.*, 2013; Gilbert *et al.*, 2013; 2014; Konermann *et al.*, 2013; Maeder *et al.*, 2013; Mali *et al.*, 2013a; Perez-Pinera *et al.*, 2013; Qi *et al.*, 2013; Tanenbaum *et al.*, 2014; Zalatan *et al.*, 2015). There is now a cohort of different repressors and activators that can be recruited and allow for fine-tuned transcriptional control; however, targeting dCas9 to modulate transcription may not be as straightforward as when targeting a gene for mutagenesis. DNA target positioning relative to the transcriptional start site, accessibility of Cas9 to the DNA, and residency time are issues. Several studies have identified promoter regions possessing low relative histone density as optimal regions for targeting dCas9, presumably because Cas9 encounters less competition for promoter binding and can bind and remain associated longer (Gilbert *et al.*, 2014; Hinz *et al.*, 2015; Horlbeck *et al.*, 2016; Isaac *et al.*, 2016).

CRISPR/Cas9 can also be used to generate custom genetic balancers. These allow lethal mutations to be easily maintained, assuming the gene is not haploinsufficient. Simultaneously targeting Cas9 to two different chromosomes can result in a reciprocal chromosomal translocation (Chen *et al.*, 2015; Iwata *et al.*, 2016). With precise targeting, these translocations can produce custom-engineered genetic balancers for mutations that previously were not covered by balancers. This is especially applicable to less well-studied nematode species, which may lack balancers altogether.

An intriguing new application of CRISPR technology can trace cell lineages (McKenna *et al.*, 2016). The Cas9/guide complex can introduce unique mutations into cells as the organism develops. As Cas9 cleaves, repair usually results in small insertions or deletions within the targeted spacer sequence. Without an intact spacer, Cas9 cannot retarget the same genomic site and the unique Cas9 edit is passed down in subsequent cell divisions to all descendants. Following which cells share unique edits provides a precise lineage map within the organism that is amenable to high-throughput analysis.

There have been numerous other Cas9 technological advances in mammalian systems that are primed for future use in nematodes. These include fluorescently tagging dCas9 to visualize DNA dynamics, to target and track RNAs or to target RNAs for cleavage (Horlbeck *et al.*, 2016; Nelles *et al.*, 2016; O'Connell *et al.*, 2014; Tanenbaum *et al.*, 2014; Wang *et al.*, 2016). Additionally, dCas9 has been fused to activation-induced cytidine deaminases to perform efficient local mutagenesis, generating point mutations, rather than indels (Komor *et al.*, 2016; Nishida *et al.*, 2016).

1.5 Conclusion

Targetable nucleases have revolutionized the field of genome engineering in a relatively short time. The general rules elucidated for targeting nucleic acids have democratized the field, making it universally accessible. And the inherent portability of programmed nucleases has opened the door to custom genomic modifications in species where tools previously did not exist. To begin, all one needs is a sequenced genome.

REFERENCES

Anders C, Niewoehner O, Duerst A, Jinek M. 2014. Structural basis of PAM-dependent target DNA recognition by the Cas9 endonuclease. *Nature* **513**: 569–573.

Arribere JA, Bell RT, Fu BXH, *et al.* 2014. Efficient marker-free recovery of custom genetic modifications with CRISPR/Cas9 in *Caenorhabditis elegans*. *Genetics* **198**: 837–846.

Bell RT, Fu BXH, Fire AZ. 2016. Cas9 variants expand the target repertoire in *Caenorhabditis elegans*. *Genetics* **202**: 381–388.

Boch J, Scholze H, Schornack S, *et al.* 2009. Breaking the code of DNA binding specificity of TAL-type III effectors. *Science* **326**: 1509–1512.

Bogdanove AJ, Schornack S, Lahaye T. 2010. TAL effectors: finding plant genes for disease and defense. *Curr Opin Plant Biol* **13**: 394–401.

Boulin T, Bessereau J-L. 2007. Mos1-mediated insertional mutagenesis in *Caenorhabditis elegans*. *Nat Protoc* **2**: 1276–1287.

Brenner S. 1974. The genetics of *Caenorhabditis elegans*. *Genetics* **77**: 71–94.

Carroll D. 2014. Genome engineering with targetable nucleases. *Annu Rev Biochem* **83**: 409–439.

Cencic R, Miura H, Malina A, *et al.* 2014. Protospacer adjacent motif (PAM)-distal sequences engage CRISPR Cas9 DNA target cleavage. *PLoS One* **9**: e109213.

Chen B, Gilbert LA, Cimini BA, *et al.* 2013a. Dynamic imaging of genomic loci in living human cells by an optimized CRISPR/Cas system. *Cell* **155**: 1479–1491.

Chen C, Fenk LA, de Bono M. 2013b. Efficient genome editing in *Caenorhabditis elegans* by CRISPR-targeted homologous recombination. *Nucleic Acids Res* **41**: e193.

Chen X, Li M, Feng X, Guang S. 2015. Targeted chromosomal translocations and essential gene knockout using CRISPR/Cas9 technology in *Caenorhabditis elegans*. *Genetics* **201**: 1295–1306.

Cheng AW, Wang H, Yang H, *et al.* 2013. Multiplexed activation of endogenous genes by CRISPR-on, an RNA-guided transcriptional activator system. *Cell Res* **23**: 1163–1171.

Chiu H, Schwartz HT, Antoshechkin I, Sternberg PW. 2013. Transgene-free genome editing in *Caenorhabditis elegans* using CRISPR-Cas. *Genetics* **195**: 1167–1171.

Cho SW, Lee J, Carroll D, Kim J-S, Lee J. 2013. Heritable gene knockout in *Caenorhabditis elegans* by direct injection of Cas9–sgRNA ribonucleoproteins. *Genetics* **195**: 1177–1180.

Cong L, Ran FA, Cox D, *et al.* 2013. Multiplex genome engineering using CRISPR/Cas systems. *Science* **339**: 819–823.

Conradt B, Xue D. 2005. Programmed cell death. *Wormbook* www.wormbook.org/chapters/www_programcelldeath/programcelldeath.html 1–13. (Accessed September 30, 2016).

C. elegans Sequencing Consortium 1998. Genome sequence of the nematode *C. elegans*: a platform for investigating biology. *Science* **282**: 2012–2018.

Dickinson DJ, Pani AM, Heppert JK, Higgins CD, Goldstein B. 2015. Streamlined genome engineering with a self-excising drug selection cassette. *Genetics* **200**: 1035–1049.

Dickinson DJ, Ward JD, Reiner DJ, Goldstein B. 2013. Engineering the *Caenorhabditis elegans* genome using Cas9-triggered homologous recombination. *Nat Methods* **10**: 1028–1034.

Doench JG, Fusi N, Sullender M, *et al.* 2016. Optimized sgRNA design to maximize activity and minimize off-target effects of CRISPR-Cas9. *Nat Biotechnol* **34**: 184–191.

Doench JG, Hartenian E, Graham DB, *et al.* 2014. Rational design of highly active sgRNAs for CRISPR-Cas9-mediated gene inactivation. *Nat Biotechnol* **32**: 1262–1267.

Dorman JB, Albinder B, Shroyer T, Kenyon C. 1995. The age-1 and daf-2 genes function in a common pathway to control the lifespan of *Caenorhabditis elegans*. *Genetics* **141**: 1399–1406.

Doudna JA, Charpentier E. 2014. Genome editing: the new frontier of genome engineering with CRISPR-Cas9. *Science* **346**: 1258096.

Edgley M, D'Souza A, Moulder G, *et al.* 2002. Improved detection of small deletions in complex pools of DNA. *Nucleic Acids Res* **30**: e52–e52.

Ellis HM, Horvitz HR. 1986. Genetic control of programmed cell death in the nematode *C. elegans*. *Cell* **44**: 817–829.

Ellis RE, Jacobson DM, Horvitz HR. 1991. Genes required for the engulfment of cell corpses during programmed cell death in *Caenorhabditis elegans*. *Genetics* **129**: 79–94.

Farboud B, Meyer BJ. 2015. Dramatic enhancement of genome editing by CRISPR/Cas9 through improved guide RNA design. *Genetics* **199**: 959–971.

Félix M-A, Braendle C, Cutter AD. 2014. A streamlined system for species diagnosis in *Caenorhabditis* (Nematoda: Rhabditidae) with name designations for 15 distinct biological species. *PLoS One* **9**: e94723.

Fire A, Xu S, Montgomery MK, *et al.* 1998. Potent and specific genetic interference by double-stranded RNA in *Caenorhabditis elegans*. *Nature* **391**: 806–811.

Friedland AE, Tzur YB, Esvelt KM, *et al.* 2013. Heritable genome editing in *C. elegans* via a CRISPR-Cas9 system. *Nat Methods* **10**: 741–743.

Frøkjær-Jensen C, Davis MW, Ailion M, Jorgensen EM. 2012. Improved Mos1-mediated transgenesis in *C. elegans*. *Nat Methods* **9**: 117–118.

Frøkjær-Jensen C, Davis MW, Hollopeter G, *et al.* 2010. Targeted gene deletions in *C. elegans* using transposon excision. *Nat Methods* **7**: 451–453.

Frøkjær-Jensen C, Davis WM, Hopkins CE, *et al.* 2008. Single-copy insertion of transgenes in *Caenorhabditis elegans*. *Nat Genet* **40**: 1375–1383.

Frøkjaer-Jensen C, Davis MW, Sarov M, *et al.* 2014. Random and targeted transgene insertion in *C. elegans* using a modified Mosl transposon. *Nat Methods* **11**: 529–534.

Fruscoloni P, Zamboni M, Panetta G, De Paolis A, Tocchini-Valentini GP. 1995. Mutational analysis of the transcription start site of the yeast tRNA(Leu3) gene. *Nucleic Acids Res* **23**: 2914–2918.

Fu BXH, Hansen LL, Artiles KL, Nonet ML, Fire AZ. 2014. Landscape of target: guide homology effects on Cas9-mediated cleavage. *Nucleic Acids Res* **42**: 13778–13787.

Gagnon JA, Valen E, Thyme SB, *et al.* 2014. Efficient mutagenesis by Cas9 protein-mediated oligonucleotide insertion and large-scale assessment of single-guide RNAs. *PLoS One* **9**: e98186.

Garneau JE, Dupuis M-È, Villion M, *et al.* 2010. The CRISPR/Cas bacterial immune system cleaves bacteriophage and plasmid DNA. *Nature* **468**: 67–71.

Gilbert LA, Horlbeck MA, Adamson B, *et al.* 2014. Genome-scale CRISPR-mediated control of gene repression and activation. *Cell* **159**: 647–661.

Gilbert LA, Larson MH, Morsut L, *et al.* 2013. CRISPR-mediated modular RNA-guided regulation of transcription in eukaryotes. *Cell* **154**: 442–451.

Hengartner MO, Ellis R, Horvitz R. 1992. *Caenorhabditis elegans* gene ced-9 protects cells from programmed cell death. *Nature* **356**: 494–499.

Hinz JM, Laughery MF, Wyrick JJ. 2015. Nucleosomes inhibit cas9 endonuclease activity in vitro. *Biochemistry (Mosc)* **54**: 7063–7066.

Horlbeck MA, Witkowsky LB, Guglielmi B, *et al.* 2016. Nucleosomes impede Cas9 access to DNA in vivo and in vitro. *eLife* **5**.

Hsu PD, Lander ES, Zhang F. 2014. Development and applications of CRISPR-Cas9 for genome engineering. *Cell* **157**: 1262–1278.

Hwang WY, Fu Y, Reyon D, *et al.* 2013a. Heritable and precise zebrafish genome editing using a CRISPR-Cas system. *PLoS One* **8**: e68708.

Hwang WY, Fu Y, Reyon D, *et al.* 2013b. Efficient genome editing in zebrafish using a CRISPR-Cas system. *Nat Biotechnol* **31**: 227–229.

Isaac RS, Jiang F, Doudna JA, *et al.* 2016. Nucleosome breathing and remodeling constrain CRISPR-Cas9 function. *eLife* **5**.

Iwata S, Yoshina S, Suehiro Y, Hori S, Mitani S. 2016. Engineering new balancer chromosomes in *C. elegans* via CRISPR/Cas9. *Sci Rep* **6**: 33840.

Jansen G, Hazendonk E, Thijssen KL, Plasterk RHA. 1997. Reverse genetics by chemical mutagenesis in *Caenorhabditis elegans*. *Nat Genet* **17**: 119–121.

Jinek M, Chylinski K, Fonfara I, *et al.* 2012. A programmable dual-RNA-guided DNA endonuclease in adaptive bacterial immunity. *Science* **337**: 816–821.

Katic I, Großhans H. 2013. Targeted heritable mutation and gene conversion by Cas9-CRISPR in *Caenorhabditis elegans*. *Genetics* **195**: 1173–1176.

Katic I, Xu L, Ciosk R. 2015. CRISPR/Cas9 genome editing in *Caenorhabditis elegans*: evaluation of templates for homology-mediated repair and knock-ins by homology-independent DNA repair. *G3 Genes, Genomes, Genetics* **5**: 1649–1656.

Kent T, Mateos-Gomez PA, Sfeir A, Pomerantz RT. 2016. Polymerase θ is a robust terminal transferase that oscillates between three different mechanisms during end-joining. *eLife* **5**.

Kenyon C, Chang J, Gensch E, Rudner A, Tabtiang R. 1993. A *C. elegans* mutant that lives twice as long as wild type. *Nature* **366**: 461–464.

Kim H, Ishidate T, Ghanta KS, Seth M, *et al.* 2014. A co-CRISPR strategy for efficient genome editing in *Caenorhabditis elegans*. *Genetics* **197**: 1069–1080.

Kimura KD, Tissenbaum HA, Liu Y, Ruvkun G. 1997. daf-2, an insulin receptor-like gene that regulates longevity and diapause in *Caenorhabditis elegans*. *Science* **277**: 942–946.

Kiontke KC, Félix M-A, Ailion M, *et al.* 2011. A phylogeny and molecular barcodes for *Caenorhabditis*, with numerous new species from rotting fruits. *BMC Evol Biol* **11**: 339.

Kleinstiver BP, Prew MS, Tsai SQ, *et al.* 2015. Engineered CRISPR-Cas9 nucleases with altered PAM specificities. *Nature* **523**: 481–485.

Komor AC, Kim YB, Packer MS, Zuris JA, Liu DR. 2016. Programmable editing of a target base in genomic DNA without double-stranded DNA cleavage. *Nature* **533**: 420–424.

Konermann S, Brigham MD, Trevino AE, *et al.* 2013. Optical control of mammalian endogenous transcription and epigenetic states. *Nature* **500**: 472–476.

Lee RC, Feinbaum RL, Ambros V. 1993. The *C. elegans* heterochronic gene lin-4 encodes small RNAs with antisense complementarity to lin-14. *Cell* **75**: 843–854.

Li W, Yi P, Ou G. 2015. Somatic CRISPR-Cas9-induced mutations reveal roles of embryonically essential dynein chains in *Caenorhabditis elegans* cilia. *J Cell Biol* **208**: 683–692.

Lin K, Dorman JB, Rodan A, Kenyon C. 1997. daf-16: an HNF-3/forkhead family member that can function to double the life-span of *Caenorhabditis elegans*. *Science* **278**: 1319–1322.

Liu H, Wei Z, Dominguez A, *et al.* 2015. CRISPR-ERA: a comprehensive design tool for CRISPR-mediated gene editing, repression and activation. *Bioinforma Oxf Engl* **31**: 3676–3678.

Lo T-W, Pickle CS, Lin S, *et al.* 2013. Precise and heritable genome editing in evolutionarily diverse nematodes using TALENs and CRISPR/Cas9 to engineer insertions and deletions. *Genetics* **195**: 331–348.

Maeder ML, Linder SJ, Cascio VM, *et al.* 2013. CRISPR RNA-guided activation of endogenous human genes. *Nat Methods* **10**: 977–979.

Mali P, Aach J, Stranges PB, *et al.* 2013a. Cas9 transcriptional activators for target specificity screening and paired nickases for cooperative genome engineering. *Nat Biotechnol* **31**: 833–838.

Mali P, Esvelt KM, Church GM. 2013b. Cas9 as a versatile tool for engineering biology. *Nat Methods* **10**: 957–963.

Markov GV, Meyer JM, Panda O, *et al.* 2016. Functional conservation and divergence of daf-22 paralogs in *Pristionchus pacificus* dauer development. *Mol Biol Evol* **33**: 2506–2514.

McKenna A, Findlay GM, Gagnon JA, *et al.* 2016. Whole-organism lineage tracing by combinatorial and cumulative genome editing. *Science* **353**: aaf7907.

Mello CC, Kramer JM, Stinchcomb D, Ambros V. 1991. Efficient gene transfer in *C. elegans*: extrachromosomal maintenance and integration of transforming sequences. *EMBO J* **10**: 3959–3970.

Mitani S. 1995. Genetic regulation of mec-3 gene expression implicated in the specification of the mechanosensory neuron cell types in *Caenorhabditis elegans*. *Dev Growth Differ* **37**: 551–557.

Mojica FJM, Díez-Villaseñor C, García-Martínez J, Almendros C. 2009. Short motif sequences determine the targets of the prokaryotic CRISPR defence system. *Microbiol Read Engl* **155**: 733–740.

Montague TG, Cruz JM, Gagnon JA, Church GM, Valen E. 2014. CHOPCHOP: a CRISPR/Cas9 and TALEN web tool for genome editing. *Nucleic Acids Res* **42**: W401–W407.

Moreno-Mateos MA, Vejnar CE, Beaudoin J-D, *et al.* 2015. CRISPRscan: designing highly efficient sgRNAs for CRISPR/Cas9 targeting in vivo. *Nat Methods* **12**: 982–988.

Morris JZ, Tissenbaum HA, Ruvkun G. 1996. A phosphatidylinositol-3-OH kinase family member regulating longevity and diapause in *Caenorhabditis elegans*. *Nature* **382**: 536–539.

Morton J, Davis MW, Jorgensen EM, Carroll D. 2006. Induction and repair of zinc-finger nuclease-targeted double-strand breaks in *Caenorhabditis elegans* somatic cells. *Proc Natl Acad Sci USA* **103**: 16370–16375.

Moscou MJ, Bogdanove AJ. 2009. A simple cipher governs DNA recognition by TAL effectors. *Science* **326**: 1501.

Nelles DA, Fang MY, O'Connell MR, *et al.* 2016. Programmable RNA tracking in live cells with CRISPR/Cas9. *Cell* **165**: 488–496.

Nishida K, Arazoe T, Yachie N, *et al.* 2016. Targeted nucleotide editing using hybrid prokaryotic and vertebrate adaptive immune systems. *Science* **353**: aaf8729.

O'Connell MR, Oakes BL, Sternberg SH, *et al.* 2014. Programmable RNA recognition and cleavage by CRISPR/Cas9. *Nature* **516**: 263–266.

Ogg S, Paradis S, Gottlieb S, *et al.* 1997. The Fork head transcription factor DAF-16 transduces insulin-like metabolic and longevity signals in *C. elegans*. *Nature* **389**: 994–999.

Pabo CO, Peisach E, Grant RA. 2001. Design and selection of novel Cys2His2 zinc finger proteins. *Annu Rev Biochem* **70**: 313–340.

Paix A, Folkmann A, Rasoloson D, Seydoux G. 2015. High efficiency, homology-directed genome editing in *Caenorhabditis elegans* using CRISPR-Cas9 ribonucleoprotein complexes. *Genetics* **201**: 47–54.

Paix A, Wang Y, Smith HE, *et al.* 2014. Scalable and versatile genome editing using linear DNAs with microhomology to Cas9 sites in *Caenorhabditis elegans*. *Genetics* **198**: 1347–1356.

Pattanayak V, Lin S, Guilinger JP, *et al.* 2013. High-throughput profiling of off-target DNA cleavage reveals RNA-programmed Cas9 nuclease specificity. *Nat Biotechnol* **31**: 839–843.

Pavletich NP, Pabo CO. 1991. Zinc finger-DNA recognition: crystal structure of a Zif268-DNA complex at 2.1 A. *Science* **252**: 809–817.

Perez-Pinera P, Kocak DD, Vockley CM, *et al.* 2013. RNA-guided gene activation by CRISPR-Cas9-based transcription factors. *Nat Methods* **10**: 973–976.

Praitis V, Casey E, Collar D, Austin J. 2001. Creation of low-copy integrated transgenic lines in *Caenorhabditis elegans*. *Genetics* **157**: 1217–1226.

Qi LS, Larson MH, Gilbert LA, *et al.* 2013. Repurposing CRISPR as an RNA-guided platform for sequence-specific control of gene expression. *Cell* **152**: 1173–1183.

Reinhart BJ, Slack FJ, Basson M, *et al.* 2000. The 21-nucleotide let-7 RNA regulates developmental timing in *Caenorhabditis elegans. Nature* **403**: 901–906.

Ren X, Yang Z, Xu J, *et al.* 2014. Enhanced specificity and efficiency of the CRISPR/Cas9 system with optimized sgRNA parameters in *Drosophila. Cell Rep* **9**: 1151–1162.

Richardson CD, Ray GJ, DeWitt MA, Curie GL, Corn JE. 2016. Enhancing homology-directed genome editing by catalytically active and inactive CRISPR-Cas9 using asymmetric donor DNA. *Nat Biotechnol* **34**: 339–344.

Sapranauskas R, Gasiunas G, Fremaux C, *et al.* 2011. The *Streptococcus thermophilus* CRISPR/Cas system provides immunity in *Escherichia coli. Nucleic Acids Res* **39**: 9275–9282.

Shen Z, Zhang X, Chai Y, *et al.* 2014. Conditional knockouts generated by engineered CRISPR-Cas9 endonuclease reveal the roles of coronin in *C. elegans* neural development. *Dev Cell* **30**: 625–636.

Shmakov S, Abudayyeh OO, Makarova KS, *et al.* 2015. Discovery and functional characterization of diverse class 2 CRISPR-Cas systems. *Mol Cell* **60**: 385–397.

Sternberg SH, Redding S, Jinek M, Greene EC, Doudna JA. 2014. DNA interrogation by the CRISPR RNA-guided endonuclease Cas9. *Nature* **507**: 62–67.

Sulston JE, Schierenberg E, White JG, Thomson JN. 1983. The embryonic cell lineage of the nematode *Caenorhabditis elegans. Dev Biol* **100**: 64–119.

Tanenbaum ME, Gilbert LA, Qi LS, Weissman JS, Vale RD. 2014. A protein-tagging system for signal amplification in gene expression and fluorescence imaging. *Cell* **159**: 635–646.

Thompson O, Edgley M, Strasbourger P, *et al.* 2013. The Million Mutation Project: a new approach to genetics in *Caenorhabditis elegans. Genome Res* **23**: 1749–1762.

Thyme SB, Schier AF. 2016. Polq-mediated end joining is essential for surviving DNA double-strand breaks during early zebrafish development. *Cell Rep* **15**: 1611–1613.

Tian D, Diao M, Jiang Y, *et al.* 2015. Anillin regulates neuronal migration and neurite growth by linking RhoG to the actin cytoskeleton. *Curr Biol* **25**: 1135–1145.

Tsai SQ, Zheng Z, Nguyen NT, *et al.* 2015. GUIDE-seq enables genome-wide profiling of off-target cleavage by CRISPR-Cas nucleases. *Nat Biotechnol* **33**: 187–197.

Urnov FD, Rebar EJ, Holmes MC, Zhang HS, Gregory PD. 2010. Genome editing with engineered zinc finger nucleases. *Nat Rev Genet* **11**: 636–646.

van Schendel R, Roerink SF, Portegijs V, van den Heuvel S, Tijsterman M. 2015. Polymerase Θ is a key driver of genome evolution and of CRISPR/Cas9-mediated mutagenesis. *Nat Commun* **6**: 7394.

Waaijers S, Portegijs V, Kerver J, *et al.* 2013. CRISPR/Cas9-targeted mutagenesis in *Caenorhabditis elegans. Genetics* **195**: 1187–1191.

Wang S, Su J-H, Zhang F, Zhuang X. 2016. An RNA-aptamer-based two-color CRISPR labeling system. *Sci Rep* **6**: 26857.

Wang T, Wei JJ, Sabatini DM, Lander ES. 2014. Genetic screens in human cells using the CRISPR-Cas9 system. *Science* **343**: 80–84.

Ward JD. 2015. Rapid and precise engineering of the *Caenorhabditis elegans* genome with lethal mutation co-conversion and inactivation of NHEJ repair. *Genetics* **199**: 363–377.

Wei Q, Shen Y, Chen X, Shifman Y, Ellis RE. 2014. Rapid creation of forward-genetics tools for *C. briggsae* using TALENs: lessons for nonmodel organisms. *Mol Biol Evol* **31**: 468–473.

Witte H, Moreno E, Rödelsperger C, *et al.* 2015. Gene inactivation using the CRISPR/Cas9 system in the nematode *Pristionchus pacificus. Dev Genes Evol* **225**: 55–62.

Wood AJ, Lo T-W, Zeitler B, *et al.* 2011. Targeted genome editing across species using ZFNs and TALENs. *Science* **333**: 307.

Wu X, Scott DA, Kriz AJ, *et al.* 2014. Genome-wide binding of the CRISPR endonuclease Cas9 in mammalian cells. *Nat Biotechnol* **32**: 670–676.

Zalatan JG, Lee ME, Almeida R, *et al.* 2015. Engineering complex synthetic transcriptional programs with CRISPR RNA scaffolds. *Cell* **160**: 339–350.

Zetsche B, Gootenberg JS, Abudayyeh OO, *et al.* 2015. Cpf1 is a single RNA-guided endonuclease of a class 2 CRISPR-Cas system. *Cell* **163**: 759–771.

Zhang L, Ward JD, Cheng Z, Dernburg AF. 2015. The auxin-inducible degradation (AID) system enables versatile conditional protein depletion in *C. elegans*. *Dev Camb Engl* **142**: 4374–4384.

Zhao P, Zhang Z, Ke H, Yue Y, Xue D. 2014. Oligonucleotide-based targeted gene editing in *C. elegans* via the CRISPR/Cas9 system. *Cell Res* **24**: 247–250.

Zhao P, Zhang Z, Lv X, *et al.* 2016. One-step homozygosity in precise gene editing by an improved CRISPR/Cas9 system. *Cell Res* **26**: 633–636.

2 Unbiased Detection of Off-target Cleavage by CRISPR/Cas9 and TALENs Using Integration-defective Lentiviral Vectors

Youjun Wu and Jiing-Kuan Yee

2.1 Introduction

The rapid progress in the development of gene editing tools to manipulate genomes from different organisms proves invaluable for studying basic biology and human diseases. These gene editing tools, comprising site-specific endonucleases, are able to modify a complex genome almost at will by creating DNA double-strand breaks (DSBs) at specific genomic loci. DSBs are repaired through one of the two DNA repair mechanisms that normally function in cells: non-homologous end joining (NHEJ) and homology-directed repair (HDR) (Figure 2.1; Lieber, 2010). NHEJ is mediated by ligation of the broken DNA ends without a repair template. Since the repair mechanism of NHEJ is not precise, errors in the form of small insertions or deletions (indels) occur frequently. If the DSB occurs in a coding region, the indel frequently leads to an altered reading frame and gene inactivation. By contrast, HDR relies on a donor template that shares sequence homology with the intended genomic target for DNA repair. The donor template can either be the homologous sister chromatid or alternatively from an exogenous DNA that shares sequence homology with the targeted genomic site and contains the intended DNA modification (Shrivastav *et al.*, 2008; Wyman and Kanaar, 2006). Although HDR occurs normally in mammalian cells at a frequency of roughly once per 10^5 to 10^7 cells, the frequency can be increased by several orders of magnitude through the introduction of a DSB near the genomic site targeted for modification (Gaj *et al.*, 2013; Weinstock and Jasin, 2006).

Development of engineered endonucleases with tailored sequence specificities allows the introduction of DSBs at designated genomic sites to stimulate HDR (Choulika *et al.*, 1995). The ease of assembly of these endonucleases has revolutionized our capacity for gene mutation, gene insertion, transcription modulation (Gaj *et al.*, 2013), live genome imaging (Chen *et al.*, 2013), RNA tracking (Nelles *et al.*, 2016) and genomic screening (Zhou *et al.*, 2014). Up to five tailored endonucleases have been developed to create site-specific DSBs. They include meganuclease derived from mobile genetic elements or introns (Silva *et al.*, 2011), zinc-finger nuclease (ZFN) derived from eukaryotic Cys_2-His_2 zinc-finger

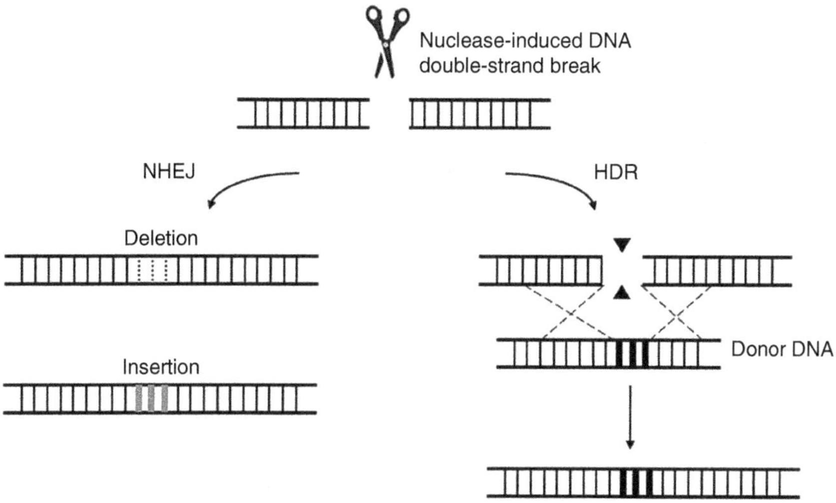

Figure 2.1 Two repair pathways for DNA double-strand breaks (DSBs). Nuclease-mediated DSBs can be repaired by non-homologous end joining (NHEJ) or homology-directed repair (HDR) pathway. NHEJ often generates small insertions or deletions (indels), while HDR generates precise gene mutations or insertions in the presence of a donor DNA template carrying homologous sequences flanking the nuclease cleavage site.

proteins covalently linked to the nuclease domain of the type IIS restriction enzyme Fok1 (Kim *et al.*, 1996), transcription activator-like effector nuclease derived from transcription activator-like effector (TALE) of bacteria *Xanthomonas* linked with the Fok1 nuclease domain (Christian *et al.*, 2010), clustered regularly interspaced short palindromic repeat (CRISPR) with CRISPR-associated Cas9 nuclease from *Streptococcus pyogenes* (spCas9) (Jinek *et al.*, 2012) and more recently the argonaute protein from *Natronobacterium gregoryi* (NgAgo) with a 5′-phosphorylated single-stranded guide DNA (gDNA; Gao *et al.*, 2016). Meganuclease, ZFN and TALEN interact with their genomic targets via protein–DNA interactions. CRISPR/Cas9 and NgAgo-gDNA form Watson–Crick base pairings between the targeted genomic sequence and the single guide RNA (sgRNA) or the gDNA, respectively. Target recognition by CRISPR/Cas9 also depends on a conserved dinucleotide-containing protospacer adjacent motif (PAM) in the genomic target adjacent to the sgRNA binding site. By substituting the 5′ 20 bp sequence in the sgRNA that forms base pairing with the genomic target, the Cas9 nuclease can cleave virtually any genomic site with an adjacent PAM sequence (Mojica *et al.*, 2009; Shah *et al.*, 2013). In contrast, the NgAgo-gDNA system does not require a PAM. The extensive experience with CRISPR/Cas9, the ease of retargeting Cas9, its highly efficient nuclease activity and the capacity for multiplexed target recognition make CRISPR/Cas9 one of the most popular systems for carrying out gene editing in many different organisms. However, these advantages of CRISPR/Cas9 may be replaced by the NgAgo-gDNA system because of the lack of a requirement for PAM with NgAgo-gDNA. Lack of target sequence preference for gene editing with NgAgo-gDNA may also represent another advantage of this gene editing system over CRISPR/Cas9 (Gao *et al.*, 2016). Nevertheless, more data needs to be gathered, especially for NgAgo-gDNA, to determine whether the on-target cleavage efficiency and the off-target effect of these two systems are equivalent.

One of the most important concerns in gene editing is the cleavage of unintended genomic targets by the engineered nuclease. Cleavage by an engineered nuclease in unintended genes or *cis*-regulatory elements followed by NHEJ can cause aberrant gene expression and alter the cellular activities. This risk raises the safety concern of applying gene editing in disease treatment in particular. Taking CRISPR/Cas9 as an example, the Cas9 protein clearly tolerates mismatches throughout the pairing between the sgRNA and its genomic target, depending on the number, position and distribution of the mismatches (Lin *et al.*, 2014). These factors have been incorporated to create several algorithms for predicting potential off-target sites of a given sgRNA. Since the epigenetic status of a genomic site can affect its accessibility to CRISPR/Cas9 and is not considered in these web-based algorithms, it is not known what fraction of the predicted off-target sites is actually recognized and cleaved by the engineered nuclease (Hsu *et al.*, 2013; Wu *et al.*, 2014). The accessibility of a genomic site may also be altered, depending on the proliferation and differentiation status of the target cell. Thus, computational design programs can help finding target sites with minimal off-target cleavage, but empirical testing in appropriate cell types will likely be required to ensure optimal specificity performance. The risk of an off-target effect is a concern for interpreting the phenotype derived from gene editing, which can be addressed by several strategies: (1) Similar phenotypes are generated with multiple sgRNAs targeting the same genomic region. It is highly unlikely that multiple sgRNAs with different off-target effects will generate the same phenotype. (2) Introduction of the wild-type gene to complement the lost gene activity can reverse the phenotype of the mutant cells. (3) Secondary gene editing to correct the mutation reverses the phenotype. These strategies, however, are not applicable in the context of using gene edited primary cell populations for therapeutic treatment, as these cells cannot proliferate indefinitely in culture. Whole genomic sequencing (WGS) of gene edited cell populations represents an unbiased approach to detect off-target sites (Smith *et al.*, 2014; Veres *et al.*, 2014). However, WGS currently lacks the sensitivity to detect rare off-target cleavage events in large cell populations, and is labor-intensive and expensive. It is therefore important to seek other more sensitive, genome-wide and unbiased approaches to identify off-target cleavages by an engineered nuclease.

Lentiviral vectors are widely used in gene transfer and gene therapy due to their highly efficient infectivity of mammalian cells and stable integration of the delivered gene into the host genome (Vigna and Naldini, 2000). However, the vector system is limited by semi-random integration into active host genes that can cause insertional mutagenesis (Hacein-Bey-Abina *et al.*, 2003). This drawback has been overcome by the development of an integration-defective lentiviral vector (IDLV) containing a mutant integrase (IN) that supports viral reverse transcription and virion morphogenesis, but can no longer mediate viral DNA integration into the host genome (Wanisch and Yanez-Munoz, 2009). In IDLV transduced cells, the linear vector genome remains episomal and is lost with extended cell division (Pierson *et al.*, 2002). In non-dividing cells or slowly dividing cells, the vector genome can persist extensively (Saenz *et al.*, 2004). Gabriel *et al.* first showed that the IDLV genome present in the nuclei of transduced cells could be incorporated into DSBs by NHEJ, thereby stably tagging those DSBs generated by DNA damaging agents, including ZFNs (Gabriel *et al.*, 2011). They mapped the IDLV-tagged sites and showed that most of IDLVs were incorporated less than 60 bp away from the ZFN cleavage site. This was consistent with earlier studies showing that small indels derived from NHEJ were less than 60 bp from the nuclease cleavage site (Honma *et al.*, 2007) The off-target site was determined based on two criteria: (1) two or more independent IDLV tagging events occurred within any 500 bp genomic fragment; (2) within this genomic fragment, a sequence that shares sequence homology with the on-target site could be identified. They termed this exercise clustered integration site (CLIS) analysis. This analysis is based on the hypothesis that only specific cleavage by an engineered nuclease will allow repeated IDLV incorporation into the same genomic region. Transient ZFN expression followed by PCR amplification of those IDLV-tagged off-target sites and pyrosequencing validated the off-target sites predicted by the CLIS analysis. This study therefore provides a strategy for unbiased and genome-wide mapping of off-target sites generated by an engineered nuclease. Osborn *et al.* used the same IDLV strategy to identify three off-target sites generated by a TALEN pair specific for the gene encoding type VII collagen in HEK293 cells (Osborn *et al.*, 2013). In this chapter, we describe the use of this strategy to mark DSBs generated by TALEN and CRISPR/Cas9 followed by mapping of these off-target sites in human cell lines. We also discuss some factors that could influence off-target site detection and the mismatch tolerance of CRISPR/Cas9 (Wang *et al.*, 2015).

2.2 Methodology

2.2.1 Transfection of CRISPR/Cas9 and TALEN Expression Plasmids

HEK293T cells and HT1080 cells were maintained in high-glucose DMEM (Life Technologies, Grand Island, NY) supplemented with 10% characterized fetal bovine serum (Life Technologies), 2 mM L-glutamine, 100 U/mL penicillin and 100 mg/mL streptomycin (Life Technologies). Cells were seeded in 48-well plates at 40% confluence and transfected with 150 ng TALEN plasmids or co-transfected with 67 ng CRISPR and 200ng Cas9 expression plasmids using Lipofectamine 2000 (ThermoFisher, Waltham, MA). After 40 h, the genomic DNA was extracted with Epicentre QuickExtract solution (Epicentre Biotechnologies, Madison, WI).

2.2.2 Surveyor Assay

This assay relies on the formation of heteroduplex DNA between mutant and wild-type alleles (Guschin *et al.*, 2010). The mismatched nucleotides can be recognized by the Surveyor nuclease and cleaved. As DSBs from an engineered nuclease induce indel formation due to NHEJ, a PCR-amplified genomic fragment spanning the indel, after denaturation and renaturation, can be recognized and cleaved by the Surveyor enzyme (Figure 2.2). The efficiency of an engineered nuclease is calculated based on the fraction of cleaved fragments. To carry out the assay, the nuclease expression plasmid was first transfected into the target cells and the genomic DNA prepared 40 h after transfection. The genomic region spanning the nuclease cleavage site was PCR-amplified using Hotstar Taq DNA polymerase following the manufacturer's instructions (Qiagen, Valencia, CA). PCR conditions were as follows: initial DNA denaturation for 15 min at 95 °C; 35 PCR cycles at 94 °C for 30 s, 60 °C for 30 s and 72 °C for 30 s; final extension for 5 min at 72 °C. The PCR product was denatured and renatured using the following conditions: 95 °C for 10 min, 95 °C to 85 °C ramping at −2 °C/s, 85 °C to 25 °C at −0.3 °C/s, and 25 °C for 1 min. The renatured products were treated with the Surveyor enzyme and resolved on 2% agarose gel. Densitometry was performed and the cleavage efficiency was quantified using image J. The indel percentage was calculated by the following equations:

indel formation (%) = 1 – [(1 – fraction cleaved) $^{1/2}$] fraction cleaved = (sum of cleaved band intensities)/(sum of the cleaved and parental band intensities)

2.2.3 IDLV Production and Transduction

To generate IDLV, HEK293T cells were transfected with pC-HELP, pCMV-G, and pHIV7/pgk-puro. The packaging plasmid, pC-HELP, contains a mutation in the HIV-1 *IN* gene that abolishes the catalytic activity of the integrase. The envelope protein expression plasmid, pCMV-G, contains the gene encoding the glycoprotein of vesicular stomatitis virus, controlled by the immediate early promoter of cytomegalovirus. The lentiviral construct, pHIV7/pgk-puro, contains the puromycin-resistant (pur^R) gene controlled by the phosphoglycerate kinase (pgk) promoter. Infectious IDLV was harvested 40 h after transfection, concentrated in polyethylene glycol, and titered on HT1080 cells for puromycin resistance (2 μg/mL). Although IDLV is non-integrating, its linear genome in the nucleus can be incorporated into naturally occurring DSBs in the host genome via the mechanism of NHEJ. IDLV titers determined by this approach therefore represent relative infectious titers rather than absolute titers. To generate a pool of pur^R clones, approximately 4×10^6 HEK293T cells or HT 1080 cells were transduced with IDLV (3000 random integration equivalents)16 h after the transfection of the expression plasmid for the engineered nuclease. Transduced cells were subjected to puromycin selection at a concentration of 2 μg/mL 40 h after the transduction. After two weeks in selection, the genomic DNA from pools of pur^R colonies was prepared and subjected to non-restrictive linear amplification-mediated polymerase chain reaction (nrLAM-PCR) analysis.

2.2.4 nrLAM-PCR

IDLV-tagged DSBs were detected by nrLAM-PCR according to the protocol described by Paruzynski *et al.*, using primers specific for the long terminal repeat (LTR) of IDLV and linker primers (Paruzynski *et al.*, 2010; Figure 2.3). A LTR primer

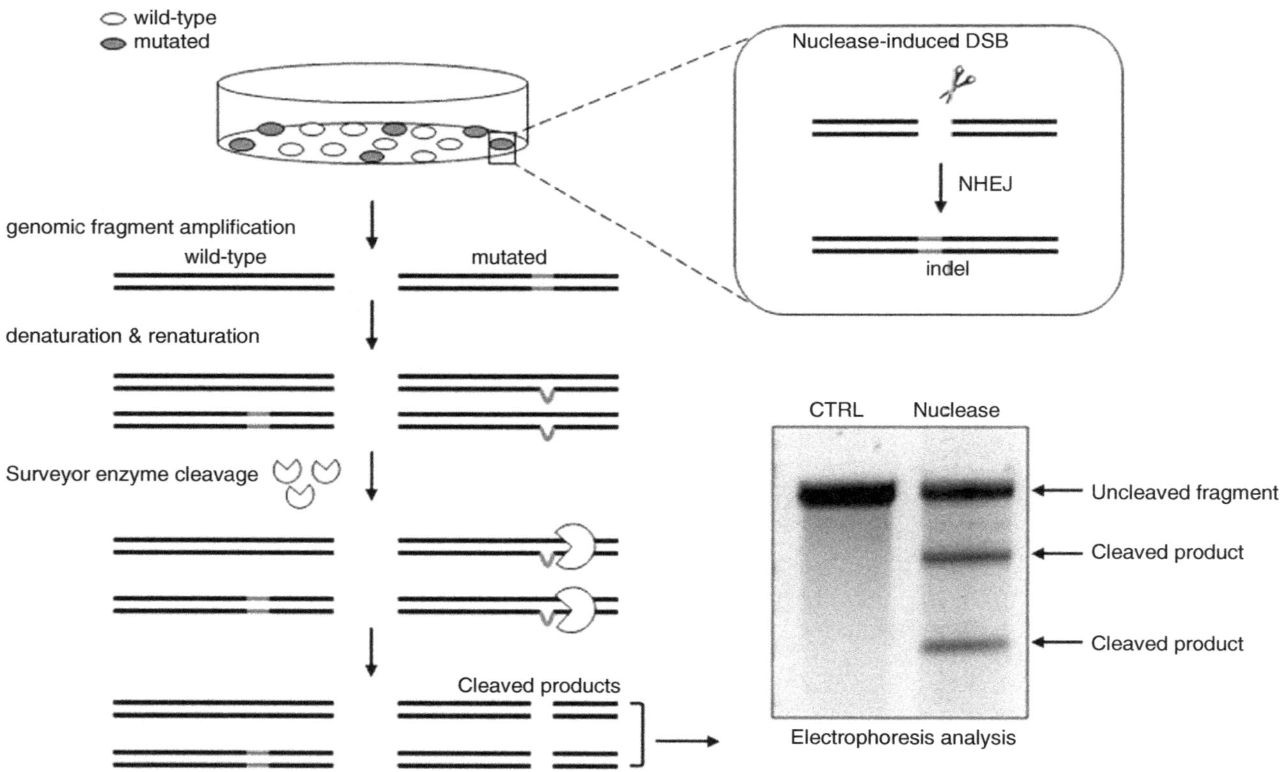

Figure 2.2 Detection of the nuclease-induced indel by the Surveyor assay. Genomic fragments spanning indel loci are amplified, the PCR products are allowed to denature and renature, followed by the digestion of the Surveyor enzyme, which recognizes and cleaves mismatches. The cleaved DNA products are visualized by gel electrophoresis and quantified. Arrows indicate the uncleaved and cleaved DNA fragments. Cells without nuclease expression are set as the negative control.

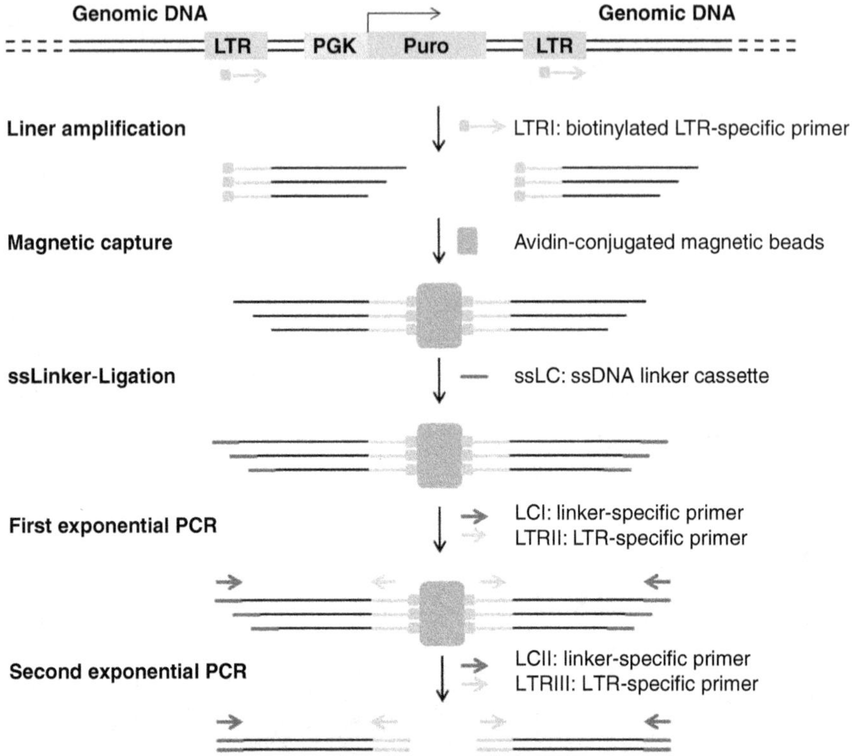

Figure 2.3 Scheme of nrLAM-PCR-mediated detection of IDLV-tagged DSBs. The process of nrLAM-PCR is initiated with linear amplification of vector-genomic DNA junctions using a 5′-biotinylated vector-specific primer (LTRI). The PCR product is captured by streptavidin-conjugated magnetic beads. The captured product is ligated to a single-strand linker cassette (ssLC) so that subsequent exponential PCR can be carried out using linker-specific primer LCI and vector-specific primer LTRII. After magnetic capture, the product is subjected to a second exponential PCR using linker- (LCII) and vector-specific primers (LTRIII). (A black-and-white version of this figure will appear in some formats. For the color version, please refer to the plate section.)

was biotinylated and used first to carry out a linear amplification step on the host genomic DNA adjacent to the incorporated IDLV genome. This step specifically amplifies the vector-genome junction. The amplified single-stranded fragments were purified by streptavidin conjugated to magnetic beads. The isolated single-stranded DNA fragments were ligated to a linker oligodeoxynucleotide, followed by exponential nested PCR and high-throughput sequencing to identify the tagged IDLV genomic sites.

2.2.5 Deep-sequencing Analysis of IDLV-tagged Genomic Sites

Before deep sequencing, the nrLAM-PCR product was verified by an Agilent bioanalyzer, quantified using qPCR and analyzed by a paired-end read flow cell for cluster generation after dilution to a final concentration of 10 pM. Paired-end reads were performed using an Illumina HiSeq2500 instrument with 5′ adaptor and 3′ adaptor primers. The reads that contained the two adaptors were mapped to

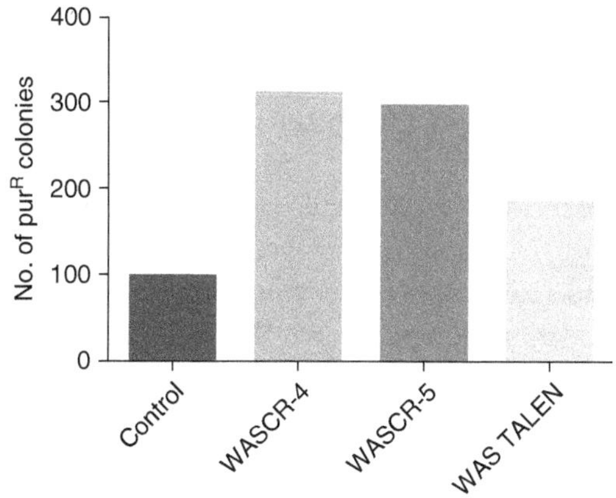

Figure 2.4 Donor template delivery via IDLV transduction increases the frequency of the pur^R colony. HEK293T cells were transduced with IDLV 16 hours after transfection with the CRISPR/Cas9 or TALEN expression plasmids as indicated. Transduced cells were subjected to puromycin selection 40 hours after IDLV transduction. Puromycin-resistant colonies were scored after two weeks. Cells transduced with IDLV only without the nuclease expression plasmids were used as the negative control.

Novoalign v2.07 and analyzed by the Bioconductor package "CHIPpeakAnno" to identify IDLV-tagged genomic sites. Sites with more than 20 reads were defined as specific IDLV tagging sites. The potential sites of nuclease cleavage were filtered out by CLIS analysis with two criteria: (1) two or more independent IDLV tagging events occurred within any 500 bp genomic fragment, (2) within this genomic fragment, a sequence that shares sequence homology with the on-target site could be identified.

2.3 Results and Discussion

2.3.1 DSBs Increase IDLV Genomic Tagging

To determine the targeting specificity of an engineered nuclease, we evaluated two genetic loci: genes encoding the Wiskott–Aldrich Syndrome (WAS) protein and tyrosine aminotransferase (TAT). For each gene, we designed three CRISPRs and a TALEN pair targeting the same genomic region. The cleavage efficiency of these engineered nucleases was determined in HEK293T cells by the Surveyor assay. Once the on-target cleavage by these engineered nucleases was confirmed, the expression plasmids for the nuclease were introduced into HEK293T cells through DNA transfection. An IDLV with the puromycin-resistance (pur^R) gene was used to transduce the HEK293T cells 16 h after transfection. Incorporation of the IDLV by on- and off-target cleavages led to the emergence of pur^R colonies upon selection. As shown in our study, inclusion of CRISPR/Cas9 or TALEN increased the number of pur^R colonies by two- to threefold when compared with the control without the engineered nuclease (Figure 2.4). This result is consistent with the hypothesis that site-specific cleavage by these nucleases increases the frequency of IDLV incorporation into the DSBs.

2.3.2 Analysis of IDLV Incorporation by nrLAM-PCR and Deep Sequencing

The IDLV-tagged DSBs were selectively amplified by nrLAM-PCR from ~3000 pooled pur^R colonies and subjected to high-throughput sequencing. We observed that ~80% of the mapped IDLV-tagged genomic sites had less than 20 reads and none of these sites appeared in a study repeat. We also used the CLIS analysis to determine whether additional IDLV incorporation events could be detected near these sites. We found that no other IDLV incorporation event took place within 500 bp of these sites. In addition, no extensive sequence homology with the on-target sequence could be detected within 500 bp of these IDLV-tagged genomic sites. Based on these findings, these sites were considered random incorporation of the IDLV caused by naturally occurring DSBs instead of those generated by the engineered nuclease. As expected, the CLIS analysis showed multiple IDLV incorporations with more than 20 reads near the on-target site. IDLV integration usually occurred within 60 bp of the on-target cleavage site, consistent with the notion that most of NHEJ-mediated events occurred within 60 bp of the DSB (Honma *et al.*, 2007). The CLIS analysis also identified additional cleavage sites with two of the three CRISPRs for the *WAS* gene and one of the three CRISPRs for the *TAT* gene. These off-target sites not only harbored multiple IDLV incorporations, with more than 20 reads for each incorporation, they also exhibited sequence homology with the on-target site. To confirm that these off-target sites marked by IDLV did indeed result from cleavage by the site-specific nuclease, we transfected the expression plasmids for the CRISPR of interest and Cas9 into HEK293T cells and directly measured the nuclease cleavage efficiency of each mapped off-target site using the Surveyor assay and deep sequencing. These two different assays confirmed that the IDLV-tagged genomic sites represented genuine off-target sites generated by the cleavage of CRISPR/Cas9. Based on these analyses, we estimated that the detection limit for indel formation was approximately 0.5%. As a comparison, the indel formation at the on-target site was >10%. Using the same IDLV tagging strategy, we were not able to detect any off-target site in HEK293T cells transfected with the TALEN expression plasmids for the *WAS* and the *TAT* genes. To ensure that this is not unique to the two TALEN pairs we examined, we used the same strategy to analyze two additional TALEN pairs targeting *ZRSR2* and *miR-9-2* genes, respectively. No off-target site could be detected with these two additional TALEN pairs in HEK293T cells. These studies suggest that TALEN may have improved specificity over CRISPR/Cas9 for target recognition, although off-target effects with TALEN have been reported in a few cases (Guilinger *et al.*, 2014a; Osborn *et al.*, 2013).

2.3.3 Evaluation of the Sensitivity of the IDLV Tagging Strategy

Although the IDLV tagging assay detects indel formation at a frequency as low as 0.5%, it is not clear whether this assay detects all the off-target sites. To address this issue, we used *in silico* off-target site prediction software, plus additional selection criteria to generate a list of potential off-target sites. The criteria for predicting a potential off-target site include: (1) a PAM sequence at the 3′ end of the non-target strand, (2) no more than one mismatch between sgRNA and the off-target site from positions 1 to 10 proximal to the PAM sequence, (3) when there is one mismatch from positions 1 to 10, no more than two mismatches from positions 11 to 20 between the sgRNA and the off-target site, (4) if there is a complete match from positions 1 to 10, four mismatches in total from positions 11 to 20 are allowed. These selection criteria generated a list of 29 potential off-target sites for WASCR-4, one of the three CRISPRs specific for the *WAS* gene. The list includes six of the seven off-target sites mapped by the IDLV tagging strategy. Only one off-target site, in the *MycN* gene, with six mismatches from positions 11 to 20, is not covered by the list. To determine whether the other 23 potential off-target sites that the IDLV tagging strategy failed to identify were cleaved by WASCR-4, HEK293T cells were co-transfected with WASCR-4 and Cas9 expression plasmids and the genomic regions spanning these potential off-target sites were PCR-amplified and subjected to deep sequencing. Out of the 23 potential off-target sites, 5 exhibited indel formation at a frequency ranging from 0.04–0.6%. This result suggests that the lower limit for IDLV to detect DSBs is around 0.5% and those off-target sites with an indel frequency lower than 0.5% cannot be consistently identified by the IDLV strategy. We carried out a similar study with another*WAS*-specific CRISPR, WASCR-5. Deep sequencing failed to detect any cleavage in the 22 potential off-target sites tested. With limited data sets, it therefore remains difficult at present to derive any general rule to predict off-target cleavage by a given CRISPR based entirely on the primary sequence homology between a potential off-target site and the on-target site.

To verify the reliability of the IDLV tagging strategy, we applied this assay on a well-characterized CRISPR specific for the *VEGFA* gene site 3. A previous study confirmed the presence of a minimum of seven off-target sites with this CRISPR (Fu *et al.*, 2013). We transfected HEK293T cells with the CRISPR and Cas9 expression plasmids, followed by IDLV transduction and selection for pur^R colonies. Our analysis showed that six of the seven previously mapped off-target sites were identified by IDLV tagging. Only a site at the *OT3-17* locus failed to be identified by IDLV and this site had the lowest indel frequency based on the previous study (Fu *et al.*, 2013). However, the IDLV strategy identified two additional low-frequency off-target sites near the *LEPREL2* and *LINC00887* loci. Together, our study demonstrates that IDLV consistently identifies all of the off-target sites with an indel frequency of 1% or higher. However, with the experimental conditions we used, detection of off-target sites with indel frequencies lower than 1% is sporadic and inconsistent.

2.3.4 Factors Influencing Off-target Detection

We next evaluated factors that might affect off-target site detection. We measured the effect of the pool size of the pur^R colony on off-target site detection. The above-mentioned study used a pool of ~3000 HEK293T colonies. We repeated the study with WASCR-4 and collected pools of ~500 and ~15 000 colonies. The CLIS analysis carried out with the 500 colony pool identified only five of the seven off-target sites. The two sites with the lowest indel frequency failed to be detected in this colony pool. With the ~15 000 colony pool, all of the seven off-target sites were repeatedly identified. Moreover, four additional off-target sites were identified in this pool. These four sites have the least numbers of mapped IDLV incorporation events and higher numbers of nucleotide mismatches when compared with the on-target site. This is consistent with the notion that the four new sites represent those with lower nuclease cleavage efficiencies and can only be detected when the colony pool size is increased. These data suggests that the sensitivity of the IDLV tagging assay for off-target site detection can be further increased with an enlarged colony pool size. We also evaluated the effect of nuclease concentration on the off-target effect. It has been reported that reduction of the CRISPR/Cas9 concentration could reduce the off-target effect (Fu *et al.*, 2013; Hsu *et al.*, 2013). We used an IDLV tagging assay to map off-target sites in HEK293T cells transfected with threefold lower WASCR-4 and Cas9 expression plasmids. Among the seven off-target sites mapped previously, two sites with the lowest indel frequency failed to be detected, suggesting that the dosage of an engineered nuclease could affect off-target site detection. Taken together, the data demonstrates that the sensitivity of the IDLV tagging assay for off-target detection can be modulated by multiple factors, and modulation of these factors should further improve the detection sensitivity, ideally, to be near the level of deep sequencing.

2.3.5 Effect of Bulge Formation on CRISPR/Cas9-mediated Cleavage

The imperfect complementarity between the on- and off-target sequences suggests that some mismatches can be tolerated by the CRISPR/Cas9 system. Previous studies have shown that if the mismatches exceeded five base pairs, efficient cleavage by CRISPR/Cas9 would be severely compromised or even completely abolished (Fu *et al.*, 2013; Hsu *et al.*, 2013). However, two WASCR-4-derived off-target sites in the *MycN* and *SH2D1A* genes exhibit six and thirteen mismatches, respectively, when compared with the on-target sequence. These two loci represent genuine off-target sites as deep sequencing confirmed the formation of indels by WASCR-4. This finding is therefore inconsistent with the notion that mismatches over five base pairs would abolish Cas9 cleavage. We noticed, however, that if the base "A" at position 13 distal to the PAM sequence in the sgRNA for *MycN* was skipped, leading to a bulge formation, the pairing between the sgRNA and the off-target site would have only two instead of six mismatches (Figure 2.5). Similarly, if the base "U" at position 1 in the sgRNA for *SH2D1A* was skipped, the mismatches between the sgRNA and the off-target site were reduced to only two instead of thirteen mismatches. Systematically skipping a single base from the 20 bases of the sgRNA followed by the Surveyor assay confirmed that bulges formed at certain positions of the base pairing between the sgRNA and the on-target sequence could be tolerated and cleaved by Cas9. These findings expand the repertoire of potential off-target sites and should be taken into consideration when designing CRISPR for gene editing.

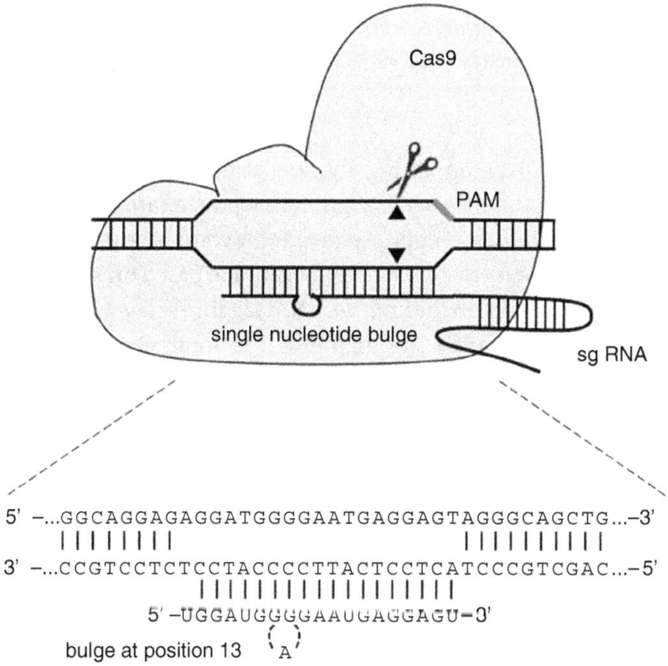

Figure 2.5 Bulge formed between the sgRNA and its genomic target can be tolerated by CRISPR/Cas9 and generate cleavage. The sequence shows that the off-target site for WASCR-4 in the *MycN* gene can form base pairing with the sgRNA with only two mismatches when the nucleotide "A" at position 13 of the sgRNA is skipped.

2.4 Concluding Remarks

Besides the IDLV tagging strategy described here, there are currently four other unbiased, genome-wide strategies to map off-target sites generated by an engineered nuclease. These strategies include: (1) genome-wide, unbiased identification of DSBs enabled by sequencing (GUIDE-seq; Tsai *et al.*, 2015). Instead of IDLV, this strategy uses double-stranded oligodeoxynucleotides (dsODNs) to tag DSBs generated by the engineered nuclease. The ODN-tagged genomic sequence is then selectively amplified by PCR and subjected to deep sequencing, (2) high-throughput, genome-wide, translocation sequencing (HTGTS; Frock *et al.*, 2015). This strategy relies on chromosome translocations between DSBs generated at on- and off-target sites. The translocation breakpoint is then mapped by LAM-PCR using primers specific for the on-target sequence and an added linker, (3) direct *in situ* break labeling, enrichment on streptavidin and next-generation sequencing (BLESS; Crosetto *et al.*, 2013). This strategy involves ligation of a biotin-labeled first linker *in situ* to DSBs in fixed nuclei to avoid DSBs generated during genomic DNA extraction. The genomic DNA is then fragmented and linker-containing DNA fragments captured by streptavidin. A second linker is added to the captured fragments, followed by PCR amplification and deep sequencing, (4) *in vitro* Cas9-digested whole-genome sequencing (digenome-seq; Kim *et al.*, 2015). This strategy first digests purified mock-transfected genomic DNA with the engineered nuclease *in vitro*. Such digestion will generate sequence reads that can be vertically aligned at the on-target and potential off-target sites. Expression of the engineered nuclease in cells will lead to indel formation at both the on- and off-target sites. Since indels are resistant to nuclease digestion, genomic DNA containing an indel at the on- and off-target sites will exhibit an altered sequence read alignment upon treatment of the isolated genomic DNA with the same nuclease *in vitro*. A computer program is then used to identify those off-target sites across the genome based on sequence read alignment. Some of the assays, including the IDLV tagging assay, have been used by multiple groups for off-target site mapping and some claim to have sensitivity near the limit of detection, as directed by deep sequencing (Frock *et al.*, 2015; Kim *et al.*, 2015; Ran *et al.*, 2015; Wang *et al.*, 2015,). However, none of them is capable of detecting all off-target sites.

Significant progress has also been made lately to modify the engineered nuclease, mainly the CRISPR/Cas9 system, to increase its targeting specificity. These improvements include: (1) Truncated sgRNAs with shorter regions of target complementarity in the range of 17–18 nucleotides in length reduce the off-target effect without sacrificing the on-target cleavage efficiency (Fu *et al.*, 2014). (2) The use of D10A mutant of spCas9(Cas9n) (Ran *et al.*, 2013). Instead of generating DSBs, Cas9n generates nicks. To generate DSBs, a pair of offset sgRNAs complementary to opposite strands of the target site is required. Since nicks are predominantly repaired through the high fidelity base excision repair, the use of Cas9n significantly reduces the off-target effect. (3) Fusion between the Fok1 nuclease domain onto a catalytically inactive Cas9 (dCas9) (Guilinger *et al.*, 2014b). Since Fok1 nuclease is only activated when the nuclease domain is dimerized, DSBs can only be generated when the genomic target is bound by two sgRNAs with defined spacing and orientation. This requirement substantially reduces the possibility of off-target cleavage. (4) Study in TALEN shows that by altering the protein residues, TALEN mutants that discriminate between on- and off-target sites can be derived (Guilinger *et al.*, 2014a). A similar approach was applied to modify the Cas9. Two different mutant Cas9 proteins containing multiple residue variants were shown to destabilize the interaction between sgRNA and the off-target site, whereas their interaction with the on-target site remains stable (Ran *et al.*, 2013; Shen *et al.*, 2014). Thus, domain modification of Cas9 may represent an effective way to reduce the off-target effect. (5) Fusion between dCas9 and a cytidine deaminase that retains the ability to be programmed with a sgRNA (Komor *et al.*, 2016). This fusion protein does not induce DSBs, but mediates the direct conversion of cytidine to uridine, thereby effecting a C to T (or G to A) substitution. This strategy effectively corrects a variety of point mutations relevant to human disease. Significantly, fewer off-target effects are expected with this strategy since no DSB is generated by the Cas9-cytidine deaminase fusion protein. In summary, the combination of unbiased, genome-wide strategies for the detection of off-target effects and the significant progress achieved in enhancing the targeting specificity of CRISPR/Cas9 constitutes the foundation for applying the powerful gene editing system effectively and safely in basic research and clinical application.

REFERENCES

Chen B, Gilbert LA, Cimini BA, *et al.* 2013. Dynamic imaging of genomic loci in living human cells by an optimized CRISPR/Cas system. *Cell* **155**: 1479–1491.

Choulika A, Perrin A, Dujon B, Nicolas JF. 1995. Induction of homologous recombination in mammalian chromosomes by using the I-Scei system of *Saccharomyces cerevisiae. Mol Cell Biol* **15**: 1968–1973.

Christian M, Cermak T, Doyle EL, *et al.* 2010. Targeting DNA double-strand breaks with TAL effector nucleases. *Genetics* **186**: 757–761.

Crosetto N, Mitra A, Silva MJ, *et al.* 2013. Nucleotide-resolution DNA double-strand break mapping by next-generation sequencing. *Nat Methods* **10**: 361–365.

Frock RL, Hu J, Meyers RM, *et al.* 2015. Genome-wide detection of DNA double-stranded breaks induced by engineered nucleases. *Nat Biotechnol* **33**: 179–186.

Fu Y, Foden JA, Khayter C, *et al.* 2013. High-frequency off-target mutagenesis induced by CRISPR-Cas nucleases in human cells. *Nat Biotechnol* **31**: 822–826.

Fu Y, Sander JD, Reyon D, *et al.* 2014. Improving CRISPR-Cas nuclease specificity using truncated guide RNAs. *Nat Biotechnol* **32**: 279–284.

Gabriel R, Lombardo A, Arens A, *et al.* 2011. An unbiased genome-wide analysis of zinc-finger nuclease specificity. *Nat Biotechnol* **29**: 816–823.

Gaj T, Gersbach CA, Barbas CF III. 2013. ZFN, TALEN, and CRISPR/Cas-based methods for genome engineering. *Trends Biotechnol* **31**: 397–405.

Gao F, Shen XZ, Jiang F, Wu Y, Han, C. 2016. DNA-guided genome editing using the Natronobacterium gregoryi Argonaute. *Nat Biotechnol* **34**: 768–773.

Guilinger JP, Pattanayak V, Reyon D, *et al.* 2014a. Broad specificity profiling of TALENs results in engineered nucleases with improved DNA-cleavage specificity. *Nat Methods* **11**: 429–435.

Guilinger JP, Thompson DB, Liu DR. 2014b. Fusion of catalytically inactive Cas9 to FokI nuclease improves the specificity of genome modification. *Nat Biotechnol* **32**: 577–582.

Guschin DY, Waite, AJ, Katibah GE, *et al.* 2010. A rapid and general assay for monitoring endogenous gene modification. *Methods Mol Biol* **649**: 247–256.

Hacein-Bey-Abina S, Von Kalle C, Schmidt M, *et al.* 2003. A serious adverse event after successful gene therapy for X-linked severe combined immunodeficiency. *N Engl J Med* **348**: 255–256.

Honma M, Sakuraba M, Koizumi T, *et al.* 2007. Non-homologous end-joining for repairing I-SceI-induced DNA double strand breaks in human cells. *DNA Repair (Amst)* **6**: 781–788.

Hsu PD, Scott DA, Weinstein JA, *et al.* 2013. DNA targeting specificity of RNA-guided Cas9 nucleases. *Nat Biotechnol* **31**: 827–832.

Jinek M, Chylinski K, Fonfara I, *et al.* 2012. A programmable dual-RNA-guided DNA endonuclease in adaptive bacterial immunity. *Science* **337**: 816–821.

Kim D, Bae S, Park J, *et al.* 2015. Digenome-seq: genome-wide profiling of CRISPR-Cas9 off-target effects in human cells. *Nat Methods* **12**: 237–243, 1 p following 243.

Kim YG, Cha J, Chandrasegaran S. 1996. Hybrid restriction enzymes: zinc finger fusions to Fok I cleavage domain. *Proc Natl Acad Sci USA* **93**: 1156–1160.

Komor AC, Kim YB, Packer MS, Zuris JA, Liu DR. 2016. Programmable editing of a target base in genomic DNA without double-stranded DNA cleavage. *Nature* **533**: 420–424.

Lieber MR. 2010. The mechanism of double-strand DNA break repair by the nonhomologous DNA end-joining pathway. *Annu Rev Biochem* **79**: 181–211.

Lin YN, Cradick TJ, Brown MT, *et al.* 2014. CRISPR/Cas9 systems have off-target activity with insertions or deletions between target DNA and guide RNA sequences. *Nucleic Acids Res* **42**: 7473–7485.

Mojica FJ, Diez-Villasenor C, Garcia-Martinez J, Almendros C. 2009. Short motif sequences determine the targets of the prokaryotic CRISPR defence system. *Microbiology* **155**: 733–740.

Nelles DA, Fang MY, O'Connell, MR, *et al.* 2016. Programmable RNA tracking in live cells with CRISPR/Cas9. *Cell* **165**: 488–496.

Osborn MJ, Starker CG, McElroy AN, *et al.* 2013. TALEN-based gene correction for epidermolysis bullosa. *Mol Ther* **21**: 1151–1159.

Paruzynski A, Arens A, Gabriel R, *et al.* 2010. Genome-wide high-throughput integrome analyses by nrLAM-PCR and next-generation sequencing. *Nat Protoc* **5**: 1379–1395.

Pierson TC, Kieffer TL, Ruff CT, *et al.* 2002. Intrinsic stability of episomal circles formed during human immunodeficiency virus type 1 replication. *J Virol* **76**: 4138–4144.

Ran FA, Cong L, Yan WX, *et al.* 2015. In vivo genome editing using *Staphylococcus aureus* Cas9. *Nature* **520**: 186–191.

Ran FA, Hsu PD, Lin CY, *et al.* 2013. Double nicking by RNA-guided CRISPR Cas9 for enhanced genome editing specificity. *Cell* **154**: 1380–1389.

Saenz DT, Loewen N, Peretz M, *et al.* 2004. Unintegrated lentivirus DNA persistence and accessibility to expression in nondividing cells: analysis with class I integrase mutants. *J Virol* **78**: 2906–2920.

Shah SA, Erdmann S, Mojica FJ, Garrett RA. 2013. Protospacer recognition motifs: mixed identities and functional diversity. *RNA Biol* **10**: 891–899.

Shen B, Zhang W, Zhang J, *et al.* 2014. Efficient genome modification by CRISPR-Cas9 nickase with minimal off-target effects. *Nat Methods* **11**: 399–402.

Shrivastav M, De Haro LP, Nickoloff JA. 2008. Regulation of DNA double-strand break repair pathway choice. *Cell Res* **18**: 134–147.

Silva G, Poirot L, Galetto R, *et al.* 2011. Meganucleases and other tools for targeted genome engineering: perspectives and challenges for gene therapy. *Curr Gene Ther* **11**: 11–27.

Smith C, Gore A, Yan W, *et al.* 2014. Whole-genome sequencing analysis reveals high specificity of CRISPR/Cas9 and TALEN-based genome editing in human iPSCs. *Cell Stem Cell* **15**: 12–13.

Tsai SQ, Zheng Z, Nguyen NT, *et al.* 2015. GUIDE-seq enables genome-wide profiling of off-target cleavage by CRISPR-Cas nucleases. *Nat Biotechnol* **33**: 187–197.

Veres A, Gosis BS, Ding Q, *et al.* 2014. Low incidence of off-target mutations in individual CRISPR-Cas9 and TALEN targeted human stem cell clones detected by whole-genome sequencing. *Cell Stem Cell* **15**: 27–30.

Vigna E, Naldini L. 2000. Lentiviral vectors: excellent tools for experimental gene transfer and promising candidates for gene therapy. *J Gene Med* **2**: 308–316.

Wang X, Wang Y, Wu X, *et al.* 2015. Unbiased detection of off-target cleavage by CRISPR-Cas9 and TALENs using integrase-defective lentiviral vectors. *Nat Biotechnol* **33**: 175–178.

Wanisch K, Yanez-Munoz RJ. 2009. Integration-deficient lentiviral vectors: a slow coming of age. *Mol Ther* **17**: 1316–1332.

Weinstock DM, Jasin M. 2006. Alternative pathways for the repair of RAG-induced DNA breaks. *Mol Cell Biol* **26**: 131–139.

Wu X, Scott DA, Kriz AJ, *et al.* 2014. Genome-wide binding of the CRISPR endonuclease Cas9 in mammalian cells. *Nat Biotechnol* **32**: 670–676.

Wyman C, Kanaar R. 2006. DNA double-strand break repair: all's well that ends well. *Annu Rev Genet* **40**: 363–383.

Zhou Y, Zhu S, Cai C, *et al.* 2014. High-throughput screening of a CRISPR/Cas9 library for functional genomics in human cells. *Nature* **509**: 487–491.

3 *In Vivo* Studies of miRNA Target Interactions Using Site-specific Genome Engineering

Andrew R. Bassett, Qianxin Wu, Quentin R.V. Ferry and Tudor Fulga

3.1 Introduction

It has become increasingly clear that miRNA-dependent gene regulation forms a critical component of the gene regulatory structure in the cell, and is important for almost every biological process, including differentiation, tissue patterning, cell division and cellular responses to their environment (Bartel, 2009; Bushati and Cohen, 2007; Esteller, 2011; Farh *et al.*, 2005; Sun and Lai, 2013). Profiling of miRNA expression (Liu *et al.*, 2008), and elegant studies of miRNA loss of function and gain of function have shown that these small ~21 nt RNA molecules provide a critical regulatory layer in both normal development (Mendell, 2005) and in many diseases, such as cancer (Croce and Calin, 2005; Mendell, 2005).

miRNA-dependent gene regulation depends on complementary base pairing of the miRNA to miRNA response elements (MREs) in a set of target mRNAs, and concomitant downregulation of the mRNAs by transcript destabilization or translational repression (Bartel, 2004; 2009; Farh *et al.*, 2005). Such targeting is surprisingly resilient to mismatches between miRNA and MRE, and analysis of known target genes has identified the "seed" region (nucleotides 2–8 of the mature miRNA) to be a key determinant of such regulation (Jinek and Doudna, 2009; Lewis *et al.*, 2005). This makes prediction of miRNA targets challenging, since a 6 nt sequence would be expected to occur at random every 4^6 bases or every 4.1 kb throughout the genome, and computational algorithms for target prediction are therefore limited by high false-positive rates (Alexiou *et al.*, 2009; Rajewsky, 2006).

A lack of appropriate experimental techniques has hindered large-scale identification and analysis of miRNA target genes and their corresponding MREs, and most currently used assays rely on indirect readouts of miRNA activity (Figure 3.1). The most widely used assay to measure the activity of an MRE is the 3′UTR reporter, whereby the genomic region containing the MRE is fused to a constitutively expressed reporter gene, and the effect of manipulating miRNA levels is analyzed by expression of the reporter gene compared to a similar construct where the MRE is absent or mutated (Jin *et al.*, 2013; Lewis *et al.*, 2003).

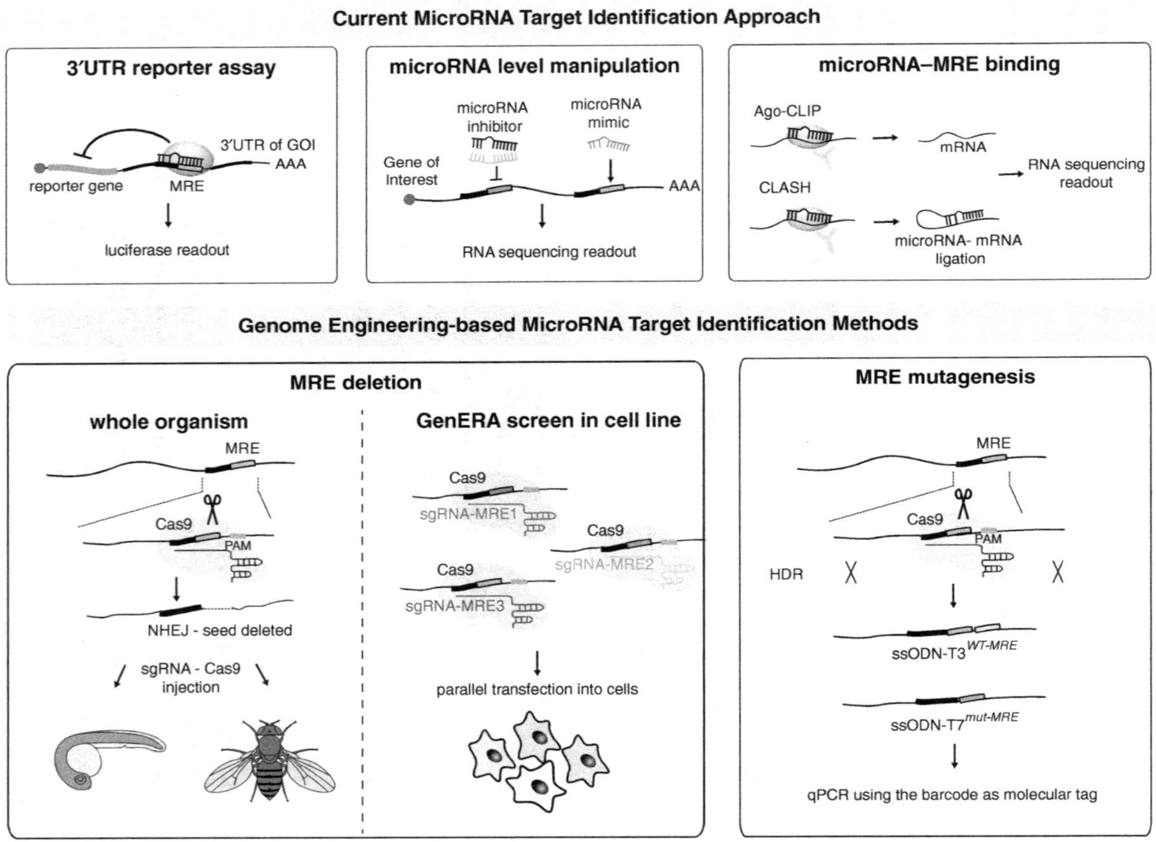

Figure 3.1 Approaches to identifying and validating miRNA targets. Current approaches include 3′UTR reporter assays where the 3′UTR of the gene of interest is coupled to a fluorescent or luminescent reporter gene. Comparison of reporter expression in the presence or absence of the MRE, or upon miRNA manipulation allows investigation of MRE activity. Targets can also be identified after manipulation of miRNA levels by miRNA inhibitors or miRNA mimics, followed by RNA-Seq to identify the resulting gene expression changes. Direct binding of miRNAs to MREs can be detected by techniques such as Ago-CLIP and CLASH that rely on crosslinking of miRNAs to their targets followed by biochemical purification of Ago-containing complexes, and subsequent analysis of bound mRNAs by RNA-Seq. Genome engineering can be employed to investigate MRE functionality by utilizing deletions resulting from inefficient NHEJ-based repair of Cas9-induced DSBs to mutate MREs in the context of a whole organism. A similar approach can be employed to screen many MREs for their functionality using a multiplexed assay in cell lines. Homology directed repair (HDR) can also be employed to make defined mutations in MREs of interest in cell lines and coupled to a molecular tag that can be used to specifically detect the effect of these changes. (A black-and-white version of this figure will appear in some formats. For the color version, please refer to the plate section.)

Whilst informative, this assay does not reflect the endogenous stoichiometry of miRNA and target mRNA, making it difficult or impossible to draw conclusions about the effect of a particular MRE in its native context.

A second widely used method for analysis of miRNA target genes is manipulation of miRNA levels by either overexpression (Pillai *et al.*, 2004; Wang, 2011), knockout (Park *et al.*, 2010) or sequestration by anti-miRs (Krutzfeldt *et al.*, 2005), followed by global gene expression profiling using techniques such as RNA-Seq (Lim *et al.*, 2005; Thomas *et al.*, 2010; Figure 3.1). Whilst one would assume that expression of true miRNA targets should be restricted under native conditions, and upregulated upon knockout, such experiments often show a large number of genes to be affected, frequently with an approximately equal number of genes being up- and downregulated (Cloonan, 2015). This observation, along with a relatively small (though often significant) enrichment of the relevant MREs within the differentially expressed genes, suggests that many of these genes are secondary or compensatory effects of the loss of miRNA-mediated regulation (Cloonan, 2015). Dissecting the direct targets of miRNA-mediated regulation therefore often relies on cross-referencing with computational prediction of MREs, with its associated false-positive rates (Alexiou *et al.*, 2009; Rajewsky, 2006).

A different approach to this problem is to look for direct physical binding of the miRNA to its target mRNA through chemical crosslinking and biochemical purification of miRNA–MRE pairs (Figure 3.1). This has been achieved by several different techniques (Hafner *et al.*, 2010; Helwak *et al.*, 2013; Thomson *et al.*, 2011), and provides important information on the set of miRNA target genes bound within a particular cell line or tissue. However, whether it is possible to extrapolate the functionality of an MRE from its ability to bind a miRNA is still debatable (Agarwal *et al.*, 2015), and therefore other assays must be employed to establish whether this is indeed the case.

Importantly, none of these techniques allow analysis of the functionality of a specific MRE *in vivo* in its native context. Target protector oligonucleotides, which bind to a specific MRE and prevent binding of the endogenous miRNA, offer one strategy to address this question (Cacchiarelli *et al.*, 2011; Choi *et al.*, 2007; Gehrke *et al.*, 2010; Staton and Giraldez, 2011). However, these assays are limited by potential off-target effects of the oligonucleotides, an ability to deliver the reagents into the appropriate cell type or tissue, and the transient nature of the suppression of MRE activity (Staton and Giraldez, 2011).

The recent availability of facile genome engineering technologies such as CRISPR/Cas9 (Cong *et al.*, 2013; Hwang *et al.*, 2013; Mali *et al.*, 2013) has provided a valuable opportunity to rapidly generate genetic mutations that affect MRE sequence and function. This enables dissection of the functionality of a single MRE *in vivo*, in its native context (Bassett *et al.*, 2014) and also offers the exciting possibility to perform larger-scale studies of multiple MREs in cell culture to establish their activity (Wu *et al.*, 2017). Here we highlight the techniques we have developed to probe MRE function in whole organisms and cultured cell lines (Bassett *et al.*, 2014; Wu *et al.*, 2017; Figure 3.1). Together, such techniques should facilitate the identification of functional MREs in specific cellular contexts, their importance in cellular and organismal function and provide rules to better predict the targets of miRNAs. Eventually, such techniques will help to establish the miRNA regulatory networks at work within particular cell types in normal development and disease states, providing an invaluable resource when seeking to manipulate such networks for therapeutic and biotechnological purposes.

3.2 Overview of Strategies

Double-strand breaks (DSBs) generated by CRISPR/Cas9 nucleases are repaired in cells predominantly by two mechanisms, non-homologous end joining (NHEJ) or homology-directed repair (HDR) (Bibikova *et al.*, 2002), both of which can be exploited to generate mutations in MRE sequences. NHEJ is achieved by ligation of the DNA termini resulting from a DSB, but is somewhat error-prone, often resulting in a variety of small insertions and deletions (indels), typically between 1 and 20 nt, at the cleavage site that can be used to delete MRE sequences (Basset *et al.*, 2013). Repair by HR uses a homologous sequence as a template for repair, and by supplying an excess of a desired template DNA, it can be used to produce defined deletions or alterations to the DNA sequence (Bibikova *et al.*, 2002). We describe MRE editing strategies using both mechanisms in whole organisms and cultured cells (Figure 3.1; Bassett *et al.*, 2014; Michaels *et al.*, 2017; Wu *et al.*, 2017).

3.2.1 MRE Deletion

In virtually any system in which CRISPR reagents can be delivered, the small indels created through imperfect NHEJ can be used to effectively delete short regulatory elements, including MREs. We have employed this strategy in *Drosophila* and zebrafish to demonstrate the functionality of MREs *in vivo* in a whole organism (Bassett *et al.*, 2014) and enabling phenotypic analysis of the resulting mutant organisms. The efficiency of this process is high enough that it can also be applied to mutate larger numbers of MREs in cell culture systems, which will allow high-throughput screening for true target genes and the analysis of entire miRNA target networks (Wu *et al.*, 2017). The natural heterogeneity in the length of indels generated from imperfect NHEJ can also be utilized to functionally dissect the critical sequence determinants of MRE-directed gene regulation (Wu *et al.*, 2017).

3.2.2 MRE Mutagenesis

DNA repair by NHEJ predominates in most cell types, and so deletion is typically the most efficient way of removing MREs. However, the homologous recombination machinery can be exploited to generate defined sequence changes to mutate MREs if there is a means of enriching those cells in which such events have happened. We describe a technique whereby a sequence tag is inserted into the genome to mark a specific MRE mutation. This allows specific detection of only the small population of cells that contain the integration by quantitative PCR

(Bassett *et al.*, 2014; Michaels *et al.*, 2017). This makes it possible to assay the effect of these defined mutations within a population of cells, providing a simple assay for MRE functionality, and a means to dissect the detailed sequence requirements of MRE function.

3.3 Methods

3.3.1 Computational Selection of sgRNA Target Sites

The specificity of the Cas9 endonuclease for a particular DNA sequence is determined by base pairing with the first 20 nt of the bound guide RNA (gRNA) (Jinek *et al.*, 2012). The only prerequisite for targeting is the presence of a protospacer adjacent motif (PAM) sequence adjacent to the targeted region. In the case of the *Streptococcus pyogenes* Cas9 protein this is NGG, although cleavage can occur at a lower efficiency with a NAG sequence (Zhang *et al.*, 2014). In order to target an MRE by either NHEJ-based indels (typically 1–20 nt) or HDR, this sequence must be present on either DNA strand within approximately 20 nt to either side of the MRE. Computational analysis of all predicted MREs in human, mouse, rat and *Drosophila* shows that this is possible for >95% of all MRE sequences. We have generated a design tool, miR-CRISPR (http://mir-crispr.molbiol.ox.ac.uk/tulga/miR-CRISPR.cgi) that allows easy design of guide RNAs to target miRNA predicted MREs (Bassett *et al.*, 2014; John *et al.*, 2005). This is searchable by gene or miRNA for individual MREs, can be used in batch mode to generate reagents for all MREs present in one gene or targeted by one miRNA, and can also be used for bespoke sequences that are defined by the user. The output provides information about distance to the MRE seed region, potential off-target sites and the sequences of primers that can be used to generate expression vectors for sgRNA expression (Bassett *et al.*, 2014).

3.3.2 MRE Deletion *In Vivo*

3.3.2.1 Strategy

The understanding of the importance and relevance of specific miRNA–target interactions on phenotype in the context of the whole organism is a critical question in the field, and one that can be addressed using genome engineering technology. Here we describe the application of CRISPR-based MRE deletion in *Drosophila* and zebrafish (Figure 3.2), but such techniques can equally be applied to other organisms in which the CRISPR components can be delivered into the germline, such as mice, rats, pigs, etc.

3.3.2.2 RNA Production and Injection

There are a number of different ways of delivering the sgRNA and Cas9 components of the CRISPR system, and the choice will largely depend on the exact nature of the system. Recombinant Cas9 protein and *in vitro* transcribed or fully synthetic guide RNAs are a simple and effective means of delivery in many cases (Kim *et al.*, 2014b). In *Drosophila*, plasmid-based and transgenic expression

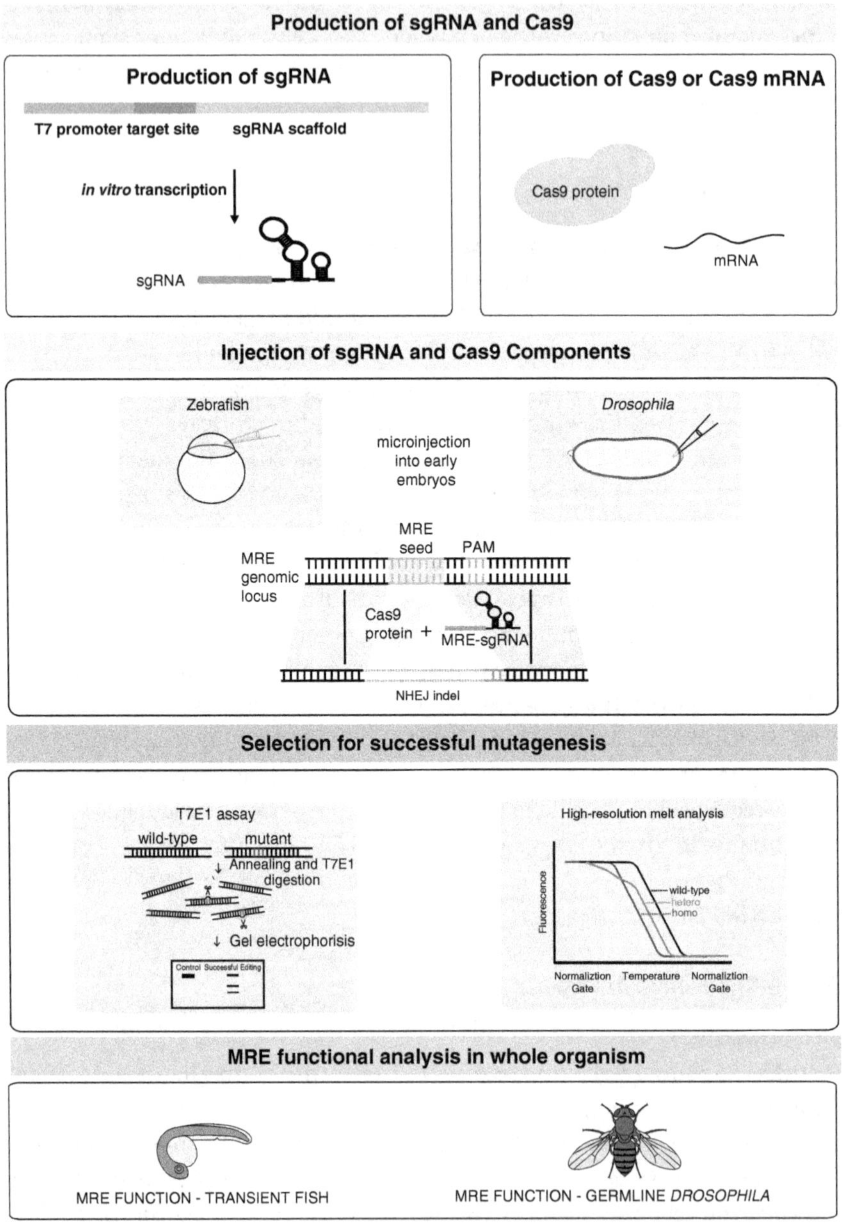

Figure 3.2 Strategy for deletion of MREs in whole organisms. sgRNAs can be produced by *in vitro* transcription using a PCR product generated from overlapping primers that contains a T7 promoter (green), followed by the specific target sequence (pink), and the sgRNA scaffold (orange). These sgRNAs are injected into early embryos of for instance zebrafish or *Drosophila* along with Cas9 mRNA or protein to induce mutagenesis at the MRE of interest. Mutations are screened by T7 endonuclease (T7E1) assay, high-resolution melt (HRM) analysis, restriction enzyme polymorphisms and DNA sequencing for those of interest, and functional analysis of the MRE deletion can be performed in the intact organism. (A black-and-white version of this figure will appear in some formats. For the color version, please refer to the plate section.)

systems have been employed to great effect and probably represent the most efficient means of generating mutant alleles (Port *et al.*, 2015).

We have used injection of *in vitro* transcribed sgRNAs and Cas9 mRNA to produce MRE deletions, which is a generally applicable technique (Figure 3.2) (Bassett *et al.*, 2014). sgRNAs are produced from a DNA template consisting of a T7 promoter, 19 nt target sequence and 80 nt common sgRNA backbone, which can be easily generated through primer extension using two overlapping DNA oligonucleotides (Bassett and Liu, 2014; Bassett *et al.*, 2013). After purification, this is used as a template for *in vitro* transcription with the Megascript T7 kit to generate sgRNAs. These sgRNAs can be mixed with Cas9 mRNA or protein, both of which can now be purchased from many suppliers. Delivery of CRISPR reagents in *Drosophila* and zebrafish, as with many other organisms, is by microinjection into early embryos, using techniques routinely used for transgenesis.

3.3.2.3 Selection

Whilst in some instances the mutagenic efficiency can be high enough to see effects in the mosaic, injected generation e.g. *lft2* in zebrafish (Bassett *et al.*, 2014; Figure 4.3), this is the exception rather than the rule and stable mutant lines normally need to be generated. Selection of those organisms containing the mutations of interest is probably the most time-consuming and labor-intensive part of the process. However, generating stable lines is important for establishing and understanding the physiological importance of MREs at all stages of development and across different tissues.

Methods to identify mutant alleles are numerous, including loss of restriction enzyme sites or CRISPR cleavage sites (Kim *et al.*, 2014a), identification of deletions by reduction of PCR product size (Carrington *et al.*, 2015), Surveyor or T7 endonuclease (T7E1) assays, high-resolution melt analysis (HRMA) (Bassett *et al.*, 2013), and direct Sanger (Brinkman *et al.*, 2014) or targeted next-generation sequencing (Bell *et al.*, 2014; Figure 3.2).

We have found HRMA-based genotyping is an effective way to identify mosaic animals. Genotypes of the subsequent generation can then be established by individual tissue PCR and sequencing prior to setting up crosses with desired mutations (Bassett *et al.*, 2014).

3.3.2.4 Analysis

Once stable lines have been produced, the consequence of MRE disruption on gene expression and phenotype during development can be analyzed by standard techniques, including the manipulation of miRNA levels. Interestingly, our analysis of the interaction of the *bantam* miRNA with an MRE in the *enabled* gene in *Drosophila* shows that whilst this MRE is functional upon overexpression of *bantam*, its loss does not cause any overt gene expression changes or functional phenotypic consequences (Bassett *et al.*, 2014). This reveals the importance of techniques enabling analysis of the functionality of MREs in their endogenous context, whilst maintaining the appropriate stoichiometry of miRNA to mRNA expression levels.

3.3.3 MRE Mutagenesis in Cell Lines

Whilst manipulation of MREs in the context of a whole organism is very valuable, it is often prohibitively expensive or time-consuming to screen large numbers of predicted MREs in this way. Cultured cell lines offer a powerful alternative system that is more amenable to screen large numbers of MREs for functionality under normal conditions or upon stimulation with defined exogenous factors. An added advantage is that a specific cell type is isolated, which allows analysis of the miRNA regulatory networks in a defined cell, independent of the surrounding tissue and cell–cell signaling.

3.3.3.1 Strategy

A major bottleneck in generating MRE deletion mutants is the clonal isolation of defined MRE mutant lines, which requires single cells to be grown to sufficient numbers to perform gene expression analyses. This is necessary since not all cells within a population will contain the same MRE mutation, or even a mutation of the MRE at all. We describe a system that circumvents this problem by supplying oligonucleotide HDR templates that link a sequence tag to an MRE mutation, and therefore allows specific analysis of the mutant cells within a mixed population (Bassett *et al.*, 2014; Figure 3.3). It also allows defined mutations within the MRE sequence to be tested rather than simple deletions, making it possible to understand the sequence determinants of MRE function.

This is achieved by generating CRISPR/Cas9 reagents cutting near to the MRE, and supplying two HDR templates simultaneously (Figure 3.3). The first mutates or deletes the MRE and integrates a T7 sequence tag, and the second maintains the MRE and integrates a T3 sequence tag. Alleles containing the T7 or T3 sequence tags can then be specifically assayed by quantitative PCR using a primer binding to either the T7 or T3 tag, and a universal gene-specific primer. Integration efficiency of each tag can be assessed by qPCR analysis of gDNA, in which the ratio of T7/T3 should be approximately 1. If there is an effect of mutation of a specific MRE, the T7 alleles should be expressed higher than the T3 alleles at the RNA level. This change can be detected if a similar analysis is performed on cDNA, and if the MRE is functional, we would expect the ratio of T7/T3 in cDNA to be greater than that observed in gDNA (Bassett *et al.*, 2014; Figure 3.3). This provides an internally controlled experiment that can be used to assess even relatively small changes in gene expression level.

3.3.3.2 Vector Construction and Transfection

As previously mentioned, there are many strategies for expressing the gRNA and Cas9 protein, including *in vitro* transcribed mRNA and sgRNAs and recombinant Cas9 protein with synthetic RNA oligonucleotides. All of these can be used, but for mammalian cell culture we typically use the pX459 vector (Addgene plasmid #48139) that expresses both Cas9 and the sgRNA, along with a puromycin resistance marker for selection (Ran *et al.*, 2013). sgRNA sequences are rapidly and simply cloned downstream of a human U6 promoter using pairs of annealed oligonucleotides which ligate into a *Bbs* I site in the vector to achieve scarless integration (Ran *et al.*, 2013).

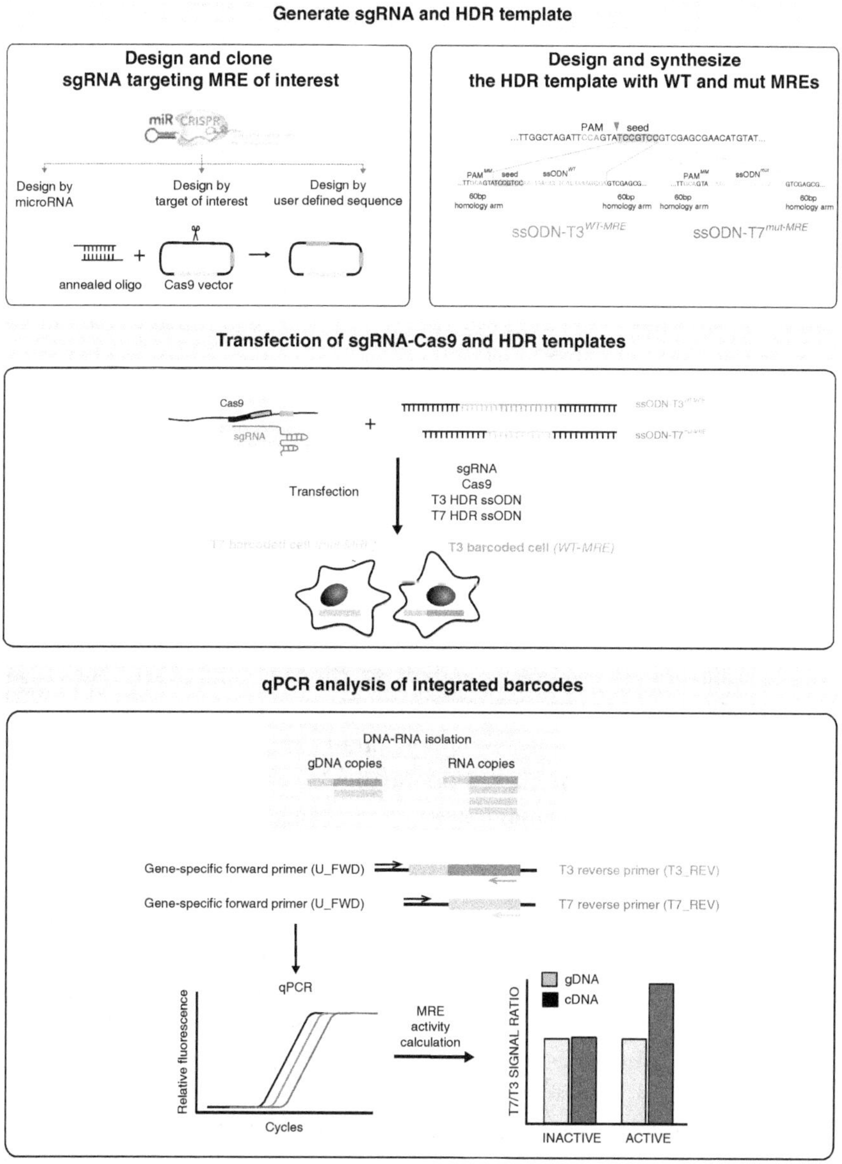

Figure 3.3 Strategy for MRE mutagenesis in cell lines. sgRNAs are designed using miR-CRISPR to target near the MRE of interest, and target sequences cloned into an appropriate expression vector (e.g. pX459). Simultaneously, two oligonucleotide HDR templates are synthesized, which contain 60 nt homology arms and either integrate a T3 sequence, whilst maintaining the MRE of interest, or a T7 sequence, whilst mutating the MRE. The CRISPR construct is co-transfected with both HDR templates into the cell line of interest, and gDNA and RNA are isolated from the pool of cells followed by reverse transcription to generate cDNA. Integration of the two HDR templates can be specifically detected by qPCR using primers specific for the T7 or T3 tag and a common gene-specific forward primer. The integration efficiency of the two HDR templates is assayed by measuring the T7/T3 ratio in gDNA and the activity of the MRE determined by an increase in this ratio when measured in the cDNA sample. (A black-and-white version of this figure will appear in some formats. For the color version, please refer to the plate section.)

Oligonucleotide HDR repair templates are designed to span the MRE of interest and CRISPR cleavage site with approximately 40–60 nt homology arms either side. Two oligos are designed, the first mutates or deletes the MRE and integrates a T7 sequence, and the second maintains the MRE, but inserts a T3 sequence (Figure 3.3). Importantly, in both cases the CRISPR target site or PAM sequence must be disrupted such that Cas9 can no longer cut once HDR has occurred, so that subsequent NHEJ events do not cause indel formation. Also, it is important to check computationally that the integrated T7 or T3 sequences do not introduce additional MREs, so that one can be sure that the effects observed are due to the MRE of interest (Bassett *et al.*, 2014). A tag-swap experiment (i.e. where the T3 sequence is associated with the MRE mutation) can be performed in parallel to control for such effects. Oligos can be ordered without any chemical modification from standard suppliers (e.g. IDT).

CRISPR vectors and the two HDR oligos are subsequently transfected into cells using standard techniques such as lipid-based transfection or electroporation, and transfected cells selected in puromycin to enrich for those containing mutations.

3.3.3.3 Analysis

Genomic DNA and RNA are extracted from each transfected well, and cDNA is prepared for each sample. Subsequently, quantitative PCR is performed using one primer specific for the tag (T7 or T3) and one for the gene of interest on the gDNA and cDNA. Ratios of T7 to T3 are calculated for both gDNA and cDNA using the difference in Ct values (ΔCt) between the T7 and T3 amplicons and transforming using $2^{-\Delta Ct}$. If the MRE has no effect, the ratio should be the same in both samples, but if deletion of the MRE causes upregulation of gene expression, the T7 to T3 ratio should be higher in cDNA than gDNA (Bassett *et al.*, 2014; Figure 3.3).

Using this technique, we have investigated the effect of three MREs in human cells that have been shown to physically interact with particular miRNAs in HEK293T cells by CLASH (Helwak *et al.*, 2013). Interestingly, only one of these MREs has an effect on gene expression when mutated in the same cell line, suggesting that other factors must play a role in regulating the functionality of specific MREs in particular cellular contexts (Bassett *et al.*, 2014).

3.3.4 High-throughput MRE Deletion in Cell Lines (GenERA)

The techniques described so far are designed to understand the functionality of individual miRNA–target interactions. However, to fully understand the network of targets regulated by miRNAs, it is important to address the function of larger sets of MREs.

3.3.4.1 Strategy

In order to achieve this, we have developed a technique which we have called GenERA (genome engineering-based evaluation of RNA regulatory element activity) that deletes candidate RNA regulatory elements such as MREs using indels generated by inefficient NHEJ (Figure 3.4; Wu *et al.*, 2017). In this case, high-throughput sequencing (HTS) of amplicons spanning the MRE in both cDNA and

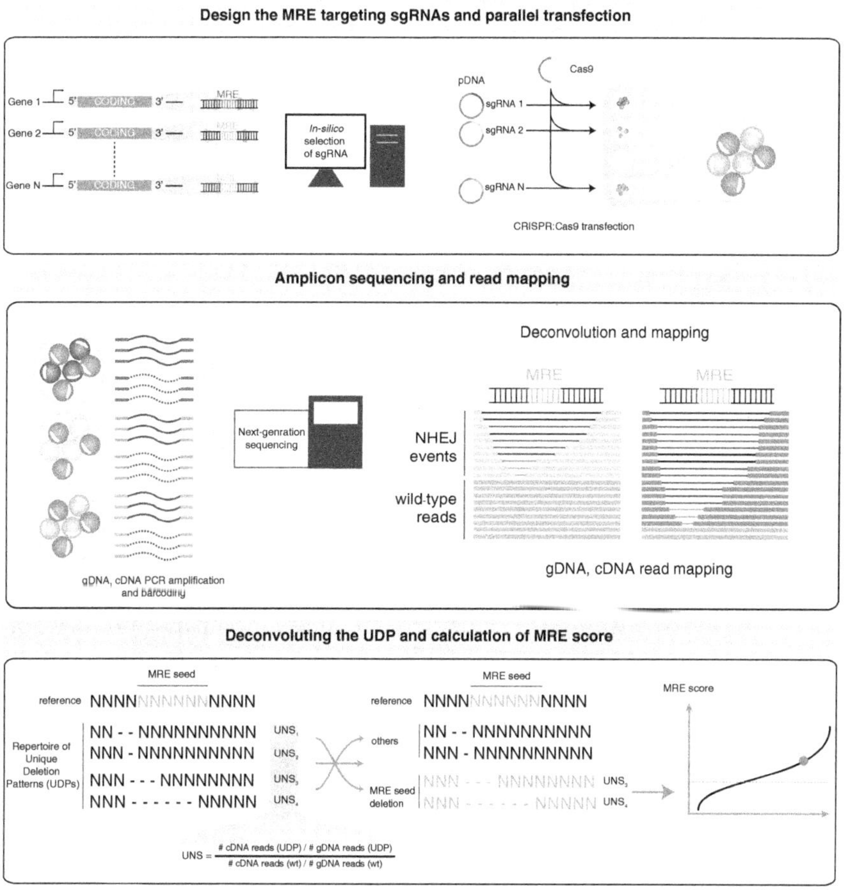

Figure 3.4 Genome engineering-based evaluation of RNA regulatory element activity (GenERA). sgRNAs are designed computationally using miR-CRISPR to target within 20 nt of MREs of interest. Oligonucleotides are cloned into sgRNA expression vectors, transfected in individual wells to generate mixed populations of mutant cells and gDNA and RNA are extracted from each well. Amplicons are designed spanning the MRE of interest, and PCR is performed on gDNA and cDNA followed by adaptor ligation, barcoding and high-throughput sequencing. Reads are mapped to each amplicon, and segregated into unique deletion profiles (UDPs). A unique normalized score (UNS) is calculated to assess the effect of each UDP on gene expression, using the ratio of cDNA to gDNA reads, normalized to WT reads. An overall MRE score can be calculated by a sum of MRE seed deleting UNSs, weighted to account for gDNA read number. (A black-and-white version of this figure will appear in some formats. For the color version, please refer to the plate section.)

gDNA is used to assess gene expression levels. A sequencing-based readout allows the mutations to be classified according to the nature of the mutation, and the ratio between cDNA and gDNA reads is compared between these classes to establish their effect on expression of the gene, in a manner analogous to the T7/T3 integration system described above. The high efficiency of NHEJ-generated indels, lack of homology constructs and HTS readout allows tens or hundreds of MREs to be assayed for their functionality. Additionally, the heterogeneity in the deletions generated can be exploited to extract useful information about the sequence requirements of the resulting gene expression changes (Wu *et al.*, 2017).

3.3.4.2 GenERA procedure

Similarly to the system of MRE mutagenesis described above, sgRNAs are designed to target near to MREs (within 20 bp), cloned into expression vectors and delivered individually to cells through lipid-based transfection. Puromycin selection for Cas9 expression is then applied for five days to remove untransfected cells and maximize mutagenic efficiency. Genomic DNA and RNA are extracted from each pool of cells, and limiting cycle PCR with amplicons spanning the MRE and CRISPR target site is performed on gDNA and cDNA. Illumina sequencing adaptors are added to the amplicons by ligation or a second round of PCR, and they are sequenced by high-throughput sequencing on a MiSeq platform to quantify gene expression (Wu *et al.*, 2017; Figure 3.4).

3.3.4.3 Data Analysis

The amplicon sequencing data is mapped to a custom genome made up of the concatenated amplicon sequences using a Burrows–Wheeler algorithm, which is able to effectively map reads containing large indels. Comparison of deletion profiles between gDNA and cDNA for a particular amplicon allows a simple assessment of MRE potency as ablation of a functional MRE should translate to an increase of mutant reads in the cDNA library.

A more thorough analysis is performed by classifying gDNA and cDNA reads for a particular amplicon into unique deletion patterns (UDP), each consisting of a single type of deletion. The impact of each UDP on transcript abundance is computed as a UDP normalized score (UNS) by comparing the corresponding cDNA to gDNA read count ratio to its wild-type counterpart. After filtering for UDPs introducing novel MREs or RNA regulatory elements that might confound the results, analysis of the sequence determinants driving changes in gene expression can be performed by monitoring variations in UNSs across the UDP repertoire. This allows a near base pair resolution dissection of the sequences necessary for changes in gene expression (Wu *et al.*, 2017). Additionally, an overall "MRE-score" can be calculated from this data by a sum of all UNSs whose corresponding deletions cover the seed region, weighted to take into account the number of gDNA reads that contribute to each UNS score. This provides an overall effect size for each MRE of interest (Wu *et al.*, 2017; Figure 3.4).

3.3.4.4 Conclusions from this Data

We have applied this technique to analyze the complete set of predicted miRNA targets for miR-184 in a *Drosophila* cell line. This shows that most canonical MREs are active in this context, but the extent to which they alter gene expression varies greatly. Machine learning techniques have shown that the magnitude of the effect of an MRE deletion depends on a series of features, including the presence of co-targeting miRNAs, expression levels and AU content. We also were able to correlate this data with a quantitative manipulation of miRNA levels followed by global gene expression profiling by RNA-Seq (unpublished data). This not only confirms that those targets with the highest MRE-scores are most strongly anticorrelated with miRNA expression, but also demonstrates that quantitative manipulation of miRNA levels is an effective means of predicting new targets that may not be identified by current algorithms.

3.4 Concluding Remarks

The application of genome engineering technology to the study of MRE functionality offers an exciting means to test the functionality of predicted miRNA–target interactions, and to gain a more thorough understanding of miRNA regulatory networks. This will aid in the definition of functional miRNA–target pairs and in understanding the role that they play in the context of individual cells or whole organisms in normal development and in disease states (Bassett *et al.*, 2014).

We have also used high-throughput MRE deletion to define functional MREs within the complete set of target genes for a single miRNA (Wu *et al.*, 2017). This information could be used to improve prediction algorithms, and provides an important data set from which to understand the complex network of gene regulatory interactions that are controlled by this miRNA. We expect that in the future, such technologies could be employed to delete the entire repertoire of MREs in human cells, to understand and model miRNA interaction networks acting in a particular cell type.

Together, we hope that this suite of techniques will provide insight into this complex yet critical layer of gene regulation and allow us to more effectively and safely manipulate its function.

REFERENCES

Agarwal V, Bell GW, Nam JW, Bartel DP. 2015. Predicting effective microRNA target sites in mammalian mRNAs. *Elife* **4**: e05005.

Alexiou P, Maragkakis M, Papadopoulos GL, Reczko M, Hatzigeorgiou AG. 2009. Lost in translation: an assessment and perspective for computational microRNA target identification. *Bioinformatics* **25**: 3049–3055.

Bartel DP. 2004. MicroRNAs: genomics, biogenesis, mechanism, and function. *Cell* **116**: 281–297.

Bartel DP. 2009. MicroRNAs: target recognition and regulatory functions. *Cell* **136**: 215–233.

Bassett A, Liu JL. 2014. CRISPR/Cas9 mediated genome engineering in *Drosophila*. *Methods* **69**: 128–136.

Bassett AR, Azzam G, Wheatley L, *et al.* 2014. Understanding functional miRNA-target interactions in vivo by site-specific genome engineering. *Nat Commun* **5**: 4640.

Bassett AR, Tibbit C, Ponting CP, Liu JL. 2013. Highly efficient targeted mutagenesis of *Drosophila* with the CRISPR/Cas9 system. *Cell Rep* **4**: 220–228.

Bell CC, Magor GW, Gillinder KR, Perkins AC. 2014. A high-throughput screening strategy for detecting CRISPR-Cas9 induced mutations using next-generation sequencing. *BMC Genomics* **15**: 1002.

Bibikova M, Golic M, Golic KG, Carroll D. 2002. Targeted chromosomal cleavage and mutagenesis in *Drosophila* using zinc-finger nucleases. *Genetics* **161**: 1169–1175.

Brinkman EK, Chen T, Amendola M, van Steensel B. 2014. Easy quantitative assessment of genome editing by sequence trace decomposition. *Nucleic Acids Res* **42**: e168.

Bushati N, Cohen SM. 2007. microRNA functions. *Annu Rev Cell Dev Biol* **23**: 175–205.

Cacchiarelli D, Incitti T, Martone J, *et al.* 2011. miR-31 modulates dystrophin expression: new implications for Duchenne muscular dystrophy therapy. *EMBO Rep* **12**: 136–141.

Carrington B, Varshney GK, Burgess SM, Sood R. 2015. CRISPR-STAT: an easy and reliable PCR-based method to evaluate target-specific sgRNA activity. *Nucleic Acids Res* **43**: e157.

Choi WY, Giraldez AJ, Schier AF. 2007. Target protectors reveal dampening and balancing of Nodal agonist and antagonist by miR-430. *Science* **318**: 271–274.

Cloonan N. 2015. Re-thinking miRNA-mRNA interactions: intertwining issues confound target discovery. *Bioessays* **37**: 379–388.

Cong L, Ran FA, Cox D, *et al.* 2013. Multiplex genome engineering using CRISPR/Cas systems. *Science* **339**: 819–823.

Croce CM, Calin GA. 2005. miRNAs, cancer, and stem cell division. *Cell* **122**: 6–7.

Esteller M. 2011. Non-coding RNAs in human disease. *Nat Rev Genet* **12**: 861–874.

Farh KK, Grimson A, Jan C, *et al.* 2005. The widespread impact of mammalian microRNAs on mRNA repression and evolution. *Science* **310**: 1817–1821.

Gehrke S, Imai Y, Sokol N, Lu B. 2010. Pathogenic LRRK2 negatively regulates microRNA-mediated translational repression. *Nature* **466**: 637–641.

Hafner M, Landthaler M, Burger L, *et al:* 2010. Transcriptome-wide identification of RNA-binding protein and microRNA target sites by PAR-CLIP. *Cell* **141**: 129–141.

Helwak A, Kudla G, Dudnakova T, Tollervey D. 2013. Mapping the human miRNA interactome by CLASH reveals frequent noncanonical binding. *Cell* **153**: 654–665.

Hwang WY, Fu Y, Reyon D, *et al.* 2013. Efficient genome editing in zebrafish using a CRISPR-Cas system. *Nat Biotechnol* **31**: 227–229.

Jin Y, Chen Z, Liu X, Zhou X. 2013. Evaluating the microRNA targeting sites by luciferase reporter gene assay. *Methods Mol Biol* **936**: 117–127.

Jinek M, Chylinski K, Fonfara I, *et al.* 2012. A programmable dual-RNA-guided DNA endonuclease in adaptive bacterial immunity. *Science* **337**: 816–821.

Jinek M, Doudna JA. 2009. A three-dimensional view of the molecular machinery of RNA interference. *Nature* **457**: 405–412.

John B, Enright AJ, Aravin A, *et al.* 2005. Human microRNA targets. *PLoS Biol* **3**: e264.

Kim JM, Kim D, Kim S, Kim JS. 2014a. Genotyping with CRISPR-Cas-derived RNA-guided endonucleases. *Nat Commun* **5**: 3157.

Kim S, Kim D, Cho SW, Kim J, Kim JS. 2014b. Highly efficient RNA-guided genome editing in human cells via delivery of purified Cas9 ribonucleoproteins. *Genome Res* **24**: 1012–1019.

Krutzfeldt J, Rajewsky N, Braich R, *et al.* 2005. Silencing of microRNAs in vivo with "antagomirs". *Nature* **438**: 685–689.

Lewis BP, Burge CB, Bartel DP. 2005. Conserved seed pairing, often flanked by adenosines, indicates that thousands of human genes are microRNA targets. *Cell* **120**: 15–20.

Lewis BP, Shih IH, Jones-Rhoades MW, *et al.* 2003. Prediction of mammalian microRNA targets. *Cell* **115**: 787–798.

Lim LP, Lau NC, Garrett-Engele P, *et al.* 2005. Microarray analysis shows that some microRNAs downregulate large numbers of target mRNAs. *Nature* **433**: 769–773.

Liu CG, Calin GA, Volinia S, Croce CM. 2008. MicroRNA expression profiling using microarrays. *Nat Protoc* **3**: 563–578.

Mali P, Yang L, Esvelt KM, *et al.* 2013. RNA-guided human genome engineering via Cas9. *Science* **339**: 823–826.

Mendell JT. 2005. MicroRNAs: critical regulators of development, cellular physiology and malignancy. *Cell Cycle* **4**: 1179–1184.

Michaels YS, Wu Q, Fulga TA. 2017. Interrogation of functional miRNA-target interactions by CRISPR/Cas9 genome engineering. *Meth Mol Biol* **1580**: 79–97.

Park CY, Choi YS, McManus MT. 2010. Analysis of microRNA knockouts in mice. *Hum Mol Genet* **19**: R169–R175.

Pillai RS, Artus CG, Filipowicz W. 2004. Tethering of human Ago proteins to mRNA mimics the miRNA-mediated repression of protein synthesis. *RNA* **10**: 1518–1525.

Port F, Muschalik N, Bullock SL. 2015. Systematic evaluation of *Drosophila* CRISPR tools reveals safe and robust alternatives to autonomous gene drives in basic research. *G3 (Bethesda)* **5**: 1493–1502.

Rajewsky N. 2006. microRNA target predictions in animals. *Nat Genet* **38**(Suppl.): S8–S13.

Ran FA, Hsu PD, Wright J, *et al.* 2013. Genome engineering using the CRISPR-Cas9 system. *Nat Protoc* **8**: 2281–2308.

Staton AA, Giraldez AJ. 2011. Use of target protector morpholinos to analyze the physiological roles of specific miRNA-mRNA pairs in vivo. *Nat Protoc* **6**: 2035–2049.

Sun K, Lai EC. 2013. Adult-specific functions of animal microRNAs. *Nat Rev Genet* **14**: 535–548.

Thomas M, Lieberman J, Lal A. 2010. Desperately seeking microRNA targets. *Nat Struct Mol Biol* **17**: 1169–1174.

Thomson DW, Bracken CP, Goodall GJ. 2011. Experimental strategies for microRNA target identification. *Nucleic Acids Res* **39**: 6845–6853.

Wang Z. 2011. The guideline of the design and validation of MiRNA mimics. *Methods Mol Biol* **676**: 211–223.

Wu Q, Ferry QRV, Michaels YS, *et al.* 2017. *In situ* functional dissection of RNA cis-regulatory elements by multiplex CRISPR-Cas9 genome engineering. *Nat Comms* **8**: 2109.

Zhang Y, Ge X, Yang F, *et al.* 2014. Comparison of non-canonical PAMs for CRISPR/Cas9-mediated DNA cleavage in human cells. *Sci Rep* **4**: 5405.

4 Don't Kill the Messenger: Employing Genome Editing to Study Regulatory RNA Interactions

Jia Jia Chan, Zhi Hao Kwok and Yvonne Tay

4.1 Introduction

In recent years, there has been an explosion of interest in RNA biology. As the DNA content in cells is largely similar, cellular identity and the distinction between diseased and healthy cells are often determined by the part of the genome that is transcriptionally active. RNA species include messenger RNA (mRNA), transfer RNA (tRNA) and ribosomal RNA (rRNA), which play key, well-defined roles in transcription and translation. Additionally, the past two decades have seen an exponential increase in non-coding RNA-focused research that has significantly expanded upon the central dogma that DNA codes for RNA which codes for proteins. Researchers have uncovered novel classes of RNA with coding-independent functions, encompassing both long non-coding RNAs and short non-coding RNAs, such as microRNA (miRNA) and Piwi-interacting RNA (piRNA; Figure 4.1). These non-coding RNAs (ncRNAs) have been shown to play diverse roles in various important cellular processes, including the regulation of development, differentiation and proliferation (Bartel, 2004; Cech and Steitz, 2014; Lee *et al.*, 1993; Wightman *et al.*, 1993).

Recent transcriptomic and bioinformatic analyses have revealed that although 70–90% of the mammalian genome is transcribed, only 1–2% of this transcriptome codes for proteins. This observation has led to a surge of interest in unraveling the potential functions and mechanisms of action of the non-coding transcriptome (Carninci *et al.*, 2005; Djebali *et al.*, 2012). The growing prominence of ncRNAs and their key roles in various key biological processes and disease development has led to the development of more versatile and efficient platforms that enable the targeting, monitoring and engineering of RNAs *in vitro* and *in vivo*. One of the most valuable technologies to date is genome editing, which has long been an inherent part of genomic research. Researchers have recently developed sophisticated RNA-targeting genome editing platforms, which are fast becoming indispensable in the study of both coding and non-coding RNAs.

This chapter will focus on genome editing methods, including zinc-finger nucleases (ZFNs), transcription activator-like effector-based nucleases (TALENs),

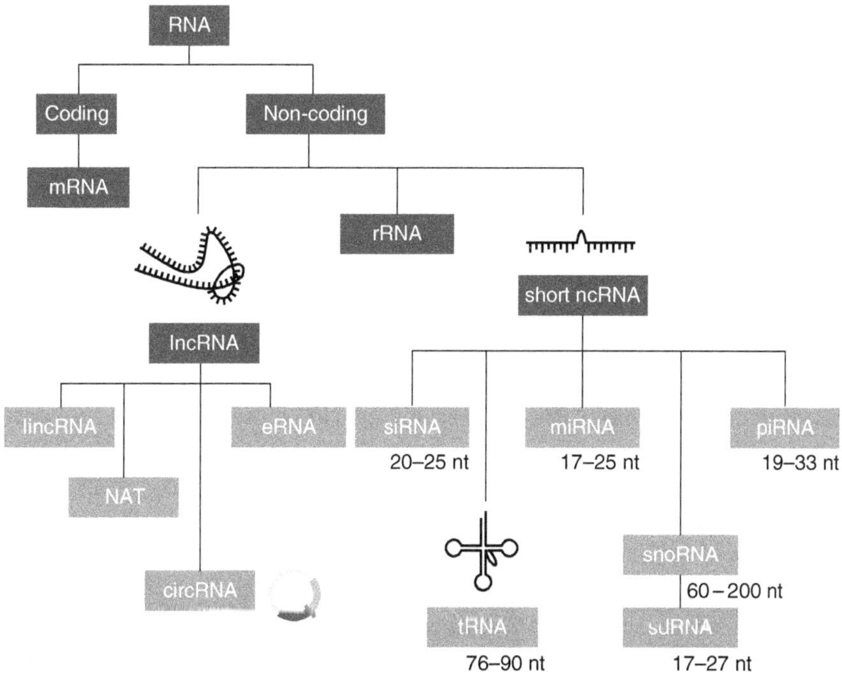

Figure 4.1 Coding and non-coding classes of RNA. LncRNAs include long intergenic RNA (lincRNA), natural antisense transcript (NAT), circular RNA (circRNA) and enhancer RNA (eRNA). Short non-coding RNAs can be further subdivided into small nucleolar RNA (snoRNA) and their subsequent sno-derived RNA (sdRNA), small interfering RNA (siRNA), microRNA (miRNA) and piwi-interacting RNA (piRNA).

and the clustered regularly interspaced short palindromic repeats (CRISPR) system that have been utilized to examine the functions and dynamics of RNA interactions. Specifically, we will focus on microRNA: competing endogenous RNA (ceRNA) interactions and describe the potential applications of genome editing platforms to unravel the intricate web of interactions in ceRNA networks and other facets of RNA biology.

4.2 miRNA and ceRNA Crosstalk

MiRNAs are a large class of small non-coding RNAs that post-transcriptionally regulate gene expression by binding to miRNA response elements (MREs) in target transcripts. MiRNA expression has been found to be dysregulated in multiple diseases including cancer, and specific miRNAs have been shown to play critical roles in disease initiation and/or progression (Calin and Croce, 2006; Calin *et al.*, 2002; Lu *et al.*, 2008; Plaisier *et al.*, 2012).

MiRNAs are first transcribed as primary miRNA transcripts in the nucleus, which are cleaved by the RNase III endonuclease *Drosha* to form ~70 nucleotide hairpin precursor miRNAs (pre-miRNAs) that are exported to the cytoplasm. Pre-miRNAs are further processed by the Dicer endonuclease to form 20–25 nucleotide mature miRNA which is incorporated into the RNA-induced silencing complex (RISC) (Bartel, 2004; Ha and Kim, 2014). In addition to regulating mRNAs via sites in their 3′UTRs (Bartel and Chen, 2004), coding regions and 5′UTRs (Hausser *et al.*, 2013, Orom *et al.*, 2008, Miranda *et al.*, 2006), miRNAs have been found to regulate various ncRNA species, such as lincRNAs and circular RNAs (Hansen *et al.*, 2013; Liu *et al.*, 2014; Memczak *et al.*, 2013). MiRNA binding can result in the degradation or translational inhibition of target transcripts, depending on the degree of complementarity between the miRNA and its MRE.

The regulation of target transcripts by miRNA binding is not unidirectional. Each miRNA can target hundreds or thousands of genes; conversely, an RNA transcript containing multiple MREs can be simultaneously targeted by many miRNAs (Salmena *et al.*, 2011). These observations led to the hypothesis that transcripts that contain MREs for the same miRNA can reciprocally co-regulate each other by competing for and sequestering these shared miRNAs, thus acting as competing endogenous RNAs (ceRNAs) (Karreth *et al.*, 2011; Tay *et al.*, 2011). Aside from protein-coding mRNAs, ncRNAs such as transcribed pseudogenes, long non-coding RNAs (lncRNAs) and circular RNA (circRNA) have also been shown to contain MREs and possess ceRNA activity (Tay *et al.*, 2014), adding another layer of complexity to miRNA:ceRNA networks.

The tumor suppressor PTEN is an example of a key protein-coding gene that is susceptible to ceRNA regulation. Studies have validated both protein-coding and non-coding PTEN ceRNAs (Karreth *et al.*, 2011; Poliseno *et al.*, 2010; Sumazin *et al.*, 2011; Tay *et al.*, 2011). Aside from the PTEN pseudogene, other pseudogenes with validated ceRNA activity include BRAFP1, KRASP1 and CYP4Z2P, which regulate the expression of their ancestral genes BRAF, KRAS and CYP4Z1, respectively; whilst lncRNAs such as HULC, HOTAIR and MALAT1 have also been shown to function as ceRNAs in different cancers (Chiu *et al.*, 2015; Karreth *et al.*, 2015; Liu *et al.*, 2014; Poliseno *et al.*, 2010; Wang *et al.*, 2010; Xiao *et al.*, 2015). These studies highlight the increasing importance of not only miRNAs, but also the coding-independent functions of both mRNA and ncRNA transcripts and hence the need for technologies to manipulate the human genome to facilitate the deconvolution of these endogenous layers of RNA crosstalk.

4.3 Current Technologies for Genome Editing and Engineering

In 1998, Fire and Mello discovered gene silencing by double-stranded RNA, or RNA interference (RNAi) (Fire *et al.*, 1998). As loss-of-function models are crucial to investigate the physiological functions of different genes, this breakthrough paved the way for a new era of RNA research. Various types of RNAi methods, such as chemically synthesized small interfering RNA (siRNA), expression vectors containing short hairpin RNA (shRNA) or endoribonuclease-prepared siRNA (esiRNA) pools generated via digestion of a specific double-stranded RNA, are now commonly used to perturb gene expression. Although RNAi is an invaluable research tool for manipulating transcript expression, limitations associated with its use include: (1) certain transcripts may be difficult to target due to factors such as secondary structure or subcellular localization, (2) off-target effects that may lead to cellular toxicity (Fedorov *et al.*, 2006) and (3) hijacking and oversaturation of the cellular RNAi machinery, which will dysregulate the activity of endogenous miRNAs (Khan *et al.*, 2009). To overcome these issues, many studies have focused on the development of complementary platforms to specifically perturb gene expression, particularly in physiological contexts. To date, these methods include programmable genome editing approaches such as ZFNs, TALENs and most recently, the CRISPR/Cas9 system (Bibikova *et al.*, 2001; Boch *et al.*, 2009; Christian *et al.*, 2010; Cong *et al.*, 2013; Jinek *et al.*, 2012; Urnov *et al.*, 2005).

4.3.1 ZFNs and TALENs

ZFNs and TALENs are chimeras of a DNA-binding domain (DBD) fused to a genetically engineered, non-specific DNA cleavage domain derived from FokI endonuclease. ZFNs consist of a nuclease fused to a zinc-finger DBD, whereas TALENs carry a transcription activator-like effector (TALE) DBD (Boch, 2011; Maeder *et al.*, 2008). These DBDs are customizable and can be engineered to target and bind to specific DNA sequences. The FokI endonuclease is a bipartite structure that is only functional as a dimer, therefore two ZFNs or TALENs must target a pair of non-palindromic DNA binding sites to allow dimerization of the nuclease domain followed by DNA cleavage at the target site to create a double-strand break (DSB) (Bitinaite *et al.*, 1998). The DSB is then repaired, either by the error-prone non-homologous end joining (NHEJ), which causes frameshift mutations and subsequent gene knockout, or homologous recombination (HR) leading to the integration of exogenous sequences into the targeted gene locus (Moehle *et al.*, 2007; Orlando *et al.*, 2010; Santiago *et al.*, 2008; Wyman and Kanaar, 2006). Despite the successful application of ZFNs and TALENs in genome editing, there are several considerations that have limited the widespread use of these two platforms. First, the specificity of zinc-finger DBDs and site-specific nucleases and associated off-target toxicity remain a major concern, particularly for ZFNs. Secondly, each target requires a custom-designed DBD, the generation of a new chimeric ZFN or TALEN, followed by an extensive screening process, which is both laborious and time-consuming. Finally, these systems have relatively low targeting efficiencies (Cade *et al.*, 2012; Carlson *et al.*, 2012; Gupta *et al.*, 2012; Kim and Kim, 2014; Kim *et al.*, 2010; 2013a; Reyon *et al.*, 2012).

4.3.2 CRISPR/Cas9

The CRISPR/Cas9 system was first discovered as a prokaryotic adaptive immune system that confers resistance to foreign pathogens (Mojica *et al.*, 2005). In 2012, it took the genome editing field by storm after independent groups reported the application of programmed CRISPR/Cas9 genome editing in human and mouse cells (Cong *et al.*, 2013; Jinek *et al.*, 2012). In contrast to ZFN and TALEN, CRISPR technology is much more versatile and robust as it uses synthetic small guide RNAs, denoted single-guide RNA (sgRNA), in concert with the Cas9 nuclease to generate DSBs. As these two components are separate entities, each DNA target requires only the design of a new sgRNA, which is much faster and more straightforward than the ZFN or TALEN method. Interestingly, the catalytically deficient mutant of Cas9 (dCas9) retains its ability to home in on its target sequence, thus functioning as a DNA-binding protein with many potential applications (Jinek *et al.*, 2012). CRISPR/Cas9, ZFN and TALEN technologies represent alternatives to the traditional *LoxP-Cre* system to introduce mutations or gene knockouts *in vivo* and studies are underway to further improve their performance by fine-tuning the specificity of the targeting domains and the efficiency of the nucleases.

4.4 Programmable RNA-targeting Genome Engineering Platforms

During the early days of ZFN, TALEN and CRISPR/Cas9 applications in gene editing, these nucleases were designed to target solely genomic DNA. However, they have been gradually adapted and developed into valuable tools to target and manipulate the sequences and expression of non-coding RNAs, which dovetails with the increasing interest in investigating the roles of coding and non-coding RNAs in various key signaling pathways and biological processes.

4.4.1 LncRNA targeting by ZFNs

ZFN was the first of the nuclease-based gene editing techniques to be adapted to target the genomic loci of non-coding RNAs. In 2011, a study reported the use of ZFN to target and silence the lncRNA MALAT1 (Gutschner *et al.*, 2011). LncRNAs are notoriously challenging to silence as: (1) they do not contain open reading frames that can be disrupted, thus small deletions are insufficient to cause complete silencing, (2) it is difficult to predict and target functionally important regions of lncRNAs and (3) lncRNAs can be transcribed from several promoters, so multiple regions need to be targeted to effectively knock down gene expression. Gutschner *et al.* harnessed ZFN technology to integrate RNA destabilizing elements (RDE) in a site-specific manner into the MALAT1 locus. As a proof-of-principle, a few different RDEs were tested, including RNaseP and RNaseZ cleavage product of MALAT1, known as mascRNA, and polyA signals from bGH and SV40. These RDEs were cloned into integration cassettes containing a CMV promoter to replace the endogenous TATA box, a GFP selectable marker and left and right homology arms that facilitate HR DNA repair. The ZFN was designed to target immediately downstream of the TATA box of the last promoter to generate a DSB, followed by HR repair that led to biallelic integration of the RDE, resulting in 1000-fold reduction in MALAT1 RNA expression. The inclusion of a GFP selectable marker allowed for FACS sorting of GFP-positive clones for subsequent analysis. Interestingly, this study also demonstrated the effective silencing of a protein-coding gene, IL2RG, using the same strategy of combining a ZFN and RDE. Moreover, the knockdown effect was 300-fold more efficient compared to RNAi-mediated gene silencing, had better specificity and fewer off-target effects.

As with all gene editing tools, there are a few considerations that should be taken into account when using this new technology. Successful gene knockdown may be affected by the transfection efficiency of the cell line of choice, the type of RDE used and the design of the ZFN. This study showed that whilst there are many RDE options, the polyA signal from bGH is universal and gave the most consistent phenotype independent of the construct and promoter in use. It is also advised that the ZFN target site should be immediately downstream of the last promoter to prevent the formation of biologically functional truncated transcripts. Additionally, the GFP marker may be substituted with other selectable markers, such as antibiotic markers. This system, if properly optimized, could prove to be a versatile and powerful tool that can be used to target non-coding RNAs, and serve as an alternative to other genome editing strategies for modifying protein-coding transcripts.

4.4.2 TALEN-based Ablation of miRNAs

Several independent studies have been carried out using TALENs that have been modified to target miRNAs, leading to the successful ablation of specific miRNA expression. TALENs are particularly suitable for targeting miRNAs as they are applicable to almost any DNA sequence, unlike ZFNs, and are thus able to target small loci with narrow targetable regions.

Kim *et al.* designed and constructed a library of 540 TALEN pairs for 274 loci that target authentic miRNAs (Kim *et al.*, 2013b). The TALENs were designed to target the mature miRNA seed region within nucleotides 1–8 and 2–7, or the Drosha processing site at the 5′ and 3′ of the pri-miRNA transcript. DSBs at the target site resulted in small insertions or deletions, known as indels. This genomic editing consequently distorted the miRNA secondary structure and inhibited Drosha processing. The study showed that all TALEN pairs designed were effective in inducing mutations. To facilitate the screening of positive clones and the efficacy of the TALENs, the authors co-expressed the TALEN constructs with a reporter plasmid containing the TALEN recognition sequence fused to a gene that encodes a surface antigen to enrich for the knockout cells. As this strategy is highly dependent on the miRNA stem-loop structures, the authors highlighted the use of the pri-miRNA secondary structure to predict the outcome of miRNA biogenesis, as well as the preferential selection of clones with the largest-sized indels to sufficiently distort the miRNA hairpins and ensure complete miRNA depletion. The TALENs generated in this study are highly specific and can selectively knock-out miRNAs with highly similar seed regions, e.g. miR-141 and miR-200c, which differ in their seed regions by only one nucleotide and target nearly identical set of genes. This platform is particularly useful to study highly similar miRNAs, where traditional tools like miRNA overexpression and antisense inhibition are not ideal due to the high probability of off-target effects and crossreactivity. It can also be used to generate double or multiple knockouts to study miRNA regulatory networks. However, this approach has its limitations, as knockout efficiency is dependent on indel sizes and mutations are often monoallelic, hence the need for single-cell derivation and a 2–4 week screening process to obtain biallelic mutated clones.

A different approach was developed using a combination of TALENs that target the miRNA seed region and a homologous recombination (HR) donor vector carrying a selectable marker, similar to the ZFN technique used to target lncRNAs (see Section 4.1) (Uhde-Stone *et al.*, 2014). As a proof-of-concept, a TALEN pair was designed to target six nucleotides in the seed region of miR-21, leading to targeted cleavage by FokI nuclease in the human HEK-293 cell line. Knockout of miR-21 then occurred via either NHEJ or HR, whereby pre-miR-21 was replaced by exogenous DNA consisting of RFP and the puromycin resistance expression cassette from the HR donor vector. The HR mechanism deleted the miRNA stem-loop, whereas NHEJ disrupted the seed region to generate efficient biallelic miRNA ablation clones. The donor vector was designed to allow two rounds of clonal selection. The two selectable markers, RFP and puromycin resistance, are flanked by LoxP sites. They facilitate the first round of selection for positive clones that can be analyzed using junction PCR, and can be removed for a second round of screening. The authors demonstrated an 87% efficiency using this TALEN design, in which almost all modifications were specific to the targeted region on miR-21, with at least 4 bp deletions or a complete change to the seed sequence. The effect of miR-21 ablation was also confirmed by an increase in the expression of the miR-21 target gene, PDCD4, accompanied by reduced proliferation and changes in global miRNA expression. A second study also used the TALEN approach to target miR-21 and characterize the effects of miR-21 knockouts both *in vitro* and *in vivo* (Chen *et al.*, 2015).

Hu *et al.* described the design of TALENs that either disrupted the miRNA seed regions or completely removed the hairpin sequences to successfully target miR-155*, miR-155, miR-146a and miR-125 in human cells (Hu *et al.*, 2013). Using a single TALEN pair that flanks the seed region of miR-155*, the authors generated both biallelic and monoallelic mutations in the targeted miRNA, whereas targeting miR-155 with a combination of two TALEN pairs that bind to each end of the miR-155 hairpin sequence produced clones that carry deletions spanning the length of pre-miR-155. They also showed that TALENs that target sequences upstream of the miRNA seed region are just as effective in disrupting the seed sequences; therefore, this approach allows for consistent and successful targeting of DNA sequences as small as eight nucleotides.

In what is possibly the first application of the TALEN system to target miRNAs in mice, Takada and colleagues optimized the microinjection of synthesized TALEN RNA without a poly(A) tail targeting different miRNAs into mouse embryos (Takada *et al.*, 2013). This resulted in a reasonable rate of mutated miRNA alleles that were successfully transmitted to the next generation, producing mmu-mir -146a, -10a and -10b knockout mice.

The TALEN system is one of the first genome engineering approaches to be adapted for miRNA ablation. Combined with the innovative design of HR donor vectors and selection strategies, this genome editing tool could be transformed into a flexible and robust platform for studies on ncRNAs.

4.4.3 RNA-guided RNA-targeting CRISPR/Cas9

Initially, CRISPR/Cas9 was largely used to facilitate gene knockout by targeting DNA. However, recent years have seen an increasing number of studies that have taken advantage of the versatility of the CRISPR/Cas9 technology and adapted it to not only successfully target the genomic loci of non-coding RNAs, but also directly target RNA transcripts for different applications.

4.4.3.1 CRISPR Interference (CRISPRi)

Recent studies found that dCas9 retains its ability to bind dsDNA. Researchers then redesigned the CRISPR/Cas9 system to target transcriptional regulation, which mimics the RNAi effect, leading to the birth of CRISPRi (Bikard *et al.*, 2013; Gilbert *et al.*, 2013; Heintze *et al.*, 2013). Bikard *et al.* engineered sgRNAs that guided dCas9 to the transcription initiation sites of different gene loci, thus enabling dCas9 to compete with RNA polymerase for binding to promoters. Additionally, dCas9 can be targeted to coding or enhancer regions, inhibiting transcription at the elongation phase or through steric hindrance that blocks the assembly of transcription factors (Bikard *et al.*, 2013; Qi *et al.*, 2013). A different study fused dCas9 to effector domains of different transcriptional repressors, such as the KRAB domain of Kox1, the CS domain of HP1α or the WRPW domain of Hes1, and used co-expressed sgRNAs to guide these fusion proteins to specific loci, resulting in highly robust and specific transcriptional repression that was validated by RNA sequencing (Gilbert *et al.*, 2013). Conversely, fusion to transcriptional activators VP16 or p65 led to elevated targeted gene expression. The study also suggested possible optimization strategies by altering the design of the fusion protein or partner, highlighting the endless potential of CRISPRi as a modular protein effector recruitment system for transcriptional regulation. Furthermore, this system allows researchers to quantitatively tune the extent of transcriptional repression through the introduction of mismatches to the sgRNA sequence, thus adjusting the complementarity with the target cognate sequence and quantitatively weakening the binding of dCas9. A further advantage of the CRISPRi platform includes its multiplexing functions, whereby the programmable RNA-guided transcriptional regulation can be used to simultaneously and reversibly repress multiple genes in a highly efficient and specific manner, with minimal off-target and cytotoxic effects (Qi *et al.*, 2013).

In a more recent study, Zalatan *et al.* modified the guide RNAs and constructed scaffold RNAs (scRNAs) that consist of a 5′ 20-base targeting sequence, a common dCas9-binding domain and 3′ protein recruitment domain (Zalatan *et al.*, 2015). Thus, a single molecule of scRNA encodes information on the target locus as well as instructions on the relevant regulatory functions, effectively linking DNA binding, protein recruitment and subsequent regulatory activities. This approach was successfully applied in yeast to reprogram cell function and fate by redirecting transcriptional programs and a complex metabolic pathway. Therefore, this platform effectively mimics the effects of CRISPRi (see Section 4.3.1), but with a greater magnitude. Moreover, the authors demonstrated the functional efficacy of this platform in human cells by targeting and activating reporter and endogenous genes in HEK293T cells. Similar to CRISPRi, this scRNA and Cas9-mediated system allows concurrent and programmable control of multiple genes, thus is a promising platform for multi-directional and high-throughput regulation to facilitate the study of complex, intertwined RNA networks.

4.4.3.2 CRISPR/Cas9 for Programmable RNA Recognition, Tracking, Knockdown and Isolation

The more recent advances in the adaptation of the CRISPR/Cas9 system for RNA targeting not only recognize RNA sequences, but enable the tracking, knockdown and isolation of specific RNAs. O'Connell *et al.* developed RNA-targeting synthetic protospacer adjacent motif (PAM)-presenting oligonucleotides, which they named PAMmers (O'Connell *et al.*, 2014). PAMmers provide the PAM sequence in *trans* as a separate DNA oligonucleotide and are engineered to result in a mismatched PAM sequence in the PAMmer–RNA hybrid, which facilitates exclusive targeting of RNA transcripts and not the corresponding genomic DNA sites that lack PAMs. With the use of PAMmers, O'Connell *et al.* observed high affinity binding of Cas9 to single-stranded RNA (ssRNA) and efficient site-specific cleavage by Cas9 only for ssRNA and RNA molecules in RNA–DNA hybrids. Interestingly, this study also showed a novel function of the RNA-programmable RNA-targeting CRISPR/Cas9 system, whereby untagged RNA transcripts could be targeted and isolated. Cas9-compatible PAMmers with 2′-OMe chemical modifications abrogated RNaseH activity and effectively isolated endogenous full-length GAPDH mRNA from HeLa cell lysate. Thus, with only a set of matching sgRNA and complementary PAMmer, this approach bypasses the need for crosslinking, costly biotinylated probes and laborious optimization of salt conditions, to enable the purification of endogenous tagless RNA transcripts.

Following this exciting breakthrough, Nelles *et al.* exploited the newly developed PAMmers to track endogenous RNAs in live cells (Nelles *et al.*, 2016). RNA-targeting Cas9 (RCas9) was engineered with a fusion fluorescent tag, such as mCherry, RFP or GFP, to allow fluorescent tracking. To prove the efficacy of this concept, the authors tracked RNA trafficking of the highly abundant *ACTB* mRNA and the less abundant *CCNA2* and *TFRC* transcripts to stress granules over time and observed significant enrichment of these targets, as well as the RCas9 signal, in stress granules during oxidative stress. This study reinforced previous observations whereby RNA targeting and binding by Cas9-sgRNA is independent of, but strengthened by PAMmers, eliminating previous concerns about the requirement for the PAM sequence being a limiting factor in the CRISPR/Cas9 system. More importantly, it showed that RNA targeting by RCas9 does not alter mRNA and protein expression. Not only is the performance of this platform comparable to fluorescence *in situ* hybridization (FISH), an established method for untagged RNA labeling, it is more advantageous than existing molecular beacons and aptamer-based RNA tracking which suffer from limitations such as delivery issues or the need to modify the target RNA sequence for tag incorporation. Finally, this is a promising platform that could be used for multiplexed targeting.

4.4.3.3 CRISPR-Display (CRISP-Disp)

In 2015, Shechner *et al.* described a unique function of the CRISPR/Cas9 system. Due to the lack of biological tools to differentiate the functions of RNA transcripts from those of the encoding DNA or peptides, a system was developed to relocate RNA transcripts to ectopic sites (Shechner *et al.*, 2015). The group exploited the CRISPR/Cas9 system to deploy large RNA cargos to specific DNA loci and termed this novel technique CRISPR-Display (CRISP-Disp). In CRISP-Disp, sgRNAs are modified to incorporate RNA cargos of various sizes up to 4.8 kb, e.g. lncRNA transcriptional activator *HOTTIP*. Subsequently, dCas9, acting as an RNA display device, deploys the attached RNA cargos to targeted genomic loci. The appended cargos could serve as scaffolds as they can extend to functional RNA domains, such as protein-binding motifs, artificial aptamers or small molecules. Additionally, these lncRNA cargos were functionally tested and were shown to successfully repress or activate the reporter, GLuc. The authors also showed that CRISP-Disp could harness the capabilities of artificial RNA devices, such as aptamers, for various applications, including cell imaging. Furthermore, they successfully demonstrated the multiplexing capacity and modular control of gene expression using one common toolkit. By binding dCas9 to multiple genomic targets and co-expressing reporter-gene-targeting sgRNA variants bearing different RNA-binding proteins, they showed that the multiplexed functions and targeting of CRISP-Disp can be individually or concomitantly controlled. Thus, CRISP-Disp is a versatile genome engineering platform with a broad functional repertoire to study the roles and regulation of diverse RNA species.

4.5 Transforming the ceRNA Field with Programmable Genome Engineering Tools

In recent years, researchers have successfully adapted the three main genome editing platforms (ZFNs, TALENs and CRISPR) to genetically alter a wide range of RNA species, as well as directly target RNA transcripts for various downstream applications, with CRISPR/Cas9 being the biggest breakthrough in the past decade. These revolutionary technologies can be harnessed to provide novel insights into multiple aspects of miRNA:ceRNA regulatory crosstalk.

First, these platforms can be used in conjunction with computational tools to study miRNA:ceRNA regulation. For example, they could be used to target genomic DNA to delete specific MREs identified via bioinformatic analysis (Figure 4.2). Interestingly, several studies have demonstrated the ability of CRISPR/Cas9 to introduce point mutations to genomic DNA in both mouse and zebrafish models (Armstrong *et al.*, 2016; Inui *et al.*, 2014). This serves as an attractive tool that can be adapted to more specifically edit MREs on targeted RNA transcripts to systematically modify miRNA binding. On the other hand, the TALEN technology has been successfully applied to knockdown miRNA expression in cells and *in vivo*, which is also useful for studying the subsequent effect on ceRNA interactions. Additionally, recent breakthroughs in the RNA-targeting and engineering technologies have opened up new avenues to target the diverse ncRNAs that are

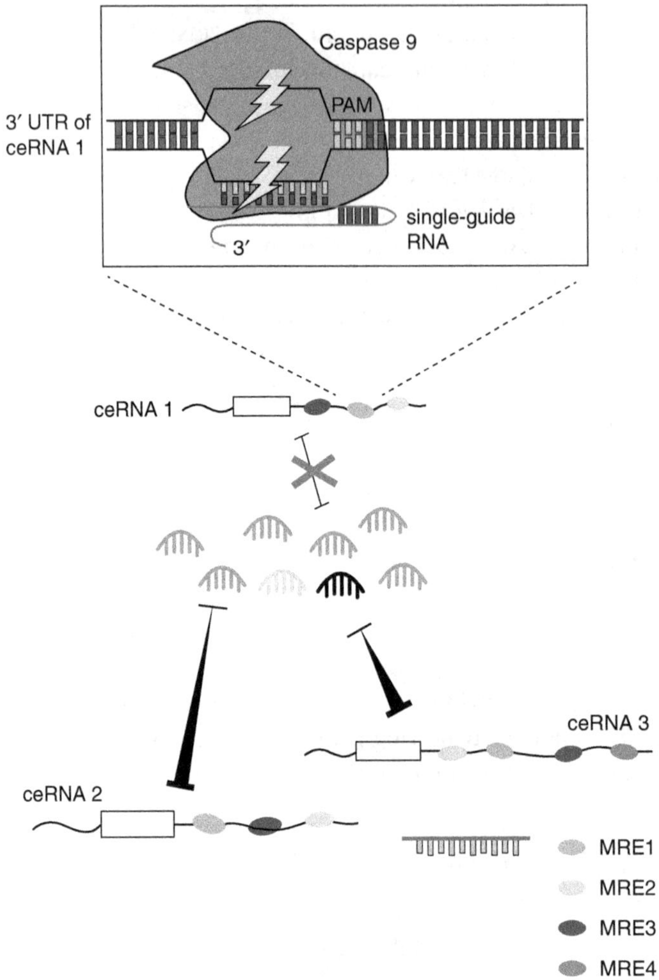

Figure 4.2 Application of the CRISPR/Cas9 system to study of microRNA:ceRNA interactions. A single-guide RNA (sgRNA) designed to target a specific genomic locus directs the Caspase 9 (Cas9) endonuclease to bind and cleave dsDNA (MRE 1 on the 3′UTR of ceRNA 1). The DSB is subsequently repaired via the NHEJ mechanism, resulting in a deletion within MRE 1, hence a reduction in miRNA binding and dysregulation of ceRNA transcript expression. Consequently, the increased pool of miRNAs can bind and alter other transcript expression (ceRNAs 2 and 3) to perturb the miRNA:ceRNA regulatory network. (A black-and-white version of this figure will appear in some formats. For the color version, please refer to the plate section.)

involved in ceRNA networks, including pseudogenes and lncRNAs, some of which have proven challenging to knockdown due to their non-coding nature.

Aside from genome editing, these new platforms have many potential applications that can be exploited for ceRNA studies. For example, the endogenous RNA tracking and isolation capabilities of the adapted CRISPR/Cas9 platform that directly targets RNA transcripts could be used in combination with RNA sequencing or mass spectrometry to facilitate the identification of novel miRNAs, ceRNAs and RNA-binding proteins (RBPs) that may be involved in ceRNA crosstalk.

Recent studies have emphasized the importance of stoichiometry in dictating the level of ceRNA activities, with reports showing that equimolarity of interacting miRNA and ceRNA molecules may result in optimal ceRNA regulation (Ala *et al.*, 2013; Bosson *et al.*, 2014). Thus, the programmable CRISPR/Cas9 system employed by Nelles *et al.* to fluorescently tag dCas9 and track targeted RNA transcripts in live cells could prove invaluable as this platform could be harnessed to quantitatively track RNA transcripts and ceRNA activity using fluorescent intensity measurements (Nelles *et al.*, 2016). With the discovery of a new RNA-guided RNA-targeting nuclease, C2c2, a class 2 type VI CRISPR effector whose catalytically inactive form is analogous to dCas9, this fluorescent system could be further developed for multiplexed targeting and tracking (Abudayyeh *et al.*, 2016). As the regulation of ceRNETs has never been studied in live cells, it could be an exciting development in the field if fluorescence-mediated RNA tracking could be expanded to exploit the Förster resonance energy transfer (FRET) phenomenon, to not only track, but also quantitatively measure interactions of ceRNAs and miRNAs in their physiological states.

In conclusion, recently developed genome engineering platforms have significantly transformed RNA-based research. These gene editing tools provide more flexible and robust approaches to precisely modify components of regulatory RNA networks (summarized in Figure 4.3), and represent a complementary, more physiologically relevant alternative to transient vector or oligo-based knockdown or overexpression studies. The ongoing evolution in these fields will serve to broaden our understanding of the crosstalk and regulatory mechanisms involved in miRNA:ceRNA regulation and help to functionally characterize the non-coding transcriptome, which could facilitate the development of novel diagnostic, prognostic and therapeutic tools, not only in cancer, but also other diseases.

Acknowledgements

We apologize to all colleagues whose work could not be cited due to space constraints. We thank Tay lab members for critical reading of the manuscript. This work was supported by the RNA Biology Center at the Cancer Science Institute of Singapore, NUS, as part of funding under the Singapore Ministry of Education's Tier 3 grants, grant number MOE2014-T3-1–006. Y.T. was supported by a Singapore National Research Foundation Fellowship and a National University of Singapore President's Assistant Professorship.

REFERENCES

Abudayyeh OO, Gootenberg JS, Konermann S, *et al.* 2016. C2c2 is a single-component programmable RNA-guided RNA-targeting CRISPR effector. *Science* **353**(6299): aaf5573

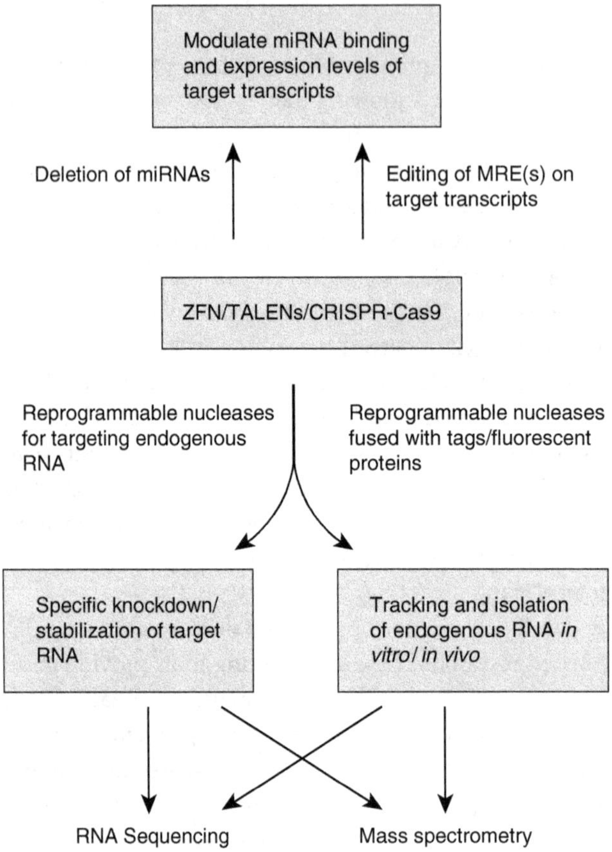

Figure 4.3 Potential applications of genome editing techniques in the study of RNA biology.

Ala U, Karreth FA, Bosia C, *et al.* 2013. Integrated transcriptional and competitive endogenous RNA networks are cross-regulated in permissive molecular environments. *Proc Natl Acad Sci USA* **110**: 7154–7159.

Armstrong GA, Liao M, You Z, *et al.* 2016. Homology directed knockin of point mutations in the zebrafish tardbp and fus genes in ALS using the CRISPR/Cas9 system. *PLoS One* **11**: e0150188.

Bartel DP. 2004. MicroRNAs: genomics, biogenesis, mechanism, and function. *Cell* **116**: 281–297.

Bartel DP, Chen CZ. 2004. Micromanagers of gene expression: the potentially widespread influence of metazoan microRNAs. *Nat Rev Genet* **5**: 396–400.

Bibikova M, Carroll D, Segal DJ, *et al.* 2001. Stimulation of homologous recombination through targeted cleavage by chimeric nucleases. *Mol Cell Biol* **21**: 289–297.

Bikard D, Jiang W, Samai P, *et al.* 2013. Programmable repression and activation of bacterial gene expression using an engineered CRISPR-Cas system. *Nucleic Acids Res* **41**: 7429–7437.

Bitinaite J, Wah DA, Aggarwal AK, Schildkraut I. 1998. FokI dimerization is required for DNA cleavage. *Proc Natl Acad Sci USA* **95**: 10570–10575.

Boch J. 2011. TALEs of genome targeting. *Nat Biotechnol* **29**: 135–136.

Boch J, Scholze H, Schornack S, *et al.* 2009. Breaking the code of DNA binding specificity of TAL-type III effectors. *Science* **326**: 1509–1512.

Bosson AD, Zamudio JR, Sharp PA. 2014. Endogenous miRNA and target concentrations determine susceptibility to potential ceRNA competition. *Mol Cell* **56**: 347–359.

Cade L, Reyon D, Hwang WY, *et al.* 2012. Highly efficient generation of heritable zebrafish gene mutations using homo- and heterodimeric TALENs. *Nucleic Acids Res* **40**: 8001–8010.

Calin GA, Croce CM. 2006. MicroRNA signatures in human cancers. *Nat Rev Cancer* **6**: 857–866.

Calin GA, Dumitru CD, Shimizu M, *et al.* 2002. Frequent deletions and down-regulation of micro-RNA genes miR15 and miR16 at 13q14 in chronic lymphocytic leukemia. *Proc Natl Acad Sci USA* **99**: 15524–15529.

Carlson DF, Tan W, Lillico SG, *et al.* 2012. Efficient TALEN-mediated gene knockout in livestock. *Proc Natl Acad Sci USA* **109**: 17382–17387.

Carninci P, Kasukawa T, Katayama S, *et al.* 2005. The transcriptional landscape of the mammalian genome. *Science* **309**: 1559–1563.

Cech TR, Steitz JA. 2014. The noncoding RNA revolution-trashing old rules to forge new ones. *Cell* **157**: 77–94.

Chen B, Chen X, Wu X, *et al.* 2015. Disruption of microRNA-21 by TALEN leads to diminished cell transformation and increased expression of cell-environment interaction genes. *Cancer Lett* **356**: 506–516.

Chiu HS, Llobet-Navas D, Yang X, *et al.* 2015. Cupid: simultaneous reconstruction of microRNA-target and ceRNA networks. *Genome Res* **25**: 257–267.

Christian M, Cermak T, Doyle EL, *et al.* 2010. Targeting DNA double-strand breaks with TAL effector nucleases. *Genetics* **186**: 757–761.

Cong L, Ran FA, Cox D, *et al.* 2013. Multiplex genome engineering using CRISPR/Cas systems. *Science* **339**: 819–823.

Djebali S, Davis CA, Merkel A, *et al.* 2012. Landscape of transcription in human cells. *Nature* **489**: 101–108.

Fedorov Y, Anderson EM, Birmingham A, *et al.* 2006. Off-target effects by siRNA can induce toxic phenotype. *RNA* **12**: 1188–1196.

Fire A, Xu S, Montgomery MK, *et al.* 1998. Potent and specific genetic interference by double-stranded RNA in *Caenorhabditis elegans*. *Nature* **391**: 806–811.

Gilbert LA, Larson MH, Morsut L, *et al.* 2013. CRISPR-mediated modular RNA-guided regulation of transcription in eukaryotes. *Cell* **154**: 442–451.

Gupta A, Christensen RG, Rayla AL, *et al.* 2012. An optimized two-finger archive for ZFN-mediated gene targeting. *Nat Methods* **9**: 588–590.

Gutschner T, Baas M, Diederichs S. 2011. Noncoding RNA gene silencing through genomic integration of RNA destabilizing elements using zinc finger nucleases. *Genome Res* **21**: 1944–1954.

Ha M, Kim VN. 2014. Regulation of microRNA biogenesis. *Nat Rev Mol Cell Biol* **15**: 509–524.

Hansen TB, Jensen TI, Clausen BH, *et al.* 2013. Natural RNA circles function as efficient microRNA sponges. *Nature* **495**: 384–388.

Hausser J, Syed AP, Bilen B, Zavolan M. 2013. Analysis of CDS-located miRNA target sites suggests that they can effectively inhibit translation. *Genome Res* **23**: 604–615.

Heintze J, Luft C, Ketteler R. 2013. A CRISPR CASe for high-throughput silencing. *Front Genet* **4**: 193.

Hu R, Wallace J, Dahlem TJ, Grunwald DJ, O'Connell RM. 2013. Targeting human microRNA genes using engineered Tal-effector nucleases (TALENs). *PLoS One* **8**: e63074.

Inui M, Miyado M, Igarashi M, *et al.* 2014. Rapid generation of mouse models with defined point mutations by the CRISPR/Cas9 system. *Sci Rep* **4**: 5396.

Jinek M, Chylinski K, Fonfara I, *et al.* 2012. A programmable dual-RNA-guided DNA endonuclease in adaptive bacterial immunity. *Science* **337**: 816–821.

Karreth FA, Reschke M, Ruocco A, *et al.* 2015. The BRAF pseudogene functions as a competitive endogenous RNA and induces lymphoma in vivo. *Cell* **161**: 319–332.

Karreth FA, Tay Y, Perna D, *et al.* 2011. In vivo identification of tumor-suppressive PTEN ceRNAs in an oncogenic BRAF-induced mouse model of melanoma. *Cell* **147**: 382–395.

Khan AA, Betel D, Miller ML, *et al.* 2009. Transfection of small RNAs globally perturbs gene regulation by endogenous microRNAs. *Nat Biotechnol* **27**: 549–555.

Kim H, Kim JS. 2014. A guide to genome engineering with programmable nucleases. *Nat Rev Genet* **15**: 321–334.

Kim JS, Lee HJ, Carroll D. 2010. Genome editing with modularly assembled zinc-finger nucleases. *Nat Methods* **7**: 91; author reply 91–92.

Kim Y, Kweon J, Kim JS. 2013a. TALENs and ZFNs are associated with different mutation signatures. *Nat Methods* **10**: 185.

Kim YK, Wee G, Park J, *et al.* 2013b. TALEN-based knockout library for human microRNAs. *Nat Struct Mol Biol* **20**: 1458–1464.

Lee RC, Feinbaum RL, Ambros V. 1993. The *C. elegans* heterochronic gene lin-4 encodes small RNAs with antisense complementarity to lin-14. *Cell* **75**: 843–854.

Liu XH, Sun M, Nie FQ, *et al.* 2014. Lnc RNA HOTAIR functions as a competing endogenous RNA to regulate HER2 expression by sponging miR-331-3p in gastric cancer. *Mol Cancer* **13**: 92.

Lu M, Zhang Q, Deng M, *et al.* 2008. An analysis of human microRNA and disease associations. *PLoS One* **3**: e3420.

Maeder ML, Thibodeau-Beganny S, Osiak A, *et al.* 2008. Rapid "open-source" engineering of customized zinc-finger nucleases for highly efficient gene modification. *Mol Cell* **31**: 294–301.

Memczak S, Jens M, Elefsinioti A, *et al.* 2013. Circular RNAs are a large class of animal RNAs with regulatory potency. *Nature* **495**: 333–338.

Miranda KC, Huynh T, Tay Y, *et al.* (2006). A pattern-based method for the identification of microRNA binding sites and their corresponding heteroduplexes. *Cell* **126**: 1203–1217.

Moehle EA, Rock JM, Lee YL, *et al.* 2007. Targeted gene addition into a specified location in the human genome using designed zinc finger nucleases. *Proc Natl Acad Sci USA* **104**: 3055–3060.

Mojica FJ, Diez-Villasenor C, Garcia-Martinez J, Soria E. 2005. Intervening sequences of regularly spaced prokaryotic repeats derive from foreign genetic elements. *J Mol Evol* **60**: 174–182.

Nelles DA, Fang MY, O'Connell MR, *et al.* 2016. Programmable RNA tracking in live cells with CRISPR/Cas9. *Cell* **165**: 488–496.

O'Connell MR, Oakes BL, Sternberg SH, *et al.* 2014. Programmable RNA recognition and cleavage by CRISPR/Cas9. *Nature* **516**: 263–266.

Orlando SJ, Santiago Y, Dekelver RC, *et al.* 2010. Zinc-finger nuclease-driven targeted integration into mammalian genomes using donors with limited chromosomal homology. *Nucleic Acids Res* **38**: e152.

Orom UA, Nielsen FC, Lund AH. 2008. MicroRNA-10a binds the 5′UTR of ribosomal protein mRNAs and enhances their translation. *Mol Cell* **30**: 460–471.

Plaisier CL, Pan M, Baliga NS. 2012. A miRNA-regulatory network explains how dysregulated miRNAs perturb oncogenic processes across diverse cancers. *Genome Res* **22**: 2302–2314.

Poliseno L, Salmena L, Zhang J, *et al.* 2010. A coding-independent function of gene and pseudogene mRNAs regulates tumour biology. *Nature* **465**: 1033–1038.

Qi LS, Larson MH, Gilbert LA, *et al.* 2013. Repurposing CRISPR as an RNA-guided platform for sequence-specific control of gene expression. *Cell* **152**: 1173–1183.

Reyon D, Tsai SQ, Khayter C, *et al.* 2012. FLASH assembly of TALENs for high-throughput genome editing. *Nat Biotechnol* **30**: 460–465.

Salmena L, Poliseno L, Tay Y, Kats L, Pandolfi PP. 2011. A ceRNA hypothesis: the Rosetta Stone of a hidden RNA language? *Cell* **146**: 353–358.

Santiago Y, Chan E, Liu PQ, *et al.* 2008. Targeted gene knockout in mammalian cells by using engineered zinc-finger nucleases. *Proc Natl Acad Sci USA* **105**: 5809–5814.

Shechner DM, Hacisuleyman E, Younger ST, Rinn JL. 2015. Multiplexable, locus-specific targeting of long RNAs with CRISPR-Display. *Nat Methods* **12**: 664–670.

Sumazin P, Yang X, Chiu HS, *et al.* 2011. An extensive microRNA-mediated network of RNA-RNA interactions regulates established oncogenic pathways in glioblastoma. *Cell* **147**: 370–381.

Takada S, Sato T, Ito Y, *et al.* 2013. Targeted gene deletion of miRNAs in mice by TALEN system. *PLoS One* **8**: e76004.

Tay Y, Kats L, Salmena L, *et al.* 2011. Coding-independent regulation of the tumor suppressor PTEN by competing endogenous mRNAs. *Cell* **147**: 344–357.

Tay Y, Rinn J, Pandolfi PP. 2014. The multilayered complexity of ceRNA crosstalk and competition. *Nature* **505**: 344–352.

Uhde-Stone C, Sarkar N, Antes T, *et al.* 2014. A TALEN-based strategy for efficient bi-allelic miRNA ablation in human cells. *RNA* **20**: 948–955.

Urnov FD, Miller JC, Lee YL, *et al.* 2005. Highly efficient endogenous human gene correction using designed zinc-finger nucleases. *Nature* **435**: 646–651.

Wang J, Liu X, Wu H, *et al.* 2010. CREB up-regulates long non-coding RNA, HULC expression through interaction with microRNA-372 in liver cancer. *Nucleic Acids Res* **38**: 5366–5383.

Wightman B, Ha I, Ruvkun G. 1993. Posttranscriptional regulation of the heterochronic gene lin-14 by lin-4 mediates temporal pattern formation in *C. elegans*. *Cell* **75**: 855–862.

Wyman C, Kanaar R. 2006. DNA double-strand break repair: all's well that ends well. *Annu Rev Genet* **40**: 363–383.

Xiao H, Tang K, Liu P, *et al.* 2015. LncRNA MALAT1 functions as a competing endogenous RNA to regulate ZEB2 expression by sponging miR-200s in clear cell kidney carcinoma. *Oncotarget* **6**: 38005–38015.

Zalatan JG, Lee ME, Almeida R, *et al.* 2015. Engineering complex synthetic transcriptional programs with CRISPR RNA scaffolds. *Cell* **160**: 339–350.

Part II

Genome Editing in Model Organisms

5 Genome Editing in the Crustacean *Daphnia magna* using CRISPR/Cas and TALEN Systems

Yasuhiko Kato, Takashi Nakanishi and Hajime Watanabe

5.1 Introduction

Daphnia, the water flea, is a model organism for ecological and toxicological genomics owing to its important position in the aquatic food chain, high phenotypic plasticity in response to environmental changes and high sensitivity to chemical compounds. This small planktonic crustacean is ubiquitously found in freshwater environments (Miner *et al.*, 2012; Piña and Barata, 2011). In addition, *Daphnia* recently became a model organism for biomedical research at the National Institutes of Health in the United States (Ebert, 2011). Expressed sequence tags (ESTs) and whole genomes are available for two species, *D. magna* (Orsini *et al.*, 2011; Watanabe *et al.*, 2005) and *D. pulex* (Colbourne *et al.*, 2011), while transcriptome analyses have also been conducted (Eads *et al.*, 2007; Watanabe *et al.*, 2007). These advancements allow the use of *Daphnia* as a tool for analyzing genomic responses to stressors that may have harmful effects on humans.

Programmable nucleases are promising tools for targeted gene knockout and knockin (Carroll, 2014; Gaj *et al.*, 2013). They can be divided into two categories: (1) custom-designed artificial nucleases, such as zinc-finger nucleases (ZFNs) and transcription activator-like effector nucleases (TALENs), and (2) RNA-guided nucleases based on CRISPR (clustered regularly interspaced short palindromic repeats)/Cas (CRISPR-associated) systems. Both ZFNs and TALENs are fusion proteins consisting of a designable DNA-binding domain and a FokI nuclease domain. Compared with ZFNs, which harbor a DNA-binding domain that consists of three or four zinc fingers, TALENs have three advantages in targeted mutagenesis: (1) higher DNA-binding specificity, (2) fewer off-target effects and (3) easier construction of DNA-binding domains (Pan *et al.*, 2013). Because the FokI domain needs to be dimerized in order to exert its nuclease activity, the generation of double-strand breaks at a target site requires the co-expression of two TALENs that bind to proximal DNA sites and to each other. Recent studies reported that heterodimerization of the modified FokI domains, ELD and KKR, reduces TALEN cytotoxicity and increases nuclease activity (Cade *et al.*, 2012; Nakajima and Yaoita, 2013).

The CRISPR/Cas system is a recently developed tool for induction of targeted mutagenesis. In this system, a CRISPR RNA (crRNA) interacts with a transactivating CRISPR RNA (tracrRNA), forming the tracrRNA:crRNA duplex. This acts as a guide RNA (gRNA) that directs the endonuclease Cas9 to its cognate target DNA, where Cas9 induces double-strand breaks (DSBs) (Jinek *et al.*, 2012). Recent studies reported that a single chimeric RNA, formed by the fusion of a crRNA and a tracrRNA, can also function as a gRNA (Jinek *et al.*, 2012). The number of possible targets is limited by the fact that the system requires ~20 bp target sequences with a protospacer adjacent motif (PAM; 5′-NGG-3′) at their 3′ end (Jinek *et al.*, 2012). To date, co-expression of a customized gRNA with the Cas9 nuclease has been used in various organisms for targeted mutagenesis.

Cleavage by the programmable nucleases and repair of the cleaved site by non-homologous end joining (NHEJ) or homologous recombination (HR) pathways are key steps for genome editing. In knockout experiments, NHEJ repair works through incorporation of random insertions or deletions (indels) at the genomic target site. In knockin experiments, the HR pathway has been mainly used for the precise integration of foreign genes via homologous recombination between the genome and donor DNA containing the gene(s) of interest. Recently, TALEN-mediated knockin of plasmid DNA via the NHEJ pathway was reported in studies using zebrafish as a model (Auer *et al.*, 2014). This knockin method requires concurrent cleavage of the injected plasmid DNA and the genome DNA at the targeted site. It allows the knockin of exogenous DNA fragments that are much longer than is possible with other approaches, and has the added benefits of simplicity, flexibility and high efficiency.

As a target for genome editing in *Daphnia magna*, we chose a functionally conserved regulator of eye development, the *D. magna eyeless* gene (*Dma-ey*). *Eyeless* is a mammalian *pax6* homolog originally found in *Drosophila* (Quiring *et al.*, 1994). *Eyeless* mutants show abnormal eye morphogenesis, resulting in complete or partial loss of eyes as well as mutations in the mouse *pax6* gene (Hill *et al.*, 1991). In the current study, we demonstrate that both the TALEN and the CRISPR/Cas systems can introduce heritable mutations into the endogenous *eyeless* gene of *D. magna*. In addition, we demonstrate that TALEN-mediated knockin of the GFP-expressing plasmid DNA sequence was achieved via NHEJ.

5.2 Methodology

A brief overview of the methods used in the studies described in this chapter can be found below.

5.2.1 *Daphnia* Strains

The used *D. magna* strain (NIES clone) was obtained from the National Institute for Environmental Studies (NIES; Tsukuba, Japan), and cultured under laboratory conditions for many generations as previously described (Törner *et al.*, 2014).

5.2.2 Construction of Cas9 and gRNA Expression Vectors

To generate the Cas9 expression vector pCS-Dmavas-Cas9, the Cas9 ORF was amplified using the plasmid MLM3613 (#42251; Addgene, Cambridge, MA, USA). DNA fragments of the 5′ and 3′ UTRs of the *D. magna vasa* (*Dmavas*) gene (Accession: AB193324.1) were obtained by PCR using cDNAs of the NIES strain. The PCR products were simultaneously cloned into the downstream region of the SP6 promoter of pCS2+ vecto (Turner and Weintraub, 1994) using the In-Fusion PCR cloning kit (Clontech Laboratories, Mountain View, CA, USA). To generate the gRNA expression vector pDR274-Dma-ey, pairs of *Dma-ey* targeting oligonucleotides for gRNA-1 and gRNA-2 were annealed and then ligated into the linearized pDR274 vector (Addgene #42250; Nakanishi *et al.*, 2014).

5.2.3 Construction of TALEN expression vectors

The TALEN target sites were identified using TAL Effector Nucleotide Targeter (TALE-NT) v2.0 (Doyle *et al.*, 2012) with the following parameters: (1) a spacer length of 15 to 20 nucleotides, (2) a repeat array length of 15 to 20 nucleotides, (3) NN for G substitute and (4) T at position 0, according to the guidelines of Streubel *et al.* (2012).

Construction of TALEN expression vectors was divided into three steps. First, ey1-TALEN assemblies of the RVD-containing repeats were performed using the Golden Gate approach (Cermak *et al.*, 2011). Once assembled, the repeats were cloned into a pT3TS destination vector with the GoldyTALEN (Addgene #38142) backbone, resulting in the creation of pT3TS-ey1-TALEN-left and -right. In the second step, vectors for the synthesis of ey1-TALEN mRNAs to be injected into *D. magna* eggs were generated by amplifying and cloning ey1-TALEN-left and -right coding sequences in pCS-Dmavas-Cas9 between the 5′ UTR and 3′ UTR of *Dmavas*, resulting in the construction of the homodimeric TALEN expression vectors pCS-Dmavas-ey1-TALEN-left and pCS-Dmavas-ey1-TALEN-right. Third, the QuickChange Lightening Multi Site-Directed Mutagenesis Kit (Agilent Technologies, Cedar Creek, TX, USA) was used to introduce the Q486E, I499L and N496D mutations to ey1-TALEN-left, and the E490K, I538K and H537K mutations to ey1-TALEN-right; this resulted in the construction of the heterodimeric TALEN expression vectors pCS-Dmavas-ey1-TALEN-ELD and pCS-Dmavas-ey1-TALEN-KKR.

The homodimeric ey2-TALEN expression vectors, pT3TS-ey2-TALEN-left and pT3TS-ey2-TALEN-right, as well as homodimeric ey1-TALEN expression vectors, were also created. The regions encoding RVD-containing repeats were cloned into pCS-Dmavas-ey1-TALEN-ELD and pCS-Dmavas-ey1-TALEN-KKR, resulting in the construction of heterodimeric ey2-TALEN expression vectors, pCS-Dmavas-ey2-TALEN-ELD and pCS-Dmavas-ey2-TALEN-KKR, respectively (Naitou *et al.*, 2015).

5.2.4 Construction of Donor Plasmid

To generate the donor plasmid for the knockin experiments, the 4xEcRE region of the 4xEcRE-H2B-GFP plasmid(Asada *et al.*, 2014) was replaced with four tandem repeats of the juvenile hormone response element (JHRE) of the *Krüppel homolog 1* gene from *Tribolium castaneum* (Kayukawa *et al.*, 2013). The fluorescent reporter gene *H2B-GFP* encodes a GFP protein fused with *D. magna* histone H2B (Nakanishi *et al.*, 2016).

5.2.5 *In Vitro* RNA Synthesis

For the synthesis of Cas9 mRNA, templates with a T7 promoter were amplified by PCR from the pCS-Dmavas-Cas9. For the synthesis of gRNAs, DraI-digested pDR274-Dma-ey vectors were prepared. For the synthesis of TALEN mRNAs, the TALEN expression vectors were linearized with Acc65I endonuclease. To synthesize capped RNA, each template was subjected to *in vitro* transcription with the mMessage mMachine T7 kit (Life Technologies, Carlsbad, CA, USA). Poly(A) tails were attached to Cas9 and TALEN mRNAs by using a Poly(A) Tailing Kit (Life Technologies), following the manufacturer's instructions. The *in vitro* synthesized RNAs were column-purified using mini Quick Spin RNA columns (Roche diagnostics GmbH, Mannheim, Germany), followed by phenol/chloroform extraction, ethanol precipitation and dissolution in DNase/RNase-free water (Life Technologies; Nakanishi *et al.*, 2014).

5.2.6 Microinjection

Eggs were obtained from *D. magna* and placed in ice-cold M4 medium containing 80 mM sucrose (M4-sucrose). In addition to the synthesized RNAs, each injection contained 1 mM of the water-soluble fluorescent dye Lucifer Yellow (LY; Life Technologies), which was used to quantify the injection volume 1 h after microinjection and distinguish eggs injected with RNA from unmodified eggs (Törner *et al.*, 2014). Injection volume was approximately 0.3 nL. Injected eggs were transferred into 96-well plates at a concentration of one egg per well, and incubated for the appropriate time (Kato *et al.*, 2011).

5.2.7 Genotyping

A standard phenol/chloroform protocol for mouse-tail DNA extraction was modified and used for this study. Targeted genome regions were amplified from extracted genomic DNA, then cloned and sequenced. For inverse PCR, approximately 2 µg of genomic DNA from transgenic animals was digested with NcoI and circularized by ligation for 24 h at 4 °C. PCR was performed using primers that amplify the junction regions between the integrated plasmid and the surrounding genome (Nakanishi *et al.*, 2016).

5.3 Results and Discussion

5.3.1 Target Sites of CRISPR/Cas and TALENs

To test whether the CRISPR/Cas and TALEN systems can induce targeted mutagenesis in *D. magna, Dma-ey* was chosen as the targeted locus (Figure 5.1), because *Dma-ey* RNAi induced the screenable phenotype "deformed eye," in which compound eyes are located at the inner side and cannot build a precise spherical shape (Figure 5.2A; Nakanishi *et al.*, 2014). We attempted to introduce indel mutations

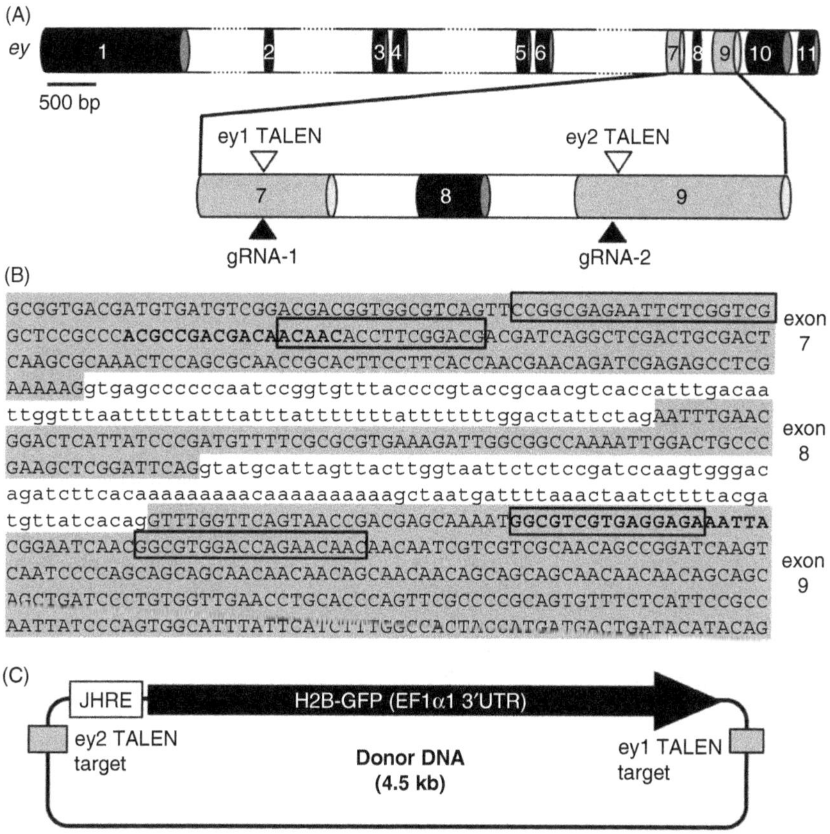

Figure 5.1 Molecular reagents used in knockin and knockout experiments. (A) Exon-intron structure of the *eyeless* (*ey*) gene in *D. magna*. A predicted structure of *ey* is schematically represented. Exons and introns are shown in black and white, respectively, with the exception of ey1-TALEN targeting exon 7 (gray) and ey2-TALEN targeting exon 9 (gray). White arrowheads indicate ey1-TALEN and ey2-TALEN cleavage sites. Black arrowheads indicate gRNA-targeting sites. (B) Nucleotide sequence of exons 7, 8 and 9. Exons and introns are shown in gray and white, respectively. Guide RNA-binding sites are written in bold font. TALEN-binding sites are enclosed by squares. (C) Schematic representation of donor DNA. This figure was drawn based on a previously published paper (Nakanishi *et al.*, 2016).

into exons 7, 8 and 9 (Figure 5.1A) that code for the homeobox domain, because *ey*-deficient *Drosophila* flies were found to harbor mutated *ey* alleles lacking all C-terminal domains, including the homeobox domain (Kronhamn *et al.*, 2002). Thus, we hypothesized that *Dma-ey*-deficient daphniids can be generated by inducing frameshift mutations at gene regions coding parts of the homeobox domain closer to the N-terminus.

We used two gRNAs, gRNA-1 and gRNA-2, which were designed to bind the sense strand at exon 7 and the antisense strand at exon 9, respectively (Figure 5.1B). The binding sites of ey1-TALEN and ey2-TALEN partially overlapped with those of gRNA-1 and gRNA-2, respectively (Figure 5.1B). Because the homodimer of the wild-type FokI domain was shown to confer severe toxicity to various organisms, including *D. magna* (Cade *et al.*, 2012; Naitou *et al.*, 2015), heterodimeric TALENs were used by introducing ELD/KKR mutations into the FokI domains of the homodimeric ey1-TALEN pair.

5.3.2 Strategy to Test CRISPR/Cas and TALEN Systems in Asexual *D. magna*

The activity of Crispr/Cas and TALEN systems in *D. magna* was examined in asexual females. In laboratory culture conditions, *Daphnia* produces female-only offspring by asexual reproduction. Seven days after birth, it starts to produce asexual eggs and continues to lay eggs every 2.5 days. Eggs were immediately collected after ovulation and cooled on ice to prevent hardening of egg membranes. Before the initiation of cell division, we injected the synthesized RNAs into the eggs (G0 eggs), and incubated each injected egg in a well of a 96-well plate until it became a juvenile. In G0 juveniles, the eye phenotype was observed to check the introduction of *eyeless* mutations in somatic cells. In G0 adults, founder animals that produced G1 juveniles with eye deformity were counted to calculate the efficiency of heritable mutagenesis. The G1 juveniles that were produced by the founder animals were subjected to genotyping to investigate whether the mutations had been introduced into the target site.

5.3.3 *Eyeless* Disruption by CRISPR/Cas

To examine whether the CRISPR/Cas system works in *D. magna*, Cas9 mRNA and gRNAs were injected into eggs. Because a previous report suggested that co-injection of multiple gRNAs increased mutation efficiency (Gratz *et al.*, 2013), we co-injected two gRNAs (gRNA-1 and gRNA-2) together with the Cas9 mRNA. This Cas9 mRNA contains the untranslated regions (UTRs) of the *Dmavas* gene, which is exclusively expressed in *D. magna* germ cells (Sagawa *et al.*, 2005). We tested three different concentrations of Cas9 mRNA (500, 1000 and 2000 ng/μL) and two of the gRNAs (50 and 150 ng/μL each; Table 5.1). At the first instar juvenile stage, 18–47% of the surviving juveniles developed deformed eye phenotypes without significant dose dependency, suggesting that the CRISPR/Cas system worked in *D. magna* (Table 5.1).

Founder animals represented 2.6–8.2% of surviving adults, excluding the case when 50 ng/μL each of gRNA-1 and gRNA-2 were injected with 2000 ng/μL Cas9 mRNA (Table 5.1 and Figure 5.2A). The highest efficiency was achieved by injecting 50 ng/μL each of gRNA-1 and gRNA-2 with 500 ng/μL Cas9 mRNA. Of the 14 founder animals, 13 produced deformed-eye progeny that died within a week. The lethality of *Dma-ey* knockout mutations was consistent with the previous report on *Drosophila* (Kronhamn *et al.*, 2002). However, the remaining founder animal produced viable deformed eye progeny and the phenotype was observed over generations.

To investigate how indel mutations were introduced into the *Dma-ey* locus, we performed genotyping of deformed eye G1 progeny from four different founder lines before they died, and from the one founder line that produced healthy progeny. All of the non-viable G1 progenies had biallelic indel mutations around the gRNA-targeted sites (Figure 5.3), and coded for proteins lacking regions

Table 5.1 Mutation frequencies induced by microinjection using various concentrations of Cas9 mRNA/gRNA mix and TALEN mRNAs

RNA concentration (ng/μL)		Injected embryos	Juveniles: Surviving	Juveniles: Deformed eye	Adults: Surviving	Adults: Founder lines
gRNA-1, −2 50 each	Cas9 500	77	59/77 (77%)	28/59 (47%)	49/77 (64%)	[a]4/49 (8.2%)
	Cas9 1000	121	90/121 (74%)	16/90 (18%)	81/121 (67%)	5/81 (6.2%)
	Cas9 2000	113	75/113 (66%)	29/75 (39%)	61/113 (54%)	0/61 (0%)
gRNA-1, −2 150 each	*Cas9* 500	98	70/98 (71%)	19/70 (27%)	60/98 (61%)	2/60 (3.3%)
	Cas9 1,000	86	67/86 (78%)	20/67 (30%)	60/86 (70%)	[b]2/60 (3.3%)
	Cas9 2,000	64	39/64 (61%)	8/39 (21%)	38/64 (59%)	[b]1/38 (2.6%)
ey1-TALEN-ELD, KKR 500 each		78	62/78 (79%)	26/62 (42%)	[a]NA	NA
ey1-TALEN-ELD, KKR 250 each		95	76/95 (80%)	17/76 (22%)	NA	NA
ey1-TALEN-ELD, KKR 125 each		116	86/116 (74%)	24/86 (28%)	NA	NA
ey2-TALEN-ELD, KKR 250 each		105	50/105 (48%)	39/50 (78%)	33/105 (33%)	[b]6/33 (18%)
ey2-TALEN-ELD, KKR 125 each		46	19/46 (41%)	10/19 (53%)	9/46 (20%)	2/9 (22%)

[a] The first number denotes the number of mutant lines that were subjected to sequencing of the *Dma-ey* loci.

[b] NA: not analyzed.

spanning from the homeobox domain to the C-terminus (Figure 5.3A; m1 to m4). Interestingly, the viable mutant harbored one *Dma-ey* allele that had lost an 877 bp region that included the entire homeobox, but it retained the correct reading frame at the part coding for the remaining C-terminal region (Figure 5.3A; m5), whereas the other allele had a single one-nucleotide deletion. This might account for the difference in the mutants' viability, as discussed previously (Nakanishi *et al.*, 2014). Thus, we concluded that the CRISPR/Cas system works not only in somatic cells, but also in germline cells, and can thus induce heritable mutations at the target site.

5.3.4 *Eyeless* Disruption by TALENs

To examine whether the TALEN system works in *D. magna*, ey1-TALEN mRNAs were first injected into G0 eggs. Three different concentrations (125, 250 and 500 ng/μL) of each of the heterodimeric ey1-TALEN-left and -right mRNAs were used, resulting in respective development rates of 74%, 80% and 79%, while 28%, 22%

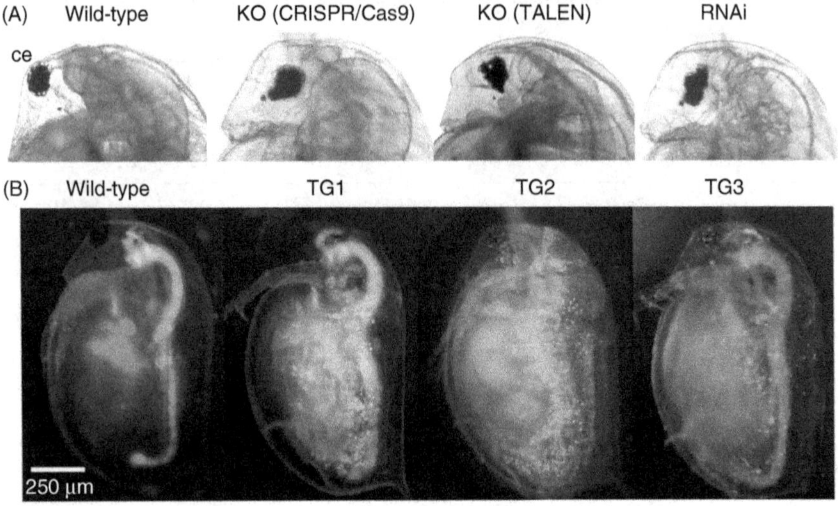

Figure 5.2 Phenotypes of knockout (A) and knockin animals (B). Each image was originally used in previous papers (Nakanishi et al., 2014; 2015; 2016). (A black-and-white version of this figure will appear in some formats. For the color version, please refer to the plate section.)

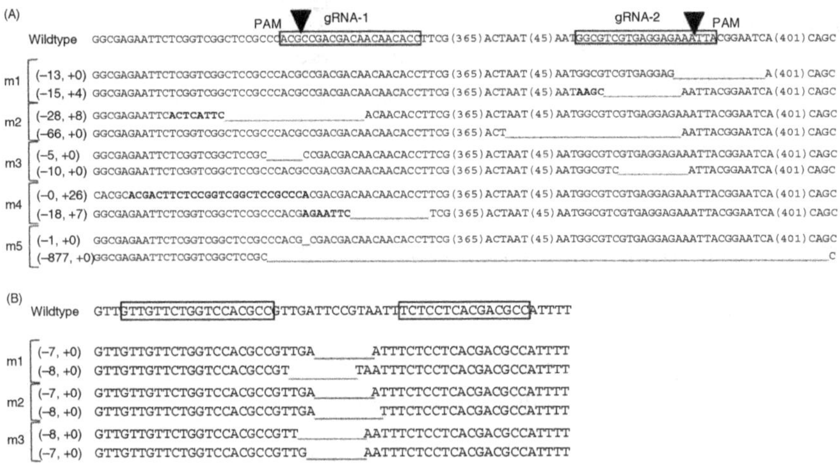

Figure 5.3 Genotype of knockout animals generated by the CRISPR/Cas (A) and TALEN systems (B). Binding sites of gRNAs and TALENs are enclosed by squares. Cas9 cleavage sites are indicated with black arrowheads. In each mutant sequence, deletions are indicated by underbars, while insertions by bold letters. Sequence data were originally used in previous papers (Nakanishi *et al.*, 2014; 2015).

and 42% of the surviving juveniles, respectively, showed abnormal eye morphology (Table 5.1). The phenotype was similar to the deformed eye phenotype induced by CRISPR/Cas-mediated *ey* knockout (Figure 5.2A), suggesting that the heterodimeric *ey1*-TALENs could be used in *D. magna*.

To provide additional evidence, the experiment was repeated with two concentrations (125 and 250 ng/μL) of each of the heterodimeric ey2-TALEN-left and -right mRNAs. The respective development rates of the injected embryos were 41% and 48%, while 53% and 78% of the surviving juveniles, respectively, showed abnormal eye morphology (Table 5.1). The higher occurrence of phenotypic change by ey2-TALEN compared to ey1-TALEN prompted us to analyze the phenotypes and genotypes of the G1 juveniles. Founder G0 animals represented 18–22% of the surviving adults (Table 5.1). All of the founder G0 animals produced non-viable progeny that died within a week. We sequenced the TALEN-targeted site in the genomes of G1 progenies from three founder G0 animals. All of the G1 progenies had biallelic indel mutations, resulting in frameshift mutations in both alleles (Figure 5.3B). Thus, we concluded that heterodimeric TALENs can induce heritable mutations at the endogenous *ey* locus (Naitou *et al.*, 2015).

5.3.5 Comparison of knockout efficiency between CRISPR/Cas and TALEN systems

Both gRNAs and heterodimeric TALENs were targeted at the homeobox region of the *ey* gene, allowing us to compare the efficiency of TALEN and CRISPR/Cas9 systems in introducing mutations (Table 5.1). There was no apparent difference between the two systems in the timing of the onset of the effects of mutagenesis, because both systems worked during embryogenesis and led to production of juveniles with abnormal eye morphology. The somatic and heritable mutagenesis efficiencies of TALENs were higher than those of the CRISPR/Cas9 system, although optimal concentrations of gRNAs and Cas9 mRNAs need to be further examined. Therefore, for the targeted mutagenesis of a single gene in *D. magna*, TALENs might be more reliable and efficient than CRISPR/Cas9. However, for multiple-target mutagenesis, the CRISPR/Cas9 system would be more suitable because of the simplicity of gRNA design.

5.3.6 Construction of the Donor Plasmid for TALEN-mediated Knockin

The higher heritable mutagenesis efficiency of TALENs compared to the CRISPR/Cas9 system prompted us to test TALEN-mediated knockin of plasmid DNA by NHEJ in *D. magna*. To confirm the genomic integration and the germline transmission of plasmid DNA, donor plasmid DNA was constructed to encode a histone H2B-GFP fusion protein with nuclear localization, which facilitates its visualization in individual cells (Figure 5.1C) (Kanda *et al.*, 1998; Kato *et al.*, 2012). Four tandem repeats of the juvenile hormone responsive element (JHRE) from the promoter region of the *Krüppel homolog1* gene from the beetle *Tribolium castaneum* (Kayukawa *et al.*, 2013) were inserted upstream of *H2B-GFP*. In addition, the 127 and 173 bp regions that contained the ey2-TALEN and ey1-TALEN target sites were inserted upstream and downstream, respectively, in the JHRE/H2B-GFP reporter gene construct (Figure 5.1C).

5.3.7 Knockin of Plasmid DNA by TALEN Systems

We first tested ey1-TALEN-mediated knockin. Donor plasmid DNA was co-injected into 48 eggs with 500 ng/μL of ey1-TALEN mRNA, which would expectedly lead to concurrent cleavage of the genomic *ey* locus and the donor plasmid.

Of the 38 surviving animals, one founder produced fluorescent juveniles with normal eyes (Figure 5.2B; TG1). These juveniles were cultured and maintained as a potentially transgenic line (TG1).

The suitability of ey2-TALEN for knockin was also tested by co-injecting donor plasmid DNA with ey2-TALEN mRNA into 117 eggs. Of the 60 surviving animals, two founders produced fluorescent juveniles (Figure 5.2B; TG2 and TG3). As was also the case with TG1, one founder produced juveniles that did not show any eye deformities. The juveniles were cultured and maintained as a potentially transgenic line (TG2). In contrast, the other founder only produced juveniles with eye deformities. These juveniles had smaller compound eyes that were located at the inner side (Nakanishi *et al.*, 2016), which led to death before sexual maturation. Some offspring from this potential transgenic line (TG3) were collected before death and underwent genotyping.

To identify the integration site of the donor plasmid in the potentially transgenic lines TG1 and TG2, we performed inverse PCR. In TG1, two copies of the *H2B-GFP* gene were tandemly integrated into the genome, at the target site of the *ey* locus. There were no indel mutations at the 5′ and 3′ junctions (Figure 5.4A). The other allele possibly did not harbor any deleterious mutation, as this line did not show eye deformity. In TG2, we found that the insertion was not mapped to the *ey* locus but to the *abba* (another B-box affiliate) locus. Moreover, there was a 12 bp deletion and an 8 bp indel mutation at the 5′ and 3′ junctions, respectively (Figure 5.4B). Because TG2 also produced juveniles with normal compound eyes and the plasmid was found to be integrated into the *abba* locus of these transgenic animals, TALEN would not lead to deleterious mutations in both alleles of the *ey* locus. Integration into the *abba* locus did not appear to be a result of TALEN off-target activity, as discussed previously (Nakanishi *et al.*, 2016). Instead, another mechanism, namely random TALEN-independent integration of the donor DNA into the genome, might have occurred at this locus.

As members of the TG3 line die during the juvenile stage, we were unable to raise a sufficient number of animals to obtain the quantity of genomic DNA required for inverse PCR. Instead, we performed genomic PCR to confirm whether the donor plasmid DNA inserted into the target site of the *ey* locus. We found that two plasmids, the first of which harbored a large deletion, had tandemly integrated at the 5′ junction, whereas the 3′ junction was mapped to the target site of the ey2-TALEN, and contained a 6 bp deletion and an 8 bp indel mutation, suggesting that NHEJ repair had occurred at this site (Figure 5.4C). Unexpectedly, the 5′ junction was located at the ey1-TALEN-target site in exon 7 and did not have any indel mutations, although this transgenic animal had been created with ey2-TALEN, which cleaves exon 9. Because our donor plasmid contained sequences from exon 7, we hypothesised that the 5′ end of the transgene had been ligated via homologous recombination between the integrated donor plasmid-originating exon 7 sequence and the endogenous *eyeless* gene. In addition, TG3 mutants were expected to also harbor deleterious mutations in the other *ey* allele, which would explain its non-viability.

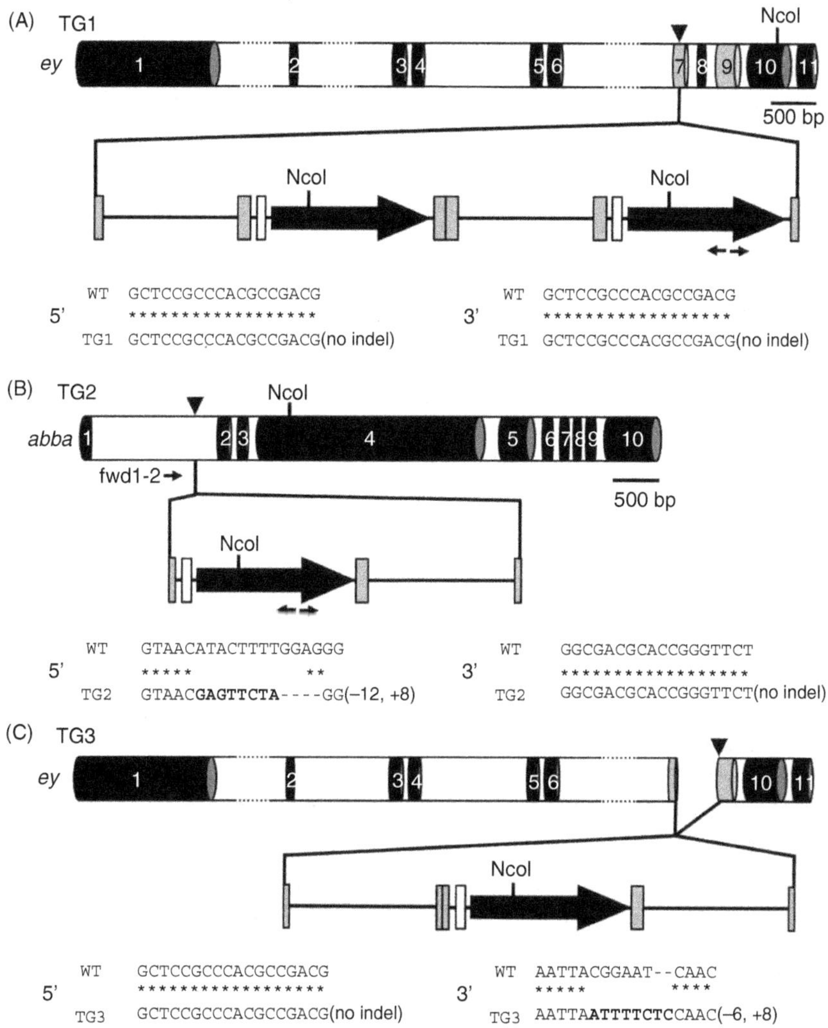

Figure 5.4 Genotypes of transgenic lines TG1 (A), TG2 (B) and TG3 (C). A complete schematic is shown for each of the three transgenic lines produced in this study (TG1, TG2, and TG3). In all three lines, the top image shows the complete structure of the endogenous locus into which donor DNA was integrated, the middle image presents the structure of the integrated plasmid in the genome and the bottom image displays the sequences of the junctions between the transgene and the surrounding genome. NcoI sites used for inverse PCR are shown above each gene. Primer binding sites for inverse PCR are shown with small black arrows, TALEN targets with gray boxes and the *H2B-GFP* reporter gene with a large black arrow. Arrowheads indicate the ey1- and ey2-TALEN cleavage sites. This figure was drawn based on a previously published paper (Nakanishi *et al.*, 2016).

5.4 Conclusion

We successfully performed knockout of the endogenous *eyeless* gene using the CRISPR/Cas and TALEN systems in *D. magna*. TALEN-mediated knockin through a homology-independent DNA repair pathway was also established. In addition, we recently reported that TALEN-mediated homologous recombination can be performed in this species (Nakanishi *et al.*, 2015). These trans-generational genetic manipulations could prove to be important tools that would take advantage of the available EST and genomic sequences to pursue genetic studies in *Daphnia*, complementing the two transient manipulation methods we previously developed, namely RNA interference and overexpression by injection of RNAs into eggs. Genome editing and engineering could lead to a better understanding of the ecology, evolution and molecular biology mechanisms of this important model organism.

REFERENCES

Asada M, Kato Y, Matsuura T, Watanabe H. 2014. Visualization of ecdysteroid activity using a reporter gene in the crustacean, *Daphnia*. *Mar Environ Res* **93**: 118–122.

Auer TO, Duroure K, De Cian A, Concordet JP, Del Bene F. 2014. Highly efficient CRISPR/Cas9-mediated knock-in in zebrafish by homology-independent DNA repair. *Genome Res* **24**: 142–153.

Cade L, Reyon D, Hwang WY, *et al.* 2012. Highly efficient generation of heritable zebrafish gene mutations using homo- and heterodimeric TALENs. *Nucleic Acids Res* **40**: 8001–8010.

Carroll D. 2014. Genome engineering with targetable nucleases. *Annu Rev Biochem* **83**: 409–439.

Cermak T, Doyle EL, Christian M, *et al.* 2011. Efficient design and assembly of custom TALEN and other TAL effector-based constructs for DNA targeting. *Nucleic Acids Res* **39**: e82.

Colbourne JK, Pfrender ME, Gilbert D, *et al.* 2011. The ecoresponsive genome of *Daphnia pulex*. *Science* **331**: 555–561.

Doyle EL, Booher NJ, Standage DS, *et al.* 2012. TAL Effector-Nucleotide Targeter (TALE-NT) 2.0: tools for TAL effector design and target prediction. *Nucleic Acids Res* **40**: W117–W122.

Eads BD, Colbourne JK, Bohuski E, Andrews J. 2007. Profiling sex-biased gene expression during parthenogenetic reproduction in *Daphnia pulex*. *BMC Genomics* **8**: 464.

Ebert D. 2011. A genome for the environment. *Science* **331**: 539–540.

Gaj T, Gersbach CA, Barbas CF. 2013. ZFN, TALEN, and CRISPR/Cas-based methods for genome engineering. *Trends Biotechnol* **31**: 397–405.

Gratz SJ, Cummings AM, Nguyen JN, *et al.* 2013. Genome engineering of *Drosophila* with the CRISPR RNA-guided Cas9 nuclease. *Genetics* **194**: 1029–1035.

Hill RE, Favor J, Hogan BL, *et al.* 1991. Mouse small eye results from mutations in a paired-like homeobox-containing gene. *Nature* **354**: 522–525.

Jinek M, Chylinski K, Fonfara I, *et al.* 2012. A programmable dual-RNA-guided DNA endonuclease in adaptive bacterial immunity. *Science* **337**: 816–821.

Kanda T, Sullivan KF, Wahl GM. 1998. Histone-GFP fusion protein enables sensitive analysis of chromosome dynamics in living mammalian cells. *Curr Biol* **8**: 377–385.

Kato Y, Matsuura T, Watanabe H. 2012. Genomic integration and germline transmission of plasmid injected into crustacean *Daphnia magna* eggs. *PLoS One* **7**: e45318.

Kato Y, Shiga Y, Kobayashi K, *et al.* 2011. Development of an RNA interference method in the cladoceran crustacean *Daphnia magna*. *Dev Genes Evol* **220**: 337–345.

Kayukawa T, Tateishi K, Shinoda T. 2013. Establishment of a versatile cell line for juvenile hormone signaling analysis in *Tribolium castaneum*. *Sci Rep* **3**: 1570.

Kronhamn J, Frei E, Daube M, *et al.* 2002. Headless flies produced by mutations in the paralogous Pax6 genes eyeless and twin of eyeless. *Development* **129**: 1015–1026.

Miner BE, De Meester L, Pfrender ME, Lampert W, Hairston NG. 2012. Linking genes to communities and ecosystems: *Daphnia* as an ecogenomic model. *Proc Biol Sci* **279**: 1873–1882.

Naitou A, Kato Y, Nakanishi T, Matsuura T, Watanabe H. 2015. Heterodimeric TALENs induce targeted heritable mutations in the crustacean *Daphnia magna*. *Biol Open* **4**: 364–369.

Nakajima K, Yaoita Y. 2013. Comparison of TALEN scaffolds in *Xenopus tropicalis*. *Biol Open* **2**: 1364–1370.

Nakanishi T, Kato Y, Matsuura T, Watanabe H. 2014. CRISPR/Cas-mediated targeted mutagenesis in *Daphnia magna*. *PLoS One* **9**: e98363.

Nakanishi T, Kato Y, Matsuura T, Watanabe H. 2015. TALEN-mediated homologous recombination in *Daphnia magna*. *Sci Rep* **5**: 18312.

Nakanishi T, Kato Y, Matsuura T, Watanabe H. 2016. TALEN-mediated knock-in via non-homologous end joining in the crustacean *Daphnia magna*. *Sci Rep* **6**: 36252.

Orsini L, Decaestecker E, De Meester L, Pfrender ME, Colbourne JK. 2011. Genomics in the ecological arena. *Biol Lett* **7**: 2–3.

Pan Y, Xiao L, Li ASS, *et al.* 2013. Biological and biomedical applications of engineered nucleases. *Mol Biotechnol* **55**: 54–62.

Piña B, Barata C. 2011. A genomic and ecotoxicological perspective of DNA array studies in aquatic environmental risk assessment. *Aquat Toxicol* **105**: 40–49.

Quiring R, Walldorf U, Kloter U, Gehring WJ. 1994. Homology of the eyeless gene of *Drosophila* to the Small eye gene in mice and Aniridia in humans. *Science* **265**: 785–789.

Sagawa K, Yamagata H, Shiga Y. 2005. Exploring embryonic germ line development in the water flea, *Daphnia magna*, by zinc-finger-containing VASA as a marker. *Gene Expr Patterns* **5**: 669–678.

Streubel J, Blucher C, Landgraf A, Boch J. 2012. TAL effector RVD specificities and efficiencies. *Nat Biotechnol* **30**: 593–595.

Törner K, Nakanishi T, Matsuura T, Kato Y, Watanabe H. 2014. Optimization of mRNA design for protein expression in the crustacean *Daphnia magna*. *Mol Genet Genomics* **289**: 707–715.

Turner DL, Weintraub H. 1994. Expression of achaete-scute homolog-3 in xenopus-embryos converts ectodermal cells to a neural fate. *Genes Dev* **8**: 1434–1447.

Watanabe H, Takahashi E, Nakamura Y, *et al.* 2007. Development of a *Daphnia magna* DNA microarray for evaluating the toxicity of environmental chemicals. *Environ Toxicol Chem* **26**: 669–676.

Watanabe H, Tatarazako N, Oda S, *et al.* 2005. Analysis of expressed sequence tags of the water flea *Daphnia magna*. *Genome* **48**: 606–609.

6 Leapfrogging: A Method for Targeting Genome Editing to the Germline

Ira L. Blitz

6.1 Introduction

CRISPR/Cas9-based genome editing is transforming biological research by facilitating DNA alterations in organisms that were previously not readily genetically modifiable. The African clawed frog *Xenopus laevis*, not traditionally a system used by geneticists, has been an important model organism for over half a century and has contributed greatly to investigations in fields including molecular and cell biology, developmental biology, physiology, endocrinology, neurobiology, evolution and immunology, to name a few. In recent years, the closely related species *Xenopus tropicalis* has been exploited because it is diploid, unlike its allotetraploid "cousin" *X. laevis*, and has a gene content similar to human and mouse (Blitz *et al.*, 2017; Hellsten *et al.*, 2010). The diploidy of *X. tropicalis* motivated forward genetic screens, identifying numerous mutant phenotypes, and genetic mapping strategies identified causative genes (Abu-Daya *et al.*, 2009; Chung *et al.*, 2016; Geach and Zimmerman, 2010; Goda *et al.*, 2006; Grammer *et al.*, 2005; Nakayama *et al.*, 2017; Noramly *et al.*, 2005). The CRISPR/Cas9 "revolution" in genome editing, combined with the complete genome sequences of both *Xenopus tropicalis* (Hellsten *et al.*, 2010) and *laevis* (Sessions *et al.*, 2016), is now concentrating interest in the research community on creating targeted mutations and gene modifications in both *Xenopus* species (Aslan *et al.*, 2017; Blitz *et al.*, 2013; Guo *et al.*, 2014; Jaffe *et al.*, 2016; Nakayama *et al.*, 2013; Naleade *et al.*, 2014; Wang *et al.*, 2015).

Despite these advances in genome editing, the creation of lines bearing mutations in critical genes is laborious. The high efficiency of CRISPR/Cas9 mutagenesis is often sufficient to produce mosaic F0 animals containing biallelic mutations in most cells, allowing for F0 phenotypic analyses (Bhattacharya *et al.*, 2015; Blitz *et al.*, 2013; Guo *et al.*, 2014; Nakayama *et al.*, 2013). However, mutant lines will still be necessary for many applications. First, mosaic F0 animals typically display partial loss-of-function phenotypes (Blitz *et al.*, 2013; 2016; Guo *et al.*, 2014; Nakayama *et al.*, 2013), but complete LOF will be required to fully understand the contribution of genes to phenotypes. In addition, maternal effect genes are expected to be completely refractory to F0 analyses because their wild-type gene products produced during oogenesis will already be deposited in the mature oocyte. Indeed, a large portion of *Xenopus* mRNAs and proteins found in unfertilized eggs persist long into early embryogenesis (Owens *et al.*, 2016; Peshkin *et al.*, 2015). Mutant lines will be valuable for studying maternal loss-of-function phenotypes for many genes that contribute to early development.

However, creating mutant lines becomes complicated if a gene of interest is critical for survival to breeding age. Efficient CRISPR/Cas9 mutagenesis results in animals containing a large contribution of cells with biallelic mutations, resulting in lethality before mutant alleles can be propagated to the next generation. The typical approach for dealing with lethality is to reduce the level of mutagenesis, titrating to permit survival to breeding age. However, this can result in a later, time-intensive, screening of F0 animals for germline transmission of mutant alleles, and requires the researcher to maintain more animals in order to ensure the successful recovery of mutant germ cells. Once germline transmission has been confirmed, phenotypic characterization can then be accomplished by creating F1 heterozygotes, which are then intercrossed to produce homozygous recessive F2 animals. This multi-generation breeding scheme is time-consuming and laborious, which led us to seek alternative methods to circumvent lethality.

One solution is to target mutagenesis specifically in the germline. Thus, animals are spared from mutations deleterious to somatic cells and therefore are viable, yet can deliver defective alleles via their mutant germ cells. We have developed a rapid and inexpensive method, called leapfrogging, to target primordial germ cells very early in embryogenesis (Blitz *et al.*, 2016). While strategies exist for delivering Cas9 and sgRNAs to the germline, including transgenic approaches, transfection of isolated adult gonadal stem cells, or partial enrichment of Cas9 mRNA in germ cells (reviewed in Blitz *et al.*, 2016), leapfrogging relies on high-efficiency, biallelic, whole-embryo mutagenesis followed by transplantation of primordial germ cells (PGCs) into wild-type sibling embryos at an early embryonic stage. Early transplantation has the benefit of avoiding the effects of somatic mutations that impair development of the embryo. We showed that leapfrogging is an efficient means of creating mutations in *Xenopus tropicalis*, and suggested that this approach can be applied to other organisms where similar manipulations can be devised (Blitz *et al.*, 2016).

Here I provide a detailed set of instructions for leapfrogging in *Xenopus* that can be employed in any laboratory that houses *Xenopus*, with no special equipment required beyond that needed for routine molecular biological techniques and a standard set-up for performing embryo microinjections.

6.2 Methodology

The initial steps in leapfrogging are very similar to standard CRISPR/Cas9-based mutagenesis in *Xenopus* (Nakayama *et al.*, 2014). Suitable target sites for Cas9 cleavage must be identified within the gene of interest. Careful choice of target sites is critical, not only to guarantee efficient Cas9 cleavage, but also to ensure that all mutations produced by double-strand break repair are likely to result in loss of function (LOF) for the gene product. Delivery of Cas9 and sgRNAs into *Xenopus* entails straightforward

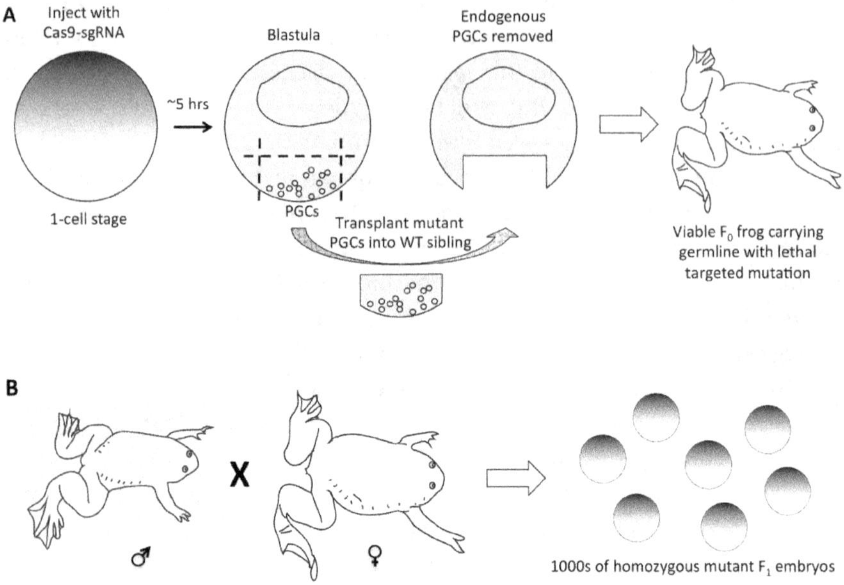

Figure 6.1 The strategy of leapfrogging. (A) Whole embryos are efficiently mutagenized with CRISPR/Cas9 (or TALENs) by microinjection of Cas9-sgRNA into embryos at the one-cell stage. Embryos are incubated until blastula stage, and a region of the vegetal pole that contains PGCs is transplanted from the mutagenized donor embryo to an uninjected sibling. This recipient embryo had its PGC removed as part of the transplantation procedure, and therefore it has undergone a germline replacement with mutant cells. As the somatic tissues of these F0 embryos are wild-type, they are viable, robust and can be grown to sexual maturity. (B) Intercrossing F0 animals produced by leapfrogging results in F1 embryos that are homozygous mutant and hence will display LOF phenotypes. Since one *Xenopus* female produces 1000s of eggs in a single day, one can obtain 1000s of mutant embryos for subsequent analyses.

and well-established microinjection techniques. Titration of Cas9-sgRNA in microinjections is first used to optimize the dosing of these components to achieve efficient mutagenesis at the target site. Once it is determined that Cas9-sgRNAs injections are optimal, one is ready to perform PGC transplantation.

PGC-transplant donor embryos are injected with Cas9-sgRNA and several hours later PGCs are transplanted from the vegetal pole of blastula-stage donor embryos into sibling "recipient" embryos that have had their PGCs removed (Figure 6.1A). Engraftment is rapid and efficient, and "leapfrogged" embryos are then grown to sexual maturity for downstream analyses.

6.2.1 sgRNA Design and Microinjection of Cas9-sgRNA into Embryos

Numerous web-based CRISPR target identification tools exist to assist in choosing appropriate sgRNA sequences. I direct the reader to our published methods review on performing CRISPR/Cas9 mutagenesis in *Xenopus* (Nakayama *et al.*, 2014) for basic information on the design of sgRNAs. A list of CRISPR design websites interfacing with the *Xenopus* genome sequence can be found on Xenbase's genome editing support webpage (www.xenbase.org/other/static/CRISPR.jsp). We have most recently been relying on either CRISPRDirect (https://crispr.dbcls.jp/) or

CRISPRScan (www.crisprscan.org/) primarily because these offer simple user interfaces. Most sgRNAs we have designed using these, and other sites (e.g. Zifit and ChopChop), have worked efficiently in *X. tropicalis*. We rely on PCR-based synthesis to produce DNA templates for T7 Megascript (ThermoFisher Scientific) *in vitro* transcription reactions to make sgRNAs and the details are also described in Nakayama *et al.* (2014).

For leapfrogging, we have been using commercially available Cas9 protein (PNA Bio., Inc.), but Cas9 mRNA (using a human codon-optimized ORF) induces mutations with similar efficiency (Blitz, Fish and Cho, unpublished observations; see also Fig. 17.4 of Nakayama *et al.*, 2014; Shigeta *et al.*, 2016). When using Cas9 protein, we preassemble Cas9-sgRNA complexes *in vitro* according to the manufacturer's instructions. Typical doses for Cas9 protein microinjection are 1–1.5 ng/embryo together with ~250 pg of sgRNA, but in some cases higher doses of sgRNA, up to 1–1.5 ng, might be needed. If using codon-optimized Cas9 mRNA, one might need as much as 1 ng to achieve optimal mutagenesis (see Table 17.1 of Nakayama *et al.*, 2014 for more details on the use of different codon-optimized versions of Cas9 mRNA in *X. tropicalis* embryos).

Female frogs are boosted with 100 units of human chorionic gonadotropin (Chorulon) in the morning and typically within 3–4 hours they begin to lay eggs. *In vitro* fertilization is performed according to Ogino *et al.* (2006) and at 10 minutes post-flooding of the dish with 1/9X Marc's modified Ringer's solution (MMR), the embryos are dejellied in 3% cysteine (free base), made in 1/9X MMR (pH adjusted to ~7.8–8.0 with NaOH). Embryos are then washed multiple times in fresh 1/9X MMR and transferred to an agarose-coated dish and maintained in 0.1X MMR at room temperature until injections are completed.

Cas9-sgRNA complexes (or Cas9 mRNA plus sgRNA) are microinjected into a single site in the animal pole at the one-cell stage, ~15–20 minutes post-fertilization (defined as the time of addition of testis homogenate). We inject embryos in 60 mm Petri plates containing a submerged agarose bed made in 1X MMR, without inclusion of ficoll. Recipes for these common frog embryo media can be found in Sive *et al.* (2000). After ~15 minutes of injection site healing in 1X solution, the embryos are transferred to agarose-coated plates containing 1/9X MMR. Injected *X. tropicalis* embryos and their uninjected siblings are transferred to a 25 °C incubator for culturing to late blastula stages. I monitor embryos for their progression through cleavage and blastula stages as embryos will be ready for transplantation early in blastula stage 9, which is approximately 5 hours post-fertilization (hpf).

6.2.2 Preparation for PGC Transplantation

While the embryos are developing one can prepare the workstation for PGC transplantation. The following materials are needed:

- A dissecting stereomicroscope fitted with a fiber optic light source.
- A pair of sharp Dumont #5 forceps (Fine Science Tools). These are needed for removal of the vitelline envelope.

- Eyebrow hair knives and hair loops for embryo microsurgery (see Sive *et al.*, 2007 for preparation of these simple tools).
- Two 60 mm Petri dishes, each containing a ~5 mm thick bed of 1% agarose prepared in 0.3X MMR. ~3.5–4 mm deep wells are created in the agarose, using a mold created from 12 wells of a 96-well PCR plate. These dishes are used for surgeries. Dishes are flooded with 0.3X MMR supplemented with 50 μg gentamycin sulfate/mL.
- A 24-well plate coated with a thin layer of 1% agarose made in 1/9X MMR. Wells are flooded with 1/9X MMR supplemented with 50 μg gentamycin sulfate/mL. This dish will be used for overnight culturing, a single embryo per well, after the PGC transplants have engrafted.

6.2.3 PGC Transplantation

Once embryos reach stage 9 (Nieuwkoop and Faber, 1994), typically ~4.5 hpf at ~25 °C, they are moved from the incubator to the benchtop to equilibrate to room temperature. I typically perform transplantation beginning at ~5 hpf ("stage 9.25"), during the 1.5 hour time window until the dorsal blastopore lip just becomes visible at stage 10, which in our hands occurs at ~6.5 hpf in *X. tropicalis*. To perform transplants, one Cas9-sgRNA injected embryo and one uninjected sibling are transferred to a 60 mm dish containing 0.3X MMR containing gentamycin 50 μg/ML (1X MMR is required for transplantation if using *X. leavis*) and an agarose bed. Each agarose dish is prepared with depressions to temporarily hold embryos (Figure 6.2A) while the transplants heal into place. Careful attention is necessary to keep track of which embryo is to be the graft donor and which will be the recipient. Their vitelline envelopes are manually removed using forceps, taking care to minimize damage to the embryos. Each embryo is then manually rotated such that their vegetal poles are visible. Starting with the uninjected embryo that is to be the graft recipient, four shallow incisions into the vegetal pole are made using the tip of an eyebrow hair knife to create a square (Figure 6.2B). The hair loop is used to maintain control of the embryo's position. After the initial incisions are made, they are repeated to deepen the cuts to ~1/3 to 1/2 of the depth to the blastocoel floor. This is followed by a single horizontal cut to release the vegetal fragment bearing the PGCs. This dissected vegetal fragment from the graft recipient, approximately 0.4–0.45 mm on a side and 0.25–0.3 mm deep (*X. tropicalis* embryos are ~0.8 mm in diameter; Crowder *et al.*, 2015), doesn't play a further role in the procedure and is discarded. Most or all of the recipient embryo's PGCs have been removed and this embryo is ready to receive a PGC graft from a donor embryo.

Working quickly, a similar-sized PGC explant is then excised from the Cas9-sgRNA injected donor embryo. This fragment is then gently moved into place in the opening created in the recipient embryo, using the eyebrow hair knife, with assistance from the hair loop if necessary, to carry it into position. The PGC-bearing fragment is inserted into the recipient embryo, placing the deeper cell layers into the opening (Figure 6.1A), and maintaining the original external surface of the explant on the outside of the recipient embryo. However, the dorsal–ventral and left–right orientations of the graft, with respect to the recipient embryo's axes, does not appear to be an

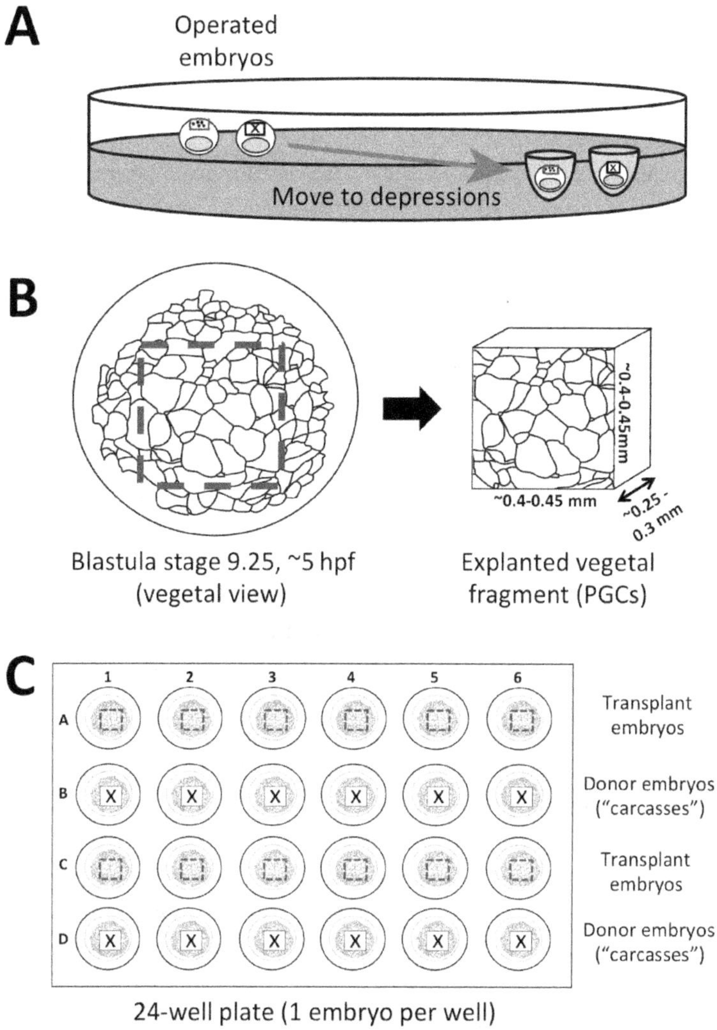

Figure 6.2 How to perform leapfrog transplants. (A) Transplantations are performed in a dish containing a thick bed of agarose and 0.3X MMR, with depressions to hold embryos while they heal post-surgery. (B) An illustration of the vegetal pole of a *Xenopus tropicalis* blastula stage embryo at 5 hpf, to show the size of vegetal blastomeres. The blue box indicates the approximate size and boundary of the region of the explanted/transplanted tissue. (C) Once transplants have healed into place, the embryos are moved to a 24-well agarose-coated dish (vegetal pole up), containing 1/9X MMR to develop overnight. Transplant donor embryos are also moved to this dish and each pair of donor and recipient embryos is tracked as the donor embryos are used as proxies to assess the mutagenesis efficiency in the transplanted PGCs carried by the "leapfrogged" embryos. (A black-and-white version of this figure will appear in some formats. For the color version, please refer to the plate section.)

important consideration and is not critical for successful development and survival of the embryo. Following insertion of the graft, the shaft of the eyebrow hair knife is used to gently press it into place.

The donor embryo is gently moved to one depression in the agarose using the eyebrow hair knife to slide it along the agarose surface. Once placed in the well, its open vegetal wound is positioned facing bulk solution to heal closed, unimpeded. Similarly, the embryo receiving the PGC graft is moved to an adjacent well and, likewise, its vegetal pole is positioned to face the bulk solution to allow for healing without pressure or friction from the agarose below. The donor embryo is not discarded because it will serve an important purpose: as a proxy for assessing the efficacy of mutagenesis in the transplanted PGCs (see Section 6.2.5 below).

Another pair of donor and recipient embryos are transferred from their respective dishes and subsequent rounds of transplantation are performed. I perform transplantations until gastrulation begins, to minimize disruptions to gastrulation movements. It is also unclear whether one needs to stop transplantation at early gastrulation, or how long one can continue into gastrulation (see Section 6.3 below). In *X. tropicalis*, in the ~1.5 hours from early stage 9 (~5 hpf) until the appearance of the bottle cells of the dorsal blastopore at stage 10 (~6.5 hpf), an experienced embryo surgeon can expect to complete 12–24 transplants.

6.2.4 Post-surgical Husbandry of Embryos

After sufficient wound healing has occurred, ~20–30 minutes, embryos should gently be transferred into a fresh dish. To keep track of individual donor–recipient pairs, I use a glass Pasteur pipette to transfer individual embryos to wells of a 24-well dish containing 1/9X MMR (+50 μg/mL gentamycin) that has been precoated with 1% agarose (Figure 6.2C). Embryos are positioned vegetal pole facing away from the agarose surface and the dish is moved to a 25 °C incubator for overnight culture. Keeping embryos separate for the first day also serves to protect healthy embryos from lysed cells from embryos that might die overnight. Uninjected control embryos and excess unused Cas9-sgRNA injected embryos are also maintained in a separate dish to assess overall embryo viability and appearance of phenotypes in the injected embryos to assess efficacy of the mutagenesis.

The next day, the embryos have reached early tailbud stages. Donor embryos, lacking a portion of their vegetal pole material, typically show a narrower ventral endoderm region, with some showing larger-scale abnormalities arising from morphogenesis defects. Donor embryos are individually transferred to 0.2 mL PCR strip tubes, most of the solution is removed, and they are homogenized in 100 μL of lysis solution (50 mM Tris [pH 8.8], 1 mM EDTA, 0.5% Tween 20 and 200 μg proteinase K/mL) by pipetting up and down, and incubated 2 h to overnight at 56 °C (Nakayama *et al.*, 2014). Proteinase K is heat inactivated by incubation at 90–95 °C for 10 min and the lysates can be stored at –20 °C, or used immediately for PCR to assess mutagenesis efficiency (see below).

Embryos receiving transplants can be maintained in the 12-well plate for a short time (~3 days total), but it is recommended that they be moved to larger confines where their growth will be less limited by alterations to the medium due to metabolic activity (e.g. pH). I typically move surviving embryos to clean wells in a 6-well agarose-coated plate containing 1/9 X MMR (and gentamycin) on day 2.

6.2.5 Assessing the Efficacy of PGC Mutagenesis using Donor Embryo Carcasses

It is advantageous to determine the efficacy of CRISPR/Cas9 mutagenesis in the embryos as early as possible, to avoid wasting effort growing animals unnecessarily. Once the tadpoles are transferred to an aquatic modular flow system (~1.5–2 weeks post-fertilization), less hands-on care is needed. Since success of mutagenesis of the PGCs in the surviving embryos cannot be directly assessed until the animals reach sexual maturity, the carcasses are used as proxies for the transplanted PGCs (Blitz *et al.*, 2016). PCR amplification of small genomic fragments, 200–800 bp in length, containing the CRISPR target site, serve as starting material to assess the population of DNA molecules. I use a touchdown PCR scheme outlined in Nakayama *et al.* (2014), which has been effective in amplifying specific fragments from genomic DNA, usually with minimal contaminating fragments. The amount of lysate needed in such reactions varies based on the cell number, and thus DNA content, of different stages of embryogenesis. For tailbud-stage embryos, typically 1 μL of lysate can be used directly in a 25 μL PCR reaction (Nakayama *et al.*, 2014).

Thus far we have relied on a simple approach to assessing mutagenesis efficiency in single embryos by directly sequencing the population of PCR products ("DSP assays"; Nakayama *et al.*, 2014), which requires a minimum of effort as DNA samples simply need to be purified over spin columns designed to remove PCR reaction components and then mixed with oligos and sent off to companies offering Sanger sequencing services. Typically one of the oligos used in the initial amplification is sufficient for sequencing from one end. The extent of "scrambling" of the sequence peaks, beginning at or near the predicted site of the double-strand break, has been used as a rough measure of the efficiency of mutagenesis. When lightly mutagenized, wild-type peaks can be seen overlapping scrambled peaks that represent different nucleotides at the same position (e.g., see Fig. S1 of Blitz *et al.*, 2016). More precise quantitation of the extent of mutagenesis indicated by these traces can be performed using the web tool Tracking of Indels by Decomposition (TIDE; Brinkman *et al.*, 2014), which has been shown to predict mutagenesis efficiencies as well as high-throughput sequencing assays (Sentmanat *et al.*, 2018).

6.2.6 Assessment of Mutant Phenotypes in F1 Embryos

Once animals receiving leapfrog transplants are grown to sexual maturity they can be used to determine germline transmission rates and evaluate LOF phenotypes. Raising tadpoles to adulthood is not covered in the current chapter, but more information on *Xenopus tropicalis* tadpole and adult animal care can be found on the National *Xenopus* Resource's website (www.mbl.edu/xenopus/files/2016/06/Xenopus-Tadpole.pdf and www.mbl.edu/xenopus/files/2016/06/Xenopus-Tropicalis.pdf) and on the European *Xenopus* Resource Center's website (https://xenopusresource.org/husbandry). In our hands, males typically reach sexual maturity beginning around 3–4 months post-fertilization, while females typically require 6–8 months. Once mature males become available one can outcross these (by stimulated "natural" mating) to assess the frequency of germline transmission of mutant alleles. Since the F1 embryos derived from outcrossing to wild-type results in heterozygotes, sequences obtained using the DSP assay contain both wild-type and mutant alleles from individual embryos.

Therefore, the mutant sequence can be read directly by "subtracting" the wild-type peaks when examining the graphical output from a Sanger sequencing run.

Once a male(s) has been identified, it can be used in intercrosses with females as they become sexually mature, without first assessing their germline transmission rates. This approach has the advantage of saving considerable time because once a female has laid eggs (for example, in an outcross to assess germline transmission rates), she will need a few months recovery time before her next egg laying.

The percentage of embryos showing phenotypes is dependent on the percentage of each parent's germline that contains mutant alleles (except in the case of maternal gene products, where the female germline is the sole provider of the mutated mRNA/protein). Once phenotypes have been assessed, it is important to correlate phenotype with genotype to demonstrate that the aberrant phenotype observed is due to carriage of the mutation. This becomes challenging if relying on DSP assays or most other genotyping techniques because both parent's germlines will be mosaic, and therefore one doesn't know, in advance, the sequence of either of the parental alleles in any individual embryo. Thus far, we have relied on Sanger sequencing of subcloned DNA molecules from individual embryos to assess the combination of alleles present in each embryo (Blitz *et al.*, 2016). One way around this problem would be to create two sets of leapfrogged animals – with animals from one set containing mutations at a different targeted site from animals within the second set. Intercrosses between animals from these two groups would produce LOF phenotypes (if target sites were well designed), and would result in heterozygosity in the sequence readouts at each of the two targeted sites. Thus, DSP assays could be used to easily determine the sequences of each mutant allele within individual embryos.

6.3 Discussion

6.3.1 On the Value of Mutant Lines

While F0 CRISPR/Cas9 analyses will be a valuable tool for studying embryogenesis in *Xenopus*, mutant lines are needed and the leapfrogging method we have developed will accelerate production of LOF embryos for phenotypic analysis. We have observed that phenotypes of both *gsc* and *tyr* in F0 animals are, on average, less severe than LOF in F1 genetic mutants, where every cell of the animal is homozygous mutant (Blitz *et al.*, 2013; 2016). Thus, F0 CRISPR/Cas9 analysis should be considered a "knockdown" analysis, which will be subject to some of the same criticisms associated with morpholino (MO) knockdowns (Teboul *et al.*, 2017). As with MOs, some sgRNAs will be less efficient at targeting the gene of interest than others, and in some cases knockdown phenotypes will be significantly less severe than a complete genetic LOF phenotype. Thus the full extent of a gene's contribution to a phenotype is likely to require moving from F0 analysis to more standard genetic LOF analysis.

For researchers working on questions in early embryogenesis, maternal deposited mRNAs and proteins are impervious to F0 CRISPR/Cas9 knockdowns because the F0 mutated genes will not be expressed until after initiation of zygotic transcription. Wild-type maternal products will already be present in the CRISPR/Cas9-injected embryos and therefore LOF phenotypes will not be obtained until these are depleted later in development. Leapfrogging is a valuable approach to making the mutant lines needed to circumvent these problems.

6.3.2 Careful Selection of Target Site Location is Critical for High-efficiency Phenotypic Analysis using Leapfrogging

A major design feature in making CRISPR/Cas9 mutants via leapfrogging is the location of the target site with respect to the protein coding sequence of a gene. If a target site is chosen simply based on being in a 5′ location within an ORF, as many CRISPR/Cas9 mutagenesis strategies employ, then it's likely that on average 2/3 of mutations will be out-of-frame indels, but 1/3 will be in-frame mutations (Varshney *et al.*, 2015). These in-frame mutations will most often encode functional proteins and thus we might expect ~1/3 of gametes *from each parent* might not be LOF alleles. F0 intercrosses under these circumstances would produce homozygous LOF in only 1/9 of the embryos. To maximize the number of gametes carrying an LOF allele, it is suggested that targeting should be designed within DNA sequence coding for a recognizable protein domain. Thus, even the 1/3 in-frame mutations might be expected to cause misfolding of the protein domain and loss of its function. Careful examination of atomic-level structural details, when available, can further minimize targeting within a loop that might tolerate more sequence variability. If a domain is targeted that is more likely to lead to complete LOF for the gene (Shi *et al.*, 2015). Therefore, all progeny derived from mutant alleles will carry a complete LOF phenotype. An example of this behavior can be found in Blitz *et al.* (2016) for targeting of the *gsc* gene. One phenotypically mutant embryo, which was found to be a compound heterozygote (i.e. containing two different mutant alleles), carried a single amino acid deletion within the Gsc DNA binding domain that was targeted. The deleted amino acid was previously shown to be important for DNA binding activity in other homeodomain proteins, providing a proof of principle for this strategy. Therefore, careful consideration of the location of target sites is highly recommended.

6.3.3 Future Improvements and Directions for Leapfrogging

While leapfrogging is currently being used in a number of laboratories to accelerate genetic studies in *Xenopus*, there remains room for both improvement of the method, and expansion of how it can be used. We are currently attempting to increase the speed of successful transplantation so that more transplants can be made in the short time window available. We are ussing 0.3X MMR to perform transplantations in *X. tropicalis*, but have found that higher ionic conditions (1X MMR) are needed to perform similar transplants in *X. laevis* (Blitz *et al.*, 2016). We have thus far not tested whether this higher strength solution would lead to higher efficiencies in *X. tropicalis* embryos.

We found that a subset of animals produced by leapfrogging have germlines that are mosaic, containing a mixed population of both mutant and wild-type alleles. This observation was hypothesized to be due to incomplete removal of wild-type PGCs from embryos receiving a mutagenized germline. We are currently testing strategies for more efficent removal of these PGCs. If PGC development can be efficiently disrupted, then any remaining PGCs after removal of vegetal tissue from these embryos will fail to populate the germline provided by the grafted mutant tissue.[1]

Another advance that could be introduced would be the exploration of the efficiency of creating deletions between two CRISPR target sites flanking a gene. While small deletions can be detected (Nakayama *et al.*, 2013), it is currently unclear how efficient this process is, or how large an interval can be excised. Instead of targeting protein domains as outlined above, one could have higher certainty that a LOF phenotype can be obtained by whole-gene deletion. The benefit of gene deletion would also be to avoid any possibility that point mutations might produce dominant negative, or possibly neomorphic, mutations. Gene deletion would ensure creation of complete LOF mutations. This would become the standard for creating mutants. If efficient enough, leapfrogging would then provide the vehicle for speeding up the breeding process, only requiring a single generation to obtain phenotypes, while also circumventing the lethality associated with loss of an essential gene.

Another use of leapfrogging would be to speed the production of multi-gene knockout animals. In many organisms the creation of animals that are homozygous mutant at two genes requires the intercrossing of animals heterozygous at both genes (*gene a*$^{+/-}$; *gene b*$^{+/-}$), to create progeny with both genes knocked out (*gene a*$^{-/-}$; *gene b*$^{-/-}$), which would occur on average in 1 in 16 embryos. In F0 *Xenopus tropicalis* embryos co-injection of multiple sgRNAs (or in one case TALENs) leads to simultaneous knockout of two or more genes (Guo *et al.*, 2014; Liu *et al.*, 2016; Naert *et al.*, 2016; Van Nieuwenhuysen *et al.*, 2015), providing a proof of principle that, if of sufficient efficiency, this can result in multi-gene-knockout gametes, and thus embryos. Such an approach would be especially valuable in cases where two or more genes are thought to be functionally redundant and therefore obtaining knockout phenotypes would require complex mutants. Leapfrogging will assist in making such genetic combinations available in the F1 generation, perhaps in 100% of embryos. We are currently performing experiments to test these ideas.

6.4 Concluding Remarks

CRISPR/Cas9-mediated mutagenesis strategies are revolutionizing biological research, with the greatest impact being in organisms where there has been no efficient method for introducing modifications to the genome. CRISPR/Cas9 is making *Xenopus* a more tractable genetic model that, when combined with more classical embryological methods, makes this system very attractive for understanding early developmental events. Adapting the germ cell transplantation work of Blackler (Blackler, 1960; Blackler and Fischberg, 1961) to blastula stages, and combining this approach with high-efficiency mutagenesis of the PGC donor embryos, allowed us to develop leapfrogging as a method to overcome lethal mutations. Animals bearing such mutations in their germline can survive to sexual maturity and pass these alleles to their F1 progeny for phenotypic studies. A side benefit of this method is that intercrossing F0 leapfrogged animals allows for phenotypic analysis in one generation (F1), thereby skipping a full generation time for analysis of the effects of genomic alterations on biological processes.

The next step in developing genome editing in the *Xenopus* system is gene transplacement, the precise (or mildly imprecise) targeted replacement of genes or non-coding elements with modified sequences. Methods utilizing homologous recombination, or homology-directed repair, will have to be advanced in *Xenopus* to accomplish such a task in an efficient manner. Leapfrogging could aid in the speedy production of animals, especially when the introduced change impairs the viability of the animal.

Acknowledgements

I would like to thank Ken Cho for his enthusiasm, critical discussions, and ongoing support. This work was performed with the support of a grant, 5R21HD080684-02, from the National Institute of Child Health and Human Development.

NOTE

1. I refer the reader to a video demonstration of PGC transplantation associated with the following publication: Blitz IL. 2018. Primordial germ cell transplantation for CRISPR/Cas9-based leapfrogging in *Xenopus*. *J Vis Exp* **132**: e56035.

REFERENCES

Abu-Daya A, Sater AK, Wells DE, Mohun TJ, Zimmerman LB. 2009. Absence of heartbeat in the *Xenopus tropicalis* mutation muzak is caused by a nonsense mutation in cardiac myosin myh6. *Dev Biol* **336**: 20–29.

Aslan Y, Tadjuidje E, Zorn AM, Cha SW. 2017. High-efficiency no-mosaic CRISPR-mediated knock-in and indel mutation in F0 *Xenopus*. *Development* **144**: 2852–2858.

Bhattacharya D, Marfo CA, Li D, Lane M, Khokha MK. 2015. CRISPR/Cas9: an inexpensive, efficient loss of function tool to screen human disease genes in *Xenopus*. *Dev Biol* **408**: 196–204.

Blackler AW. 1960. Transfer of germ-cells in *Xenopus laevis*. *Nature* **185**: 859–860.

Blackler AW, Fischberg M. 1961. Transfer of primordial germ-cells in *Xenopus laevis*. *J Embryol Exp Morph* **9**: 634–641.

Blitz IL, Biesinger J, Xie X, Cho KW. 2013. Biallelic genome modification in F0 *Xenopus tropicalis* embryos using the CRISPR/Cas system. *Genesis* **51**: 827–834.

Blitz IL, Fish MB, Cho KW. 2016. Leapfrogging: primordial germ cell transplantation permits recovery of CRISPR/Cas9-induced mutations in essential genes. *Development* **143**: 2868–2875.

Blitz IL, Pawaiso KD, Patsushev I, *et al.* 2017 A catalogue of *Xenopus tropicalis* transcription factors and their regional expression in the early gastrola stage embryo. *Dev Biol* **426**: 409–417.

Brinkman EK, Chen T, Amendola M, van Steensel B. 2014. Easy quantitative assessment of genome editing by sequence trace decomposition. *Nucleic Acids Res* **42**:e168.

Chung HA, Medina-Ruiz S, Harland RM. 2016. Sp8 regulates inner ear development. *Proc Natl Acad Sci USA* **111**: 6329–6334.

Crowder ME, Strzelecka M, Wilbur JD, *et al.* 2015. A comparative analysis of spindle morphometrics across metazoans. *Curr Biol* **25**: 1542–1550.

Geach TJ, Zimmerman LB. 2010. Paralysis and delayed Z-disc formation in the *Xenopus tropicalis* unc45b mutant dicky ticker. *BMC Dev Biol* **10**: 75.

Goda T, Abu-Daya A, Carruthers S, *et al.* 2006. Genetic screens for mutations affecting development of *Xenopus tropicalis*. *PLoS Genet* **2**: e91.

Grammer TC, Khokha MK, Lane MA, Lam K, Harland RM. 2005. Identification of mutants in inbred *Xenopus tropicalis*. *Mech Dev* **122**: 263–272.

Guo X, Zhang T, Hu Z, *et al.* 2014. Efficient RNA/Cas9-mediated genome editing in *Xenopus tropicalis*. *Development* **141**: 707–714.

Hellsten U, Harland RM, Gilchrist MJ, *et al.* 2010. The genome of the Western clawed frog *Xenopus tropicalis*. *Science* **328**: 633–636.

Jaffe KM, Grimes DT, Schottenfeld-Roames J, *et al.* 2016. c21orf59/kurly controls both cilia mobility and polarization. *Cell Rep* **14**: 1841–1849.

Liu Z, Cheng TT, Shi Z, *et al.* 2016. Efficient genome editing of genes involved in neural crest development using the CRISPR/Cas9 system in *Xenopus* embryos. *Cell Biosci* **6**: 22.

Naert T, Colpaert R, Van Nieuwenhuysen T, *et al.* 2016. CRISPR/Cas9 mediated knockout of rb1 and rbl1 leads to rapid and penetrant retinoblastoma development in *Xenopus tropicalis*. *Sci Rep* **6**: 35264.

Nakade S, Tsubota T, Sakame Y, *et al.* 2014. Microhomology-mediated end-joining-dependent integration of donor DNA in cells and animals using TALENs and CRISPR/Cas9. *Nat Commonun* **20**: 5560.

Nakayama T, Blitz IL, Fish MB, *et al.* 2014. Cas9-based genome editing in *Xenopus tropicalis*. *Methods Enzymol* **546**: 355–375.

Nakayama T, Fish MB, Fisher M, *et al.* 2013. Simple and efficient CRISPR/Cas9-mediated targeted mutagenesis in *Xenopus tropicalis*. *Genesis* **51**: 835–843.

Nakayama T, Nakajima K, Cox A, *et al.* 2017. No privacy, a *Xenopus tropicalis* mutant, is a model of human Hermansky-Pudlak Syndrome and allows visualization of internal organogenesis during tadpole development. *Dev Biol* **426**(2): 472–486.

Nieuwkoop and Faber. 1994. *Normal Table of Xenopus laevis (Daudin): A Systematical and Chronological Survey of the Development from the Fertilized Egg till the End of Metomorphosis*. New York: Garland Publishing Inc.

Noramly S, Zimmerman L, Cox A, *et al.* 2005. A gynogenetic screen to isolate naturally occurring recessive mutations in *Xenopus tropicalis*. *Mech Dev* **122**: 273–287.

Ogino H, McConnell WB, Grainger RM. 2006. High-throughput transgenesis in *Xenopus* using I-SceI meganuclease. *Nat Protoc* **1**: 1703–1710.

Owens ND, Blitz IL, Lane MA, *et al.* 2016. Measuring absolute RNA copy numbers at high temporal resolution reveals transcriptome kinetics in development. *Cell Rep* **14**: 632–647.

Peshkin L, Wühr M, Pearl E, *et al.* 2015. On the relationship of protein and mRNA dynamics in vertebrate embryonic development. *Dev Cell* **35**: 383–394.

Sentmanat MF, Peters ST, Florian CP, Connelly JP, Pruett-Miller SM. 2018. A survey of validation strategies for CRISPR-Cas9 editing. *Sci Rep* **8**: 888.

Session AM, Uno Y, Kwon T, *et al.* 2016. Genome evolution in the allotetraploid frog *Xenopus laevis*. *Nature*. **538**: 336–343.

Shi J, Wang E, Milazzo JP, *et al.* 2015. Discovery of cancer drug targets by CRISPR-Cas9 screening of protein domains. *Nat Biotechnol* **33**: 661–667.

Shigeta M, Sakane Y, Iida M, *et al.* 2016. Rapid and efficient analysis of gene function using CRISPR-Cas9 in *Xenopus tropicalis* founders. *Genes Cells* **21**: 755–771.

Sive HL, Grainger RM, Harland RM. 2000. *Early Development of Xenopus laevis: A Laboratory Manual*. Cold Spring Harbor, NY: Cold Spring Harbor Laboratory Press.

Sive HL, Grainger RM, Harland RM. 2007. Embryo dissection and micromanipulation tools. *CSH Protoc*: pdb.top7.

Teboul L, Murray SA, Nolan PM. 2017. Phenotyping first-generation genome editing mutants: a new standard? *Mamm Genome* **28**(7–8): 377–382.

Van Nieuwenhuysen T, Naert T, Tran HT, *et al.* 2015. TALEN-mediated apc mutation in *Xenopus tropicalis* phenocopies familial adenomatous polyposis. *Oncoscience* **2**: 555–566.

Varshney GK, Pei W, LaFave MC, *et al.* 2015. High-throughput gene targeting and phenotyping in zebrafish using CRISPR/Cas9. *Genome Res* **25**: 1030–1042.

Wang F, Shi Z, Cui Y, *et al.* 2015. Targeted gene disruption in *Xenopus laevis* using CRISPR/Cas9. *Cell Biosci* **5**: 15.

7 Genome Editing with Desired Mutations (Knockin) with CRISPR in Model Organisms

Chunxin Wang and Yingjie Wu

7.1 Introduction

The advent of CRISPR technology opened the door for every research laboratory to routinely manipulate genes of interest with ease. While the specificity of CRISPR has been largely addressed with improved off-target prediction tools and rationally engineered Cas9 enzymes, the efficacy of CRISPR single-guide sgRNA remains a daunting task to be tackled. In addition, knocking-in desired mutations/tags with precision still faces challenges of low frequency. For genetic studies, null alleles made by gene knockouts and hyper- or hypo-alleles made by knockins of point mutations complement each other, and together provide a full spectrum of gene functions or modes of action. In this chapter, we emphasize the importance of rational design of effective sgRNAs with available online tools, and discuss the pros and cons of ssODN (single-strand oligodeoxynucleotide) vs. plasmid as a donor. We also describe unconventional, but efficient approaches in certain species to employ non-homologous end joining (NHEJ)-mediated integration of reporters. Depending on the purpose of the study and the model species, readers are advised to choose different knockin strategies to optimize efficiency and the labor involved.

7.2 How to Design an Effective sgRNA

Double-strand DNA breaks (DSBs) generated by CRISPR/Cas9 can be repaired predominantly by NHEJ, whereas knockin often requires precise homologous recombination (HR)-dependent repair (HDR). The efficiency of obtaining correct knockin events is much lower than that of simple knockouts. For instance, Gonzalez *et al.* reported 2–3% HDR rates versus 13–49% indel rates in human cells (González *et al.*, 2014). Therefore, it is imperative to start with highly effective sgRNAs to boost HDR frequency.

sgRNA efficacy is no doubt essential for the success of genome-wide CRISPR library screens: the more low-cutting or no-cutting sgRNAs in the library, the less the coverage of the library, which ultimately leads to fewer positive hits.

Therefore, several groups who have conducted large-scale CRISPR library screens naturally investigated the characteristics of a large number of effective and ineffective sgRNAs to identify certain features that may help to predict the efficacy of sgRNAs. This will not only dramatically improve the quality of future libraries, but also help individual labs to cut down cost and time by avoiding making ineffective sgRNAs.

Based on sgRNAs across ~1400 genomic loci in several human cell lines, Chari *et al.* found that at position 20, A/G is most favored in high active sgRNAs (Chari *et al.*, 2015). The strong preference of G at position 20 is confirmed by a different study (Doench *et al.*, 2014). In contrast, Liu *et al.* found that a "C" is more favored at position 20 and an "A" is least favored (Liu *et al.*, 2016). Therefore, consensus is still lacking on what common sequence features constitute effective sgRNAs. Interestingly, sgRNAs containing NGG at the 3′ end right before the PAM motif exhibited high mutagenic activity in *C. elegans* (Farboud and Meyer, 2015). However, it is possible that such a key determining factor is only applicable to *C. elegans*. As seen in Table 7.1, over 30% sgRNAs containing 3′ GG that target human genes are still inactive.

While many groups have developed online design tools for prediction of effective sgRNAs, CRISPOR (http://crispor.org), developed by Haeussler *et al.* (2016), is the most user-friendly and retrieves scores from eight different scoring models with off-target sites and scores displayed clearly. We ran 45 sgRNAs (previously tested in our labs) with CRISPOR and listed the predicted efficacy scores in Table 7.1. The italicized scores indicate against the prediction, and the scores in bold correlate well with experimental confirmation. It is interesting to see that certain scoring models are best for avoiding false negatives, whereas others are superior in not producing false positives. Given the overall correlation between prediction and experimental data, Fusi/Doench's and Wang's models produce the highest correlation rates (above 70%).

We originally exclusively used the CRISPR design tool by Feng Zhang at MIT, http://crispr.mit.edu, which only predicts off-target probability, not sgRNA efficacy. However, our original picks always focused on guide sequences that contained a restriction site near the 3 bp upstream of the PAM motif for easy monitoring of sgRNA activity by digestion of PCR products. For some unknown reason, our success rate (around 80%) in finding effective guide sequences was very similar to the sgRNA scorer by Fusi/Doench we used subsequently. We did confirm that all of our ineffective sgRNAs have one or more disfavored bases at those strongly biased positions.

7.3 How to Make Donor DNA

Depending on what needs to be knocked-in to the genomic locus (a point mutation or a small tag such as FLAG or LoxP sites, or big fluorescent protein/luciferase reporters), different strategies can be employed to make the donor DNA, which could be ssODN or plasmids. Also, depending on where it is knocked-in (N-terminal, C-terminal or internal of target genes to make fusion proteins, or simply

Table 7.1 CRISPR prediction by various online tools on CRISPOR vs. experimental confirmation

	Fusi	Chari	Xu	Doench	Wang	Mor.-Mateos	Housden	Experimental
GTCCATGGATTGAATTCCAC *AGG*	*52*	**46**	**0**	**25**	**45**	*52*	**3**	x
GTCATCCTTCACGTAGGACA *TGG*	*57*	**37**	**0.2**	**7**	*66*	**27**	*8*	x
GAGAATCGCCTGAACCCAGG *AGG*	*66*	*99*	**0.4**	**44**	*84*	*61*	*6*	x
ATTTGGGAGGCCGAGGCAGG *TGG*	**48**	*97*	*0.7*	**5**	*68*	*71*	*7*	x
CTTTGGGAGGCTGAGGCCGG *TGG*	**49**	*92*	*0.9*	**3**	*70*	*86*	*7*	x
TTCACACACAGCTGGCAGTG *TGG*	*65*	*94*	**–0.3**	**1**	**40**	**43**	**4**	x
TGAGACCAGCCTGGCCAACA *TGG*	*59*	*71*	**–0.3**	**3**	*50*	*55*	*7*	x
CCTACTAAAGGAGAAAGATA *CGG*	*60*	*88*	**0.3**	**24**	*53*	**37**	**4**	x
GTGGGTGGAGCGGCCATGCT *GGG*	**38**	*56*	**0**	**2**	**40**	*88*	*5*	x
ACAAAGGCTCGCATAGCTGC TGG	**47**	**29**	*0.5*	**8**	*74*	**19**	*7*	x
GTACGCAGCCGGCGCTCCGT *GGG*	**42**	*73*	**0**	**13**	*66*	*57*	*6*	x
GACTCTCAATGTTAAAGGTA GGG	**54**	**56**	*0*	*16*	**69**	*32*	*3*	+
GACCTGGCCGACTACCTGCA *CGG*	**59**	**62**	*–0.2*	**57**	**58**	**58**	*4*	+
ATTATTTTAAGAAAAAGAAG *TGG*	**57**	**81**	**0.9**	*38*	**81**	*19*	*4*	++
TGGCTGCATGTCCAGGAAGG *AGG*	**61**	**93**	*0.4*	*14*	**70**	**84**	*4*	++
CTTTGGGAAGCCGAGGCGGG *TGG*	**53**	**97**	**0.7**	*7*	**65**	**71**	**6**	++
GGAAGAAGCGGAGACGGTTA *GGG*	*48*	**73**	*–0.2*	*28*	**64**	**69**	**5**	++
GTCAATCCCTTCTACGGCCA *GGG*	**54**	*35*	*0.3*	*49*	**66**	**50**	*3*	++
TATAAAAATTCATATTAATG *TGG*	**54**	**74**	*0*	*11*	**51**	*16*	*3*	++
AAAATGGATAACAGATCAGT TGG	**53**	*47*	*0.2*	*7*	**78**	*31*	*4*	++
GTTTGATGTGCATTCCCTGG AGG	**69**	**92**	**0.6**	**65**	**76**	**68**	**6**	++
GCAGATTCATCCCCCCAGCC AGG	**54**	**64**	*–0.1*	*24*	**68**	**52**	*3*	++
ACGCATCTTGAATGTGACAA AGG	**65**	**57**	*0.4*	*19*	**60**	**70**	*3*	++
GTCATATAGAGATGTCATGA *AGG*	**62**	**64**	*–0.1*	*6*	**65**	*27*	*3*	++

Continued

Table 7.1 (cont.)

	Fusi	Chari	Xu	Doench	Wang	Mor.-Mateos	Housden	Experimental
CTGGCAGCGGGTTGAGAACG *TGG*	**54**	*31*	*0.3*	*8*	**53**	**61**	**5**	+++
CGTGTCAAGGAACAAGCCTG *AGG*	**69**	**96**	**0.9**	*41*	**78**	**66**	**6**	+++
GACCCCCAGGAGTGTGACGG *TGG*	**64**	**91**	*0.4*	*10*	**86**	**65**	*4*	+++
GCAGCCACTTGCCTGGGACC *TGG*	*44*	*45*	*–0*	*10*	**66**	*28*	**6**	+++
TGGGGGCTAACAGCAGCTGC *AGG*	*42*	*49*	**0.7**	*5*	**66**	*42*	**5**	+++
GATAAACTCGTTCCGCAGCT *GGG*	**63**	**63**	*–0.4*	**54**	*46*	*43*	*4*	+++
GGTAGATAAGTAGAGGCTAG AGG	**63**	**96**	*0*	*24*	**65**	*40*	**5**	+++
GGGATTTTTGTATCCACAGG *AGG*	**59**	**83**	**0.5**	*26*	**84**	**57**	*4*	+++
ATTTAAATAAAGAAGAAGGG *AGG*	**65**	**95**	**0.9**	**63**	**79**	*45*	*4*	+++
TTTACAGATGAGGGCAACTG *AGG*	**59**	**93**	*0.3*	*31*	**59**	**57**	**5**	+++
AAACTGGAACATGTAGTTAG *AGG*	*45*	**71**	*0.4*	*8*	**57**	*45*	**9**	+++
AAGTGTAATATGAATTAAAG *TGG*	*43*	**59**	*0.2*	*17*	**68**	*38*	*3*	+++
GATCTGTAATCACTATGTGT *AGG*	**70**	*39*	*0.4*	*31*	**76**	*29*	*4*	+++
TCATCACTGATTTATACGTG TGG	**61**	**62**	*0*	*19*	**62**	*2*	**5**	+++
ATTTCCCTCAACTGAGACCA *CGG*	**59**	**56**	**0.6**	*37*	**64**	**53**	*4*	+++
AAGCCCCATGTGTTCACTGT GGG	**64**	*28*	*–0.1*	*19*	**50**	**71**	**6**	+++
GGGTGAGCTTCACTGCAGTC AGG	**51**	**57**	*–0.1*	*13*	**52**	**51**	**6**	+++
TGTGCTCTCCAATGCCCGGC TGG	*44*	**72**	*–0.2*	*26*	**66**	*41*	**7**	+++
CGGGGATGCTAGCATGGCTA AGG	**56**	*17*	*0.2*	*10*	**69**	**62**	*4*	+++
CAGCACCTCCCTGTTGTGCA *CGG*	**59**	*25*	*–0.2*	*26*	*31*	**54**	**8**	+++
TGTTTGGAATGACGACAGTA *TGG*	**49**	**83**	*–0.2*	*13*	**58**	**50**	*3*	+++
False positive	55%	73%	27%	**0%**	73%	64%	73%	
False negative	18%	26%	74%	**88%**	6%	44%	56%	
Overall correlation	**73%**	62%	36%	33%	**78%**	51%	40%	

Note: Values in bold indicate that prediction correlates with experimental confirmation

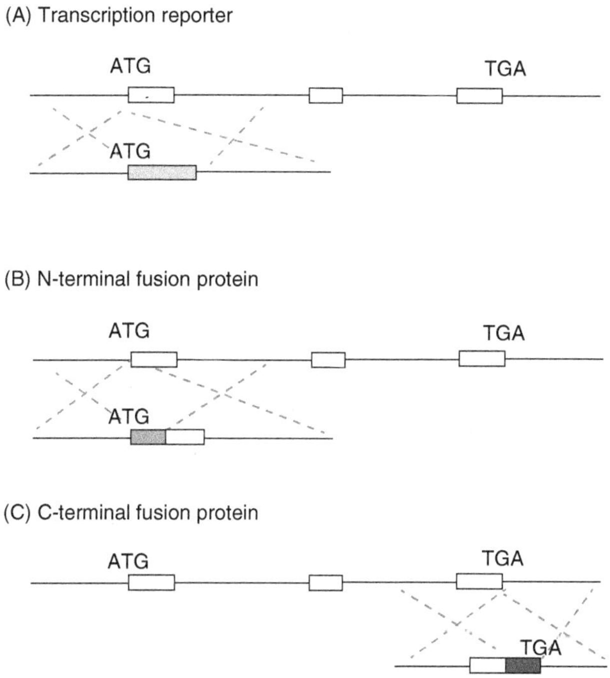

Figure 7.1 Knockin of a reporter or a tag at different genomic locations. (A) A transcription reporter does not need seamless integration and selection can be used to enrich the targeting events. The reporter can be linked to a P2A-puro marker instead of a separate promoter-driven puro marker to simplify the donor construction. (B) For N-terminal fusion of target proteins, GFP or other tags such as HA/Myc/His should be seamlessly integrated to avoid any complications. A selection marker can be used under the condition that it won't interfere with protein structure. For instance, it can be placed into the intron region driven by its own promoter in the opposite orientation of the target gene. (C) For C-terminal fusion protein, the tag can be fused with a P2A-puro marker. In some cases, when P2A interferes with the tagged protein function, the puro marker can be driven with its own promoter.

as a transcriptional reporter), seamless recombination or inclusion of selection markers for enrichment have to be taken into account (Figure 7.1). Most importantly, one should pay extreme attention to ensure in-frame integration, intact splicing and no disruption of gene function when desired.

7.3.1 ssODN Donor

ssODN is the first choice as the donor as it is so simple to make. In addition, several reports have shown that ssODN is more efficient than using donor plasmids for HDR-mediate knockin (Chen *et al.*, 2011a; Yoshimi *et al.*, 2014).

When using ssODN, the total length and the length of homology arms (HA) of the oligos are of major concern. Too short may not provide efficient targeting, too long may not be cost-effective and error-prone from synthesis. For instance, Chen *et al.* observed a drop in HDR efficiency when HA decreased from 40 bp to 30 bp (Chen *et al.*, 2011b). Surprisingly, Yang *et al.* found that longer ssODN (170 bp) showed lower efficiency of HDR than the shorter one (50–90 bp) (Yang *et al.*, 2013).

One reason might be that the long ssODN may not be stable, or may be prone to secondary structures, both decreasing the amount of donors effectively available for gene editing. As for which strand should be targeted by ssODN, Hwang *et al.* found that the strand bias could be gene-locus specific (Hwang *et al.*, 2013).

The ssODN can be easily degraded inside the cells. Hence, increasing the stability of ssODN will likely enhance HDR efficiency. Renaud *et al.* found that using phosphorothioate (PS)-modified oligonucleotides strongly enhances the genome editing efficiency of ssODN in cultured cells, and rat and mouse embryos (Renaud *et al.*, 2016). They reported that using PS-modified ssODN allowed a tenfold insertion rate increase (53% with PS-modified vs. 3% for unmodified).

The latest advance in ssODN strategy uses asymmetric ssODN. Cas9 asymmetrically releases the 3′ end of the cleaved DNA strand (PAM-distal non-target strand) after cleavage, but before complete dissociation with the other three strands (see Figure 7.2; Richardson *et al.*, 2016). This strand is free to anneal to complementary challenge DNA and can be extruded from the Cas9-DNA complex. Therefore, ssODN donor complementary to this first released non-target strand could jump-start the pairing to boost HDR frequency. Indeed, they showed that asymmetric ssODN identical to the target stand exhibits HDR frequencies up to 2.6-fold greater than ssODN identical to the non-target strand. Asymmetric ssODN overlapping the Cas9 cutting site with 36 bp on the distal side and 91bp on the proximal side of the PAM seems to be optimal and leads to 57 ± 5% HDR frequency (Figure 7.2).

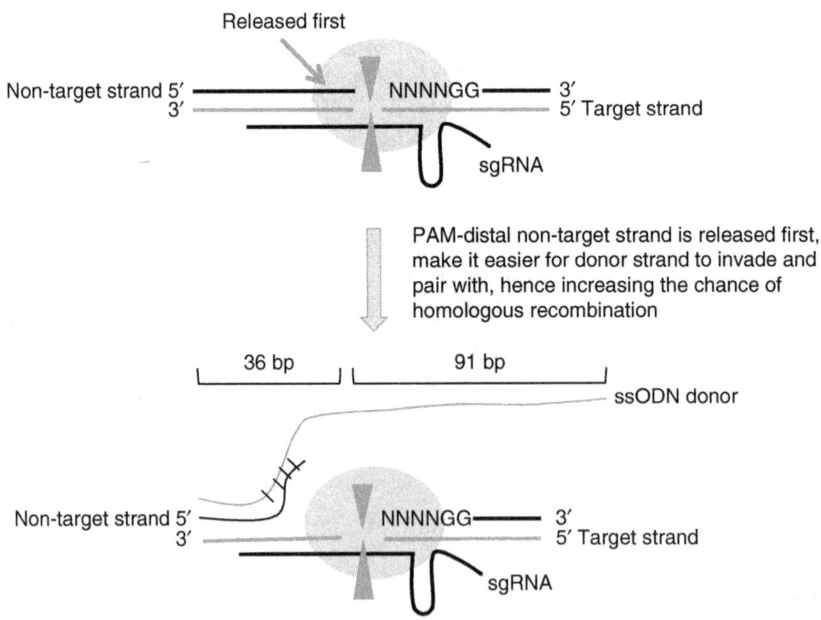

Figure 7.2 Asymmetric ssODN enhances HDR frequency. Cas9, sgRNA and the target dsDNA form a complex. Then Cas9 cuts the dsDNA to make a DSB 3 bp upstream of the PAM motif (NGG). The non-target strand 3′ end, distal to PAM, is released first while the other three strands remain associated with Cas9. Asymmetric ssODN with 36 bp overlapping with the non-target strand 3′ end of the DSB and 91 bp overlapping with the PAM proximal strand exhibits maximal HDR frequency.

While most research focuses on how to increase HDR rate with ssODN, some has also tried to address how far away a mutation can be from the CRISPR cutting site. Given the short homology arms (HA) of ssODN, it is not surprising that the optimal window is only about 10 bp distance from the cutting site. Paquet *et al.* similarly found a clear monotonic inverse relationship between the rate of mutation incorporation and distance from the cleavage site (Paquet *et al.*, 2016). With a cut-to-mutation distance of 2 bp, the HDR rate is near 100%, which drops to below 23.5% and 0.3%, respectively, when the distance is 23 or 45 bp. This is reasonable as NHEJ repair often causes more small deletions than large ones. During HDR, only the portion of the repair template overlapping the deletion may be used. Therefore, the further away the mutation is from the cleavage site, the less chance there is for it to be incorporated.

7.3.2 Plasmid Donor

The plasmid donor is more daunting and labor-intensive to construct than ssODN, but can provide many advantages. The major advantage is that it allows a powerful antibiotic selection to greatly enrich the HDR events. For instance, Park *et al.* used double selection markers in the donor DNA to knockin a destabilization domain (Park *et al.*, 2014). With only 800 bp homolog arms on each side, 7 of 8 double positive clones are real knockins. In a separate report, with a selection marker in the donor DNA, over 90% drug-resistant clones are real knockins (Merkle *et al.*, 2015). More impressively, Dickinson *et al.* reported streamlined genome editing employing a self-excising drug selection cassette (SEC) together with a dominant phenotype marker in *C. elegans* to allow minimal hands-on labor (Dickinson *et al.*, 2015). In combination with fluorescent tags, a fluorescent protein fusion, a transcriptional reporter, and a strong loss-of-function allele for any gene of interest can be obtained in a single injection step. Such flexibility and complexity of knockin cannot be achieved with ssODN.

When using a plasmid donor, the length of the homolog arm (HA) is a major factor to consider. We normally use 500–1000 bp HAs. Petrezselyova *et al.* compared 300 bp HA with 2–5 kb HA and found that the targeting efficiency is similar (Petrezselyova *et al.*, 2015). However, with shorter HAs, a greater frequency of undesired mutations at the junction is observed. This happens even with a selection marker on the targeting donor vector. However, others found that homology arms longer than 1 kb could increase gene-targeting efficiency by two- to fourfold (Chu *et al.*, 2015).

The difficulty with a plasmid donor containing both a fluorescent marker and a selection marker is that the two markers might interfere with each other. Zhu *et al.* reported that with a P2A-GFP-PGK-Puro donor, only 2.8% puromycin-resistant clones are correct knockin clones (Zhu *et al.*, 2015). Surprisingly, GFP expression is not detected until the puro marker is removed by Cre. Therefore, one must be cautious when using extra sequences/elements to boost HDR frequency to ensure that there is no interference with what you want to achieve.

7.4 NHEJ-based Integration

HDR is not only less frequent than NHEJ, but also occurs only during the S and G2 phases (Maruyama *et al.*, 2015). In contrast, NHEJ occurs throughout the cell cycle, allowing more chances of repair at all cell cycle stages. Therefore, effort has been made to employ NHEJ for precision integration of reporters, especially in species like zebrafish. For instance, NHEJ is tenfold more active than HR in early zebrafish development (Auer *et al.*, 2014). As a result, integration of a reporter by homology-independent DNA repair has an efficiency of up to 30% (compared to 1.5% with HR-mediated knockin). The caveat is that the insertion is imprecise and only one-third is in-frame. However, adding an sgRNA cutting site in the donor plasmid DNA to linearize the donor can largely enhance the insertion rate. Since zebrafish strains display a large extent of polymorphism in their DNA sequences that are not identical to the GeneBank reference sequences, researchers often need to sequence the gene of interest and its related targeting region of parent animals for precise targeting. This actually makes NHEJ a better choice for optimizing knockin reporters in zebrafish (Li *et al.*, 2015), eliminating time-consuming screening and genotyping of parent animals.

Another way to improve NHEJ-mediated knockin is to add short homologous sequences (10–40 bp) flanking the genomic target locus on the donor plasmid DNA. In addition, Cas9 cleavage sites can be introduced at both sides of the reporter (as well as the short homology arms) to linearize the donor DNA and avoid unnecessary integration of vector backbone sequences. Using this method, Hisano *et al.* achieved precise in-frame integration of exogenous DNA without backbone vector sequences (Hisano *et al.*, 2015). Although the introduction of HA only slightly increased the frequency of integration events (0 bp, 53%, compared with 10–40 bp, 77–85%), the precise integration event is largely enhanced (0% for 0 bp, 60% for 10 bp).

Similarly, He *et al.* detected a high frequency around 15–20% of reporter insertion when the single-cut donor was used in human cells (He *et al.*, 2016). However, when the double-cut donor vector was used to remove unwanted vector backbone sequence, only 27% of clones were correct at both junctions. In addition, since the double-cut donor plasmid will produce two fragments, the vector backbone fragment might compete for genomic integration of the reporter or the two fragments could be integrated into the two alleles at the same time. Therefore, NHEJ-mediated knockin is non-directional and non-selective, and raises serious concerns on using this strategy for knockin in human cells. The balance between high efficiency and lower accuracy must be calculated based on the purpose of applications. In most cases, obtaining a correctly knocked-in reporter can outweigh the unexpected alterations in the other allele. This is especially true when HDR-mediated reporter integration is at a very low frequency (for example, below 0.4% in H1 human ESCs; He *et al.*, 2016). Moreover, since HDR only occurs during G2-S phase, non-dividing cells such as neurons are not suitable for HDR-mediated knockin. In addition, He *et al.* found that for silenced or inactive loci, NHEJ-mediated knockin is 10–20-fold more efficient than the HDR-based approach.

7.5 How to Enhance the Knockin Rate?

One way to enhance knockin efficiency is to increase the homologous recombination rate by stimulating expression of homologous recombination proteins. RS-1 is a small molecule that can induce homologous recombination protein RAD51. RS-1 can enhance HDR three- to sixfold, depending on the locus and transfection method (Pinder *et al.*, 2015). Song *et al.* independently reported a similar two- to fivefold enhancement of HDR at different loci by RS-1 (Song *et al.*, 2016), whereas the NHEJ inhibitor SCR7 had minimal effects, in contrast to previous reports. For instance, Maruyama *et al.* reported that SCR7 treatment increased the efficiency of HDR-mediated genome editing in mammalian cell lines and in mice for four genes examined, up to 19-fold (Maruyama *et al.*, 2015).

On the other hand, suppressing NHEJ will enhance HDR and knockin efficiency. Suppression of KU70 and DNA ligase IV, the NHEJ key molecules, promotes the efficiency of HDR four- to fivefold (Chu *et al.*, 2015). The adenovirus 4 (Ad4) E1B55K and E4orf6 proteins mediate the ubiquitination and proteasomal degradation of DNA ligase IV. When co-expressed with the Cas9 system, E1B55K and E4orf6 improved the efficiency of HDR up to eightfold in both human and mouse cell lines.

Interestingly, limited low-dose irradiation (LDI) using either γ-ray or x-ray exposure (0.4 Gy) significantly enhanced HR frequency up to 90%, which is tenfold more than with no radiation control (Hatada *et al.*, 2015). Not only do the positive clones increase threefold, but correctly targeted clones increase 5–10-fold. Since no increased mutagenesis was induced until the higher 4.0-Gy dose of radiation was administered, the lower doses of radiation can be safely used for routine stimulation of HDR.

Another simple way to enhance knockin is to sort high-expressing Cas9/sgRNA cells by FACS (Duda *et al.*, 2014), but with caution as it also correspondingly increases off-target activity. We routinely sort GFP-positive cells (co-transfect with Cas9/sgRNA or use all-in-one vector containing GFP) or do puromycin selection to enrich transfected cells, which is the easiest way to increase targeting rate.

7.6 How to Avoid Unwanted Mutations

In our experience, we have noticed that compared to TALEN, CRISPR/Cas9 tends to produce more homozygous indels than heterozygous alleles, which has been found by several other studies as well (Dow *et al.*, 2015; Platt *et al.*, 2014). For instance, Dow showed that over 90% of targeted clones are homozygous for several genes they target, and nearly 50% of clones carry mutations in all four alleles. Although this feature is ideal for making knockout clones, it causes unwanted mutations for knockin. For instance, over 37–50% of knockin clones were found to contain unwanted indels on both targeted and untargeted alleles (Merkle *et al.*, 2015). This is because the high nuclease activity of CRISPR/Cas9 will keep cutting the wild-type protospacer sequence until sufficient indels have been introduced by NHEJ to prevent further targeting by CRISPR. Therefore, physically destroying CRISPR target sites could effectively eliminate on-target indels to avoid unwanted mutations. Indeed, less than 0.05–0.5% knockins were found to harbor the unwanted mutations after the CRISPR sites were destroyed (Merkle *et al.*, 2015).

Recently, Paquet *et al.* reported a clever method for accurate genome editing, termed "CORRECT (consecutive re-guide or re-Cas steps to erase CRISPR/Cas-blocked targets)"(Figure 7.3; Paquet *et al.*, 2016). Basically, they first introduced blocking mutations in the ssODN donor to dramatically increase HDR accuracy. Without blocking mutations, they found that only 6 to 35% of reads that incorporated expected pathogenic mutations had accurate HDR. The presence of a blocking mutation increased HDR accuracy 2–10-fold in both iPS cells and HEK293 cells. However, the blocking mutation itself sometimes is undesired and needs to be removed. Paquet *et al.* designed two approaches for this purpose: re-guide and re-Cas. The re-guide blocking mutation (near the 3′ end of sgRNA) will interfere with guide RNA targeting and the re-Cas mutation (inside the PAM motif) will block PAM detection, but can be recognized by a specificity-modified Cas9 (for instance, NGCG, target of the recently described VRER-Cas9). For re-guide, a separate re-sgRNA targeting the modified sequence is used with a new ssODN that no longer contains the original blocking mutation. For re-Cas, the modified PAM is targeted with the Cas9 variant and a new ssODN as well.

7.7 Dual sgRNAs Enhance Knockin Efficiency and Accuracy

It is not surprising that most genome editing has been done with a single sgRNA, as researchers tend to avoid extra work. More evidence suggests that dual sgRNAs,

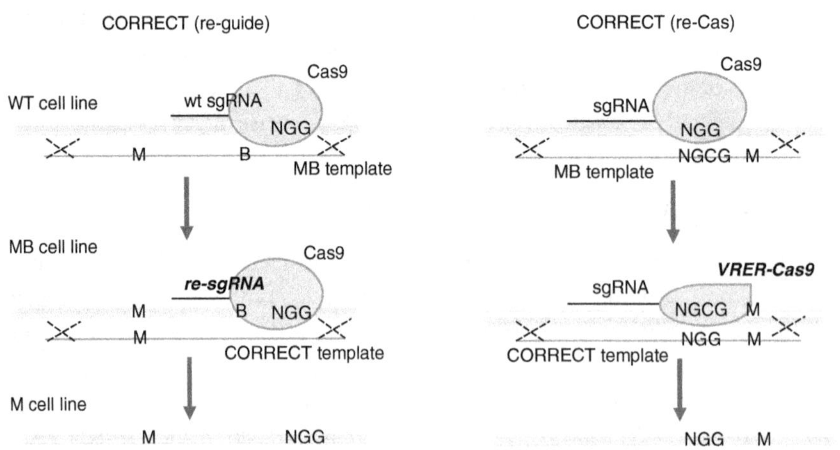

Figure 7.3 CORRECT targeting. Blocking mutation (B) and mutagenic mutation (the intended point mutation) are included in the ssODN (MB template). For CORRECT (re-Cas), the blocking mutation is in the PAM motif; in this case, a "G" to "C" change. The MB cell line is first generated with wt sgRNA and Cas9, which is then used for retargeting with re-sgRNA or VRER-Cas9, together with the CORRECT ssODN template to make the final M cell line. In many cases when a blocking mutation (mostly silent) is not interfering with gene function, the MB cell line can be used safely without going through a second targeting round to make the CORRECT M cell line.

which double the cloning work of making sgRNAs and testing sgRNA efficacy, seems to be more beneficial in the end. Chen *et al.* reported that using two sgRNAs to insert 5′ and 3′ FRT sites is necessary for generating correctly targeted homozygous hESC lines (Chen *et al.*, 2015). Even with puromycin selection, using a single 5′ sgRNA, the targeting efficiency was extremely low (less than 1.5%), and there were no homozygous clones. When using only the 3′ sgRNA, none of the homozygous clones contained the 5′ FRT site. In contrast, the FRT sites were correctly inserted in both alleles in most of the clones from the group using both 5′ and 3′ sgRNAs.

With two sgRNAs, it is also possible to make a conditional knockout (CKO) in a single step. For example, Flemr and Buhler (2015) used a homologous recombination reporter to enhance the biallelic integration of LoxP sites to improve CKO generation. In this situation, two LoxP sites flanking the targeting exon can be introduced by ssODNs. Together with the two sgRNAs near the LoxP integration sites, homozygous knockin of two LoxP sites can be generated in a single targeting step.

7.8 Concluding Remarks

There are increasing reports of the use of a new CRISPR nuclease, Cpf1, for genome editing that exhibits similar editing activity but fewer off-target effects than Cas9 (Kleinstiver *et al.*, 2016; Zetsche *et al.*, 2015). As Cpf1 generates staggered sticky ends, unlike the blunt ends produced by Cas9, such a unique feature could allow knockin without HDR via NHEJ (like a ligation event) in non-dividing cells. In addition, the Cpf1 cleavage site is far away from the PAM motif at the 5′ end of the protospacer, the indels caused by Cpf1 may not interfere with repeated cutting, which may increase the chances of new DNA being incorporated at that site. Unfortunately, one year has passed since the first report of Cpf1 and we have not seen any study reporting its use for knockin purposes, but many hope for a breakthrough soon.

For gene therapy or biomedical application of CRISPR technology, it is crucial to precisely knockin desired mutations or a wild-type copy to correct mutations in patients with no off-site targeting. To avoid any potential unwanted alterations to the genome, many approaches and reagents such as chemicals or gene silencing/overexpression that enhance the HDR rate are not suitable. In these cases, accuracy is the most important consideration factor to design any knockin strategies. In contrast, basic research and routine laboratory work with genome editing do not require such high precision as long as proper controls are included. In these cases, increasing HDR frequency can vastly lower the labor cost for screening correctly targeted knockin clones. Therefore, accuracy vs. frequency can be balanced to meet your particular research purpose. In all cases, the final knockin clones should be completely sequenced to record any alterations at both the 5′ and 3′ junction sites of integration. It is very true that one method/strategy may triumph over others in certain cell types, and there is no golden rule applicable to all situations. Given the rapid progress made in genome editing every day, it is certain that we will simultaneously achieve both extreme knockin accuracy and high frequency in the near future.

Acknowledgement

We'd like to thank Dr. Shireen Sarraf for the comments and editing. Chunxin Wang is supported in part by the Intramural Research Program of the National Institute of Neurological Disorders and Stroke (NINDS), National Institutes of Health (NIH).

REFERENCES

Auer TO, Duroure K, De Cian A, Concordet J-P, Del Bene F. 2014. Highly efficient CRISPR/Cas9-mediated knock-in in zebrafish by homology-independent DNA repair. *Genome Res* **24**(1): 142–153.

Chari R, Mali P, Moosburner M, Church GM. 2015. Unraveling CRISPR-Cas9 genome engineering parameters via a library-on-library approach. *Nat Methods* **12**(9): 823–826.

Chen F, Pruett-Miller SM, Huang Y, *et al.* 2011a. High-frequency genome editing using ssDNA oligonucleotides with zinc-finger nucleases. *Nat Methods* **8**(9): 753–755.

Chen F, Pruett-Miller SM, Huang Y, *et al.* 2011b. High-frequency genome editing using ssDNA oligonucleotides with zinc-finger nucleases. *Nat Methods* **8**(9): 753–755.

Chen Y, Cao J, Xiong M, *et al.* 2015. Engineering human stem cell lines with inducible gene knockout using CRISPR/Cas9. *Cell Stem Cell* **17**(2): 233–244.

Chu VT, Weber T, Wefers B, *et al.* 2015. Increasing the efficiency of homology-directed repair for CRISPR-Cas9-induced precise gene editing in mammalian cells. *Nat Biotechnol* **33**(5): 543–548.

Dickinson DJ, Pani AM, Heppert JK, Higgins CD, Goldstein B. 2015. Streamlined genome engineering with a self-excising drug selection cassette. *Genetics* **200**(4): 1035–1049.

Doench JG, Hartenian E, Graham DB, *et al.* 2014. Rational design of highly active sgRNAs for CRISPR-Cas9–mediated gene inactivation. *Nat Biotechnol* **32**(12): 1262–1267.

Dow LE, Fisher J, O'Rourke KP, *et al.* 2015. Inducible in vivo genome editing with CRISPR-Cas9. *Nat Biotechnol* **33**(4): 390–394.

Duda K, Lonowski LA, Kofoed-Nielsen M, *et al.* 2014. High-efficiency genome editing via 2A-coupled co-expression of fluorescent proteins and zinc finger nucleases or CRISPR/Cas9 nickase pairs. *Nucl Acids Res* **42**(10): e84.

Farboud B, Meyer BJ. 2015. Dramatic enhancement of genome editing by CRISPR/Cas9 through improved guide RNA design. *Genetics* **199**(4): 959–971.

Flemr M, Bühler M. 2015. Single-step generation of conditional knockout mouse embryonic stem cells. *Cell Rep* **12**(4): 709–716.

González F, Zhu Z, Shi Z-D, *et al.* 2014. An iCRISPR platform for rapid, multiplexable, and inducible genome editing in human pluripotent stem cells. *Cell Stem Cell* **15**(2): 215–226.

Haeussler M, Schönig K, Eckert H, *et al.* 2016. Evaluation of off-target and on-target scoring algorithms and integration into the guide RNA selection tool CRISPOR. *Genome Biol* **17**(1): 148.

Hatada S, Subramanian A, Mandefro B, *et al.* 2015. Low-dose irradiation enhances gene targeting in human pluripotent stem cells. *Stem Cells Trans Med* **4**(9): 998–1010.

He X, Tan C, Wang F, *et al.* 2016. Knock-in of large reporter genes in human cells via CRISPR/Cas9-induced homology-dependent and independent DNA repair. *Nucleic Acids Res* **44**(9): e85.

Hisano Y, Sakuma T, Nakade S, *et al.* 2015. Precise in-frame integration of exogenous DNA mediated by CRISPR/Cas9 system in zebrafish. *Scientific Rep* **5**: 8841.

Hwang WY, Fu Y, Reyon D, *et al.* 2013. Heritable and precise zebrafish genome editing using a CRISPR-Cas system. *PLoS One* **8**(7): e68708.

Kleinstiver BP, Tsai SQ, Prew MS, *et al.* (2016). Genome-wide specificities of CRISPR-Cas Cpf1 nucleases in human cells. *Nat Biotechnol* **34**(8): 869–874.

Li J, Zhang B-B, Ren Y-G, *et al.* 2015. Intron targeting-mediated and endogenous gene integrity-maintaining knockin in zebrafish using the CRISPR/Cas9 system. *Cell Res* **25**(5): 634–637.

Liu X, Homma A, Sayadi J, *et al.* 2016. Sequence features associated with the cleavage efficiency of CRISPR/Cas9 system. *Scientific Rep* **6**: 19675.

Maruyama T, Dougan SK, Truttmann MC, *et al.* 2015. Increasing the efficiency of precise genome editing with CRISPR-Cas9 by inhibition of nonhomologous end joining. *Nat Biotechnol* **33**(5): 538–542.

Merkle FT, Neuhausser WM, Santos D, *et al.* 2015. Efficient CRISPR-Cas9-mediated generation of knockin human pluripotent stem cells lacking undesired mutations at the targeted locus. *Cell Rep* **11**(6): 875–883.

Paquet D, Kwart D, Chen A, *et al.* 2016. Efficient introduction of specific homozygous and heterozygous mutations using CRISPR/Cas9. *Nature* **533**(7601): 125–129.

Park A, Won ST, Pentecost M, Bartkowski W, Lee B. 2014. CRISPR/Cas9 allows efficient and complete knock-in of a destabilization domain-tagged essential protein in a human cell line, allowing rapid knockdown of protein function. *PLoS One* **9**(4): e95101.

Petrezselyova S, Kinsky S, Truban D, *et al.* 2015. Homology arms of targeting vectors for gene insertions and CRISPR/Cas9 technology: size does not matter; quality control of targeted clones does. *Cell Mol Biol Lett* **20**(5): 773–787.

Pinder J, Salsman J, Dellaire G. 2015. Nuclear domain "knock-in" screen for the evaluation and identification of small molecule enhancers of CRISPR-based genome editing. *Nucleic Acids Res* **43**(19): 9379–9392.

Platt RJ, Chen S, Zhou Y, *et al.* 2014. CRISPR-Cas9 knockin mice for genome editing and cancer modeling. *Cell* **159**(2): 440–455.

Renaud J-B, Boix C, Charpentier M, *et al.* 2016. Improved genome editing efficiency and flexibility using modified oligonucleotides with TALEN and CRISPR-Cas9 nucleases. *Cell Rep* **14**(9): 2263–2272.

Richardson CD, Ray GJ, DeWitt MA, Curie GL, Corn JE. 2016. Enhancing homology-directed genome editing by catalytically active and inactive CRISPR-Cas9 using asymmetric donor DNA. *Nat Biotechnol* **34**(3): 339–344.

Song J, Yang D, Xu J, *et al.* 2016. RS-1 enhances CRISPR/Cas9- and TALEN-mediated knock-in efficiency. *Nat Comm* **7**: 10548.

Yang L, Guell M, Byrne S, *et al.* 2013. Optimization of scarless human stem cell genome editing. *Nucleic Acids Res* **41**(19): 9049–9061.

Yoshimi K, Kaneko T, Voigt B, Mashimo T. 2014. Allele-specific genome editing and correction of disease-associated phenotypes in rats using the CRISPR–Cas platform. *Nat Comm* **5**: 4240.

Zetsche B, Gootenberg JS, Abudayyeh OO, *et al.* 2015. Cpf1 is a single RNA-guided endonuclease of a class 2 CRISPR-Cas system. *Cell* **163**(3): 759–771.

Zhu Z, Verma N, González F, Shi Z-D, Huangfu D. 2015. A CRISPR/Cas-mediated selection-free knockin strategy in human embryonic stem cells. *Stem Cell Rep* **4**(6): 1103–1111.

8 Genetically Engineered Pig Models for Human Diseases using ZFNs, TALENs and CRISPR/Cas9

Martin F. Berthelsen, Martin K. Thomsen and Yonglun Luo

8.1 Introduction

This year marks the 33rd anniversary of the first transgenic pigs generated by microinjection (Hammer *et al.*, 1985), the 22nd anniversary of the birth of Dolly the sheep, generated by somatic cell nuclear transfer (SCNT) or cloning (Campbell, 1996), the 16th anniversary of the first α-1,3-galactosyltransferase knockout pig generated by conventional gene targeting technology and SCNT (Lai *et al.*, 2002), and the 11th anniversary of simplifying the SCNT technology for pig cloning, called handmade cloning (HMC) (Du *et al.*, 2007; Kragh *et al.*, 2005). The growing interest in generating genetically modified pigs is owing to their similarities with humans, with regard to their anatomic structure, organ size, genetics, physiology, etc. The primary aim of this great venture in generating genetically modified pigs is to provide a better animal model for studying the pathogenesis of human diseases, such as neurodegenerative diseases, atherosclerosis, cancers and diabetes. Well-established and -characterized pig models can be further used for drug testing, therapeutic/surgical practice, etc.

Although there is a great need for pig models of human diseases, the development of good models that can fully replicate human diseases and their pathological processes has progressed very slowly over the last two decades. One major challenge relates to the relatively long gastrulation and maturation time of pigs, which makes the development and characterization of these pig models expensive and impractical. Thus, most currently generated genetically modified pigs are based on well-established small animal models, such as rodents. Another major challenge is the generation of multiplexed genetically modified pigs.

During recent years, the development of programmable DNA endonucleases has greatly revolutionized life sciences and biomedical research, including the generation of genetically modified pigs. In this chapter, we will focus on providing an update of the pros and cons of using pigs as research model organisms, the commonly used methods and the application of programmable DNA nucleases including ZFNs, TALENs and CRISPR/Cas9 in generating genetically modified pigs, as well as the clinical impacts of these pig models. For a more detailed review of what kinds of genetically modified pig models have been generated for studying human diseases, please refer to the recent review articles from us and many other groups (Cooper *et al.*, 2016; Flisikowska *et al.*, 2014; Holm *et al.*, 2016; Luo *et al.*, 2012; Niemann and Petersen, 2016; Shim *et al.*, 2016; Wolf *et al.*, 2014; Yao *et al.*, 2016).

8.2 Pigs as Research Model Organisms

Our understanding of human diseases largely originates from studies of relatively few animal model organisms, include *Drosophila melanogaster, Caenorhabditis elegans, Danio rerio,* canines and rodents, as they function as links between basic research and clinical application. Although all these model organisms have their advantages, pigs do have closer resemblance to humans when looking at immunology, anatomy, physiology, metabolism, size, longevity and genetics. Non-human primates do surpass pigs in all the aforementioned parameters; however, non-human primates are more expensive to work with and it is frequently more difficult to get permission for animal experiments on non-human primates due to ethical concerns. Rodents are considered the best of the frequently used animals as model organisms for human diseases, therefore, the focus of this section is on the comparison of rodents vs. pigs as model organisms (Table 8.1). In addition to the seven aspects mentioned above, most research conducted in rodents is conducted in inbred lines, where the obtained results may be less trustworthy. It must also be kept in mind that porcine models are not perfect, but they do show advantages over other model animals.

Compared to humans, pigs are very similar with regard to chemokines, cytokines, defense proteins, immunoglobulin light chain repertoire, κ:λ expression, and T-cell antigen receptor variable β and γδ repertories and loci rearrangements (Butler *et al.*, 2009). Moreover, being non-altricial, in contrast to primates and rodents, together with contagion similarity to humans make pigs good models of the development of the immune system and immune responses (Butler *et al.*, 2009). Furthermore, the high similarity between the two immune systems makes porcine cancer models trustworthy, as the immune system has a fundamental function in the progression of human cancers (Finn, 2012).

Although all mammals reveal a very similar anatomy, pigs demonstrate a higher degree of resemblance to humans than rodents. For example, both pig and human brains are gyrencephalic, whereas rodents have a lissenceplalic brain (Holm and West, 1994; Pillay and Manger, 2007; Pond *et al.*, 2000; Thibault and Margulies, 1998).

With regard to the metabolic system, pigs and humans transport most of the cholesterol in low-density lipoproteins, whereas mice carry the majority of their cholesterol in high-density lipoprotein, making them less attractive as models for lipid metabolism (Bergen and Mersmann, 2005). There are, however, also areas where pigs and humans differ in their metabolism, for example, the pig mainly produces indoleacetylglycine from nutritional tryptophan, whereas mice and humans produce indoxyl sulfate (Merrifield *et al.*, 2011). Another aspect in

Table 8.1 Comparison of humans vs. pigs and rodents

	Humans	Pigs (minipigs)	Rodents (mice)	Reference
Immune system				
APOBEC3F	Present	Present	Absent	Puente *et al.*, 2003
CCL14,23,3L1	Present	Present	Absent	Barreiro and Quintana-Murci, 2010; Mestas and Hughes, 2004
CCL16	Present	Present	Pseudogene	Barreiro and Quintana-Murci, 2010
MHC class II-DQB2	Pseudogene	Pseudogene	Present	Shinkai *et al.*, 2002; Weaver *et al.*, 2007
ICAM3	Present	Present	Absent	Vaughan *et al.*, 2000
NLRP7,8,11,13	Present	Present	Absent	Hagemann *et al.*, 2004
TLR10	Present	Present	Absent	Leeb and Müller, 2004; Parker and Porter, 2004
Tlr1,12,13	Absent	Absent	Present	Gerner *et al.*, 2009
TNFRSF10A,6B	Present	Present	Absent	Bufler *et al.*, 2002
Size				
Body weight	62 kg	30–35 kg	15.5–23.9 g	Berry, 1970; Ellegaard Göttingen Minipigs, 2010; Walpole *et al.*, 2012
Anatomy				
Brain	Gyrencephalic	Gyrencephalic	Lissenceplalic	Holm and West, 1994; Pillay and Manger, 2007; Pond *et al.*, 2000; Thibault and Margulies, 1998
Skin	Sparse hair and adherent	Sparse hair and adherent	Furry and loose	Davidson, 1998
Eye	Photoreceptors abundant	Photoreceptors abundant	Photoreceptors sparse	Gregory-Evans and Weleber, 1997
Metabolism				
Chlosterol transport	LDL	LDL	HDL	Bergen and Mersmann, 2005
Tryptophan metabolism	Indoxyl sulfate	Indole acetylglycine	Indoxyl sulfate	Merrifield *et al.*, 2011
Physiology				
Core temperature	37 °C	38.8 °C	36.9 °C	Ingram and Legge, 1970; Talan, 1984
Heart rate	60–100 ppm	70–120 bpm	450–750 bpm	American Heart Association, 2016; Reece, 2005

Continued

Table 8.1 (cont.)

	Humans	Pigs (minipigs)	Rodents (mice)	Reference
Genetics				
5′-untranslated regions	0.073[a]	0.133[a]	0.224[a]	Wernersson *et al.*, 2005
3′-untranslated regions	0.070[a]	0.154[a]	0.230[a]	Wernersson *et al.*, 2005
Exon	0.047[a]	0.090[a]	0.132[a]	Wernersson *et al.*, 2005
Intron	0.094[a]	0.189[a]	0.305[a]	Wernersson *et al.*, 2005
Intergenic	0.092[a]	0.181[a]	0.290[a]	Wernersson *et al.*, 2005
miRNA	0.005[a]	0.009[a]	0.025[a]	Wernersson *et al.*, 2005
Longevity				
Lifespan	70–80 years	15–20 years	2–3 years	Berry, 1970; Ellegaard Göttingen Minipigs, 2010; Roser, 2016

[a] Distance to most recent common ancestor.

which the pig does not surpass the mouse as a model organism is in body temperature. The pig has a core body temperature of 38.8 °C (Ingram and Legge, 1970) whereas mice have a core body temperature of 36.9 °C (Talan, 1984), which is closer to the 37 °C human core body temperature.

The size of the pig organs and body demonstrates a much greater resemblance to the size of humans than rodents, therefore the pig once again is the obvious choice when attempting to generate human disease models. Moreover, the smaller size of the rodent's organs makes it harder for surgeons to practice organ and tumor resections on rodent cancer models; here the pig's greater size makes tumor resection training more akin to what they would experience in humans. Furthermore, it is possible to collected biopsies from a pig without euthanizing it, allowing disease progression to be followed. The large size of the adult pig can, however, make handling challenging and therefore pigs such as the Yucatan miniature pig and the Göttingen minipig have been generated, making handling of the pig models more convenient (Swindle *et al.*, 1994).

The 2–3 year lifespan of rodents (Berry, 1970) compared to the 70–80 year lifespan of humans makes assessment and use of data between rodents and humans challenging. The lifespan of miniature pigs, at 15–20 years, although not equivalent to that of humans, enables better longevity studies. However, the short generation time and cost of rodents make them desirable models compared to pigs.

It has been concluded that the pig genome shares a greater similarity with the human genome than that of rodents. An examination of the 3′-untranslated regions, 5′-untranslated regions, exonic, intronic, intergenic and miRNA regions were analyzed and all regions revealed a higher homology between pigs and humans than for mice and humans (Wernersson *et al.*, 2005). The same conclusion was reached when the full-length cDNA alignment of the three species were analyzed (Jørgensen *et al.*, 2005).

In conclusion, the pig is a potent model organism for bridging basic research and clinical application, as it is possible to conduct long-term studies of disease progression, either in combination with treatment or without, due to the overlap in drug metabolism. Moreover, the overlap in immunology, physiology and genetics makes the pig a good model for biological mechanisms of diseases.

8.3 Production of Genetically Modified Pigs

Many different cloning techniques are available for the production of transgenic animals, including pigs. In this section, a short description of the most commonly used methods for generation of genetically modified pigs is given and their advantages and disadvantages are pointed out. Five methods: microinjection, sperm-mediated gene transfer, viral vector-mediated gene transfer, embryonic stem cell-mediated gene transfer and somatic cell nuclear transfer are presented; their principles are schematically depicted in Figure 8.1.

8.3.1 Microinjection

DNA microinjection, originally performed by Hammer *et al.* (1985), is the most commonly used technique for the transfer of genes into animals. As shown in Figure 8.1A, it is carried out by delivering a gene-of-interest (GOI), as linearized DNA, into the cell or directly into the male pronucleus of fertilized eggs from superovulated donors with a microinjection needle. This is followed by *in vitro* development of the embryo to the appropriate embryonic state and finally transfer of the embryonic cells to a surrogate mother (Figure 8.1A; Nottle *et al.*, 2001). With DNA microinjection, the GOI(s) is randomly integrated into the genome of the cell, which leads to non-identical offspring, in addition to the risk of disrupting essential genes. The main advantages of microinjection are its suitability for a broad spectrum of species and good results when performed by trained scientists. However, most of the founder animals are chimeric and require a second round of breeding to generate germline-mutated animals. This issue has proved impractical and expensive for pigs, which have a gestation period of 112–114 days and require eight months to become mature. Compared to DNA, protein and mRNA have been more favored for microinjection. This method has been one of the most popular for generating gene knockout pigs using CRISPR/Cas9 technology.

8.3.2 Sperm-mediated Gene Transfer

Sperm-mediated gene transfer (SMGT), first described by Lavitrano *et al.* (1989), is a technique that exploits the incompletely understood fact that sperm cells are able to take up free-floating DNA from the surrounding medium and thereby function as transporters of exogenous DNA in addition to their own genome (Figure 8.1B). SMGT is carried out by suspending seminal plasma-free sperm

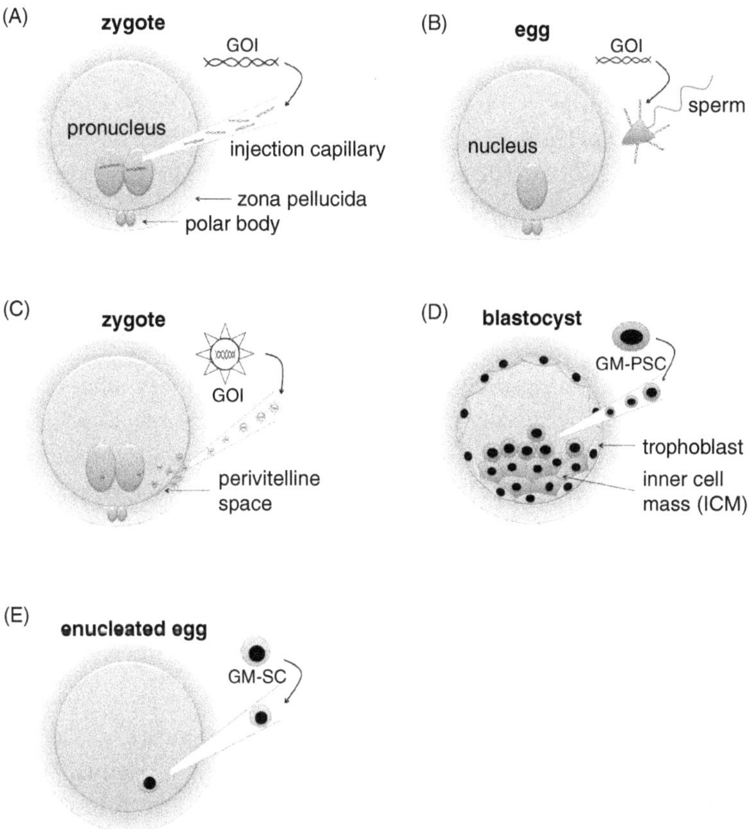

Figure 8.1 Schematic illustration of five commonly used methods to generate genetically modified animals. (A) Microinjection. GOI, gene of interest. (B) Sperm-mediated gene transfer. (C) Viral vector-mediated gene transfer. (D) Embryonic stem cell-mediated gene transfer. GM-PSC, genetically modified pluripotent stem cells. (E) Somatic cell nuclear transfer. GM-SC, genetically modified somatic cells. (A black-and-white version of this figure will appear in some formats. For the color version, please refer to the plate section.)

cells in an appropriate medium, followed by incubation with the GOI(s), whereby the DNA is taken up by the cells; DNA uptake can be enhanced by the addition of liposomes, electroporation or viral-mediated gene transfer (Farre *et al.*, 1999). Subsequently, the sperm is used for *in vitro* fertilization of oocytes followed by transplantation into a surrogate mother. SMGT has been suggested as an alternative to microinjection because it is easier to perform. As observed with microinjection, the end result of SMGT is random integration of the GOI and non-identical offspring. The success rate of this technique is inconsistent.

8.3.3 Viral Vector-mediated Gene Transfer

Retroviruses, including lentivirus, are viruses that can transduce almost all kinds of cells with high efficiency, including porcine fibroblasts and zygotes. The natural form of these viruses is capable of reversely transcribing their RNA genetic material and can thereby transfer the genome into that of an animal cell, which produces more virus particles that can infect new animal cells (Baltimore, 1970). Retroviral particles have therefore been used to introduce GOIs into the genomes of model organisms; this is done by exchanging the viral genes required for replication with GOIs. The viral vectors are produced in a packaging cell line, where the missing proteins are delivered *in trans*. When the virus particles have been produced in the packaging cell line, the virus particles that encompass the GOI are cultured with zona free zygotes (Figure 8.1C). If lentivirus is used, it is injected directly into the perivitelline space of a zygote or an oocyte, since lentiviruses are capable of infecting both dividing and non-dividing cells (Sakoda *et al.*, 2007). High transgenic efficiencies have been observed with lentivirus, making lentiviral transgenesis a good candidate for the production of transgenic pigs (Whitelaw *et al.*, 2008). However, there are some major disadvantages to this method. The two main drawbacks are random integration and limited cargo capacity, reported as up to 8 kilobases (kb); however, in practice both low efficiencies and titers are observed when the GOI is bigger than 6 kb (Whitelaw *et al.*, 2008). Another concern when using this method is the potential risk of human transmission and thus infection of researchers working with the viral vector.

8.3.4 Embryonic Stem Cell-mediated Gene Transfer

Pluripotent stem cells are cells that are capable of self-renewal and give rise to any cells in the body. The fact that totipotent stem cells are able to differentiate into every single cell type in the organism together with their unlimited passage and proliferation abilities can be exploited to generate animal model organisms (Gertsenstein *et al.*, 2002). Embryo-derived pluripotent stem cells have been the major cell source for generating genetically modified mice. Model organisms are generated by isolating embryonic stem cells from a donor animal followed by *in vitro* growth, where the GOI can be introduced into the genome, for example through HR, where electroporation of the GOI in linearized form is the preferred method of DNA delivery. Fusing the manipulated cell to the host embryos generates chimeric transgenic animals (Figure 8.1D). This method is well established in mice, but since no true embryonic stem cells have been found in pigs, these cannot be produced by this approach. An alternative is to generate induced pluripotent stem cells (iPSCs) and a few studies have shown that porcine iPSCs can be derived from primary porcine fibroblasts (Ao *et al.*, 2014; Ji *et al.*, 2013; Xie *et al.*, 2014). However, there is only one study showing that chimeric and germline-transmittable pigs can be generated using porcine iPSCs. Furthermore, other studies have demonstrated that the current iPSC generation and maintenance conditions are still not optimized for porcine iPSCs, or for porcine embryonic stem cells, with extremely low efficiency when used for somatic cell nuclear transfer (Fan *et al.*, 2013) and failure to activate endogenous pluripotency genes, such as SSEA-3 and -4 (Ezashi *et al.*, 2009), and NANOG (Choi *et al.*, 2016). The technical challenges in porcine pluripotent stem cell technology have thus greatly hampered its application in the generation of genetically modified pigs.

8.3.5 Somatic Cell Nuclear Transfer

The lack of porcine embryonic stem cells has led to the requirement for an alternative method to produce porcine model animals. The best alternative to date is somatic cell nuclear transfer (SCNT), which has a low success rate, with only 2% of transferred cloned embryos become viable piglets (Liu *et al.*, 2015). Transferring the nucleus from a genetically manipulated somatic cell, most often a fetal fibroblast isolated from either kidney or ear biopsies, to an enucleated donor oocyte generates a transgenic zygote, which can be introduced to the uterus of a surrogate mother that then produces transgenic offspring (Figure 8.1E). SCNT is the technique by which the world-famous sheep, Dolly, was cloned (Wilmut *et al.*, 1997).

Traditionally, two alternative routes were used to clone using SCNT. The cloning was achieved by enucleating an oocyte with micromanipulators and then transferring the donor nucleus, either directly into the cell cytoplasm or by subzonal injection, and then internalizing by electrofusion. Later a third method emerged termed handmade cloning (HMC) (Vajta *et al.*, 2005). Similar to the two original methods, HMC relies on electrofusion, but no micromanipulators are needed and the zona pellucida is removed. The zona-pellucida-free approach makes it possible to cryopreserve the produced embryos (Vajta and Callesen, 2012).

The first step in HMC is to force the nucleus donor cells into the G_0-resting phase through serum starvation. Then the cumulus cells are removed from the oocytes by aspiration in the medium and the zona pellucida is removed with pronase. The karyoplast, making up one-third of the oocyte, is separated from the cytoplast with a scalpel and only the cytoplast is used for cloning. The cytoplasts and the genetically engineered cells are then placed in an electropulser, where the manipulated cell is fused with a cytoplast when an electric pulse is applied. A second cytoplast is fused with the first pair to obtain a suitable volume for embryo development. Thus the transgenic nucleus is integrated into the newly formed zygote. The electropulse also functions as an activator of the resting zygotes. Finally, the reconstructed zygotes are grown to the blastula/blastocyst stage *in vitro* and transplanted into a surrogate mother. This makes HMC simpler, faster and less costly in terms of equipment, and, not least, equally or more efficient compared to traditional SCNT (Vajta *et al.*, 2005). In contrast to the above-mentioned cloning methods, SCNT has the advantage of generating non-chimeric model organisms; hence it is the only method where the offspring are complete knockouts.

8.4 Genome Editing in Pigs

Generation of genetically modified pigs can be achieved by several approaches. The first approach is transgenesis, in which a transgene is randomly inserted in the genome. Pigs generated using this method are generally referred to as transgenic. This strategy is quite fast and efficient, but normally accompanied by several disadvantages, such as transgene silencing, copy number variations and undesired insertion-mediated mutagenesis.

In this section, we will focus on the approaches that can be used to more specifically modify the porcine genome, known as targeted gene editing. Several methods have been developed for targeted genome engineering and the most popular are described in chronical order.

8.4.1 Homologous Recombination (HR)

Discovering that genes could be knocked-in by conventional homologous recombination (HR) enabled scientists to modify the genomes of model animals at will, making the field of genetic engineering possible. This discovery and its potential in biomedical research earned Capecchi, Evans, and Smities the Nobel Prize in 2007.

The introduced construct is integrated into the genome with the aid of long nucleotide arms that flank the construct; these nucleotide arms are identical to the loci in the genome where the knockin is intended.

HR was the initial path to generate knockin and knockout model animals with precise genetic modifications. This can be accomplished by designing a double-stranded DNA sequence with flanking arms identical to the desired position in the genome. Both functional and non-functional forms of the GOI can be used, thereby producing a knockin or a knockout of the targeted gene. Apart from the GOI, the targeting vector must also contain selection markers that enable both positive and negative selection of gene-targeted cells, as well as subsequent excision of the selection markers (Figure 8.2).

Precisely where between the identical flanking arms and the genomic target the recombination occurs is entirely decided by the individual cell. The replaced wild-type allele and non-integrated transgenic constructs in the transfected cells are lost by dilution and degradation in the dividing cells. The precise mechanism behind the replacement of the wild-type allele is not understood. Production of model organisms by HR is very efficient in yeast and mice. Although it also is achievable in pigs, the long reproduction time makes this method unattractive, especially when producing knockout models.

If random integration occurs, the non-homologous recombined cells can survive positive selection, but if the thymidine kinase gene has been added to the construct as a negative selection marker, ganciclovir can be used to kill cells that contain the gene. This allows for selection of cells that have undergone HR by growing them in medium with the appropriate antibiotic to select for recombination and ganciclovir to kill cells that have successfully undergone non-homologous recombination.

To produce a knockout animal, both alleles of the GOI have to be replaced by a non-functional construct. This is most often achieved by breeding animals that have had one of their alleles knocked out. It can also be accomplished by adding two different positive selection markers to the gene construct; if the animal is doubly resistant both of its alleles have been replaced.

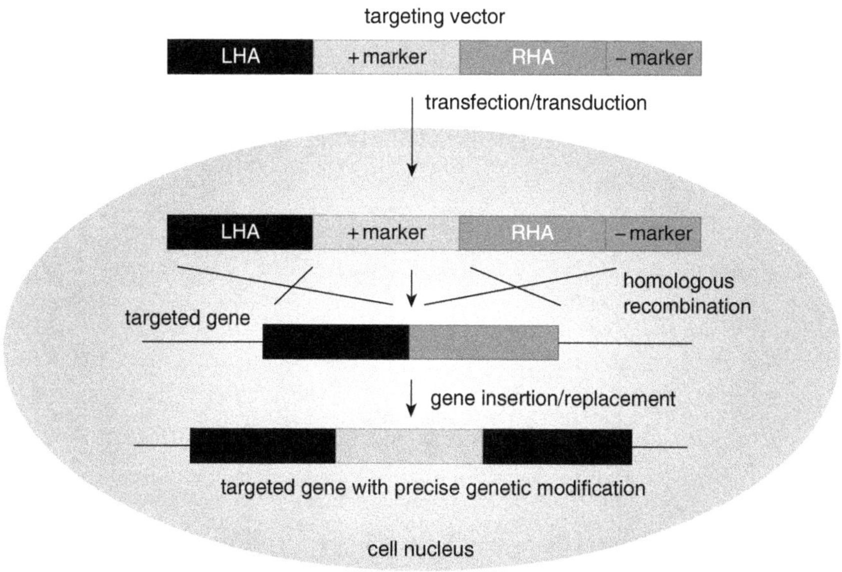

Figure 8.2 Schematic illustration of conventional gene-targeting approach. A targeting vector containing a left homology arm (LHA), positive selection marker (+ marker), right homology arm and negative selection marker (– marker) is delivered into the cell. In the cell nucleus, homologous recombination between the targeting vector and the targeted gene will lead to inversion of the positive selection marker in the targeted locus. This approach has been the most traditional strategy for conventional gene knockout or knockin.

The workflow of producing transgenic animals by conventional HR in pigs is as depicted in Figure 8.3. Primary fibroblasts are established from fetal or newborn piglet tissues, such as ear or kidney, and grown in fibroblast medium. Since fibroblasts have a very limited proliferation capacity, we have previously developed a system that allows more sustained culture and manipulation of porcine fibroblasts *in vitro* (Luo *et al.*, 2011). Cells that carry the correct genetic modifications are used as donor cells for SCNT to generate genetically modified pigs.

This method allows various modifications, e.g., reporter, knockin, knockout, conditional alleles, etc. However, due to the low efficiency of HR, the method is time-consuming and labor-intensive, inefficient, requires long homology arms and drug selection, and biallelic targeting is difficult to achieve. Due to these drawbacks, faster and more efficient techniques have been developed, such as ZFN, TALEN and CRISPR.

8.4.2 Programmable DNA Nucleases

The concept of harnessing programmable DNA nucleases for genome editing is based on the discovery that all cells and organisms have intrinsic DNA damage-repair systems. In mammalian cells, an unrepaired double-stranded DNA break (DSB) is lethal. To repair DSBs, mammalian cells have developed two major

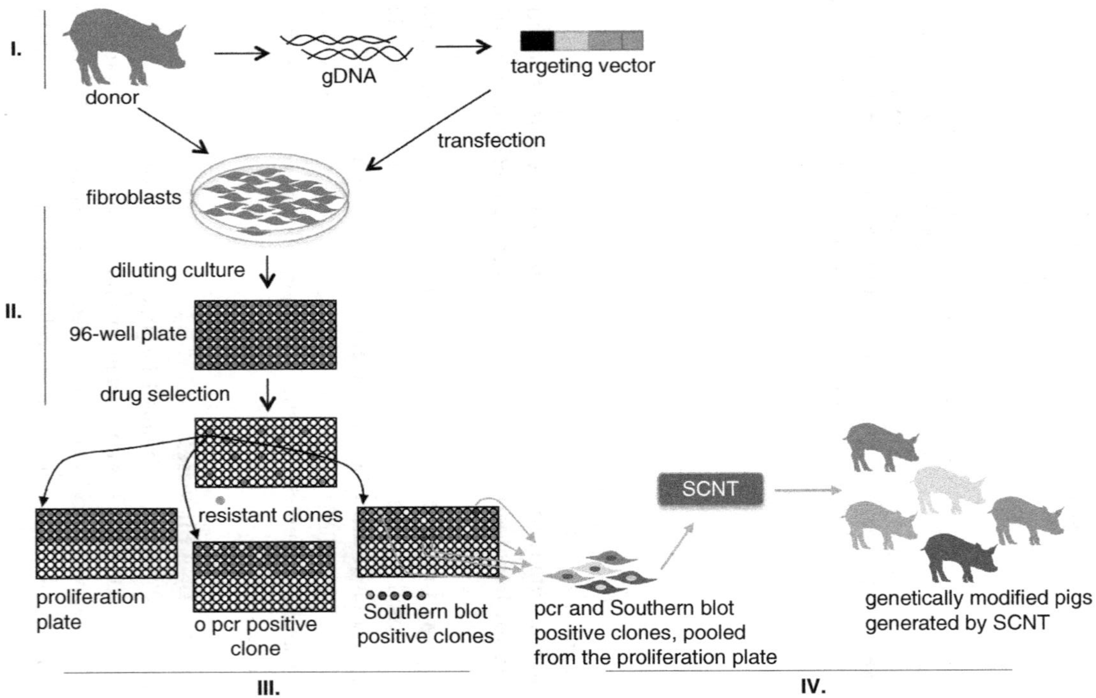

Figure 8.3 Workflow for generating genetically modified pigs by SCNT. (I) An isogenetic targeting vector is generated from the genomic DNA of the targeted cells/or cell donors. (II) The targeting vector is delivered into the fibroblasts by transfection/transduction, followed by approximately two weeks of drug selection based on the positive and negative markers presented in the targeting vector. (III) PCR and Southern-blot-based genotyping is applied to the drug-resistant clones. (IV) Positive clones, as analyzed by PCR and Southern blot are pooled as donors for SCNT. See Luo *et al.* (2012) and Li *et al.* (2015).

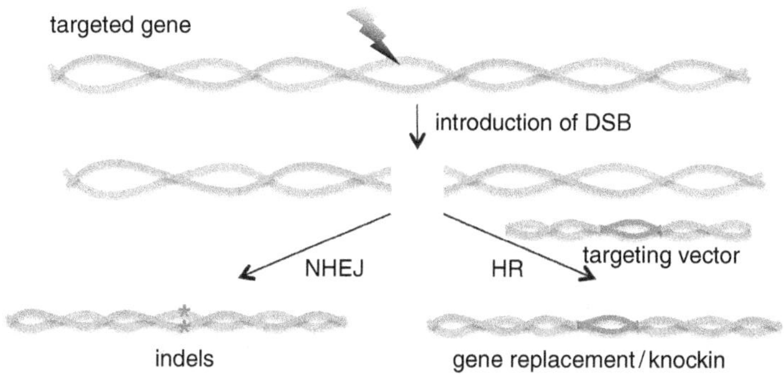

Figure 8.4 Schematic illustration of DSB-triggered genome editing. The introduction of double-stranded DNA breaks (DSBs) to genomic DNA in mammalian cells by programble DNA nuclease or other factors will trigger the DSB repair pathways: non-homologous end joining (NHEJ) and homologous recombination (HR). The majority of DSB events in mammalian cells, including pigs, are repaired by NHEJ, which leads to unmodified or small deletions or insertions, known as indels. Homologous recombination results in targeted insertion or replacement if a donor targeting DNA is used as the repair template, or unmodified repair if the homologous sister chromosome is used.

pathways: non-homologous end joining (NHEJ) and HR (Figure 8.4). Compared to the HR pathway, NHEJ is error-prone, but is the most frequently used pathway for DSB repair. This has greatly facilitated loss-of-function genetic modifications such as gene knockout, but has hampered other modifications such as knockin, gene replacement or gene tagging. Compared to the conventional HR approach, studies have observed the introduction of DBSs to a user-defined genomic locus and enhanced genetic modifications at this specific locus. Thus, during the last two decades, scientists have spent great effort to generate programmable and artificial DNA nucleases that can introduce DSBs to a given genomic locus. In the following section, we will discuss the three major classes of programmable DNA nucleases that have been harnessed for genome editing in pigs.

8.4.2.1 Zinc-finger Nucleases

Although originally intended to replace or complement restriction enzymes (Kim *et al.*, 1996), zinc-finger nucleases (ZFNs) became the first breakthrough in genetically engineered model animals owing to their ability to generate insertions and deletions (indels) in desired loci within the genome. These indels lead to frameshifts in the DNA sequences of genes and thus to the introduction of premature stop codons, resulting in knockout of the mutated genes.

ZFNs are chimeric proteins consisting of the FokI endonuclease fused to the DNA-binding domain of a modified zinc-finger motif. In order to obtain a catalytically functional ZFN two FokI hetero-monomers must dimerize. This dimerization is achieved by binding ZFN monomers to the sense and antisense strands of the locus the researcher wishes to cleave. When the zinc-finger motifs have bound to the DNA helix, the FokI dimer is able to induce a DSB. As different zinc-finger motifs can be designed to bind to different nucleotide triplets, they can be constructed to bind to specific loci in the genome.

The DSB is repaired in the cell by either NHEJ or HR. As DSBs are most often repaired by the error-prone NHEJ method, this leads to the introduction of indels and thus knockout of the GOI. Not only genes, but also non-coding, long non-coding RNA and microRNAs can be knocked out.

HR can be used for knockin if an HR template with the right flanking arms is supplied. Compared to other methods of introducing mutations, such as time-consuming and expensive ionizing radiation and chemical mutagenesis, which also introduces undesired random mutations, ZFNs are highly efficient and specific. Moreover, when using ZFNs there is no need to establish stem cell cultures, which have not yet been identified in pigs; thus ZFNs can be used to produce any desired model organism, including pigs.

The drawbacks of this method are that it is challenging to obtain highly efficient ZFN with low off-target effects (cytotoxicity) and low efficiency compared to TALENs and CRISPR. Moreover, ZFNs are restricted by a relatively low number of target sites (approximately one per 300 base pairs) and a preference for guanine-rich target sites.

8.4.2.2 TALENs

As a type of ZFN, transcription activator-like effector nucleases (TALENs) use the FokI endonuclease to mediate DSBs and rely on dimerization of FokI to obtain a catalytically active nuclease. However, the DNA-binding domain does differ from ZFNs. In TALENs it is the transcription activator-like effectors that bind to the DNA. These domains bind single nucleotides, making TALENs much more versatile as they can be designed for any possible DNA sequence. As with ZFNs, TALENs are not limited by the requirement to establish embryonic stem cultures to produce transgenic animals.

The primary weaknesses of TALENs are off-target DSBs and toxic effects (Mussolino *et al.*, 2011), and the difficulty in assembling and delivering such large and repeat-prone sequences. Finally, TALENs cannot cleave methylated cytosines. Table 8.2 offers examples of gene-modified pigs produced using TALENs.

Table 8.2 Examples of porcine models produced by TALENs

Genes	Delivery	Form	Target cells	Reference
GGTA1	Transfection	Plasmid	Fibroblasts	Kang *et al.*, 2016b
MSTN	Transfection	Plasmid	Fibroblasts	Rao *et al.*, 2016
RAG2	Electroporation	Plasmid	Fibroblasts	Lee *et al.*, 2014
GGTA1	Co-transfection	Plasmid	Fibroblasts	Xin *et al.*, 2013
DAZL and APC	Transfection	RNA	Fibroblasts	Tan *et al.*, 2013
LDLR and DMD	Transfection	Plasmid	Fibroblasts	Carlson *et al.*, 2012

8.4.2.3 CRISPR

In contrast to ZFNs and TALENs, clustered regularly interspaced short palindromic repeat (CRISPR) systems are composed of two separate components: the CRISPR-associated protein 9 (Cas9) endonuclease and a programmable small guide RNA (gRNA), that as the name implies, guides the Cas9 protein to specified genetic loci via Watson–Crick base pairing (Jinek *et al.*, 2012). This separation of the two components opens up new ways to engineer model organisms by delivering them at different time points, as well as multiplexity by delivering one Cas9 with several gRNAs. As with ZFNs and TALENs, gene-specific modifications are generated via DSBs, created by supplementing an HR template. The efficiency can be enhanced by the stimulation of the HR pathway over NHEJ by only inducing nicks in the genome, which can be accomplished using the modified Cas9 protein, D10ACas9 nickase (Shen *et al.*, 2014). The advantages of the CRISPR system over the two other nuclease systems are the ease and speed of preparing the system.

Although CRISPR is an extremely promising system, there are still some disadvantages, which include risk of off-target cleavage and difficulty in delivering the Cas9 due to its relatively large size. The risk of off-target cleavage can be significantly reduced by using the modified D10A Cas9 nickase that only is capable of inducing single-strand breaks (Ran *et al.*, 2013). Here, two adjacent target sites are used to induce nicks in both strands of the DNA double helix, which leads to a DSB. If the gRNAs have potential off-targets these will only be nicked once as the chance of both gRNAs having the off-target site close enough to generate a DSB is negligible. Furthermore, several other approaches, such as using truncated gRNAs and neutralizing the positively charged residues in the Cas9 protein that interact with DNA have been developed to improve specificity. Several smaller Cas9 orthologs with smaller size have been discovered to enhance viral delivery. Finally, persistent expression of the Cas9 protein might stimulate host immune responses, which leads to elimination of cells expressing Cas9 (Wang *et al.*, 2015). This can be overcome by only expressing the protein transiently by either transfection/electroporation of cultured cells or viral transduction of live animals. Delivery by non-integrating viral vectors could be a good way to generate cancer models, as the mutation will not be present through embryogenesis and thereby effect the development of the piglet. Moreover, this model also resembles human cancers better, as the majority of cancers first occur in adult humans. Table 8.3 offers examples of gene-modified pigs produced by CRISPR.

8.5 Clinical Impact of Pig Models

The difference in brain structure makes it difficult to study neurodegenerative diseases such as Alzheimer's and Huntington's disease in rodents, which is better carried out in pigs (Kragh *et al.*, 2009; Yang *et al.*, 2010). Porcine disease models are also preferred to rodents when making eye disease models. The smaller rodent eye

Table 8.3 Examples of porcine models produced by CRISPR

Genes	Delivery	Form	Target cells	Reference
IL2RG	MI	RNA	Zygotes	Kang *et al.*, 2016a
SOX10	MI	RNA	Zygotes	Zhou *et al.*, 2016
PARKIN, DJ-1 and PINK1	MI	RNA	Zygotes	Wang *et al.*, 2016
RUNX3	Co-transfection	Plasmid	Fibroblasts	Kang *et al.*, 2016c
OCT4	Transfection	Plasmid	Fibroblasts	Lai *et al.*, 2016
GGAT and CMAH	Co-transfection	Plasmid	Liver-derived cells	Butler *et al.*, 2016
RAG2 and IL2RG	MI	RNA	Zygotes	Lei *et al.*, 2016
hALB	MI	RNA and plasmid	Zygotes	Peng *et al.*, 2015
PERVs	Transfection	Plasmid	PK15	Yang *et al.*, 2015
IgM	Transfection	Plasmid	Fibroblasts	Chen *et al.*, 2015
BMP15	Transfection	Plasmid	Fibroblasts	He *et al.*, 2015
OCT4	MI	RNA	Zygotes	Kwon *et al.*, 2015
GGTA1, CMAH and β4GalNT2	Co-transfection	Plasmid	Liver-derived cells	Estrada *et al.*, 2015
Npc1l1	MI	RNA	Zygote	Wang *et al.*, 2015
TYR, PARK2 and PINK1	Transfection	Plasmid	Fibroblasts	Zhou *et al.*, 2015
GGTA1, CMAH and iGb3S	Co-transfection	Plasmid	Liver-derived cells	Li *et al.*, 2015
CD163 and CD1D	Transfection or MI	Plasmid	Fibroblasts	Whitworth *et al.*, 2014
VWF	MI	RNA	Zygotes	Hai *et al.*, 2014

contains a much lower number of photoreceptors that also are differently distributed than in humans. The pig eye, on the other hand, has a high resemblance to humans (Gregory-Evans and Weleber, 1997), making it possible to study eye diseases such as retinitis pigmentosa that cannot be performed in rodents (Kraft *et al.*, 2005; Petters *et al.*, 1997).

The correspondence between the pig and human gastrointestinal system and metabolism can be used to generate porcine models of diabetes. Expression of a dominant-negative isoform of the incretin glucose-dependent insulinotropic polypeptide receptor revealed prediabetic conditions similar to the human disease, whereas transgenic mice developed surprisingly severe diabetes (Renner *et al.*, 2010). Moreover, a porcine model of the autoimmune type 1 diabetes mellitus has been generated that has offered new insights to the disease (Lee *et al.*, 2010). A drug metabolism study on cinobufagin metabolism revealed stronger correspondence between pigs and humans than between mice and humans (Ma *et al.*, 2011).

From a physiological angle, the porcine model also outcompetes the rodent models when mimicking human diseases, e.g. a porcine model of cystic fibrosis that developed symptoms approaching those observed in humans has been made, which cannot be generated in rodent models (Rogers *et al.*, 2008). Moreover, the correspondence in heart rate and cardiac size between pigs and humans compared with the small size and high heart rate observed in mice makes pigs a preferred model for heart diseases (Marchant *et al.*, 2001; Stubhan *et al.*, 2008; Uechi *et al.*, 1998; Zerhouni *et al.*, 1988). Cardiac dysfunction is one of the main causes of non-injury deaths in Western countries. A porcine model of heterotopic heart transplantation was used to conclude that endothelial injury at insertion accelerates endothelial dysfunction as a result of rejection after heart transplantation leading to cardiac allograft vasculopathy (El-Hamamsy *et al.*, 2005). Another porcine cardiac study validated that transplanting of bone marrow stromal differentiated cells into a myocardial infarct region formed cardiac-like tissue, induced angiogenesis, inhibited thinning and expansion of the infarct region, and improved contractile function (Tomita *et al.*, 2002). A third porcine model validated the reducing effect of exenatide on myocardial infarct size (Timmers *et al.*, 2009). Exenatide is now being used in patients with myocardial infarction (Woo *et al.*, 2013).

Some cancer models have also been generated in pigs; these include a model of human familial adenomatous polyposis syndrome. The pig was produced by introducing premature codons into the *APC* gene. This was carried out by electroporation of linearized vector DNA into mesenchymal stem cells, which were used for SCNT to produce the pigs. As observed in humans with familial adenomatous polyposis syndrome, the pig developed numerous benign polyps (Flisikowska *et al.*, 2012). Another porcine cancer model mimicking the human Li–Fraumeni syndrome was generated by introducing a dominant-negative missense mutation in the pig tumor suppressor gene $TP53^{R167H}$. This was carried out by transducing fetal fibroblasts with recombinant adeno-associated virus, followed by SCNT for the production of the transgenic piglets. The piglets developed numerous synchronous tumors throughout the body, as observed in humans with Li–Fraumeni syndrome (Sieren *et al.*, 2014). A third porcine cancer model, which harbors a Cre-LoxP inducible $TP53^{\mathrm{R167H}}$ and $KRAS^{\mathrm{G12D}}$ cassette that can be activated by viral delivery of the Cre recombinase, has been generated. The piglets were produced by co-transfection of linearized mutant *TP53* and *KRAS* into fetal fibroblasts through electroporation, followed by SCNT for the production of the transgenic piglets. Upon viral Cre intramuscular, subcutaneous and intra-testicular injection, tumorgenesis occurred in all three locations (Adam *et al.*, 2007).

Due to the similar size of pigs and humans, pigs can, unlike rodents, be used to investigate the technical feasibility and safety of invasive approaches in connection with disease treatment. For example, pigs have been used to validate an endoscopic transgastric approach and endoscopic gastrojejunostomy with survival (Kalloo *et al.*, 2004; Kantsevoy *et al.*, 2005). These methods could, for obvious reasons, not have been validated in small rodents.

The similarity in skin anatomy has made pigs a valuable model for both normal and pathological wound-healing studies, with more than 1500 articles on the pathophysiology of different wound types based on porcine models published (Seaton *et al.*, 2015). These models include studies of all four main wound categories: incisional wounds, partial-thickness excisional wounds, full-thickness excisional wounds and burn wounds (Seaton *et al.*, 2015). Some of the best-validated studies on wound-healing pathologies have involved chronic non-healing wounds (Hyodo *et al.*, 1995; Roy *et al.*, 2009), diabetic wounds (Velander *et al.*, 2008), infected wounds (Breuing *et al.*, 2003), burns (Wang *et al.*, 2008) and hypertrophic scars (Cuttle *et al.*, 2006; Zhu *et al.*, 2003).

An important take-home message on the use of animals as model organisms for human diseases is that animal suffering should always be kept to an absolute minimum. The three Rs (replacement, reduction and refinement) must always be applied when designing any animal experiment.

REFERENCES

Adam SJ, Rund LA, Kuzmuk KN, *et al.* 2007. Genetic induction of tumorigenesis in swine. *Oncogene* **26**(7): 1038–1045.

American Heart Association. 2016. Target heart rates. www.heart.org/HEARTORG/HealthyLiving/PhysicalActivity/FitnessBasics/Target-Heart-Rates_UCM_434341_Article.jsp#.WfncRohpyM8 (accessed November 2017).

Ao Y, Mich-Basso JD, Lin B, Yang L. 2014. High efficient differentiation of functional hepatocytes from porcine induced pluripotent stem cells. *PLoS One* **9**(6): 1–11.

Baltimore D. 1970. RNA-dependent DNA polymerase in virions of RNA tumour viruses. *Nature* **226**(5252): 1209–1211.

Barreiro LB, Quintana-Murci L. 2010. From evolutionary genetics to human immunology: how selection shapes host defence genes. *Nat Rev Genet* **11**(1): 17–30.

Bergen WG, Mersmann HJ. 2005. Comparative aspects of lipid metabolism: impact on contemporary research and use of animal models. *J Nutr* **135**(11): 2499–2502.

Berry RJ 1970. The natural history of the house mouse. *Field Studies* **3**: 219–262.

Breuing K, Kaplan S, Liu P, Onderdonk AB, Eriksson E. 2003. Wound fluid bacterial levels exceed tissue bacterial counts in controlled porcine partial-thickness burn infections. *Plast Reconstruct Surg* **111**(2): 781–788.

Bufler P, Azam T, Gamboni-Robertson F, *et al.* 2002. A complex of the IL-1 homologue IL-1F7b and IL-18-binding protein reduces IL-18 activity. *Proc Natl Acad Sci USA* **99**(21): 13723–13728.

Butler JE, Lager KM, Splichal I, *et al.* 2009. The piglet as a model for B cell and immune system development. *Vet Immunol Immunopathol* **128**(1–3): 147–170.

Butler JR, Martens GR, Li P, *et al.* 2016. The fate of human platelets exposed to porcine renal endothelium: a single-pass model of platelet uptake in domestic and genetically modified porcine organs. *J Surg Res* **200**(2): 698–706.

Campbell KH, McWhir J, Ritchie WA, Wilmut I. 1996. Sheep cloned by nuclear transfer from a cultured cell line. *Nature* **380**(6569): 64–66.

Carlson DF, Tan W, Lillico SG, *et al.* 2012. Efficient TALEN-mediated gene knockout in livestock. *Proc Natl Acad Sci USA* **109**(43): 17382–17387.

Chen F, Wang Y, Yuan Y, *et al.* 2015. Generation of B cell-deficient pigs by highly efficient CRISPR/Cas9-mediated gene targeting. *J Genet Genom* **42**: 437–444.

Choi KH, Park JK, Son D, *et al.* 2016. Reactivation of endogenous genes and epigenetic remodeling are barriers for generating transgene-free induced pluripotent stem cells in pig. *PLoS One* **11**(6): 1–18.

Cooper DKC, Ekser B, Ramsoondar J, Phelps C, Ayares D. 2016. The role of genetically engineered pigs in xenotransplantation research. *J Pathol* **238**(2): 288–299.

Cuttle L, Kempf M, Phillips GE, *et al.* 2006. A porcine deep dermal partial thickness burn model with hypertrophic scarring. *Burns* **32**(7): 806–820.

Davidson JM. 1998. Animal models for wound repair. *Arch Dermatol Res* **290**(14): S1–S11.

Du Y, Kragh PM, Zhang Y, *et al.* 2007. Piglets born from handmade cloning, an innovative cloning method without micromanipulation. *Theriogenology* **68**(8): 1104–1110.

El-Hamamsy I, Stevens LM, Vanhoutte PM, Perrault LP. 2005. Injury of the coronary endothelium at implantation increases endothelial dysfunction and intimal hyperplasia after heart transplantation. *J Heart Lung Transplant* **24**(3): 251–258.

Ellegaard Göttingen Minipigs. 2010. Taking good care of Ellegaard Göttingen Minipigs. https://minipigs.dk/fileadmin/filer/pdf/Taking_good_care_of_Ellegaard_Goettingen_Minipigs_13.03.13.pdf (accessed October 2017).

Estrada J, Martens G, Li P, *et al.* 2015. Evaluation of human and non-human primate antibody binding to pig cells lacking GGTA1/CMAH/β4GalNT2 genes. *Xenotransplantation* **22**(3): 194–202.

Ezashi T, Telugu BPVL, Alexenko AP, *et al.* 2009. Derivation of induced pluripotent stem cells from pig somatic cells. *Proc Natl Acad Sci USA* **106**(27): 10993–10998.

Fan N, Chen J, Shang Z, *et al.* 2013. Piglets cloned from induced pluripotent stem cells. *Cell Res* **23**(1): 162–166.

Farre L, Rigau T, Mogas T, *et al.* 1999. Adenovirus-mediated introduction of DNA into pig sperm and offspring. *Mol Reprod Devel* **53**(2): 149–158.

Finn OJ. 2012. Immuno-oncology: understanding the function and dysfunction of the immune system in cancer. *Ann Oncol* **23**(Suppl. 8): 8–11.

Flisikowska T, Kind A, Schnieke A. 2014. Genetically modified pigs to model human diseases. *J Appl Genet* **55**(1): 53–64.

Flisikowska T, Merkl C, Landmann M, *et al.* 2012. A porcine model of familial adenomatous polyposis. *Gastroenterology* **143**(5): 1173–1177.

Gerner W, Käser T, Saalmüller A. 2009. Porcine T lymphocytes and NK cells: an update. *Devel Compar Immunol* **33**(3): 310–320.

Gertsenstein M, Lobe C, Nagy A. 2002. ES cell-mediated conditional transgenesis. *Meth Mol Biol* **185**(5): 285–307.

Gregory-Evans K, Weleber RG. 1997. An eye for an eye: new models of genetic ocular disease. *Nat Biotech* **15**: 947–948.

Hagemann S, Günther T, Dennemärker J, *et al.* 2004. The human cysteine protease cathepsin V can compensate for murine cathepsin L in mouse epidermis and hair follicles. *Euro J Cell Biol* **83**(11–12): 775–780.

Hai T, Teng F, Guo R, Li W, Zhou Q. 2014. One-step generation of knockout pigs by zygote injection of CRISPR/Cas system. *Cell Res* **24**(3): 372–375.

Hammer RE, Pursel VG, Rexroad CE, *et al.* 1985. Production of transgenic rabbits, sheep and pigs by microinjection. *Nature* **315**(6021): 680–683.

He Z, Shi X, Du B, *et al.* 2015. Highly efficient enrichment of porcine cells with deletions induced by CRISPR/Cas9 using dual fluorescence selection. *J Biotechnol* **214**: 69–74.

Holm IE, Alstrup AKO, Luo Y. 2016. Genetically modified pig models for neurodegenerative disorders. *J Pathol* **238**(2): 267–287.

Holm IE, West MJ. 1994. Hippocampus of the domestic pig: a stereological study of subdivisional volumes and neuron numbers. *Hippocampus* **4**(1): 115–125.

Hyodo A, Reger SI, Negami S, *et al.* 1995. Evaluation of a pressure sore model using monoplegic pigs. *Plast Reconstr Surg* **96**(2): 421–428.

Ingram DL, Legge KF. 1970. Variations in deep body temperature in the young unrestrained pig over the 24 hour period. *J Physiol* **210**(4): 989–998.

Ji G, Ruan W, Liu K, *et al.* 2013. Telomere reprogramming and maintenance in porcine iPS cells. *PLoS One* **8**(9): e74202.

Jinek M, Chylinski K, Fonfara I, *et al.* 2012. A programmable dual-RNA-guided DNA endonuclease in adaptive bacterial immunity. *Science* **337**(6069): 816–821.

Jørgensen FG, Hobolth A, Hornshøj H, *et al.* 2005. Comparative analysis of protein coding sequences from human, mouse and the domesticated pig. *BMC Biol* **3**: 2.

Kalloo AN, Singh VK, Jagannath SB, *et al.* 2004. Flexible transgastric peritoneoscopy: a novel approach to diagnostic and therapeutic interventions in the peritoneal cavity. *Gastrointest Endosc* **60**(1): 114–117.

Kang J-T, Cho B, Ryu J, *et al.* 2016a. Biallelic modification of IL2RG leads to severe combined immunodeficiency in pigs. *Reprod Biol Endocrinol* **14**(1): 74.

Kang J-T, Kwon DK, Park AR, *et al.* 2016b. Production of α-1,3-galactosyltransferase targeted pigs using transcription activator-like effector nuclease-mediated genome editing technology. *J Vet Sci* **17**(1): 89–96.

Kang, J-T, Ryu J, Cho B, *et al.* 2016c. Generation of RUNX3 knockout pigs using CRISPR/Cas9-mediated gene targeting. *Reprod Dom Animals* **51**(6): 970–978.

Kantsevoy SV, Jagannath SB, Niiyama H, *et al.* 2005. Endoscopic gastrojejunostomy with survival in a porcine model. *Gastrointest Endosc* **62**(2): 287–292.

Kim YG, Cha J, Chandrasegaran S. 1996. Hybrid restriction enzymes: zinc finger fusions to Fok I cleavage domain. *Proc Natl Acad Sci USA* **93**(3): 1156–1160.

Kraft TW, Allen D, Petters RM, *et al.* 2005. Altered light responses of single rod photoreceptors in transgenic pigs expressing P347 L or P347S rhodopsin. *Mol Vision* **11**: 1246–1256.

Kragh PM, Du Y, Corydon TJ, *et al.* 2005. Efficient in vitro production of porcine blastocysts by handmade cloning with a combined electrical and chemical activation. *Theriogenology* **64**(7): 1536–1545.

Kragh PM, Nielsen AL, Li J, *et al.* 2009. Hemizygous minipigs produced by random gene insertion and handmade cloning express the Alzheimer's disease-causing dominant mutation APPsw. *Transgenic Res* **18**: 545–558.

Kwon J, Namgoong S, Kim N. 2015. CRISPR/Cas9 as tool for functional study of genes involved in preimplantation embryo development. *PLoS One* **10**(3): e0120501.

Lai L, Kolber-Simonds D, Park K-W, *et al.* 2002. Production of alpha-1,3-galactosyltransferase knockout pigs by nuclear transfer cloning. *Science* **295**(5557): 1089–1092.

Lai S, Wei S, Zhao B, *et al.* 2016. Generation of knock-in pigs carrying Oct4-tdTomato Reporter through CRISPR/Cas9-mediated genome engineering. *PLoS One* **11**(1): e0146562.

Lavitrano M, Camaioni A, Fazio VM, *et al.* 1989. Sperm cells as vectors for introducing foreign DNA into eggs: genetic transformation of mice. *Cell* **57**(5): 717–723.

Lee K, Kwon D-N, Ezashi T, *et al.* 2014. Engraftment of human iPS cells and allogeneic porcine cells into pigs with inactivated RAG2 and accompanying severe combined immunodeficiency. *Proc Natl Acad Sci USA* **111**(20): 7260–7265.

Lee PY, Park SG, Kim EY, *et al.* 2010. Proteomic analysis of pancreata from mini-pigs treated with streptozotocin as type I diabetes models. *J Microbiol Biotechnol* **20**(4): 817–820.

Leeb T, Müller M. 2004. Comparative human-mouse-rat sequence analysis of the ICAM gene cluster on HSA 19p13.2 and a 185-kb porcine region from SSC 2q. *Gene* **343**(2): 239–244.

Lei S, Ryu J, Wen K, *et al.* 2016. Increased and prolonged human norovirus infection in RAG2/IL2RG deficient gnotobiotic pigs with severe combined immunodeficiency. *Sci Rep* **6**: 25222.

Li P, Estrada JL, Burlak C, *et al.* 2015. Efficient generation of genetically distinct pigs in a single pregnancy using multiplexed single-guide RNA and carbohydrate selection. *Xenotransplantation* **22**(1): 20–31.

Li X-Q, Donnelly DJ, Jensen TG (eds.). 2015. *Somatic Genome Manipulation*. New York, NY: Springer New York.

Liu Y, Li J, Løvendahl P, *et al.* 2015. In vitro manipulation techniques of porcine embryos: a meta-analysis related to transfers, pregnancies and piglets. *Reprod Fert Dev* **27**(3): 429.

Luo Y, Li J, Liu Y, *et al.* 2011. High efficiency of BRCA1 knockout using rAAV-mediated gene targeting: developing a pig model for breast cancer. *Transgen Res* **20**(5): 975–988.

Luo Y, Lin L, Bolund L, Jensen TG, Sørensen CB. 2012. Genetically modified pigs for biomedical research. *J Inherit Metabol Dis* **35**(4): 695–713.

Ma XC, Ning J, Ge GB, *et al.* 2011. Comparative metabolism of cinobufagin in liver microsomes from mouse, rat, dog, minipig, monkey, and human. *Drug Metabol Dispos* **39**(4): 675–682.

Marchant JN, Whittaker X, Broom DM. 2001. Vocalisations of the adult female domestic pig during a standard human approach test and their relationships with behavioural and heart rate measures. *Appl Animal Behav Sci* **72**(1): 23–39.

Merrifield CA, Lewis M, Claus SP, *et al.* 2011. A metabolic system-wide characterisation of the pig: a model for human physiology. *Mol BioSystems* **7**(9): 2577–2588.

Mestas J, Hughes CCW. 2004. Of mice and not men: differences between mouse and human immunology. *J Immunol* **172**(5): 2731–2738.

Mussolino C, Morbitzer R, Lütge F, *et al.* 2011. A novel TALE nuclease scaffold enables high genome editing activity in combination with low toxicity. *Nucleic Acids Res* **39**: 9283–9293.

Niemann H, Petersen B. 2016. The production of multi-transgenic pigs: update and perspectives for xenotransplantation. *Transgen Res* **25**(3): 361–374.

Nottle MB, Haskard KA, Verma PJ, *et al.* 2001. Effect of DNA concentration on transgenesis rates in mice and pigs. *Transgen Res* **10**(6): 523–531.

Parker N, Porter ACG. 2004. Identification of a novel gene family that includes the interferon-inducible human genes 6–16 and ISG12. *BMC Genom* **5**(1): 8.

Peng J, Wang Y, Jiang J, *et al.* 2015. Production of human albumin in pigs through CRISPR/Cas9-mediated knockin of human cDNA into swine albumin locus in the zygotes. *Sci Rep* **5**(April): 16705.

Petters RM, Alexander CA, Wells KD, *et al.* 1997. Genetically engineered large animal model for studying cone photoreceptor survival and degeneration in retinitis pigmentosa. *Nat Biotechnol* **15**(10): 965–970.

Pillay P, Manger PR. 2007. Order-specific quantitative patterns of cortical gyrification. *Euro J Neurosci* **25**(9): 2705–2712.

Pond WG, Boleman SL, Fiorotto ML, *et al.* 2000. Perinatal ontogeny of brain growth in the domestic pig. *Proc Soc Exp Biol Med* **223**(1): 102–108.

Puente X, Sánchez L, Christopher M, López-Otín C. 2003. Human and mouse proteases: a comparative genomic approach. *Nat Rev Genet* **4**(7): 544–558.

Ran FA, Hsu PD, Lin C-Y, *et al.* 2013. Double nicking by RNA-guided CRISPR Cas9 for enhanced genome editing specificity. *Cell* **154**(6): 1380–1389.

Rao S, Fujimura T, Matsunari H, *et al.* 2016. Efficient modification of the myostatin gene in porcine somatic cells and generation of knockout piglets. *Mol Reprod Dev* **83**(1): 61–70.

Reece WO. 2005. *Dukes' Physiology of Domestic Animals, 12th edn.* Ithaca, NY: Cornell University Press.

Renner S, Fehlings C, Herbach N, *et al.* 2010. Glucose intolerance and reduced proliferation of pancreatic β-cells in transgenic pigs with impaired glucose-dependent insulinotropic polypeptide function. *Diabetes* **59**(5): 1228–1238.

Rogers CS, Stoltz DA, Meyerholz DK, *et al.* 2008. Disruption of the CFTR gene produces a model of cystic fibrosis in newborn pigs. *Science* **321**(5897): 1837–1841.

Roser M. 2016. Life expectancy. *Our World in Data.* https://ourworldindata.org/life-expectancy/ (accessed November 2017).

Roy S, Biswas S, Khanna S, *et al.* 2009. Characterization of a preclinical model of chronic ischemic wound. *Physiol Genomics* **37**(3): 211–224.

Sakoda T, Kasahara N, Kedes L, Ohyanagi M. 2007. Lentiviral vector-mediated gene transfer to endotherial cells compared with adenoviral and retroviral vectors. *Prep Biochem Biotechnol* **37**(1): 1–11.

Seaton M, Hocking A, Gibran NS. 2015. Porcine models of cutaneous wound healing. *ILAR J* **56**(1): 127–138.

Shen B, Zhang W, Zhang J, *et al.* 2014. Efficient genome modification by CRISPR-Cas9 nickase with minimal off-target effects. *Nat Meth* **11**(4): 399–402.

Shim J, Al-Mashhadi RH, Sørensen CB, Bentzon JF. 2016. Large animal models of atherosclerosis – new tools for persistent problems in cardiovascular medicine. *J Pathol* **238**(2): 257–266.

Shinkai A, Komuta-Kunitomo M, Sato-Nakamura N, Anazawa H. 2002. N-terminal domain of eotaxin-3 is important for activation of CC chemokine receptor 3. *Prot Engineering Design Select* **15**(11): 923–929.

Sieren JC, Meyerholz DK, Wang XJ, *et al.* 2014. Development and translational imaging of a TP53 porcine tumorigenesis model. *J Clin Invest* **124**(9): 4052–4066.

Stubhan M, Markert M, Mayer K, *et al.* 2008. Evaluation of cardiovascular and ECG parameters in the normal, freely moving Göttingen minipig. *J Pharmacol Toxicol Methods* **57**(3): 202–211.

Swindle MM, Smith AC, Laber-Laird K, Dungan L. 1994. Swine in biomedical research: management and models. *ILAR J* **36**(1): 1–5.

Talan M. 1984. Body temperature of C57BL/6 J mice with age. *Exp Gerontol* **19**(1): 25–29.

Tan W, Carlson DF, Lancto CA, *et al.* 2013. Efficient nonmeiotic allele introgression in livestock using custom endonucleases. *Proc Nat Acad Sci USA* **110**(41): 16526–16531.

Thibault KL, Margulies SS. 1998. Age-dependent material properties of the porcine cerebrum: effect on pediatric inertial head injury criteria. *J Biomech* **31**(12): 1119–1126.

Timmers L, Henriques JPS, de Kleijn DPV, *et al.* 2009. Exenatide reduces infarct size and improves cardiac function in a porcine model of ischemia and reperfusion injury. *J Am Coll Cardiol* **53**(6): 501–510.

Tomita S, Mickle DAG, Weisel RD, *et al.* 2002. Improved heart function with myogenesis and angiogenesis after autologous porcine bone marrow stromal cell transplantation. *J Thorac Cardiovasc Surg* **123**(6): 1132–1140.

Uechi M, Asai K, Osaka M, *et al.* 1998. Depressed heart rate variability and arterial baroreflex in conscious transgenic mice with overexpression of cardiac Gsalpha. *Circul Res* **82**(4): 416–423.

Vajta G, Callesen H. 2012. Establishment of an efficient somatic cell nuclear transfer system for production of transgenic pigs. *Theriogenology* **77**(7): 1263–1274.

Vajta G, Kragh PM, Mtango NR, Callesen H. 2005. Hand-made cloning approach: potentials and limitations. *Reprod Fertil Dev* **17**(1–2): 97–112.

Vaughan AN, Malde P, Rogers NJ, *et al.* 2000. Porcine CTLA4-Ig lacks a MYPPPY motif, binds inefficiently to human B7 and specifically suppresses human CD4+ T cell responses costimulated by pig but not human B7. *J Immunol* **165**(6): 3175–3181.

Velander P, Theopold C, Hirsch T, *et al.* 2008. Impaired wound healing in an acute diabetic pig model and the effects of local hyperglycemia. *Wound Repair Regen* **16**(2): 288–293.

Walpole SC, Prieto-Merino D, Edwards P, *et al.* 2012. The weight of nations: an estimation of adult human biomass. *BMC Publ Health* **12**(1): 439.

Wang D, Mou H, Li S, *et al.* 2015. Adenovirus-mediated somatic genome editing of Pten by CRISPR/Cas9 in mouse liver in spite of Cas9-specific immune responses. *Human Gene Ther* **26**(7): 432–442.

Wang X, Cao C, Huang J, *et al.* 2016. One-step generation of triple gene-targeted pigs using CRISPR/Cas9 system. *Sci Rep* **6**: 20620.

Wang XQ, Kempf M, Liu PY, *et al.* 2008. Conservative surgical debridement as a burn treatment: supporting evidence from a porcine burn model. *Wound Repair Regen* **16**(6): 774–783.

Wang Y, Du Y, Shen B, *et al.* 2015. Efficient generation of gene-modified pigs via injection of zygote with Cas9/sgRNA. *Sci Rep* **5**: 8256.

Weaver BK, Bohn E, Judd BA, Gil MP, Schreiber RD. 2007. ABIN-3: a molecular basis for species divergence in interleukin-10-induced anti-inflammatory actions. *Mol Cell Biol* **27**(13): 4603–4616.

Wernersson R, Schierup MH, Jørgensen FG, *et al.* 2005. Pigs in sequence space: a 0.66X coverage pig genome survey based on shotgun sequencing. *BMC Genom* **6**: 70.

Whitelaw CBA, Lillico SG, King T. 2008. Production of transgenic farm animals by viral vector-mediated gene transfer. *Reprod Domest Animals* **43**(Suppl. 2): 355–358.

Whitworth KM, Lee K, Benne JA, *et al.* 2014. Use of the CRISPR/Cas9 system to produce genetically engineered pigs from in vitro-derived oocytes and embryos 1. *Biol Reprod* **91**: 1–13.

Wilmut I, Schnieke AE, McWhir J, Kind AJ, Campbell KH. 1997. Viable offspring derived from fetal and adult mammalian cells. *Nature* **385**(6619): 810–813.

Wolf E, Braun-Reichhart C, Streckel E, Renner S. 2014. Genetically engineered pig models for diabetes research. *Transgen Res* **23**(1): 27–38.

Woo JS, Kim W, Ha S., *et al.* 2013. Cardioprotective effects of exenatide in patients with ST-segment-elevation myocardial infarction undergoing primary percutaneous coronary intervention: results of exenatide myocardial protection in revascularization study. *Arterioscler Thrombo Vasc Biol* **33**(9): 2252–2260.

Xie B, Wang JJ, Liu S, *et al.* 2014. Positive correlation between the efficiency of induced pluripotent stem cells and the development rate of nuclear transfer embryos when the same porcine embryonic fibroblast lines are used as donor cells. *Cell Reprog* **16**(3): 206–214.

Xin J, Yang H, Fan N, *et al.* 2013. Highly efficient generation of GGTA1 biallelic knockout inbred mini-pigs with TALENs. *PLoS One* **8**(12): 1–9.

Yang D, Wang CE, Zhao B, *et al.* 2010. Expression of Huntington's disease protein results in apoptotic neurons in the brains of cloned transgenic pigs. *Human Mol Genet* **19**(20): 3983–3994.

Yang L, Güell M, Niu D, *et al.* 2015. Genome-wide inactivation of porcine endogenous retroviruses (PERVs). *Science* **350**(6264): 1101–1104.

Yao J, Huang J, Zhao J. 2016. Genome editing revolutionize the creation of genetically modified pigs for modeling human diseases. *Human Genet* **135**(9): 1093–1105.

Zerhouni EA, Parish DM, Rogers WJ, Yang A, Shapiro EP. 1988. Human heart: tagging with MR imaging – a method for noninvasive assessment of myocardial motion. *Radiology* **169**(1): 59–63.

Zhou X, Wang L, Du Y, *et al.* 2016. Efficient generation of gene-modified pigs harboring precise orthologous human mutation via CRISPR/Cas9-induced homology-directed repair in zygotes. *Human Mutat* **37**(1): 110–118.

Zhou X, Xin J, Fan N, *et al.* 2015. Generation of CRISPR/Cas9-mediated gene-targeted pigs via somatic cell nuclear transfer. *Cell Mol Life Sci* **72**(6): 1175–1184.

Zhu KQ, Engrav LH, Gibran NS, *et al.* 2003. The female, red Duroc pig as an animal model of hypertrophic scarring and the potential role of the cones of skin. *Burns* **29**(7): 649–664.

9 Gene Editing to Create Agricultural and Biomedical Swine Models

Bethany K. Redel, Chad O'Gorman, Randall S. Prather and Kristin M. Whitworth

9.1 Introduction

With an ever-growing world population, swine research in the areas of genomics and genome editing are becoming increasingly important and advanced, with a precise focus on improving production by preventing disease. Pigs are an important agricultural commodity across the world. There are an estimated 1 billion pigs in production agriculture (Robinson, 2014), as pork is the most highly consumed meat protein after fish. China is the largest producer of pork in the world, followed by the European Union and the United States. In 2012, the US hog and pig industry had sales of $22.5 billion, accounting for 6 percent of total US agriculture sales (www.agcensus.usda.gov/Publications/2012/Online_Resources/Highlights/Hog_and_Pig_Farming/). In addition to use as a protein source, new pig models are being developed for biomedical research due to the similarities between pigs and humans with regard to organ size, anatomy, physiology, metabolism, neurobiology and genome. Genome editing via site-directed nucleases, including zinc-finger nucleases (ZFNs), the transcription activator-like effector nucleases (TALENs) or clustered regularly interspaced short palindromic repeats (CRISPR)/Cas9 has resulted in an explosion of new and exciting swine models for use in both agriculture and biomedicine. This chapter will highlight a few of these gene editing technologies including ZFNs and TALENs. However, the CRISPR/Cas9 system is extremely efficient and will be the main focus of the chapter.

9.1.1 What are Site-directed Nucleases?

Site-directed nucleases, sometimes called meganucleases, are naturally occurring endo-deoxyribonucleases that have a large recognition site within a genome. Since site-directed nucleases have a high degree of efficiency and specificity, genetic engineers repurposed zinc-finger proteins, TAL-effector nuclease proteins and the CRISPR/Cas9 system for use in DNA editing of mammalian cells (Figure 9.1). Zinc-finger nucleases (ZFNs) were one of the first to be repurposed. ZFNs consist of two modular domains, a DNA-binding domain and a DNA cleavage domain (Porteus, 2008). ZFNs are designed to recognize a unique DNA sequence, which fuses to the nuclear domain resulting in a double-strand break (Urnov *et al.*, 2005). Another group of successful DNA editors are the TALENs. TALENs are also endonucleases that consist of a TAL effector DNA-binding domain (TALE) fused to a DNA cleavage domain. Both ZFNs and TALENs utilize the nuclease FokI to catalyze double-strand breaks in the DNA. TALENs bind DNA via a central domain of tandem repeats and are designed to target specific genes by changing the type of amino acid repeats to recognize specific DNA sequences (Boch *et al.*, 2009; Moscou and Bogdanove, 2009). Clustered regularly interspaced short palindromic repeats (CRISPR) use an RNA-guided nuclease Cas9 to produce double-stranded breaks in the DNA (Hwang *et al.*, 2013; Wiedenheft *et al.*, 2012). A single guide RNA (gRNA) is constructed that contains a 20 base pair region that is complementary to the target sequence and is adjacent to a protospacer motif (PAM, NGG sequence) in the genome. The gRNA can be synthesized or cloned into a plasmid that contains the CRISPR machinery (www.addgene.org/crispr/zhang/). The gRNA/protein complex can then scan the genome and Cas9 cleaves both strands of the complementary DNA (Terns and Terns, 2011; Wang *et al.*, 2011).

A successful edit from all three site-directed nuclease relies on an error in the organism's DNA repair mechanism either by non-homologous end joining repair (NHEJ) or homologous recombination (HR). A mistake in repair results in a potential insertion or deletion of at least one nucleotide (indel) which can disrupt the reading frame. If a donor DNA in the form of a targeting vector or oligonucleotide is made available, it can be used as a template for repair and result in a specific edit by HR (Figure 9.1). More detailed descriptions of how each of these site-directed nucleases function and how they are being used in DNA editing of the pig genome will be found throughout the chapter.

9.2 Zinc-finger Nucleases

Zinc-finger nucleases are a class of engineered nucleases capable of inducing site-specific double-stranded DNA breaks. This is achieved by the linkage of a restriction endonuclease cleavage domain to a sequence-specific binding domain consisting of multiple zinc-finger proteins. Due to the specificity of the zinc-finger proteins and their modular nature, several can be combined to form a peptide with very high sequence specificity.

The binding domain is composed of a series Cys_2His_2, which form(Tyr,Phe)-Xaa-Cys-Xaa2-4-Cys-Xaa3-Phe-Xaa5-Leu-Xaa2-His-Xaa3-5-His, usually in tandem arrays (Kim and Chandrasegaran, 1994). Each binds a $zinc^{2+}$ ion to form a structural domain referred to as a zinc finger. These binding proteins bind to the DNA at specific three-base sequences by inserting an α-helix into the major groove of the DNA double helix (Kim *et al.*, 1996). Due to differences in protein sequence and the resulting structure, the zinc fingers recognize and bind different three-base sequences; as well as having the ability to bind independently, each can be used modularly and linked into a peptide capable of binding DNA at a specific

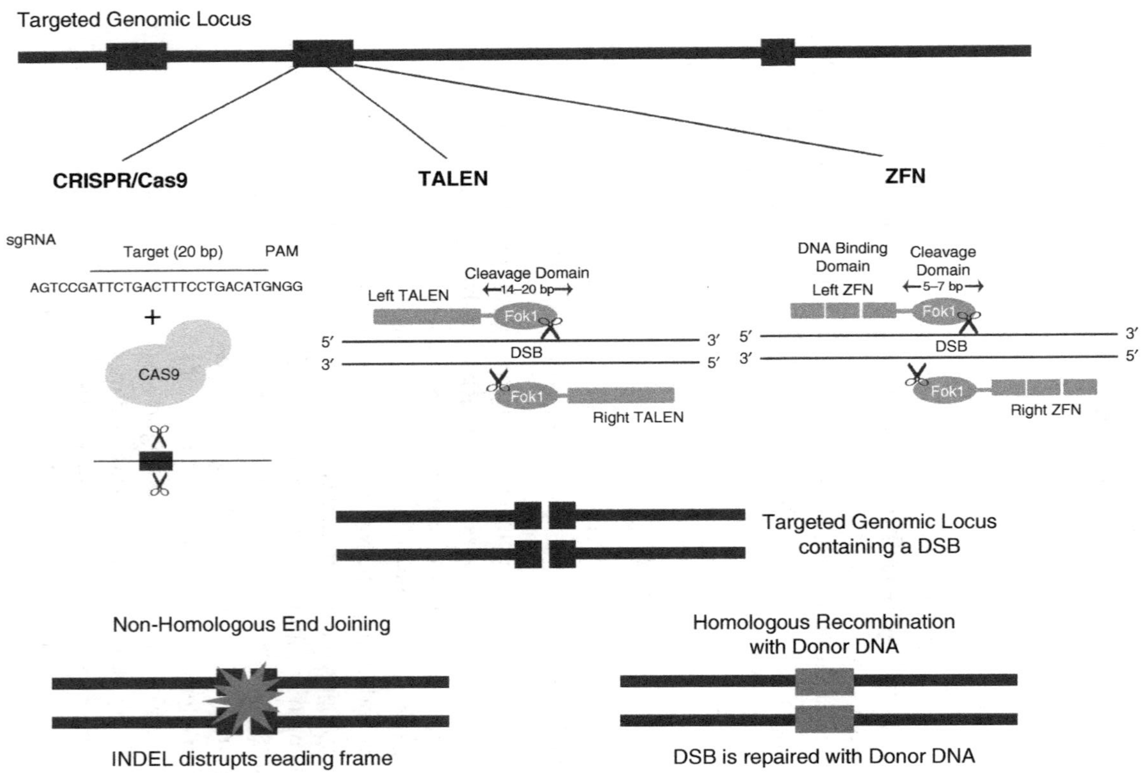

Figure 9.1 Three major groups of site-directed nucleases have been used to successfully create DNA-edited pig models: CRISPR/Cas9, TALENs and ZFN. All three site-directed nucleases use enzymes that result in a double-strand break (DSB) in the DNA and subsequent errors in DNA repair by non-homologous end joining or homologous recombination with a donor DNA (repair template).

predetermined site. One common assembly process consists of combining three separate zinc fingers that can each recognize a three base pair DNA sequence to generate a three-finger array capable of binding to a specific nine base pair target site.

The cleavage domain of each ZFN consists of the cleavage domain of a Type II restriction endonuclease, most commonly FokI. FokI, isolated from *Flavobacterium okeanokoites*, consists of an N-terminal binding domain and a C-terminal domain with non-specific DNA cleavage activity resulting in cleavage of DNA outside of the recognition site: 5′-GGATG$(N)_{9/13}$. The C-terminal cleavage domain can be linked with specific zinc-finger binding proteins to engineer hybrid endonucleases with high sequence specificity as desired by the user. By nature, it is necessary for FokI cleavage domains to dimerize in order to cleave. This, in turn makes it necessary to use ZFNs in pairs. In order to allow the two cleavage domains to dimerize, it is necessary for the two individual ZFNs to bind opposite strands of DNA with their C-terminus a certain distance apart. Commonly, linker sequences between the ZFN-binding domain and cleavage domain require each binding site to be separated by five to seven base pairs (Cathomen and Joung, 2008).

9.3 TALENs

The TALEN system of gene editing originated with the identification of TAL effector (TALE) proteins secreted by *Xanthomonas*, a species of bacteria that infects various crop plant species. The TALE proteins are capable of binding DNA as well as inducing expression of target genes. TALE proteins are composed of a central domain responsible for DNA binding, a nuclear localization signal and a domain that activates the target gene transcription (Schornack *et al.*, 2006).

The DNA-binding domain is composed of monomers, tandem repeats of 34 amino acid residues, and each of them binds one nucleotide in the target nucleotide sequence. In each monomer, two amino acids located at positions 12 and 13 are highly variable (repeat variable diresidue, RVD), these RVDs are responsible for the recognition of a specific nucleotide. Due to the simplicity of TALE proteins (one monomer – one nucleotide), as well as their high sequence specificity, TALE proteins were used to engineer a new class of nucleases (Christian *et al.*, 2010). This was accomplished by inserting the sequence encoding the DNA-binding domain of TALE into a plasmid vector used for ZFN, which resulted in the construction of an engineered nuclease with the DNA-binding domain from TALE combined with a FokI cleavage domain similar to that used for ZFN. TALENs work as pairs and their binding sites must be selected so that they are located on opposite DNA strands and are separated by a spacer sequence of 12–25 base pairs. Once in the nucleus, the TALEN nucleases bind to target sequences, and the FokI domains located at the C-terminus of a chimeric protein dimerize to cut the DNA within the spacer sequence causing a double-strand break.

9.4 CRISPR (Clustered Regulatory Interspaced Short Palindromic Repeats)/Cas

The CRISPR/Cas system is essentially a form of prokaryotic acquired immunity providing resistance to foreign genetic elements such as those present within plasmids and phages (Barrangou *et al.*, 2007; Marraffini and Sontheimer, 2008; Redman *et al.*, 2016). In the endogenous system, short regions of the unique DNA (spacers) are separated from each other by short palindromic repeats. Sequencing of bacterial genomes revealed that the spacer DNA is often homologous to the DNA of many phages and plasmids (Bolotin *et al.*, 2005; Mojica *et al.*, 2005; Pourcel *et al.*, 2005). These CRISPR elements are arranged in close proximity to *Cas* genes, the proteins of which exhibit helicase and nuclease activity. It was demonstrated that bacterial cells bearing a spacer within the CRISPR locus that is complementary to a bacteriophage DNA fragment become resistant to that phage (Marraffini and Sontheimer, 2008).

In contrast to ZFN and TALEN proteins, recognition by the CRISPR/Cas system is carried out via the complementary interaction between a non-coding RNA and the target DNA. In this system, a short gRNA is designed, consisting of a small piece of RNA sequence (about 20 bases long) designed to be complementary to the target DNA sequence, and located within a longer RNA scaffold. The RNA scaffold containing the gRNA binds to the complementary DNA. The Cas9 protein then forms a complex with the RNA scaffold along with the bound DNA. Once complexed, the Cas9 protein then cleaves both strands of DNA causing a double-strand break.

9.5 Methods for Introducing Site-directed Nucleases into Mammalian Cells and Embryos

Before the optimization of gene editing with site-directed nucleases, transgenic pigs were produced by pronuclear injection (Hammer *et al.*, 1985), retroviral transduction (Cabot *et al.*, 2001), intracytoplasmic sperm injection (ICSI)-mediated sperm transfer (Kurome *et al.*, 2006) and modification of somatic cell lines followed by somatic cell nuclear transfer (SCNT) (Dai *et al.*, 2002; Lai and Prather, 2002; Lai *et al.*, 2002a; Park *et al.*, 2001). Modification of somatic cell lines followed by SCNT continues to be an effective method for creating specific DNA edits and now is more efficient, with the introduction of site-directed nucleases providing double-stranded breaks in the DNA template. SCNT is also advantageous because the cell line can be fully characterized before SCNT, and mosaicism observed in other techniques can be prevented. ZFNs, TALENs and CRISPRs have all been used to modify pig cell lines and create edited offspring (Figure 9.2; Carlson *et al.*, 2012; Choi *et al.*, 2016; Whitworth *et al.*, 2014; Whyte *et al.*, 2011; Xin *et al.*, 2013). Methodologies for transfection of porcine cell lines have been described in detail previously (Ross *et al.*, 2010; Whitworth *et al.*, 2014). Briefly, somatic cells, most commonly fibroblasts derived from day 35 fetuses, are transfected/transduced with an exogenous DNA construct containing the site-directed nucleases and/or donor DNA of choice by a variety of methods, including electroporation (Dai *et al.*, 2002; Ramsoondar *et al.*, 2003; Watanabe *et al.*, 2005), lipid-

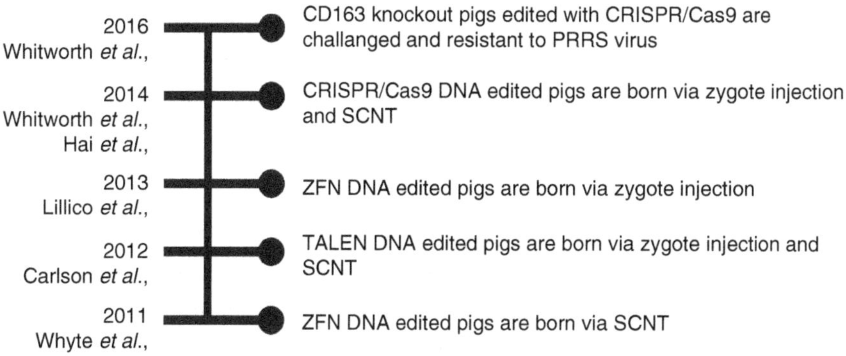

Figure 9.2 The use of site-directed nucleases has quickly been incorporated into somatic cell nuclear transfer and zygote injection protocols to create pig models.

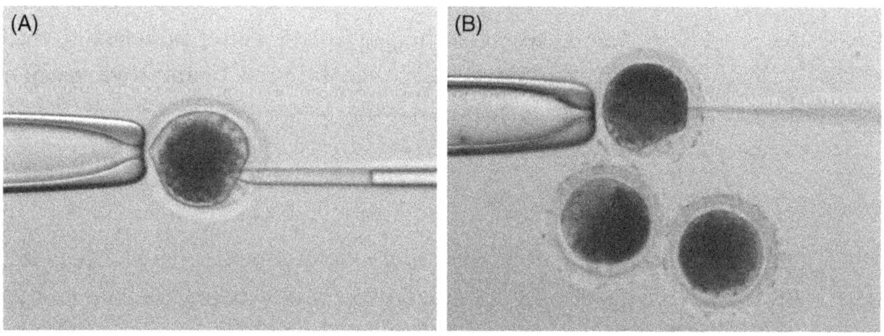

Figure 9.3 DNA editing of a somatic cell line followed by somatic cell nuclear transfer (A) and direct zygote injection of site-directed nucleases (B) results in a high rate of DNA editing in pigs. Photo credit: Joshua Benne and Stephanie Murphy.

based delivery (Hyun *et al.*, 2003) and viral delivery (Lai *et al.*, 2002b; Rogers *et al.*, 2008). After transfection/transduction, fetal fibroblasts are propagated and selected based upon genotype evaluation. Occasionally the coding sequence for a selectable marker such as neomycin or a visual marker such as enhanced green fluorescent protein (EGFP) is included to increase efficiency. After an edited or modified cell line has been characterized, piglets can be created by SCNT (Figure 9.3A).

Direct injection of mRNA encoding ZFN or TALEN (Carlson *et al.*, 2012; Lillico *et al.*, 2013) or CRISPR gRNA with Cas9 (Hai *et al.*, 2014; Whitworth *et al.*, 2014) into the cytoplasm of *in vitro* fertilized porcine zygotes is a new and powerful method that removes the reprogramming issues observed with SCNT. Optimized methods for this technique are described in the above references. Briefly, mRNA is transcribed *in vitro* from a ZFN, TALEN or CRISPR template specific to the targeted gene. mRNA encoding ZFNs, TALENs or CRISPR guide RNA with Cas9 mRNA is then microinjected into *in vitro* fertilized presumed pig zygotes (Figure 9.3B). The pig embryos are then cultured to the morula or blastocyst stage at which stage they are either genotyped for DNA edits or transferred into recipient gilts for continuation of pregnancy. Most high-impact DNA editing labs using the CRISPR/Cas9 system are reporting efficiencies of greater than 80% of resulting piglets that have both mono-allelic and biallelic DNA edits (Wang *et al.*, 2016; Whitworth *et al.*, 2014).

9.6 Gene Editing Impacts on Animal Agriculture

There is great potential to improve animal agriculture by utilizing gene editing methods. Recently pigs were created with the CRISPR/Cas9 system that contain a simple biallelic edit in the *CD163* gene (Figure 9.4). The resulting pigs were resistant to both Type 1 and Type 2 porcine reproductive and respiratory syndrome (PRRS) viral infections (Wells *et al.*, 2017; Whitworth *et al.*, 2016). Dairy cows were created with a TALEN-induced edit resulting in polled "hornless" dairy cattle (Carlson *et al.*, 2016). Swine and cattle models are also being developed to test potential resistance to African swine fever (ASF; Lillico *et al.*, 2016) and "mad cow disease" (Bevacqua *et al.*, 2016), respectively. In addition to disease resistance, agricultural models are being created to help scientists better understand viral infectivity. Such valuable information can then be used to improve vaccine development and animal treatment options.

9.7 Agricultural Swine Models

Utilizing transgenic technologies in agriculture is not a new concept. There are many readily available genetically modified plant products on the market, but utilizing the same technologies in livestock has proven to be a complicated process. One of the reasons is the lack of reliable and repeatable embryonic stem

Figure 9.4 *CD163* knockout pigs edited with CRISPR/Cas9 are challenged and resistant to PRRS virus. Photo credit: Melissa Samuel. (A black-and-white version of this figure will appear in some formats. For the color version, please refer to the plate section.)

cell lines that are commonly used to make transgenic mice. Nonetheless, creating agricultural swine models is an important tool for providing protein to an ever-growing world population. Meat products from traditional livestock species are low in omega-3 (n-3) fatty acids which have been shown to be beneficial to cardiovascular health. Pigs have been created with "healthier" ratios of omega-3 fatty acids such as linoleic acid (Lai *et al.*, 2006; Saeki *et al.*, 2004). Due to the reluctance of federal regulating agencies to approve consumption of transgenic animals and the lack of societal acceptance of consuming transgenic meat, beneficial models improving meat quality have not yet moved forward. As the use and understanding of site-directed nucleases increased, simple DNA edits in livestock species could be created without the addition of foreign DNA into the animal. A single base pair change in the right location could alter the reading frame of a targeted gene and knockout its function. Three agricultural models will be discussed in the upcoming section, including two disease models for the study of PRRS and ASF, as well as a double-muscling myostatin model (MSTN).

9.8 Disease-resistance Models

PRRS and ASF are two devastating pig diseases that have remained elusive to the development of a protective vaccine or treatment regimen. PRRS causes reproductive failure in both female and male pigs and respiratory tract illness in younger pigs. Porcine reproductive and respiratory syndrome virus (PRRSV) was first reported in the United States in 1987 (Keffaber, 1989) and later in Europe in 1990. PRRS is the most economically important swine disease in North America, Europe and Asia, costing North American producers over $600 million annually (Holtkamp *et al.*, 2013). ASF virus (ASFV) causes severe hemorrhagic fever with a high mortality rate in pigs. ASFV is endemic throughout Sub-Saharan Africa and the island of Sardinia, but in 2007, the virus produced outbreaks in the Republic of Georgia and subsequent outbreaks in the Russian Federation and Eastern Europe (Rowland *et al.*, 2008; Sanchez-Vizcaino *et al.*, 2013). When PRRSV infects large swine production systems, the only way to eradicate the virus is by depopulation of the herd, sterilization and repopulation. This has a huge economic and emotional impact on farmers and business owners. The impact of ASF can be equally devastating in developing countries where sometimes entire families groups are dependent on the meat from a few pigs that can die very quickly when infected by ASFV.

9.8.1 PRRS-resistant Pigs

Initially, PRRS infectivity models were analyzed to select candidates for editing. The widely accepted and most scientifically tested model (Van Breedam *et al.*, 2010) involves initial loose association of PRRSV with heparan sulfate. It is then proposed that the virus binds the N-terminal region of sialoadhesion (SIGLEC1, CD169) whereby it is internalized into a clathrin-coated endosome (Kreutz and Ackermann, 1996; Mercer *et al.*, 2010). In the low pH environment of the endosome, CD163 can uncoat the viral genome so that the genome can be released into the cytoplasm. Heparan sulfate is a proteoglycan that regulates a multitude of biological functions and was therefore not selected as a molecule for removal to prevent infectivity. SIGLEC1 and CD163 appeared to be the next logical choices.

To determine the role of SIGLEC1 in PRRSV infectivity, scientists utilized traditional homologous recombination techniques to knock out the *SIGLEC1* gene (*SIGLEC1*$^{-/-}$) in somatic cells and then perform SCNT to create the pigs (Prather *et al.*, 2013). The *SIGLEC1*$^{-/-}$ knockout pigs were innoculated with a North American strain of PRRSV, which resulted in productive infection. Knocking out *SIGLEC1* did not result in PRRS-resistant pigs so candidate number two, *CD163*, was chosen for editing.

CD163 is a member of the scavenger receptor cysteine-rich (SRCR) superfamily. Exogenous expression of CD163 in non-permissive cells renders them permissive (Calvert *et al.*, 2007), indicating a positive role of CD163 for infectivity of this virus. Van Gorp *et al.* (2010) further dissected the CD163 protein with a series of deletion constructs and showed that SRCR domain 5 appears to be solely responsible for viral binding (Van Gorp *et al.*, 2010). The CRISPR/Cas9 system was used to knock out the *CD163* gene by both methods described above, i.e. by editing of the fetal fibroblast cell line followed by SCNT, and direct zygote injection (Whitworth *et al.*, 2014). The resulting pigs appeared healthy, grew normally and were fertile (Whitworth *et al.*, 2016). When *CD163*$^{-/-}$ pigs were challenged with either North American or European strains of PRRSV, there were no clinical signs of infection, as measured by fever, coughing, viremia, antibody response or lung pathology (Wells *et al.*, 2017; Whitworth *et al.*, 2016). *In vitro* experiments that infected alveolar macrophages collected from *CD163*$^{-/-}$ pigs with multiple isoforms of PRRSV from both North American and European strains further confirmed resistance to infection. This was the first report of an agricultural swine model with no foreign DNA, e.g. a transgene, that showed absolute viral resistance.

Macrophages are the primary target cells for ASFV infectivity as well. Models illustrating how ASFV enters the macrophage are less defined for ASF compared to PRRS. Nonetheless, there are two candidates that are being pursued for resistance by gene editing: Sus scrofa RELA proto-oncogene, NF-kB subunit (*RELA*) and *CD163*. Domestic pigs are highly sensitive to ASF when compared to African pigs such as warthogs. Genomic analysis showed that, unlike domestic pigs, warthogs contain polymorphic sequence variation within the transactivation domains of *RELA* (Palgrave *et al.*, 2011). Researchers designed both ZFNs and TALENs to target the *RELA* locus. In one experiment, piglets were made with biallelic edits to the *RELA* coding sequence (Lillico *et al.*, 2013). In a follow-up experiment, ZFNs were used to make double-strand breaks and the template was repaired with a donor DNA that contained 251 base pairs with five base pair substitutions resulting in domestic pigs with a warthog *RELA* locus (Lillico *et al.*, 2016). Challenge experiments with this model are yet to be published. Other evidence points to CD163 as being the gatekeeper to the ASF virus. Like PRRSV, macrophages that expressed CD163 are permissive to ASF viral infection, but CD163 negative cells are not permissive (Sanchez-Torres *et al.*, 2003). Additionally, CD163 antibodies can block ASFV *in vitro*. *CD163*$^{-/-}$ pigs created for PRRSV resistance were also challenged with ASFV, but unfortunately were not resistant (Popescu *et al.*, 2016).

9.9 Utilizing Site-directed Nucleases to Improve Meat Quality

Myostatin (MSTN) is a member of the transforming growth factor-β superfamily and acts as a negative regulator of muscle growth in animals. Cattle have a naturally occurring *MSTN* knockout allele that results in the double-muscling phenotype observed at a high prevalence in certain breeds of cattle (Grobet *et al.*, 1997; Kambadur *et al.*, 1997; McPherron and Lee, 1997). *MSTN* knockout pigs were created by using the CRISPR/Cas9 system (Wang *et al.*, 2015). $MSTN^{-/-}$ pigs had an increased birth weight, with some of the pigs having an obvious intermuscular boundary with more myofiber nuclei compared to wild-type controls. Another group of $MSTN^{-/-}$ pigs were created by using the CRISPR/Cas9 system followed by removal of the selectable marker (Bi *et al.*, 2016). Muscle from these pigs also had an increased number of myofiber nuclei, larger longissimus muscle and decreased backfat thickness. The meat quality models are still in the development phases, but this does illustrate that site-directed nucleases can efficiently be used to create models with potential improvements to muscle quality.

9.10 Pig Biomedical Models of Human Disease Created by CRISPR/Cas9

Since the discovery and implementation of the CRISPR/Cas9 system, the efficiency of creating animals has accelerated and the time that it takes to produce genome-edited pigs has been reduced. There have been numerous pig models of human disease created by using this technology (Table 9.1); however, only three types of pig models used for human disease will be reviewed here.

9.11 Neurodegenerative Disorders

Neurodegenerative disorders affect hundreds of millions of people worldwide. Alzheimer's disease, Huntington's disease, amyotrophic lateral sclerosis, spinal muscle atrophy, Parkinson's disease and ataxia-telangiectasia pig models have been created and extensively reviewed by Holm *et al.* (2016). The majority of these pig models were created by using traditional transgenic approaches, either by transgene insertion or by gene disruption by vector-mediated homologous recombination. The second most common neurological disorder is Parkinson's disease (PD). Recently, CRISPR/Cas9 technology has been utilized to create two pig models of Parkinson's disease (Wang *et al.*, 2016; Zhou *et al.*, 2015). Parkinson's disease is a chronic and progressive neurodegenerative disorder that is characterized by the presence of motor symptoms such as bradykinesia, rest tremor, muscle rigidity and postural instabilities (World Health Organization; www.who.int). Pathologically, this neurodegenerative disorder is characterized by the loss of dopaminergic neurons whose cell bodies are located in the substantia nigra as well as the formation of cytoplasmic protein inclusions known as Lewy bodies (Dauer and Przedborski, 2003). The majority of these cases are sporadic; however, there are a few genes that have been shown to be responsible for hereditary

Table 9.1 CRISPR/Cas9-generated pig models

Gene	Method	Model	Reference
APC	NHEJ	Colon cancer	Tan *et al.*, 2013
DAZL	NHEJ	Spermatogenesis	Tan *et al.*, 2013
PINK1, PARK2	NHEJ	Parkinson's disease	Zhou *et al.*, 2015
PINK1, PARK2, PARK7	NHEJ	Parkinson's disease	Wang *et al.*, 2016
NPC1l1	NHEJ	Cardiovascular and metabolic diseases	Wang *et al.*, 2016
SOX10	HR	Warrdenburg disease	Zhou *et al.*, 2016
IgM	NHEJ	Production of humanized polyclonal antibodies	Chen *et al.*, 2015
ALB	HR	Human serum albumin production	Peng *et al.*, 2015
GGTA1, CMAH, CD55	NHEJ+HR	Xenotransplantation	Butler *et al.*, 2016a
GGTA1, CMAH, B4GALNT2	NHEJ	Xenotransplantation	Estrada *et al.*, 2015
GGTA1, CMAH, A3GALT2	NHEJ	Xenotransplantation	Li *et al.*, 2015
GGTA1	NHEJ	Xenotransplantation	Li *et al.*, 2015
GGTA1, CMAH	NHEJ	Xenotransplantation	Butler *et al.*, 2016b; Li *et al.*, 2015
GGTA1, A3GALT2	NHEJ	Xenotransplantation	Li *et al.*, 2015
SLA-1,-2,-3	NHEJ	Xenotransplantation	Reyes *et al.*, 2014
TYR	NHEJ	Oculocutaneous albinism type 1	Zhou *et al.*, 2015
vWD	NHEJ	Von Willebrand disease	Hai *et al.*, 2014

forms of PD (Valente *et al.*, 2004). One gene implicated in early onset PD is *PARK2*, which encodes a protein called parkin and is a component of the E3 ubiquitin ligase complex. *PINK1*, a second gene implicated in PD encodes PTEN-induced putative kinase 1. PINK1 accumulates on damaged mitochondria activating parkin's E3 ubiquitin ligase to trigger selective autophagy (Pickrell and Youle, 2015). Loss-of-function mutations in either of these two genes mediates autosomal recessive PD. Zhou and co-workers (2015) co-transfected porcine fetal fibroblasts with Cas9 and gRNA for *PINK1* and *PARK2*. Individual colonies were picked, cultured and assayed for mutations of each gene. Cell colonies with the desired mutations were used for SCNT. *PINK1* and *PARK2* double-knockout piglets had undetectable levels of PINK1 and parkin in brain cortex sections. At the time of publication, the mutant pigs were seven months and did not show any typical symptoms of PD.

A second PD pig model was created by utilizing a simultaneous approach of modifying three genes that have been implicated in hereditary early onset PD. This model incorporated a third gene, *PARK7*, which has been shown to be responsible for neuronal death due to oxidative damage (Bonifati *et al.*, 2003). These three genes, *PARK7*, *PARK2* and *PINK1*, may together exert synergistic effects on the development of autosomal recessive PD in pigs (Wang *et al.*, 2016). The CRISPR/Cas9 system was used to create piglets with biallelic modifications in all three genes. Fibroblasts from these triple-knockout pigs were depleted of PARK2 protein, which confirmed targeting of both PARK2 alleles. These piglets were born in 2016 and phenotypic results are yet to be published. Together, these authors demonstrate the CRISPR/Cas9 system is highly efficient at creating biallelic PD pig models.

9.12 Xenotransplantation Models

Currently there are 120064 people needing a life-saving organ transplant and only 9129 organ transplants were performed this year (September 2016: https://optn.transplant.hrsa.gov). The dramatic shortage in the number of donors to recipients calls us to find an alternative source for organ donors. Pigs are an attractive source as their size and physiology is comparable to that of humans. Pigs contain a galactose-α1,3-galactose (α1,3Gal) epitope on the cell surface that is not present in humans, apes or Old World monkeys (Galili *et al.*, 1988). This pig cell surface carbohydrate is recognized by pre-existing antibodies in humans and within minutes of transplantation hyperacute rejection occurs. The enzyme that is responsible for the formation of α1, 3Gal is α-galactotransferase (GGTA1). *GGTA1* knockout pigs were first reported in 2002 (Lai *et al.*, 2002a). These *GGTA1* knockout pigs had reduced α1,3Gal antigen and hyperacute rejection was not observed after organ transplant to a baboon (Kuwaki *et al.*, 2005; Yamada *et al.*, 2005); however, there are still many obstacles to overcome before pig organs can become a reality to use as donors. For example, *N*-glycolyneuraminic acid (Neu5Gc) is a well-characterized xenoantigen expressed on pig cells, but not on human cells. The enzyme cytidine monophosphate-*N*-acetylneuraminic acid hydroxylase (CMAH) allows the expression of this carbohydrate on the surface of pig cells. Knockouts of *CMAH* have been produced by using ZFNs (Kwon *et al.*, 2013; Lutz *et al.*, 2013). Since the pig cells now lack the carbohydrate and are more similar to the human cells there is no production of antibody by the host. Due to the improvement and efficiency in the production of genetically modified pigs there has been an increased effort to find a way for pig organs to be successfully used in xenotransplantation studies.

Recently, CRISPR/Cas9 technology has been employed to efficiently produce pigs with multiple modifications to attempt to overcome pig organ rejection. Li and colleagues (Li *et al.*, 2015) co-transfected liver-derived porcine fibroblast cells with plasmids expressing Cas9, and three gRNAs targeting *GGTA1, CMAH* and a putative *A3GALT2* gene. The enzyme isoglobotrihexosylceramide synthase (A3GALT2) has been shown to synthesize small amounts of α1,3Gal epitopes. Piglets were created that contained deletions or insertions in all three genes. Triple-knockout piglets had no α1,3Gal or Neu5Gc expression in kidney, liver, heart or lungs, and had an altered *A3GALT2* gene. Another group evaluated the significance of *A3GALT2* as a second source of α1,3Gal epitopes in *GGTA1* knockout animals. To determine the significance, researchers used the CRISPR/Cas9 system to target *GGTA1* and *A3GALT2* (Butler *et al.*, 2016b). This study directly compared the *GGTA1* knockout pigs to *GGTA1/A3GALT2* double-knockout pigs. There was a very small trend for decreased α1,3Gal expression in the *A3GALT2* knockout cells compared to wild-type cells, but this was not seen in the tissues. The antibody binding and cytotoxicity remained unchanged by inactivating this *A3GALT2* gene. Thus, *A3GALT2* may not contribute to antibody-mediated rejection in pig to primate or pig to human xenotransplantation.

Another study utilized CRISPR/Cas9 technology to knockout *GGTA1, CMAH* xenoantigens and also added a third gene, β1,4 *N*-acetylgalactosaminyl transferase (*B4GALNT2*; Estrada *et al.*, 2015). This enzyme catalyzes the addition of *N*-acetylgalactosamine to a sialic-acid-modified lactosamine and produces an immunogenic non-Gal glycan antibody response in the pig to baboon xenotransplantation model (Byrne *et al.*, 2014). Again, the authors co-transfected all three gRNAs and Cas9 plasmids into liver-derived cells. Transfected cells were first selected for absence of GGTA1 and then these mutant *GGTA1* cells were incubated with DBA lectin. Negative DBA cells, which don't express *B4GALNT2*, were then selected. The double-negative cells were used for SCNT and transferred to recipient sows. One *GGTA1/CMAH/B4GALNT2* knockout piglet was born. Peripheral blood mononuclear cells (PBMC) from this *GGTA1/CMAH/B4GALNT2*-deficient pig exhibited reduced human IgM and IgG binding compared to cells lacking both *GGTA1* and *CMAH*. These pigs may be able to overcome antibody-mediated rejection when transplanted into humans.

Another strategy for limiting organ rejection is by expressing human complement regulatory proteins (hCRP). One of the most commonly employed hCRP is human CD55, as it has been shown to decrease complement-mediated endothelial cell damage (Menoret *et al.*, 2004). Double- and triple-knockout pigs were generated with CRISPR/Cas9 technology for *GGTA1, CMAH* or *GGTA1, CMAH,* and *B4GALNT2*. A third type of pig was also produced by knocking out *GGTA1* and *CMAH* and adding a *CD55*-expressing cassette (*hDAF*; Butler *et al.*, 2016a). The double-knockout, triple-knockout and double-knockout with *hDAF* insertion was compared to human antibody binding from inactivated human sera. The *GGTA1, CMAH, B4GALNT2* knockout-derived pig cells had decreased human IgM and IgG antibody binding compared to the *GGTA1, CMAH* and *GGTA1, CMAH +CD55* pigs. The same was seen when looking at human antibody-mediated complement-dependent cytotoxicity to porcine cells. The triple-knockout pigs had the lowest levels of complement-mediated damage, even in the absence of hCRP. This significant finding indicated that human patients might receive a *GGTA1, CMAH, B4GALNT2* knockout pig organ transplant without the need for desensitization to avoid antibody-mediated rejection.

Porcine endogenous retroviruses present another potential threat to organ transplantation. Infection of human cells and subsequent recombination has a limited potential to result in a new virus. Another exciting application of the CRISPR/Cas9 system is the potential ability to eliminate all of these endogenous retroviruses. In one study some 60 genes were targeted in a single cell (Yang *et al.*, 2015).

9.13 Pigs as Bioreactors for Biomedical Protein Products

Pigs have been used to produce recombinant human proteins through traditional genetic engineering. The first pig to produce recombinant human protein C was created in 1992 (Velander *et al.*, 1992). Since then, a handful of other pig bioreactor models have been produced. However, with the use of the CRISPR/Cas9 system, the production of pigs that are able to produce recombinant human protein can be accelerated.

Human serum albumin (ALB) is the most abundant plasma protein found in human blood and plays an important homeostatic role in the human body. Human ALB is also prescribed to treat many diseases, but because blood is in short supply and risks, such as pathogen transmission, are associated with its use, an alternative source is important to identify. Peng and colleagues (Peng *et al.*, 2015) utilized CRISPR/Cas9 to knock in recombinant human *ALB* into the pig albumin locus immediately downstream of the start codon. The authors used gRNA that targeted the start codon region of the swine albumin locus and Cas9 mRNA, and injected those RNAs into fertilized oocytes. Sixteen live piglets containing the knockin allele were born, all of the piglets containing human ALB in their blood plasma, as recognized by antibodies specific for human albumin. Mass spectral analysis of plasma from two piglets confirmed that the ALB produced was truly human albumin. Although many others have produced efficient deletions by NHEJ, these authors present evidence that efficient HR can occur in pig zygotes.

Production of human polyclonal antibodies for use in the clinical medical field is crucial due to the dearth of human blood donors. Animal models that can produce humanized polyclonal antibodies are desperately needed. In order for animals to produce humanized antibodies, they need to have disrupted IgM genes so that these animals can be B-cell and antibody deficient. One group targeted the J_H5 region of IgM heavy chain in pigs with a Cas9-gRNA plasmid during porcine fetal fibroblast transfection (Chen *et al.*, 2015). Cells with a biallelic mutation in the J_H5 region were selected for SCNT. By targeting this region, IgM protein expression was disrupted, causing depletion of both B cells and immunoglobulins in the blood of the knockout pigs. The production of these pigs brings the researchers one step closer to being able to produce high-titer human polyclonal antibodies in pigs.

9.14 Conclusion

The technology for making genetic changes in pigs has greatly advanced with the introduction of the CRISPR/Cas9 system. Knockouts, that in some cases took years to make, are now limited only by the length of gestation in the pig (about four months). Not only can single genes be rapidly knocked out, but repair mechanisms can be used to knock in genetic elements from a few base pairs in length to full genes. The utility of these technologies is illustrated by their immediate application to both production agriculture and human medicine. The impacts are already being felt as animals resistant to disease, and others that can potentially serve as a source of organs, as models of human disease, or as a bioreactor have already been produced by using these newer technologies. The fruits of these genetic engineering technologies are only beginning to be harvested and their future applications will be transformative.

Acknowledgements

The authors wish to acknowledge funding from the National Institutes of Health via U42 OD011140-14 and Food for the 21st Century.

REFERENCES

Barrangou R, Fremaux C, Deveau H, *et al.* 2007. CRISPR provides acquired resistance against viruses in prokaryotes. *Science* **315**: 1709–1712.

Bevacqua RJ, Fernandez-Martin R, Savy V, *et al.* 2016. Efficient edition of the bovine PRNP prion gene in somatic cells and IVF embryos using the CRISPR/Cas9 system. *Theriogenology* **86**(8): 1886–1896.

Bi Y, Hua Z, Liu X, *et al.* 2016. Isozygous and selectable marker-free MSTN knockout cloned pigs generated by the combined use of CRISPR/Cas9 and Cre/LoxP. *Sci Rep* **6**: 31729.

Boch J, Scholze H, Schornack S, *et al.* 2009. Breaking the code of DNA binding specificity of TAL-type III effectors. *Science* **326**: 1509–1512.

Bolotin A, Quinquis B, Sorokin A, Ehrlich SD. 2005. Clustered regularly interspaced short palindrome repeats (CRISPRs) have spacers of extrachromosomal origin. *Microbiology* **151**: 2551–2561.

Bonifati V, Rizzu P, van Baren MJ, *et al.* 2003. Mutations in the DJ-1 gene associated with autosomal recessive early-onset parkinsonism. *Science* **299**: 256–259.

Butler JR, Martens GR, Estrada JL, *et al.* 2016a. Silencing porcine genes significantly reduces human-anti-pig cytotoxicity profiles: an alternative to direct complement regulation. *Transgenic Res* **25**(5): 751–759.

Butler JR, Skill NJ, Priestman DL, *et al.* 2016b. Silencing the porcine iGb3s gene does not affect Galalpha3 Gal levels or measures of anticipated pig-to-human and pig-to-primate acute rejection. *Xenotransplantation* **23**: 106–116.

Byrne GW, Du Z, Stalboerger P, Kogelberg H, McGregor CG. 2014. Cloning and expression of porcine beta1,4 N-acetylgalactosaminyl transferase encoding a new xenoreactive antigen. *Xenotransplantation* **21**: 543–554.

Cabot RA, Kuhholzer B, Chan AW, *et al.* 2001. Transgenic pigs produced using *in vitro* matured oocytes infected with a retroviral vector. *Anim Biotechnol* **12**(2): 205–214.

Calvert JG, Slade DE, Shields SL, *et al.* 2007. CD163 expression confers susceptibility to porcine reproductive and respiratory syndrome viruses. *J Virol* **81**: 7371–7379.

Carlson DF, Lancto CA, Zang B, *et al.* 2016. Production of hornless dairy cattle from genome-edited cell lines. *Nat Biotechnol* **34**: 479–481.

Carlson DF, Tan W, Lillico SG, *et al.* 2012. Efficient TALEN-mediated gene knockout in livestock. *Proc Natl Acad Sci USA* **109**: 17382–17387.

Cathomen T, Joung JK. 2008. Zinc-finger nucleases: the next generation emerges. *Mol Ther* **16**: 1200–1207.

Chen F, Wang Y, Yuan Y, *et al.* 2015. Generation of B cell-deficient pigs by highly efficient CRISPR/Cas9-mediated gene targeting. *J Genet Genomics* **42**(8): 437–444.

Choi YJ, Lee K, Park WJ, *et al.* 2016. Partial loss of interleukin 2 receptor gamma function in pigs provides mechanistic insights for the study of human immunodeficiency syndrome. *Oncotarget* **7**(32): 50914–50926.

Christian M, Cermak T, Doyle EL, *et al.* 2010. Targeting DNA double-strand breaks with TAL effector nucleases. *Genetics* **186**: 757–761.

Dai Y, Vaught TD, Boone J, *et al.* 2002. Targeted disruption of the alpha1,3-galactosyltransferase gene in cloned pigs. *Nat Biotechnol* **20**: 251–255.

Dauer W, Przedborski S, 2003. Parkinson's disease: mechanisms and models. *Neuron* **39**: 889–909.

Estrada JL, Martens G, Li P, *et al.* 2015. Evaluation of human and non-human primate antibody binding to pig cells lacking GGTA1/CMAH/beta4GalNT2 genes. *Xenotransplantation* **22**: 194–202.

Galili U, Shohet SB, Kobrin E, Stults CL, Macher BA. 1988. Man, apes, and Old World monkeys differ from other mammals in the expression of alpha-galactosyl epitopes on nucleated cells. *J Biol Chem* **263**: 17755–17762.

Grobet L, Martin LJ, Poncelet D, *et al.* 1997. A deletion in the bovine myostatin gene causes the double-muscled phenotype in cattle. *Nat Genet* **17**: 71–74.

Hai T, Teng F, Guo R, Li W, Zhou Q. 2014. One-step generation of knockout pigs by zygote injection of CRISPR/Cas system. *Cell Res* **24**: 372–375.

Hammer RE, Pursel VG, Rexroad CE Jr., *et al.* 1985. Production of transgenic rabbits, sheep and pigs by microinjection. *Nature* **315**: 680–683.

Holm IE, Alstrup AK, Luo Y. 2016. Genetically modified pig models for neurodegenerative disorders. *J Pathol* **238**: 267–287.

Holtkamp DJ, Kliebenstein JB, Neumann EJ, *et al.* 2013. Assessment of the economic impact of porcine reprodutive and respiratory syndrome virus on United States pork producers. *J Swine Health Prod* **21**: 72–84.

Hwang WY, Fu Y, Reyon D, *et al.* 2013. Efficient genome editing in zebrafish using a CRISPR-Cas system. *Nat Biotechnol* **31**: 227–229.

Hyun S, Lee G, Kim D, *et al.* 2003. Production of nuclear transfer-derived piglets using porcine fetal fibroblasts transfected with the enhanced green fluorescent protein. *Biol Reprod* **69**: 1060–1068.

Kambadur R, Sharma M, Smith TP, Bass JJ. 1997. Mutations in myostatin (GDF8) in double-muscled Belgian Blue and Piedmontese cattle. *Genome Res* **7**: 910–916.

Keffaber KK. 1989. Reproductive failure of unknown etiology. *Am Assoc Swine Practit Newsletter* **1**: 1–9.

Kim YG, Cha J, Chandrasegaran S. 1996. Hybrid restriction enzymes: zinc finger fusions to Fok I cleavage domain. *Proc Natl Acad Sci USA* **93**: 1156–1160.

Kim YG, Chandrasegaran S. 1994. Chimeric restriction endonuclease. *Proc Natl Acad Sci USA* **91**: 883–887.

Kreutz LC, Ackermann MR. 1996. Porcine reproductive and respiratory syndrome virus enters cells through a low pH-dependent endocytic pathway. *Virus Res* **42**(1–2): 137–147.

Kurome M, Ueda H, Tomii R, Naruse K, Nagashima H. 2006. Production of transgenic-clone pigs by the combination of ICSI-mediated gene transfer with somatic cell nuclear transfer. *Transgenic Res* **15**: 229–240.

Kuwaki K, Tseng YL, Dor FJ, *et al.* 2005. Heart transplantation in baboons using alpha1,3-galactosyltransferase gene-knockout pigs as donors: initial experience. *Nat Med* **11**: 29–31.

Kwon DN, Lee K, Kang MJ, *et al.* 2013. Production of biallelic CMP-Neu5Ac hydroxylase knock-out pigs. *Sci Rep* **3**: 1981.

Lai L, Kang JX, Li R, *et al.* 2006. Generation of cloned transgenic pigs rich in omega-3 fatty acids. *Nat Biotechnol* **24**: 435–436.

Lai L, Kolber-Simonds D, Park KW, *et al.* 2002a. Production of alpha-1,3-galactosyltransferase knockout pigs by nuclear transfer cloning. *Science* **295**: 1089–1092.

Lai L, Park KW, Cheong HT, *et al.* 2002b. Transgenic pig expressing the enhanced green fluorescent protein produced by nuclear transfer using colchicine-treated fibroblasts as donor cells. *Mol Reprod Dev* **62**: 300–306.

Lai L, Prather RS. 2002. Progress in producing knockout models for xenotransplantation by nuclear transfer. *Ann Med* **34**: 501–506.

Li P, Estrada JL, Burlak C, *et al.* 2015. Efficient generation of genetically distinct pigs in a single pregnancy using multiplexed single-guide RNA and carbohydrate selection. *Xenotransplantation* **22**: 20–31.

Lillico SG, Proudfoot C, Carlson DF, *et al.* 2013. Live pigs produced from genome edited zygotes. *Sci Rep* **3**: 2847.

Lillico SG, Proudfoot C, King TJ, *et al.* 2016. Mammalian interspecies substitution of immune modulatory alleles by genome editing. *Sci Rep* **6**: 21645.

Lutz AJ, Li P, Estrada JL, *et al.* 2013. Double knockout pigs deficient in N-glycolylneuraminic acid and galactose alpha-1,3-galactose reduce the humoral barrier to xenotransplantation. *Xenotransplantation* **20**: 27–35.

Marraffini LA, Sontheimer EJ. 2008. CRISPR interference limits horizontal gene transfer in staphylococci by targeting DNA. *Science* **322**: 1843–1845.

McPherron AC, Lee SJ, 1997. Double muscling in cattle due to mutations in the myostatin gene. *Proc Natl Acad Sci USA* **94**: 12457–12461.

Menoret S, Plat M, Blancho G, *et al.* 2004. Characterization of human CD55 and CD59 transgenic pigs and kidney xenotransplantation in the pig-to-baboon combination. *Transplantation* **77**: 1468–1471.

Mercer J, Schelhaas M, Helenius A. 2010. Virus entry by endocytosis. *Ann Rev Biochem* **79**: 803–833.

Mojica FJ, Diez-Villasenor C, Garcia-Martinez J, Soria E. 2005. Intervening sequences of regularly spaced prokaryotic repeats derive from foreign genetic elements. *J Mol Evol* **60**: 174–182.

Moscou MJ, Bogdanove AJ. 2009. A simple cipher governs DNA recognition by TAL effectors. *Science* **326**: 1501.

Palgrave CJ, Gilmour L, Lowden CS, *et al.* 2011. Species-specific variation in RELA underlies differences in NF-kappaB activity: a potential role in African swine fever pathogenesis. *J Virol* **85**: 6008–6014.

Park KW, Lai L, Cheong HT, *et al.* 2001. Developmental potential of porcine nuclear transfer embryos derived from transgenic fetal fibroblasts infected with the gene for the green fluorescent protein: comparison of different fusion/activation conditions. *Biol Reprod* **65**: 1681–1685.

Peng J, Wang Y, Jiang J, *et al.* 2015. Production of human albumin in pigs through CRISPR/Cas9-mediated knockin of human cDNA into swine albumin locus in the zygotes. *Sci Rep* **5**: 16705.

Pickrell AM, Youle RJ. 2015. The roles of PINK1, parkin, and mitochondrial fidelity in Parkinson's disease. *Neuron* **85**: 257–273.

Popescu LN, Gaudreault N, Whitworth KM, *et al.* 2016. Genetically edited pigs lacking CD163 show no resistance following infection with the African swine fever (ASF) Georgia 07 virus. *Virology* **501**: 102–106.

Porteus M. 2008. Design and testing of zinc finger nucleases for use in mammalian cells. *Methods Mol Biol* **435**: 47–61.

Pourcel C, Salvignol G, Vergnaud G. 2005. CRISPR elements in *Yersinia pestis* acquire new repeats by preferential uptake of bacteriophage DNA, and provide additional tools for evolutionary studies. *Microbiology* **151**: 653–663.

Prather RS, Rowland RR, Ewen C, *et al.* 2013. An intact sialoadhesin (Sn/SIGLEC1/CD169) is not required for attachment/internalization of the porcine reproductive and respiratory syndrome virus. *J Virol* **87**: 9538–9546.

Ramsoondar JJ, Machaty Z, Costa C, *et al.* 2003. Production of alpha 1,3-galactosyltransferase-knockout cloned pigs expressing human alpha 1,2-fucosylosyltransferase. *Biol Reprod* **69**: 437–445.

Redman M, King A, Watson C, King D. 2016. What is CRISPR/Cas9? *Arch Disease Child Edu Pract Edn* **101**: 213–215.

Reyes LM, Estrada JL, Wang ZY, *et al.* 2014. Creating class I MHC-null pigs using guide RNA and the Cas9 endonuclease. *J Immunol* **193**: 5751–5757.

Robinson TP, Wint GR, Conchedda G, *et al.* 2014. Mapping the global distribution of livestock. *PLoS One* **9**(5): e96084.

Rogers CS, Hao Y, Rokhlina T, *et al.* 2008. Production of CFTR-null and CFTR-DeltaF508 heterozygous pigs by adeno-associated virus-mediated gene targeting and somatic cell nuclear transfer. *J Clin Invest* **118**: 1571–1577.

Ross JW, Whyte JJ, Zhao J, *et al.* 2010. Optimization of square-wave electroporation for transfection of porcine fetal fibroblasts. *Transgenic Res* **19**(4): 611–620.

Rowlands RJ, Michaud V, Heath L, *et al.* 2008. African swine fever virus isolate, Georgia, 2007. *Emerg Infect Dis* **14**: 1870–1874.

Saeki K, Matsumoto K, Kinoshita M, *et al.* 2004. Functional expression of a Delta12 fatty acid desaturase gene from spinach in transgenic pigs. *Proc Natl Acad Sci USA* **101**: 6361–6366.

Sanchez-Torres C, Gomez-Puertas P, Gomez-del-Moral M, *et al.* 2003. Expression of porcine CD163 on monocytes/macrophages correlates with permissiveness to African swine fever infection. *Arch Virol* **148**: 2307–2323.

Sanchez-Vizcaino JM, Mur L, Martinez-Lopez B. 2013. African swine fever (ASF): five years around Europe. *Vet Microbiol* **165**: 45–50.

Schornack S, Meyer A, Romer P, Jordan T, Lahaye T. 2006. Gene-for-gene-mediated recognition of nuclear-targeted AvrBs3-like bacterial effector proteins. *J Plant Physiol* **163**: 256–272.

Tan W, Carlson DF, Lancto CA, *et al.* 2013. Efficient nonmeiotic allele introgression in livestock using custom endonucleases. *Proc Natl Acad Sci USA* **110**: 16526–16531.

Terns MP, Terns RM. 2011. CRISPR-based adaptive immune systems. *Curr Opin Microbiol* **14**: 321–327.

Urnov FD, Miller JC, Lee YL, *et al.* 2005. Highly efficient endogenous human gene correction using designed zinc-finger nucleases. *Nature* **435**: 646–651.

Valente EM, Abou-Sleiman PM, Caputo V, *et al.* 2004. Hereditary early-onset Parkinson's disease caused by mutations in PINK1. Science **304**: 1158–1160.

Van Breedam W, Delputte PL, Van Gorp H, *et al.* 2010. Porcine reproductive and respiratory syndrome virus entry into the porcine macrophage. *J Gen Virol* **91**: 1659–1667.

Van Gorp H, Delputte PL, Nauwynck HJ. 2010. Scavenger receptor CD163, a Jack-of-all-trades and potential target for cell-directed therapy. *Mol Immunol* **47**: 1650–1660.

Velander WH, Johnson JL, Page RL, *et al.* 1992. High-level expression of a heterologous protein in the milk of transgenic swine using the cDNA encoding human protein C. *Proc Natl Acad Sci USA* **89**: 12003–12007.

Wang K, Ouyang H, Xie Z, *et al.* 2015. Efficient generation of myostatin mutations in pigs using the CRISPR/Cas9 system. *Sci Rep* **5**: 16623.

Wang R, Preamplume G, Terns MP, Terns RM, Li H. 2011. Interaction of the Cas6 riboendonuclease with CRISPR RNAs: recognition and cleavage. *Structure* **19**: 257–264.

Wang X, Cao C, Huang J, *et al.* 2016. One-step generation of triple gene-targeted pigs using CRISPR/Cas9 system. *Sci Rep* **6**: 20620.

Watanabe S, Iwamoto M, Suzuki S, *et al.* 2005. A novel method for the production of transgenic cloned pigs: electroporation-mediated gene transfer to non-cultured cells and subsequent selection with puromycin. *Biol Reprod* **72**: 309–315.

Wells KD, Bardot R, Whitworth KM, *et al.* 2017. Substitution of porcine CD163 SRCR domain 5 with human CD163-like homolog SRCR domain 8 confers resistance of pigs to genotype 1 but not genotype 2 porcine reproductive and respiratory syndrome (PRRS) viruses. *J Virol* **91**(2): pii: e01521–16.

Whitworth KM, Lee K, Benne JA, *et al.* 2014. Use of the CRISPR/Cas9 system to produce genetically engineered pigs from in vitro-derived oocytes and embryos. *Biol Reprod* **91**: 78.

Whitworth KM, Rowland RR, Ewen CL, *et al.* 2016. Gene-edited pigs are protected from porcine reproductive and respiratory syndrome virus. *Nat Biotechnol* **34**: 20–22.

Whyte JJ, Zhao J, Wells KD, *et al.* 2011. Gene targeting with zinc finger nucleases to produce cloned eGFP knockout pigs. *Mol Reprod Dev* **78**: 2.

Wiedenheft B, Sternberg SH, Doudna JA. 2012. RNA-guided genetic silencing systems in bacteria and archaea. *Nature* **482**: 331–338.

Xin J, Yang H, Fan N, *et al.* 2013. Highly efficient generation of GGTA1 biallelic knockout inbred mini-pigs with TALENs. *PLoS One* **8**: e84250.

Yamada K, Yazawa K, Shimizu A, *et al.* 2005. Marked prolongation of porcine renal xenograft survival in baboons through the use of alpha1,3-galactosyltransferase gene-knockout donors and the cotransplantation of vascularized thymic tissue. *Nat Med* **11**: 32–34.

Yang L, Guell M, Niu D, *et al.* 2015. Genome-wide inactivation of porcine endogenous retroviruses (PERVs). *Science* **350**: 1101–1104.

Zhou X, Xin J, Fan N, *et al.* 2015. Generation of CRISPR/Cas9-mediated gene-targeted pigs via somatic cell nuclear transfer. *Cell Mol Life Sci* **72**: 1175–1184.

10 Generation of New Model Cell Lines using ssODN Knockin Donors and FACS-based Genome Editing

Francesco Niola and Morten Frödin

10.1 Introduction

Programmable nucleases like the CRISPR/Cas9 (clustered regularly interspaced short palindromic repeats/CRISPR-associated 9) system, ZFNs (zinc-finger nucleases) and TALENs (transcription activator-like effector nucleases) can generate any desired modification in the genome of higher organisms. Two major barriers, however, universally limit the efficiency of genome editing for the generation of model cell lines. The first barrier relates to getting nuclease into a sufficiently high fraction of cells in the targeted population. The second barrier relates to obtaining a sufficiently high amount of nuclease in individual cells to elicit the desired mutation event. One way to overcome these obstacles is to use efficient methods for delivery of nuclease to the cells. Examples include nucleofection or viral transduction, which are highly efficient delivery methods. A completely different approach is to isolate specifically the nuclease-containing cells from the targeted population. One advantage of such a nuclease-enrichment strategy is that it can enable efficient editing, even when no efficient delivery methods are available or practical, because enrichment can compensate for poor delivery.

Two major nuclease-enrichment strategies have been developed. One strategy uses classical drug resistance to select for nuclease-expressing cells (Frank *et al.*, 2013; Ran *et al.*, 2013). In this approach, cells are transfected with vectors that carry a drug-resistance gene and a nuclease gene, either on the same or on separate constructs. Following transfection, cells are treated with the relevant drug for a few days to kill the cells that did not receive the constructs containing the selection marker and nuclease. Drug-based nuclease enrichment is an efficient strategy that has been shown to increase genome editing efficiencies several-fold in a broad range of applications. Furthermore, most of the commercial providers of CRISPR/Cas9 reagents have incorporated this approach into their vector systems and offer Cas9, gRNA or both linked to expression of drug selection markers.

We and others have developed another major nuclease-enrichment strategy, which exploits fluorescence-activated cell sorting (FACS) to isolate the nuclease-

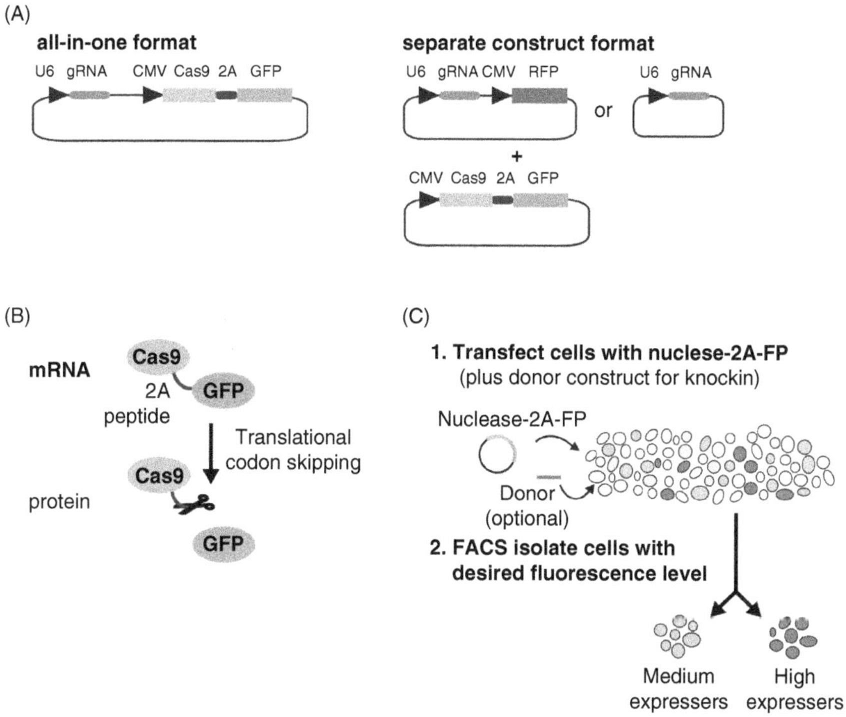

Figure 10.1 Vectors and principles of FACS-based genome editing. (A) Examples of Cas9-2A-FP vectors. (Left) All-in-one format, where Cas9-2A-FP and gRNA are expressed from the same vector. (Right) Separate construct format, where Cas9-2A-FP is expressed from one vector and gRNA from another vector that may carry a distinct FP for co-sorting or carry no marker. CMV and U6 indicate the promoters that drive expression. (B) In the Cas9-2A-FP transcript, the 2A peptide undergoes codon skipping during translation, resulting in 1:1 co-expression of nuclease and FP as separate proteins. (C) Principle of FACS-based genome editing. In the first step, cells are transfected with a nuclease-2A-FP construct. In the second step, cells with a desired level of fluorescence are isolated by FACS 1–3 days later. (A black-and-white version of this figure will appear in some formats. For the color version, please refer to the plate section.)

expressing cells. The most widely used format is based on vectors, in which the nuclease and fluorescent protein (FP) are encoded by the same transcript, but separated by a 2A peptide sequence (Figure 10.1A; Duda *et al.*, 2014; Lonowski *et al.*, 2017). The 2A sequence undergoes ribosomal codon skipping during translation (Szymczak and Vignali, 2005), resulting in 1:1 co-expression of nuclease and FP as separate entities (Figure 10.1B). One to three days after delivery, the cells that have taken up the construct will fluoresce and can be isolated by FACS, at a desired expression level (Figure 10.1C). FACS-based genome editing can efficiently tackle both of the barriers mentioned above. First, a population of 100% nuclease-expressing cells can easily be retrieved, regardless of how scarce they were in the initial targeted cell culture. Second, a population of cells that all have a sufficiently high nuclease level can easily be obtained by specifically sorting for a high fluorescence intensity. This chapter describes in detail the principles and use of FACS-based genome editing for the generation of model cell lines. This nuclease-enrichment approach has also been embraced by the commercial providers of CRISPR/Cas9 reagents, which offer a large number of different 2A-FP vectors for any delivery method, and some vectors are also available free from Addgene (see Lonowski *et al.*, 2017 for a comprehensive list).

Furthermore, we will describe the strategy to obtain knockin-modified model cell lines using single-stranded oligodeoxynucleotide (ssODN) donors and FACS-based genome editing. We originally reported that ssODNs are efficient donors for knockin of small mutations like SNPs, codon-swaps, LoxP sites or epitope tags (Chen *et al.*, 2011) and they have since become the preferred donor format for such mutations. We will also summarize the current knowledge on how to best design an ssODN donor.

10.2 Methodology

In this section, we will outline the methodological aspects of vectors for FACS-based genome editing and ssODN donor design that underlie the experiments in the Results and Discussion section.

10.2.1 Design of Nuclease-2A-FP Vectors

We originally developed and tested the principles of FACS-based genome editing by modifying Sigma's CompoZr ZFN vectors to allow 2A-coupled co-expression of a ZFN with any of the commonly used FPs (Duda *et al.*, 2014). Results using these ZFN-2A-FP vectors will be used to illustrate FACS-based knockin using ssODN donors, which will apply to any nuclease-2A-FP construct. However, because CRISPR/Cas9 has now become the editing system of choice, Cas9-2A-FP constructs will be the main focus and will be used to illustrate knockout and other aspects of FACS-based editing. We will describe results obtained with the vector pCMV-Cas9-2A-GFP (shown in Figure 10.1A, left) that works well in many cell types (Duda *et al.*, 2014). In this all-in-one vector, the gRNA can easily be swapped by a QuikChange PCR using the primers: (a) 5′-CTTGTGGAAAGGACGAAACACCG(N_{20})GTTTTAGAGCTAGAAATAGCAAGTTAA-3′ and (b) 5′-TTAACTTGCTATTTCTAGCTCTAAAAC(N_{20})CGGTGTTTCGTCCTTTCCACAAG-3′. The gRNA target sequence (N_{20}) is incorporated into primer (a) and primer (b) is the complementary sequence. The underlined nucleotides are inserted to ensure that the sgRNA will start with a guanine (in case the gRNA does not contain a 5′ guanine), which enhances U6 promotor-driven transcription. QuikChange amplification is performed under standard conditions, except that we use 16 min extension, 18 cycles and 5% DMSO to increase product yield (Duda *et al.*, 2014).

Editing by FACS for high nuclease levels has the unwanted side effect that off-target sites are also cut with higher efficiency, because the mismatch binding energy barrier may be overcome. When sorting for high Cas9 levels, it is therefore important that gRNAs have been designed such that the closest potential off-target sites have at least 2–3 mismatches, preferentially located in the conserved guanines of the PAM site (NGG) or within the 1–12 nucleotides upstream of the PAM site (the seed region), which will effectively prevent most off-target cutting.

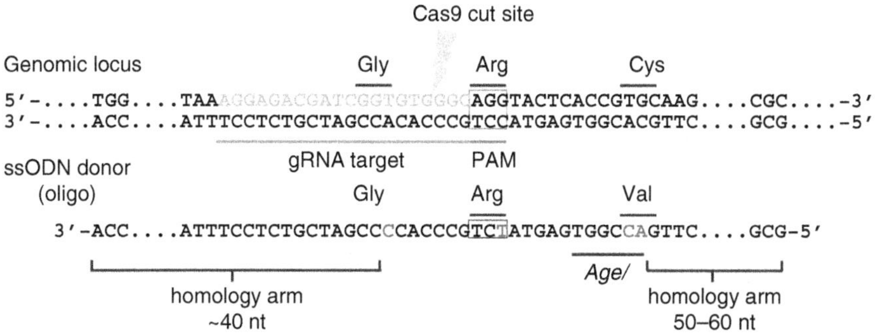

Figure 10.2 Key features of an ssODN donor for knockin editing using CRISPR/Cas9. The mutations contained in the donor are in red. The mutation-in-study is a Cys-Val codon-swap mutation, which simultaneously creates a diagnostic *AgeI* site for RFLP analysis. The gRNA target and PAM sequences are indicated, as well as the Cas9 cut site (yellow lightning). Note that the gRNA-blocking mutations are silent (Gly-Gly; Arg-Arg) and two are sufficient, since one of them mutates the PAM. Note that the effective homology arm begins at the mutation most distant to the cut site (or at the cut site itself, if there is no mutation). (A black-and-white version of this figure will appear in some formats. For the color version, please refer to the plate section.)

For design of gRNAs, we recommend the WTSI Genome Editing (WGE) algorithm (Hodgkins *et al.*, 2015; http//sanger.ac.uk/htgt/wge), which is very user-friendly. Furthermore, the closest matching potential off-target sites are listed with a clear indication of the position of mismatches and whether the off-target site is located in an exon, intron or intergenic sequence. A mouse-click on the off-target site shows its genomic location. This information can be used for rapid evaluation of whether a given off-target event is likely to occur or be a concern; off-target sites in an intergenic region, for example, typically can be ignored.

10.2.2 ssODN Donors

In Figure 10.2 are summarized all the key features of a typical ssODN (also called "oligo") donor for knockin of small mutations, which in the present fictive example is the knockin of a codon-swap (the mutation-in-study). It is of key importance to design the nuclease so as to cut as close as possible to the intended site of mutagenesis, because the efficiency of knockin of the mutation contained in the donor decreases steeply with distance from the DNA break. We typically aim for < 20 bp, but have succeeded in knocking in mutations located ~100 bp from the cut site. Critically, the donor must span the nuclease cut site (Chen *et al.*, 2011). The exact mechanism by which ssODNs are genomically integrated at the cut site is unclear, but it certainly involves homology-directed repair (HDR) mechanisms and therefore a minimum length of homology sequence on either site of the cut (called a homology arm) must be present in the ssODN. For CRISPR/Cas9, it has been reported that a relatively long homology arm at the PAM side increases efficiency (Richardson *et al.*, 2016). We typically use homology arms of 50–60 nt at the PAM end and ~40 nt at the other end of the ssODN donor. For synthesis, we use Integrated DNA Technologies DNA Ultramer oligo service with

PAGE purification (https://eu.idtdna.com). Oligos are synthesized unmodified, but if homology arms are significantly shorter than outlined here, phosphorothioate modification may increase knockin frequency, presumably by preventing exonucleases from reducing homology arms to an unproductive length (Renaud *et al.*, 2016). For CRISPR/Cas9, it has been reported that ssODN donors are generally most efficient when composed of a strand that is complementary to the gRNA sequence (Richardson *et al.*, 2016; Yang *et al.*, 2013).

FACS for high nuclease levels causes prolonged presence of nucleases in the cells. We therefore include 2–3 silent mutations in the gRNA target site in order to prevent cutting and indel mutagenesis of the donor after its genomic integration (gRNA-blocking mutations). The mutations should be designed such that either the mutation-in-study or the gRNA-blocking mutations introduce (or, less optimally, remove) a restriction endonuclease site. Such a diagnostic site is used to monitor knockin editing by restriction fragment length polymorphism (RFLP) assay, which consists of genomic PCR amplification of the mutation site, followed by digestion and agarose gel analysis of the PCR product (consult WatCut for design of silent diagnostic sites: http://watcut.uwaterloo.ca/template.php?act=silent_new). It is most optimal if the mutation-in-study creates the diagnostic site, since a positive RFLP result thereby equals knockin of this mutation. If the diagnostic site is created by the gRNA-blocking mutations, positive RFLP does not equal knockin of the mutation-in-study, although it is likely to have also been knocked-in, if located <10–15 bp from the Cas9 cut site. To avoid the need for purification of PCR products, use a diagnostic enzyme that cuts reasonably well in PCR buffer (see: www.neb.com/tools-and-resources/usage-guidelines/activity-of-restriction-enzymes-in-pcr-buffers). Note that the mutations should not generate a codon that is rarely used in the edited species, as this may reduce expression of the protein (consult the Codon Usage Database: www.kazusa.or.jp/codon). Assess also whether mutations near exon–intron junctions could affect splicing, as RNA sequence motifs exist at these sites (and even elsewhere in the pre-mRNA) that are recognized by RNA-binding proteins, and their destruction may lead to the generation of alternative transcripts (Fu and Ares, 2014; consult Human Splicing Finder: www.umd.be/HSF3/).

10.2.3 Delivery of ssODNs and Nuclease-2A-FP Constructs

We performed the experiments in human K562 cells. The ssODNs and nuclease constructs were delivered to the cells using nucleofection, which is an efficient electroporation method. Typically, we nucleofected 1.5×10^6 cells using 2 µg pCMV-Cas9-2A-GFP plasmid, 3.3 µg mRNA of each ZFN-2A-FP monomer for *RSK4* and 6 µg ssODN donor. In the electroporation reaction, cells were suspended in Nucleofector Solution V and nucleofected in an Amaxa Nucleofector 2b device using program T-016, according to the manufacturer´s instructions (Lonza). Note, that the delivery method and efficiency may often not be critical due to the nuclease-enrichment step inherent to FACS-based genome editing. We and others have successfully delivered Cas9-2A-FP constructs by liposomal transfection, electroporation/nucleofection or viral transduction and in the form of RNA, plasmid or viral particles.

10.2.4 FACS Parameters

Setting up parameters and operation of the FACS instrument is often performed by core facility staff and is beyond the scope of this chapter. However, a brief guide on setting up FACS parameters in relation to genome editing is provided in Lonowski *et al.* (2017).

10.2.5 Monitoring of Genome Editing Efficiency

Knockin editing at the biallelic *RSK4* locus in clonal cell lines was assessed by RFLP analysis. Knockout editing at the triallelic *KRAS* locus in cell populations was determined by Indel Detection by Amplicon Analysis (IDAA) (Lonowski *et al.*, 2017; Yang *et al.*, 2015). This assay can determine the frequency and size of insertion or deletion (indel) mutations elicited by nucleases and is described in Section 12.2.5.

10.3 Results and Discussion

10.3.1 Distribution of Fluorescence in a Population of Cells Transfected with Cas9-2A-FP

Cells in a transfected cell population often show a wide distribution in the amount of protein expressed by individual cells from the transfected plasmid. This is illustrated by the GFP marker in K562 cells nucleofected with a pCMV-Cas9-2A-GFP all-in-one construct with gRNA targeting the *KRAS* gene (Figure 10.3A, middle). Parallel analysis of mock transfected cells determined the autofluorescence level of K562 cells in the GFP channel (Figure 10.3A, top). Thus, only the pCMV-Cas9-2A-GFP-transfected cells with fluorescence above this level showed specific expression of the GFP marker (indicated by a bar). Note that low- and medium-level GFP-expressing cells constituted the bulk of the population, whereas high-level GFP-expressing cells were much fewer (Figure 10.3A, bottom), which is a typical distribution. The difference between the lowest and highest specific GFP fluorescence level was ~500-fold. Such a difference in transfection level has a great impact on both the intended editing and non-specific effects in the targeted cell population, as will be discussed below. It must be noted that GFP does not directly mark the cells for Cas9 *expression* because the two proteins are translated as separate entities due to the 2A format, and GFP is degraded slower than Cas9. However, because of the strict 1:1 co-translation of the 2A format, GFP faithfully marks the cells for Cas9 *transfection*. Furthermore, the 1:1 coupling also ensures that the higher the fluorescence level observed in an individual cell at any given time after transfection, the higher the amount of Cas9 expressed by that cell.

10.3.2 Knockout Editing as a Function of Fluorescence Level in Cells Transfected with Cas9-2A-FP

We used FACS to isolate 20 000 cells from six cell populations with distinct and increasing levels of GFP (populations 1–6 in Figure 10.3A, middle). We cultured

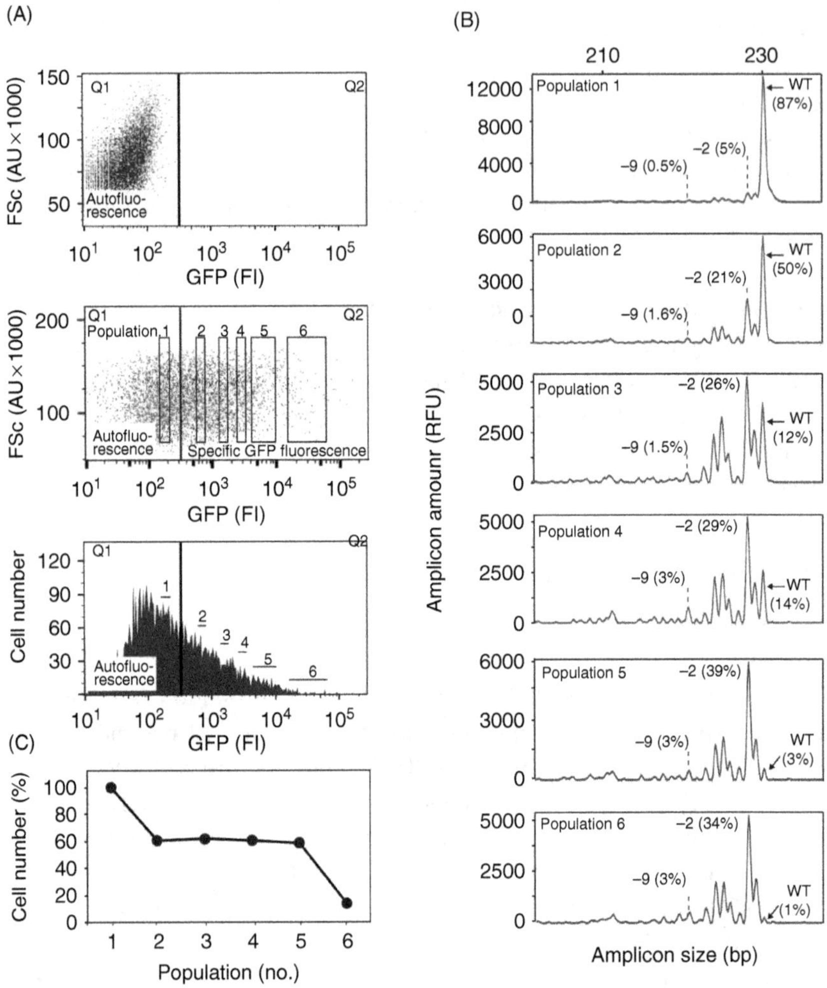

Figure 10.3 FACS-based genome editing for generation of knockout cell populations. (A) Top: FACS profile of mock-transfected K562 cells, showing the autofluorescence level of the cells in the GFP channel. Middle: FACS profile of K562 cells two days after transfection with an all-in-one pCMV-Cas9-2A-GFP construct expressing gRNA targeting the *KRAS* locus. The GFP expression level of individual cells (represented by dots) is measured as arbitrary fluorescence intensity (FI) units (*x*-axis) and cell-size as arbitrary forward scatter (FSc) units (*y*-axis). Within the pCMV-Cas9-2A-GFP-transfected cell population, six subpopulations were defined (boxed). Bottom: Another representation of the pCMV-Cas9-2A-GFP transfected cell population, showing cell number (*y*-axis) distribution at the various GFP fluorescence levels, as determined in 10 000 cells. The six distinct populations defined in the middle panel are indicated. (B) 20 000 cells from each of the six populations in (A) were bulk isolated by FACS, cultured for six days, after which an aliquot of each population was analyzed by IDAA to determine the level of indel (= knockout) mutations at the *KRAS* locus. The frequencies of wild-type and select indel alleles in the six populations are indicated. (C) Prior to IDAA, the total number of live cells in each of the six populations was determined and expressed as a percentage of population 1. (A black-and-white version of this figure will appear in some formats. For the color version, please refer to the plate section.)

these cells for six days and then analyzed the level of knockout editing. The readout for knockout was IDAA, which reveals the extent of editing, as well as the size and type of indel mutation resulting from repair of the DNA break by the non-homologous end joining (NHEJ) pathway (Figure 10.3B). The background-fluorescent population 1 showed a low level of editing. This revealed that some Cas9 had been expressed in this population and also illustrated that the cellular autofluorescence of K562 cells can mask a low level of GFP fluorescence. More importantly, at increasing levels of fluorescence, increasing levels of indel editing were observed, and in population 5, almost complete knockout editing of *KRAS* was achieved (only 3% wild-type alleles were present in this population). The results clearly show that indel editing increases in a dose-dependent manner with the cellular level of Cas9, as monitored by the GFP marker during FACS. Moreover, complete modification can even be achieved at an expression level that is >10 times lower than the maximal level. Qualitatively, the same relative distribution of indels was obtained at the various fluorescence levels; e.g. the 2 bp and 9 bp deletions are predominant and minor indels, respectively, at all fluorescence levels. We find that the dose–response relationship shown in Figure 10.3B is quite representative for a good gRNA in an all-in-one expression vector system. However, the dose–response will vary with the quality (efficiency) of the gRNA used and the fluorescence span obtained, as well as other parameters.

The ability of FACS-based genome editing to generate a pool of cells with essentially complete knockout of a gene in a single sorting can be highly advantageous. First, a population of model cells can be generated quickly for gene knockout studies. Second, it allows for the study of gene knockouts in the context of a pool of cells as opposed to a clone of cells. For some studies this is essential, since it eliminates potential confounding effects and artifacts that are derived from clonal variation among the cells of the parental, targeted cell population.

10.3.3 Balancing Desired Modifications and Undesired Effects Using FACS-based Genome Editing

Undesired effects that are linked to the transfection level will also increase when sorting for high fluorescence levels. Examples of undesired effects are off-target cutting, cell death and temporary growth arrest, all of which can be problematic depending on the extent to which they occur and the purpose of the experiment. The undesired effects are often translated into a lower cell number. An example is provided in Figure 10.3C, which assessed the total number of viable cells resulting from each of the 20 000 cells isolated from the six fluorescence levels, determined after six days of culture post-FACS and at the time of IDAA. A drop in cell number was observed between the background-fluorescent population 1 and the low-expressing population 2, showing that some factor associated with transfection significantly reduced cell number. Cell number thereafter remained stable, despite increasing transfection and on-target editing level, until population 6, where a steep drop in cell number was observed. Thus, like the Cas9 on-target editing of the *KRAS* locus, the undesired effects also increased in a dose-dependent manner. However, the dose–response relationship for undesired effects can be complex, as illustrated with the present bi-phasic case. This is not surprising, because it is the result of several factors and processes (chemical/physical transfection stress, double-stranded DNA stress, FP-toxicity, off-target cutting, etc.) that may each show distinct dose-dependency. Note that *KRAS* is fully redundant in K562 cells (Yang *et al.*, 2015), and the decrease in cell number is therefore not caused by its knockout, but by the undesired effects discussed above.

It is a key feature of FACS-based editing that a defined cell population showing an optimal balance between on-target editing efficiency and specific undesired effects can be identified and isolated for subsequent studies. Such a population could, for example, be population 5 in Figure 10.3, which showed almost complete knockout of *KRAS*, yet retained a high number of surviving cells.

It is important to note that the dose–response relationship for on-target editing and undesired effects will vary with cell type, gRNA construct, nuclease-2A-FP expression system, transfection method and other parameters. Up front in an editing project, it is therefore recommended to perform a pilot experiment like the one shown in Figure 10.3A–C that identifies the optimal fluorescence intensity level for sorting. In subsequent experiments, FACS for this level will normally produce the same level of editing and unwanted effects as the pilot experiment. This is possible because fluorescence intensity is a quite exact measure, so even if transfection efficiency varies from experiment to experiment, for example, a cell population with a specific, desirable fluorescence level can easily be retrieved each time. The same time interval between transfection and FACS must, however, be used in all experiments intended for comparison due to the distinct half-lives of GFP and Cas9, as discussed above.

10.3.4 Knockin Editing as a Function of FACS for Increasing Fluorescence Level

We tested FACS-based knockin editing using ssODN donors. A ZFN pair was constructed linking each monomer via 2A to GFP and DsRed2, respectively, and targeting a desired mutation site in *RSK4*. An ssODN was designed to knock in a codon-swap mutation and a diagnostic RFLP site on top of, and 14 nt, respectively, from the ZFN cut site in *RSK4*. Both constructs were co-nucleofected into K562 cells that were subsequently sorted for GFP/DsRed2-double fluorescence at five different levels (Figure 10.4A). Furthermore, we tested FACS-based editing for generation of clonal cell lines and therefore directly FACS-sorted cells singly into 96-well plates for clonal expansion.

As observed for knockout, ssODN knockin editing in the expanded clones increased in a dose-dependent manner with the cellular level of nuclease, as monitored by the GFP/DsRed2 markers at the time of FACS (Figure 10.4B). Furthermore, in clones derived from top nuclease expressers (population 5), nearly all clones (~95%) were knockin modified, and the majority (70%) on both alleles. Thus, although the donor is not sorted for, it does not limit the ability to obtain complete knockin. Rather, the dose-dependency is on the nuclease. One reason is that the donor is very small and therefore present in huge (50–100-fold) numerical excess of the nuclease in the nucleofection reaction with the standard amounts used for both constructs. Another reason is that the FACS method essentially sorts for the "transfectability" of the cells. Thus, cells that have the capacity to take up high levels of nuclease-2A-FP construct will normally

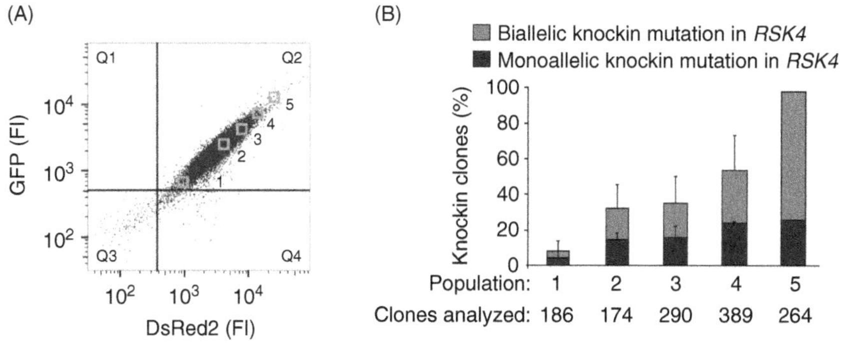

Figure 10.4 FACS-based genome editing for generation of clonal cell lines with ssODN-meditated knockin mutation. (A) FACS profile of K562 cells three days after transfection with mRNA for left ZFN-2A-GFP and right ZFN-2A-DsRed2 targeting the RSK4 locus and an ssODN donor for knockin of a Cys443Val mutation and a diagnostic *BamHI* site for RFLP analysis. The GFP and DsRed2 expression levels of individual cells (represented by dots) are measured as arbitrary fluorescence intensity (FI) units on the *y*-axis and *x*-axis, respectively. Parallel analysis of mock-transfected K562 cells was used to determine the autofluorescence level of the cells in the GFP and DsRed2 channels (not shown). Within the population with specific GFP/DsRed2 double-fluorescence (Q2), five subpopulations were defined (boxed), and single cells from each population were isolated by FACS and seeded into 96-well plates for clonal expansion. (B) After clonal expansion, each cell line was analyzed for monoallelic or biallelic knockin of the ssODN-containing mutations by RFLP analysis. The fraction of knockin modified clones is shown as a percentage on the *y*-axis and the total number of clones analyzed on the *x*-axis. The data are means +SD of three independent experiments, except for population 5, which represents one experiment. (A black-and-white version of this figure will appear in some formats. For the color version, please refer to the plate section.)

also have taken up high amounts of ssODN donor, in particular because it is much smaller than the nuclease-2A-FP construct and therefore taken up more efficiently. It is also interesting to note that, since knockin editing was almost complete in the top-expressers, very little NHEJ repair of the nuclease-induced DNA break had taken place, even though NHEJ is normally more efficient than HDR. The two repair mechanisms are competing processes and it therefore seems likely that FACS-based editing using ssODNs and sorting for high transfection levels shifts the balance of repair mechanism towards knockin repair, which is, of course, a great practical advantage.

Finally, the experiment illustrates that FACS-based editing is a very convenient and efficient method for establishing clonal lines of modified cells. Thus, the simultaneous FACS for a desired fluorescence level and sorting of cells singly into 96-well plates provide accurate and facile cloning of nuclease-expressing cells with a high probability of harboring the desired mutation.

10.3.5 Additional Uses of FACS-based Genome Editing

We and others have exploited the advantages of FACS-based editing in applications other than those illustrated here. The method is particularly well suited for isolation of rare nuclease-2A-FP transfectants. FACS instruments are very sensitive and precise and can easily extract single fluorescent cells from a large population of essentially non-fluorescent cells. We have obtained mono- as well as biallelic knockin-modified clonal cell lines at useful frequencies, even at 0.5% transfection efficiencies (Duda *et al.*, 2014). FACS-based editing can also be used to build up modification levels in a targeted cell population. Sometimes it is not possible to obtain virtually complete editing in a single sort, if, for example, the gRNA is not very active, the knockin mutagenesis is very demanding, multiple gene copies need be modified, etc. In such cases, the cells may be subjected to repeated rounds of nuclease-2A-FP delivery and FACS to gradually build up editing levels in the cell population until a high enough level has been achieved. In this scenario, a cell population with a desired fluorescence level is bulk isolated and allowed to grow for approximately a week to obtain enough cells for another round of nuclease transfection, and FP from the previous transfection to become sufficiently degraded or diluted through cell proliferation. By this approach, we have obtained cell pools with complete modification in demanding knockin (Duda *et al.*, 2014) or knockout experiments (Lonowski *et al.*, 2017).

10.3.6 Cell Viability in FACS-based Genome Editing

As mentioned above, various factors inherent to FACS-based editing can act to decrease cell viability. Several parameters, however, can be optimized to counteract this downside of the method. Importantly, the fluorescence level with the optimal balance between editing efficiency and cell survival for the cell type and nuclease construct in use should be determined and used, as outlined in Figure 10.3. Furthermore, using the larger nozzle size of the FACS instrument to decrease mechanical stress, sorting into conditioned medium and/or coated plates and sorting at the optimal temperature for the cell type used could be considered. For single-cell cloning, we often seed out many 96-well plates, since clones often grow up in only 10–20% of the wells. It is recommended to consult the literature on how to best sort the cell type of interest.

10.3.7 FACS-based vs. Drug-based Genome Editing

Both strategies can efficiently increase editing frequencies, but do have advantages relative to each other. Drug-based editing does not require a costly FACS instrument and the time and expertise to perform the sort. Furthermore, it can be applied to cell types that cannot tolerate the mechanical stress of sorting. On the other hand, only FACS allows for the controlled selection of cells with defined low, medium or high levels of nuclease. Thus, drug selection cannot specifically isolate low-nuclease expressers if, for example, a specific off-target or other event(s) is a concern in high-nuclease expressers. Similarly, it cannot specifically select for top-level transfected cells and typically therefore cannot achieve complete knockout or knockin editing in a single enrichment step.

10.4 Concluding Remarks

Our results show that FACS-based genome editing is a very efficient method for generation of model cell lines or cell pools that contain either knockout or knockin mutation of a gene of interest. For knockin of small mutations, ssODNs are the donor format of choice, as they are fully compatible with FACS-based editing, cost-efficient, and easy to design and obtain. Apart from being efficient, FACS-based genome editing also offers the possibility to balance editing efficiency and off-target/toxic effects, a feature not shared by other nuclease-enrichment strategies.

REFERENCES

Chen F, Pruett-Miller SM, Huang Y, *et al.* 2011. High frequency genome editing using ssDNA oligos and zinc finger nucleases. *Nat Methods* **8**: 753–755.

Duda K, Lonowski LA, Kofoed-Nielsen M, *et al.* 2014. High-efficiency genome editing via 2A-coupled co-expression of zinc finger nucleases or CRISPR/Cas nickase pairs. *Nucleic Acids Res* **42**(10): e84.

Frank S, Skryabin BV, Greber B. 2013. A modified TALEN-based system for robust generation of knock-out human pluripotent stem cell lines and disease models. *BMC Genomics* **14**: 773.

Fu XD, Ares M Jr. 2014. Context-dependent control of alternative splicing by RNA-binding proteins. *Nat Rev Genet* **15**(10): 689–701.

Hodgkins A, Farne A, Perera S, *et al.* 2015. WGE: a CRISPR database for genome engineering. *Bioinformatics* **31**(18): 3078–3080.

Lonowski LA, Narimatsu Y, Riaz A, *et al.* 2017. Genome editing using FACS enrichment of nuclease expressing cells and Indel Detection by Amplicon Analysis, *Nat Protocols* **12**(3): 581–603.

Ran FA, Hsu PD, Wright J, *et al.* 2013. Genome engineering using the CRISPR-Cas9 system. *Nat Protocols* **8**(11): 2281–2308.

Renaud JB, Boix C, Charpentier M, *et al.* 2016. Improved genome editing efficiency and flexibility using modified oligonucleotides with TALEN and CRISPR-Cas9 nucleases. *Cell Reports* **14**(9): 2263–2272.

Richardson CD, Ray GJ, DeWitt M, Curie GL, Corn JE. 2016. Enhancing homology-directed genome editing by catalytically active and inactive CRISPR-Cas9 using asymmetric donor DNA. *Nat Biotech* **34**: 339–344.

Szymczak AL, Vignali DA. 2005. Development of 2A peptide-based strategies in the design of multicistronic vectors. *Expert Opin Biol Ther* **5**(5): 627–638.

Yang L, Guell M, Byrne S, *et al.* 2013. Optimization of scarless human stem cell genome editing. *Nucleic Acids Res* **41**: 9049–9061.

Yang Z, Steentoft C, Hauge C, *et al.* 2015. Fast and sensitive detection of indels induced by precise gene targeting. *Nucleic Acids Res* **43**(9): e59.

Part III

Technology Development and Screening

11 CRISPR Genome Editing in Mice

Ella Paulina Thomson,[1] Ruby Emily Dawson,[1] James Nicholas Hughes and Paul Quinton Thomas

11.1 Introduction

Mice are one of the most widely used model organisms for dissecting mammalian gene function and human disease modeling. Targeted mutagenesis in mice has traditionally been performed using "gene targeting," in which mutations of interest are initially generated in embryonic stem (ES) cells by homologous recombination, which are then used to produce mutant mouse lines (Griep *et al.*, 2011). While this technique has been widely used, it is lengthy (1–2 years are often required to produce a mutant), labor-intensive, expensive and prone to failure. More recently, genome editing technologies such as zinc-finger nucleases (ZFNs) (Carbery *et al.*, 2010) and transcription activator-like effector nucleases (TALENs) (Davies *et al.*, 2013) have been successfully used to generate mutant mice via zygotic injection, although high expense and relatively cumbersome production methods have limited their use.

A paradigm shift in the field of genome editing occurred in 2013 when a group of landmark papers were published demonstrating that the clustered regularly interspaced short palindromic repeats (CRISPR) bacterial defense mechanism could be used to introduce specific mutations at precise locations within eukaryotic cells (Cong *et al.*, 2013; Jinek *et al.*, 2012; Mali *et al.*, 2013). This sparked enormous interest from the scientific community, and soon many laboratories around the world were using this system to modify a diverse variety of organisms, including plants (Xing *et al.*, 2014), yeast (DiCarlo *et al.*, 2013), zebrafish (Hwang *et al.*, 2013), fruit flies (Basset *et al.*, 2013) and monkeys (Niu *et al.*, 2014), among others (Onuma *et al.*, 2016; Wang *et al.*, 2016). This chapter focuses on the development and application of CRISPR mutagenesis in mice.

Over the past few years, CRISPR has been used extensively by our research group, allowing the generation of a variety of mutations in mice, including floxed alleles, null alleles, enhancer deletions, epitope FLAG-TAG insertions and promoter modifications. Based on our experience with this technique, we also provide insights into which strategies are the most efficient, as well as highlighting significant, but less well-known issues that can arise.

11.2 System Components and Activity

Establishing the mechanism by which the CRISPR system cleaves specific sequences in bacteria provided the groundwork for its rapid development as a genome editing tool in eukaryotes (Bolotin *et al.*, 2005; Brouns *et al.*, 2008; Cong *et al.*, 2013; Jinek *et al.*, 2012; Mali *et al.* 2013; Sorek *et al.*, 2013). The endonuclease in the CRISPR system is Cas9, which can be directed to bind and cleave specific genomic target sites through complementary base pairing of its single-guide RNA (sgRNA) to the DNA (Brouns *et al.*, 2008). By modifying the sequence of the sgRNA, Cas9 can be directed to precise locations within the genome, causing a double-strand break (DSB) in the DNA, the repair of which can be used to produce a variety of mutation types (Cong *et al.*, 2013). The flexibility and practicability of this approach provides a significant advantage over other genome editing platforms and has contributed to the massive uptake of CRISPR across the research community.

Despite its utility, it is important to note that the CRISPR–Cas9 system cannot be used to target every sequence in the genome due to the absolute requirement for a protospacer adjacent motif (PAM) sequence immediately 3′ to the targeting region of the sgRNA (Sternberg *et al.*, 2014). For the most widely used Cas9 variant (from *Streptococcus pyogenes*), the PAM corresponds to an NGG sequence, which on average will occur every 8 bp across the genome. Structural studies indicate that Cas9 binds directly to the PAM followed by DNA strand separation and "sampling" of the neighboring sequence for complementarity by the sgRNA (Sternberg *et al.*, 2014). Accordingly, mutation of the PAM will completely abolish cleavage by the Cas9, regardless of whether or not the sgRNA is a perfect match to the adjacent sequence (Jinek *et al.*, 2012; Sternberg *et al.*, 2014).

The mechanism by which Cas9 cleaves DNA is reasonably well understood and is covered in some detail in other chapters. Briefly, Cas9 has two active nuclease sites: the RuvC domain at the amino terminus and the HNH domain in the central region (Nishimasu *et al.*, 2014). Both of these active sites generate site-specific nicks on opposing DNA strands to create a blunt-ended double-strand break 3–4 nucleotides upstream of the PAM site (Jinek *et al.*, 2012). Repair of the DSB by the cell's endogenous repair machinery can result in the generation of alleles with the desired functional consequence (e.g. loss of function). Unlike traditional gene targeting, CRISPR mutagenesis permits directed modification of the zygotic genome through direct injection of sgRNA(s) designed to target the location of interest along with Cas9 mRNA or protein. Surviving zygotes are then transferred into pseudopregnant females for development to term when they can be genotyped to identify mutations of interest. Thus, remarkably, CRISPR mutagenesis permits mutant founder mice to be generated within three weeks (summarized in Figure 11.1)

11.3 Repair Mechanisms

CRISPR mutagenesis, like all forms of genome editing, relies on exploitation of the cell's endogenous repair mechanisms that ordinarily operate to protect the cell

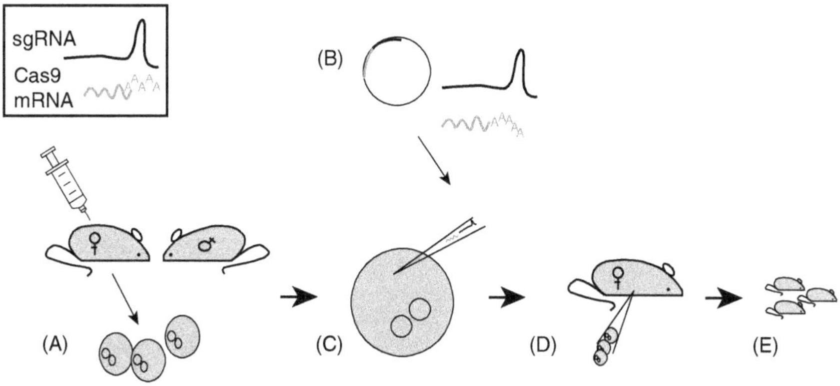

Figure 11.1 The process of using CRISPR in mice. (A) Female mice are superovulated by the injection of hormones, PMSG and HCG, and bred with males to obtain 20–30 zygotes per female. Cas9 mRNA and sgRNA are prepared through *in vitro* transcription from a plasmid template (B). CRISPR reagents are then microinjected into the cytoplasm or pronucleus of zygotes (C). Injected zygotes are transferred to pseudopregnant females that have been bred with vasectomized males (D). After three weeks gestation, mutant founder mice are born (E).

against DNA damage from sources such as replication error, fragile sites and ionizing radiation (Mehta and Haber, 2014). These repair mechanisms include canonical non-homologous end joining (cNHEJ), microhomology-mediated end joining (MMEJ), single-strand annealing (SSA) and homology-directed repair (HDR). Utilization of a particular repair mechanism is determined by a number of factors, such as cell cycle stage and flanking sequence, and, depending on the desired outcome, each of these can be exploited to generate specific types of mutations in mice, as outlined below and summarized in Figure 11.2 (Deriano and Roth, 2013).

11.3.1 Non-homologous End Joining

While cNHEJ repair is sensitive, rapid and precise, it can also result in mutations and can therefore be harnessed in CRISPR mutagenesis. This repair mechanism ligates and rejoins two blunt DNA ends without the use of a homologous template and with limited resection or restoration of nucleotides around the break (Davis and Chen, 2013; Dueva and Iliakis, 2013; Lieber, 1999; Matsuzaki *et al.*, 2012). The commonly held belief among users of CRISPR technology that NHEJ is highly error-prone is unlikely to be true (Bétermier *et al.*, 2014). This misconception arises in part because of the difficulty in measuring the fidelity of repair and is exacerbated by the fact that creation of DSBs and their repair will inevitably end only when a mistake is made. For example, in the context of CRISPR mutagenesis, precise repair will result in restoration of the target sequence, meaning that sequential rounds of cutting and repair will continue until misrepair destroys either the sgRNA-binding sequence or the PAM (Bétermier *et al.*, 2014). For this reason, CRISPR mutagenesis in the absence of a repair template invariably results in small insertions or deletions of nucleotides (indels) at the cut site. This property has been used extensively to generate frame-shifting loss-of-function mutations (Carbery *et al.*, 2010; Shen *et al.*, 2013).

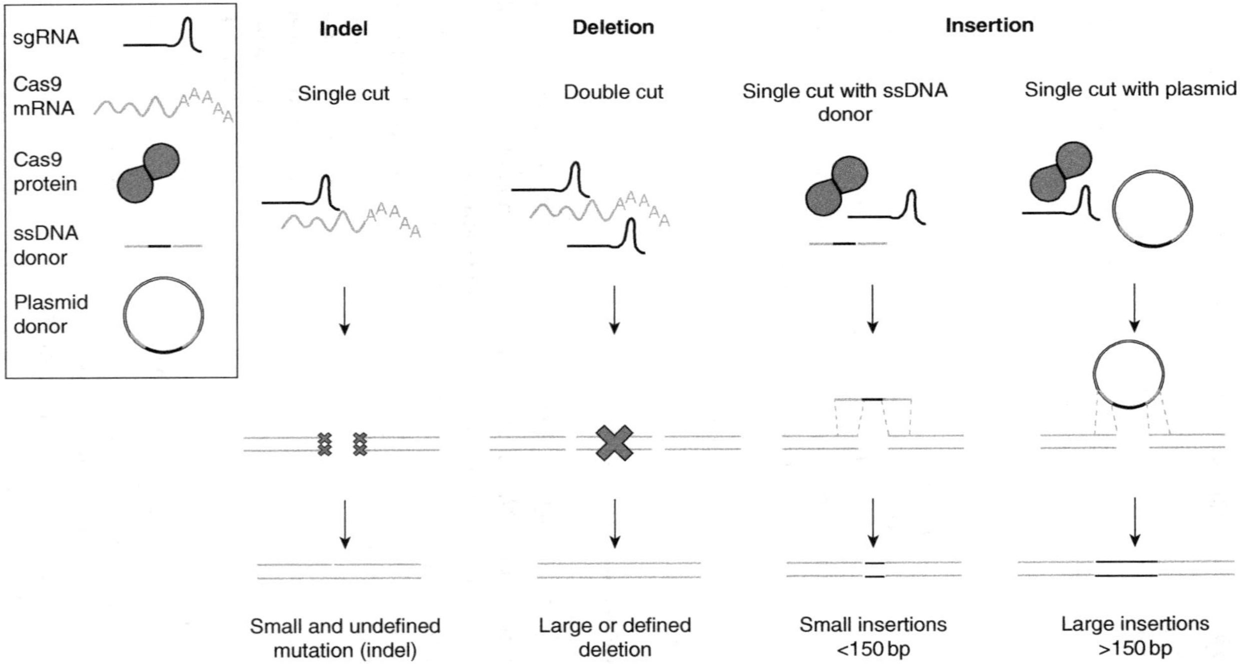

Figure 11.2 CRISPR can be used to generate a range of mutations in mice. Depending on the desired mutation, different CRISPR reagents are used to create indels and defined deletions through NHEJ or insertions through HDR. (A black-and-white version of this figure will appear in some formats. For the color version, please refer to the plate section.)

11.3.2 MMEJ and SSA (Alternative NHEJ)

Microhomology-mediated end joining (MMEJ) and single-strand annealing (SSA), which are subsets of the alternative NHEJ pathway (aNHEJ), are inherently error-prone, unlike cNHEJ (Dueva and Illakis, 2013; Simsek and Jasin 2010). Using enzymes distinct from cNHEJ, aNHEJ pathways repair DSBs by catalyzing the joining of DNA ends after resection, revealing small regions of homology on the two strands (Chiruvella *et al.* 2013; Crespan *et al.* 2012). Base pairing of the homologous regions is followed by trimming, extension and ligation to complete the repair, which has the effect of generating defined small deletions to the sequence (Crespan *et al.*, 2012; McVey and Lee, 2008; Roth and Wilson, 1986). SSA involves the annealing of larger repeat sequences, whereas MMEJ involves much shorter sequences of homology (5–25 bp; Decottignies, 2013). Because MMEJ and SSA rely on complementary base pairing of two homologous regions, the resultant deletions are predictable, allowing them to be used to our advantage in CRISPR genome editing.

11.3.3 Homology-directed Repair

Homology-directed repair (HDR) operates in competition with NHEJ to repair DSBs. The main difference between these mechanisms is that HDR relies on a repair template that is used to restore resected sequences at the break site and therefore usually results in precise repair (Jasin and Rothstein, 2013). Cells can undergo HDR when a homologous sequence of DNA is present, usually the homologous chromosome or sister chromatid (Deriano and Roth, 2013; Heyer *et al.*, 2010; Mao *et al.*, 2009). Importantly, exogenous DNA that is introduced to the cell can be copied into the genome to introduce specific alterations using this repair, by incorporating flanking regions of homology, which is a method that has historically been used in traditional gene-targeting methods. However the efficiency of this approach is very low, as millions of ES cells are typically needed in order to select for these rare recombination events (Hall *et al.*, 2009). However, when Cas9 is used to create the DSB, a remarkable increase in efficiency of template incorporation is seen, making it possible to directly modify mouse zygotes and avoid the use of ES cells entirely (Yang *et al.*, 2014).

It should be noted that all of the discussed repair mechanisms work in competition with each other, and greater understanding of these pathways should lead to an improved ability to control the type of repair used and hence the efficiency of a desired mutational outcome.

11.4 Harnessing Endogenous Repair Mechanism for Desired Mutational Outcomes

11.4.1 Single-cut Gene Disruption

The simplest way to utilize CRISPR in model organisms is to generate non-functional copies of genes following indel mutations caused by imprecise NHEJ, MMEJ or SSA. When the mutation causes a frameshift, this can result in a null allele due to loss of functional protein domain(s) (Isken and Maquat, 2007). Being able to create null mice in an efficient and reliable way is especially useful when none are commercially available on the background of choice (Barbaric *et al.*, 2007).

In 2013, Shen *et al.* published the first mutations to disrupt gene function in mice using a single-guide CRISPR strategy when they targeted a GFP transgene. They were successful in detecting mutations in 2 out of 12 founder mice. Since this study, many examples of single-cut indel mutants have been published, with mutations being found in almost all alleles (Qin *et al.*, 2015; Yasue *et al.*, 2014). In addition, Wang and colleagues (2013) and subsequent studies have demonstrated that this approach can be multiplexed to simultaneously generate mutations in multiple genes using a single zygotic injection. The utility of this approach was nicely demonstrated in a recent study of acute myeloid leukemia, in which different combinations of genes were inactivated simultaneously to allow for more accurate disease modeling (Heckl *et al.*, 2014). The use of CRISPR to generate null mice is now routine in research and is already becoming a leading genome editing technique (Singh *et al.*, 2015).

In our experience, generating single-cut null (frameshift) mice has been relatively simple. We use a strategy targeting an early exon to ensure incorporation of a stop codon soon after the translation start site. Although we have been successful in generating null mouse lines using the single-cut method, one of the disadvantages of this approach is the generation of in-frame deletions that occur in approximately one-third of mutated alleles, which can allow the resultant protein to retain significant function. In an attempt to overcome this problem, we have since moved away from this approach to a double-cut method, which also has the added benefit of easy genotyping (discussed in detail below).

11.4.2 Genotyping Single-cut Mutants

Given that single-cut mutations are generally small insertions or small deletions, they must be screened using a method that does not rely on discrimination based on mutation size. Assays that involve heteroduplex formation can be used to detect small changes to the sequence. Amplification around the mutated region will generate two or more PCR products with at least one base mismatch, which, after melting and annealing, will form DNA heteroduplexes. These heteroduplexes can either be visualized on a polyacrylamide gel, as they migrate much more slowly than homoduplexes, or they can be detected through restriction fragment length polymorphism (RFLP) in a T7E1 assay, where heteroduplexes are cleaved to form two distinct products (Shen *et al.*, 2013). These mutation screening techniques allow specific samples to be selected for sequence analysis.

Sequencing chromatograms from founder mice frequently contain overlapping distinct sequence reads from the cleavage site due to the presence of different mutations on multiple alleles. These can sometimes be difficult to unravel, particularly if more than two alleles are present as a result of Cas9 endonuclease activity post-zygotically. However, if necessary, mutant alleles can be discriminated by cloning of individual sequences. When the size difference between the two products is sufficient, the easiest method is to gel purify the individual alleles and sequence separately. Importantly, we have found that standard PCR screening using products of 200–400 bp will often miss alleles that result from larger deletions due to loss of the primer binding sequences. For this reason, it is important to out-cross founders before starting any experimental work to ensure that only the allele(s) of interest is present in the mutant line.

11.4.3 Double-cut Deletions

While frameshift mutations are very useful and relatively easy to make, the size and functional impact of deletions generated by a single cut can be difficult to predict. To overcome this issue, a double-cut strategy can be employed to generate deletions across a predetermined genomic region. The double-cut approach uses two sgRNAs flanking the interval of interest, and, after generation of the DSBs, rejoining of the blunt ends through NHEJ results in deletion of the intervening sequence (Fujii *et al.*, 2013; Mali *et al.*, 2013). Mice with specific deletions up to 10 kb were first reported by Fujii *et al.* (2013) and since, even larger deletions (65 kb) have been generated using this approach (Zhang *et al.*, 2015). Our group has adopted a double-cut strategy to delete enhancer regions, exons (around 200 bp) and whole genes (up to 9 kb). Exon deletions are now our favored approach to generate non-functional alleles by targeting cuts flanking a relatively short exon early in the transcript that contains a number of nucleotides not divisible by three. This approach can also be extended to other untranslated regions of the genome such as intronic and intergenic regions, promoters and UTRs.

11.4.4 Genotyping Double-cut Mutants

When there is a larger deletion or insertion being made, it is relatively easy to screen for the desired mutation using a PCR-based method, where a size difference will be apparent on an agarose gel. However, it is important to remember that whilst this is an efficient process, the DBS is not always at the precise predicted location and therefore can be slightly larger than intended due to NHEJ-mediated imperfect repair. For this reason, it is recommended to sequence all potential founders to determine the exact mutation.

The importance of screening for regions of homology or microhomology when designing the sgRNAs was emphasized to us when we observed a high frequency of identical small deletions in multiple mice generated from cleavage in a region of microhomology. This phenomenon has been reported previously in the literature in the context of CRISPR (Wang *et al.*, 2013) and indeed other gene editing reagents, such as TALENs (Tesson *et al.*, 2011). Microhomology can be used to predict precise deletion events, which, in theory, could be used to the advantage of the researcher (Bae *et al.*, 2014). However, these mechanisms can also result in a nasty surprise after mice have been generated, if repeat sequences are not identified during sgRNA design. Widely used current design tools do not always take this into account, but there are online tools available which can be used to screen guides for microhomology independently (Bae *et al.*, 2014; Haeussler *et al.*, 2016).

11.4.5 Point Mutations and Small Insertions

By using HDR-mediated CRISPR mutagenesis, it is now possible to introduce specific point mutations and small insertions such as epitope tags (e.g. HA-FLAG) relatively cheaply and rapidly compared to traditional methods (Wang *et al.*, 2013). These mutations are most easily achieved using a single-stranded oligonucleotide (ssDNA) donor that contains regions of homology (50–80 bp) to the endogenous locus flanking the donor sequence to be inserted at the cut site (Lee and Lloyd, 2014; Nakagawa *et al.*, 2016; Wang *et al.*, 2015; Yang *et al.*, 2014). There is ongoing optimization of the HDR-mediated CRISPR method to increase its low efficiency, including changes to homology arm length and silencing or inhibition of genes involved in NHEJ (Chu *et al.*, 2015; Maruyama *et al.*, 2015; Nakagawa *et al.*, 2016; Vartak and Raghavan, 2015). Substitution of a single base can also be used to generate silent mutations, which can destroy the relevant PAM site to prevent recutting after donor sequence insertion. From a practical standpoint we have found it useful to either destroy or introduce a restriction site as a silent mutation to aid in genotyping of founders.

Although we have been successful in generating 12 mouse lines with a variety of HDR-mediated insertions, in all cases NHEJ repair predominated, resulting in a relatively low efficiency of the desired insertion. In an attempt to increase the frequency of NHEJ repair events, we trialled co-injection of the DNA ligase IV inhibitor SCR7, which was previously shown to repress NHEJ and increase HDR efficiency (Chu *et al.*, 2015; Maruyama *et al.*, 2015; Vartak and Raghavan, 2015), although it did not increase insertion rates in our hands. Recently, it has been hypothesized in the literature that low HDR efficiency may be due to the donor oligonucleotides being degraded by the cell prior to the Cas9 mRNA being translated into functional protein (Ménoret *et al.*, 2015). We have found that injection of Cas9 protein rather than the mRNA directly into the zygote increases insertion efficiency as previously shown in the literature (Ménoret *et al.*, 2015). For the same reason, using phosphorothioate-modified oligos has also proven to be a useful approach to promote HDR insertion (Renaud *et al.*, 2016).

11.4.6 Generating Reporter and Floxed Conditional Mouse Lines

Using CRISPR to create more complex mutations, such as reporter gene insertions and floxed alleles, is possible, but generally much less efficient than the deletions and HDR-mediated small insertions described above. However, these modifications are becoming easier to generate using emerging CRISPR technologies, although strategies for longer or more complex insertions such as reporter genes (e.g. EGFP and LacZ) or two LoxP sequences are still being optimized. Whilst single-stranded oligonucleotides are an inexpensive and accessible option to insert small sequences into the mouse genome, often a much larger insert is required (Mali *et al.*, 2013; Yang *et al.*, 2013; 2014). Inserting longer sequences can be achieved using a plasmid donor template containing the insertion sequence flanked by large homology arms greater than 800 bp (Zhu *et al.*, 2015). This approach is similar in concept to targeting vectors that have been used in ES cells for over 30 years, although there is no requirement for a drug selection cassette for zygotic injections. To generate a large insertion, sgRNAs targeting the genomic region between the homology arms are injected along with the plasmid donor template (in linearized or circular form). The long homology arms provided by the donor template facilitate repair of the DSB via homologous recombination, thereby resulting in targeted insertion of the intervening sequence. It is important to note that the efficiency of this strategy is relatively low (8–10%) (Nakagawa *et al.*, 2016; Yang *et al.*, 2013). We have used this approach to generate targeted insertions of 750 bp–4 kb at three loci and, consistent with published reports, our insertion rate is relatively low (2.5–12.5%).

Floxed conditional alleles are probably the most challenging type of mutation to generate. Several alternative approaches have been used successfully, although all remain relatively inefficient. The first published approach used an oligonucleotide HDR repair strategy to simultaneously insert LoxP sequences either side of an exon of interest (Bishop *et al.*, 2016; Yang *et al.*, 2013). Using this strategy, we recently generated a floxed allele containing LoxP sites spaced 2 kb apart through co-injection of gRNAs and single-stranded oligonucleotides repair templates with 60 bp homology arms. Of the 11 founder pups born, only one had both the LoxP sequences inserted on the same allele, while the majority of the cut sites generated indels or single LoxP insertions and one case of deletion of the 2 kb region between the cuts. This relatively low efficiency is consistent with reports from the literature (Maruyama *et al.*, 2015; Nakagawa *et al.*, 2016; Yang *et al.*, 2013). The alternative approach utilizes a dsDNA (plasmid) repair template (identical in sequence to the intended floxed allele) that includes long homology arms (typically >1 kb). The floxed allele is generated by sgRNA-mediated cutting at one or both of the LoxP insertion sites, followed by homologous recombination repair. This approach has been used to insert LoxP sites separated by 71–600 bp albeit with low efficiency (2.7–16%) (Maruyama *et al.*, 2015; Yang *et al.*, 2013). A dual sgRNA Cas9-nickase approach has also been successfully used but with similarly low efficiency (2.7%) (Ran *et al.*, 2013; see below).

11.5 The CRISPR Toolbox

Since 2013, when the first mutant mouse lines were generated using the CRISPR system (Shen *et al.*, 2013; Wang *et al.*, 2013), the number of tools available to researchers has rapidly expanded to include modified versions of the wild-type *S. pyogenes* Cas9 (wt *Sp*Cas9) with altered PAM specificity, Cas9 proteins from a range of bacterial species and new programmable endonuclease enzymes (summarized in Table 11.1). This section focuses on the ever-increasing "CRISPR Toolbox" and some of the ways these new enzymes can be implemented to enable increased flexibility of target site selection and reduce off-target cutting.

11.5.1 Off-target Reduction

One of the enduring concerns of CRISPR mutagenesis is the occurrence of off-target endonuclease activity. Whilst breeding is a simple way to eliminate these from a mutant line, it can be time-consuming and expensive to backcross to be

Table 11.1 A summary of endonucleases that can be used in CRISPR genome editing

Name	Species	Engineered property	Endonuclease	PAM	Cells	Mice	Reference
SpCas9	*S. pyogenes*	N/A	Cas9	NGG	Yes	Yes	Cong *et al.*, 2013; Shen *et al.*, 2013
SpCas9n	*S. pyogenes*	Off-target minimization	Cas9	NGG	Yes	Yes	Ran *et al.*, 2013
SpCas9 VQR	*S. pyogenes*	PAM variant	Cas9	NGAN/NGNG	Yes	No	Kleinstiver *et al.*, 2015
SpCas9 EQR	*S. pyogenes*	PAM variant	Cas9	NGAG	Yes	No	Kleinstiver *et al.*, 2015
SpCas9 VRER	*S. pyogenes*	PAM variant	Cas9	NGCG	Yes	No	Kleinstiver *et al.*, 2015
eSpCas9	*S. pyogenes*	High fidelity	Cas9	NGG	Yes	No	Slaymaker *et al.*, 2016
SpCas9-HF1	*S. pyogenes*	High fidelity	Cas9	NGG	Yes	No	Kleinstiver *et al.*, 2016
SaCas9	*S. aureus*	N/A	Cas9	NGRRT/NGRRN	Yes	Yes	Ran *et al.*, 2013
KKH SaCas9	*S. aureus*	PAM variant	Cas9	NNNRRT	Yes	No	Kleinstiver *et al.*, 2015
AsCpf1	*Acidaminococcus* sp.	N/A	Cpf1	TTN	Yes	Yes	Zetsche *et al.*, 2015; Kim *et al.*, 2016
LbCpf1	*Lachnospiraceae*	N/A	Cpf1	TTN	Yes	Yes	Zetsche *et al.*, 2015; Kim *et al.*, 2016
dCas9-Fok1	*S. pyogenes*	Off-target minimization	FokI	NGG	Yes	Yes	Guilinger *et al.*, 2014; Hara *et al.*, 2015
NmCas9	*N. meningitidis*	PAM variant	Cas9	NNNNGATT	Yes	No	Lee *et al.*, 2016
StCas9	*S. thermophilus*	PAM variant	Cas9	NNAGAAW	Yes	No	Müller *et al.*, 2016

sure all off-targets have been segregated away. Targeted resequencing of potential off-target sites in founder mice indicates that sgRNA binding sites that differ by three or more nucleotides from the on-target sequence are very unlikely to be mutated (Haeussler *et al.*, 2016). Various online tools have been developed to predict where the most likely off-target cuts will occur and guide sequences can often be designed to avoid those within coding sequences (Haeussler *et al.*, 2016). In cases where there is limited flexibility in sgRNA targeting (e.g. generating a point mutation via HDR-mediated repair), potentially problematic off-target sgRNAs can now be paired with modified endonucleases that are able to minimize this occurrence (discussed below).

One of the first published solutions to off-target mutagenesis was the "dual nickase" strategy (Ran *et al.*, 2013). This approach uses a modified Cas9 protein containing an inactivating mutation in the RuvC domain that functions as a nickase. To create a mutation at the site of interest, a pair of sgRNAs (spaced by <100 bp) are used to nick opposite strands to create a dsDNA with a large 5′ overhang (Ran *et al.*, 2013). Through repair of this staggered break it is possible to generate indels or localized targeted modifications such as point mutations via HDR using a ssDNA repair template. Off-target nicks generated by a single sgRNA/Cas9 complex will be repaired efficiently by the high fidelity base excision repair method (Dianov and Hübscher, 2013), greatly minimizing the chances of unwanted mutations elsewhere in the genome. The dual nickase approach has been shown to be efficient in both cell lines and mouse models without any apparent reduction in on-target DNA cutting (Cheng *et al.*, 2014; Ran *et al.*, 2013).

Off-target cleavage has also been addressed by fusing an endonuclease "dead" Cas9 protein (dCas9) with a dimer-dependent FokI endonuclease domain to create fCas9. The FokI endonuclease domain has been used extensively in zinc-finger nuclease and TALEN genome editing, where pairs of DNA-binding proteins bind to adjacent sequences to enable generation of a double-strand break at the target site. A similar principle is used in the context of fCas9; two sgRNAs are designed to target regions 15–20 bp apart, allowing the two FokI modules to create the DSB in the intervening "spacer" region (Guilinger *et al.*, 2014). In human cell lines this approach was shown to have 140-fold higher specificity for target sites compared to wt *Sp*Cas9 (Guilinger *et al.*, 2014), and this has also been shown to be an effective strategy for use in mouse models, with similar efficiencies to Cas9 being reported (Hara *et al.*, 2015). However, a disadvantage of this approach is that regions that are able to be targeted via this method occur on average once every 34 bp, compared to every 8 bp for a single sgRNA strategy with wt *Sp*Cas9 (Guilinger *et al.*, 2014).

Structure-guided rational engineering has also been used to increase the fidelity of Cas9. Two different versions of Cas9 recently published were eSpCas9 (Slaymaker *et al.*, 2016) and SpCas9 HF-1 (Kleinstiver *et al.*, 2016). It was shown that these mutants are able to disproportionately decrease off-target binding with respect to on-target binding and hence improve specificity without compromising on-target activity in human cell lines, although this has not yet been tested in a mouse model.

11.5.2 PAM Variants

PAM sites for wt SpCas9 (NGG) are relatively common throughout genomes, although there are regions where it can be difficult to find a suitable sgRNA, such as A-T-rich regions, or in regions of highly repetitive sequence. Identification of Cas9 proteins that bind to different PAM sequences will therefore provide increased flexibility in target site identification.

To generate SpCas9 enzymes that recognize alternative PAM sequences, Kleinstiver *et al.* (2015) used a directed evolution approach targeting residues within the PAM-recognizing region of the Cas9 enzyme. This approach yielded a number of Cas9 mutants with altered PAM specificities: SpCas9 VQR, EQR and VRER, which recognize NGAN/NGNG, NGAG and NGCG, respectively. As with other Cas9 mutants, these have been tested in cell lines, but have not yet been validated in mouse zygotes.

CRISPR/Cas9 defense systems are widely used in bacteria, providing considerable scope for the development of new genome editing platforms. Inspired by the success of CRISPR/Cas9 from *Streptococcus pyogenes*, Cas9 orthologs from other bacterial species including *Staphylococcus aureus, Niessera meningiditis* (Lee *et al.*, 2016) and *Streptococcus thermophilus* (Kleinstiver *et al.*, 2015; Müller *et al.*, 2016) have recently been published, each of which utilizes a different PAM sequence. So far, only the SaCas9 has been shown to work successfully in mice, although the other strains are also likely to prove effective agents for generation of mutant mice.

11.5.3 Non-Cas9 Endonuclease

Whilst Cas9 is the pioneer endonuclease used in mammalian CRISPR applications, a newly recognized endonuclease known as Cpf1 has similar properties, and can also be harnessed for genome engineering (Zetsche *et al.*, 2015). By analyzing the CRISPR systems in hundreds of bacteria to determine if other enzymes are also capable of cutting DNA, Cpf1 was found in the *Acidaminococcus* and *Lachnospiraceae* families, and has been shown to be successful at creating mutations in human cell lines and mouse models (Hur *et al.*, 2016; Kim *et al.*, 2016).

Although they both work as part of a CRISPR bacterial defense system, Cas9 and Cpf1 have different modes of action. Specifically, Cpf1 cleaves DNA asymmetrically, such that rather than forming blunt ends, sticky ends are formed, creating a 4 or 5 bp 5′ overhang (Zetsche *et al.*, 2015). In addition, Cpf1 makes a staggered cut, 23 bp upstream from the PAM on the non-targeted strand and 18 bp on the targeted strand, rather than 3 bp upstream of the PAM site (Zetsche *et al.*, 2015). Due to the distance between the PAM site and the staggered DSB, repeated cleavage of the repaired allele can readily occur if the gRNA binding or PAM sequence is not altered. While this is tolerable for simple deletion strategies, it is essential that donor fragment integration includes a PAM-destroying silent mutation to eliminate the possibility of recutting after correct insertion.

In 2016 it was reported that the argonaute proteins from the bacterial species *Natronobacterium gregoryi* (NgAgo) could be used as ssDNA-guided genome editing agents in human cell lines (Gao *et al.*, 2016). The NgAgo protein was reported to function without the requirement for a PAM sequence, thereby bypassing one of the key limitations of existing CRISPR genome editing platforms. However, whilst this technology appeared to function efficiently in cell lines, replication studies have not yet been published. Indeed, concerns have been expressed about reproducibility of the published data. We recently compared the mutagenesis efficiency of wt SpCas9 and NgAgo in mouse zygotes (n = 29) and mouse ES cells and no evidence of mutations were detected in the NgAgo samples. Thus, the suitability of this platform for routine genome editing remains highly questionable.

11.6 Conclusion

The advent of CRISPR technology for genome editing has greatly improved the ease and affordability of creating mouse models for biological research. By harnessing the cell's endogenous repair mechanisms through this technology, researchers can now reliably and efficiently generate a range of mutations for a plethora of experimental applications. As the interest and use of CRISPR continues to escalate, it seems likely that new editing platforms and techniques will continue to be developed that will circumvent existing limitations, and inevitably increase the scope of research applications.

Acknowledgements

We would like to thank Louise Robertson and the members of the Thomas Laboratory for their help and willingness to share unpublished data.

NOTE

1. The authors contributed equally to the work.

REFERENCES

Bae S, Kweon J, Kim HS, Kim J-S. 2014. Microhomology-based choice of Cas9 nuclease target sites. *Nat Methods* **11**: 705–706.

Barbaric I, Miller G, Dear TN. 2007. Appearances can be deceiving: phenotypes of knockout mice. *Brief Funct Genomic Proteomic* **6**: 91–103.

Bassett AR, Tibbit C, Ponting CP, Liu J-L. 2013. Highly efficient targeted mutagenesis of *Drosophila* with the CRISPR/Cas9 system. *Cell Rep* **4**: 220–228.

Bétermier M, Bertrand P, Lopez BS. 2014. Is non-homologous end-joining really an inherently error-prone process? *PLoS Genet* **10**(1): e1004086.

Bishop KA, Harrington A, Kouranova E, *et al.* 2016. CRISPR/Cas9-mediated insertion of loxP sites in the mouse dock7 gene provides an effective alternative to use of targeted embryonic stem cells. *G3 (Bethesda)* **6**: 2051–2061.

Bolotin A, Quinquis B, Sorokin A, Ehrlich SD. 2005. Clustered regularly interspaced short palindrome repeats (CRISPRs) have spacers of extrachromosomal origin. *Microbiology* **151**: 2551–2561.

Brouns SJJ, Jore, MM, Lundgren M, *et al.* 2008. Small CRISPR RNAs guide antiviral defense in prokaryotes. *Science* **321**: 960–964.

Carbery ID, Ji D, Harrington A, *et al.* 2010. Targeted genome modification in mice using zinc-finger nucleases. *Genetics* **186**: 451–459.

Cheng R, Peng J, Yan Y, *et al.* 2014. Efficient gene editing in adult mouse livers via adenoviral delivery of CRISPR/Cas9. *FEBS Lett* **588**: 3954–3958.

Chiruvella KK, Liang Z, Wilson TE. 2013. Repair of double-strand breaks by end joining. *Cold Spring Harb Perspect Biol* **5**(5): a012757.

Chu VT, Weber T, Wefers B, *et al.* 2015. Increasing the efficiency of homology-directed repair for CRISPR-Cas9-induced precise gene editing in mammalian cells. *Nat Biotechnol* **33**: 543–548.

Cong L, Ran FA, Cox D, *et al.* 2013. Multiplex genome engineering using CRISPR/Cas systems. *Science* **339**: 819–823.

Crespan E, Czabany T, Maga G, Hübscher U. 2012. Microhomology-mediated DNA strand annealing and elongation by human DNA polymerases λ and β on normal and repetitive DNA sequences. *Nucleic Acids Res* **40**(12): 5577–5590.

Davies B, Davies G, Preece C, *et al.* 2013. Site specific mutation of the Zic2 locus by microinjection of TALEN mRNA in mouse CD1, C3H and C57BL/6J oocytes. *PLoS One* **8**: e60216.

Davis AJ, Chen DJ, 2013. DNA double strand break repair via non-homologous end-joining. *Transl Cancer Res* **2**: 130–143.

Decottignies A. 2013. Alternative end-joining mechanisms: a historical perspective. *Front Genet* **4**: 48.

Deriano L, Roth DB. 2013. Modernizing the nonhomologous end-joining repertoire: alternative and classical NHEJ share the stage. *Annu Rev Genet* **47**: 433–455.

Dianov GL, Hübscher, U. 2013. Mammalian base excision repair: the forgotten archangel. *Nucleic Acids Res* **41**: 3483–3490.

DiCarlo JE, Norville JE, Mali P, *et al.* 2013. Genome engineering in *Saccharomyces cerevisiae* using CRISPR-Cas systems. *Nucleic Acids Res* **41**: 4336–4343.

Dueva R, Iliakis G. 2013. Alternative pathways of non-homologous end joining (NHEJ) in genomic instability and cancer. *Transl Cancer Res* **2**: 163–177.

Fujii W, Kawasaki K, Sugiura K, Naito K. 2013. Efficient generation of large-scale genome-modified mice using gRNA and CAS9 endonuclease. *Nucleic Acids Res* **41**: e187.

Gao P, Yang H, Rajashankar KR, Huang Z, Patel DJ. 2016. Type V CRISPR-Cas Cpf1 endonuclease employs a unique mechanism for crRNA-mediated target DNA recognition. *Cell Res* **26**: 901–913.

Griep AE, John MC, Ikeda S, Ikeda A. 2011. Gene targeting in the mouse. *Methods Mol Biol* **770**: 293–312.

Guilinger JP, Thompson DB, Liu DR. 2014. Fusion of catalytically inactive Cas9 to FokI nuclease improves the specificity of genome modification. *Nat Biotechnol* **32**: 577–582.

Haeussler M, Schönig K, Eckert H. *et al.* 2016. Evaluation of off-target and on-target scoring algorithms and integration into the guide RNA selection tool CRISPOR. *Genome Biol* **17**: 148.

Hall B, Limaye A, Kulkarni AB. 2009. Overview: generation of gene knockout mice. *Curr Protoc Cell Biol* Chapter: Unit-19.1217.

Hara S, Tamano M, Yamashita S, *et al.* 2015. Generation of mutant mice via the CRISPR/Cas9 system using FokI-dCas9. *Sci Rep* **5**: 11221.

Heckl D, Kowalczyk MS, Yudovich D, *et al.* 2014. Generation of mouse models of myeloid malignancy with combinatorial genetic lesions using CRISPR-Cas9 genome editing. *Nat Biotechnol* **32**: 941–946.

Heyer W-D, Ehmsen KT, Liu J. 2010. Regulation of homologous recombination in eukaryotes. *Annu Rev Genet* **44**: 113–139.

Hur JK, Kim K, Been KW, *et al.* 2016. Targeted mutagenesis in mice by electroporation of Cpf1 ribonucleoproteins. *Nat Biotechnol* **34**(8): 807–808.

Hwang WY, Fu Y, Reyon D, *et al.* 2013. Efficient genome editing in zebrafish using a CRISPR-Cas system. *Nat Biotechnol* **31**: 227–229.

Isken O, Maquat LE. 2007. Quality control of eukaryotic mRNA: safeguarding cells from abnormal mRNA function. *Genes Dev* **21**: 1833–3856.

Jasin M, Rothstein R. 2013. Repair of strand breaks by homologous recombination. *Cold Spring Harb Perspect Biol* **5**: a012740.

Jinek M, Chylinski K, Fonfara I, *et al.* 2012. A programmable dual-RNA-guided DNA endonuclease in adaptive bacterial immunity. *Science* **337**: 816–821.

Kim D, Kim J, Hur JK, *et al.* 2016. Genome-wide analysis reveals specificities of Cpf1 endonucleases in human cells. *Nat Biotechnol* **34**: 863–868.

Kleinstiver BP, Pattanayak V, Prew MS, *et al.* 2016. High-fidelity CRISPR–Cas9 nucleases with no detectable genome-wide off-target effects. *Nature* **529**: 490–495.

Kleinstiver BP, Prew MS, Tsai SQ, *et al.* 2015. Engineered CRISPR-Cas9 nucleases with altered PAM specificities. *Nature* **523**: 481–485.

Lee AY-F, Lloyd KCK. 2014. Conditional targeting of Ispd using paired Cas9 nickase and a single DNA template in mice. *FEBS Open Bio* **4**: 637–642.

Lee CM, Cradick TJ, Bao G. 2016. The *Neisseria meningitidis* CRISPR-Cas9 system enables specific genome editing in mammalian cells. *Mol Ther J Am Soc Gene Ther* **24**: 645–654.

Lieber MR. 1999. The biochemistry and biological significance of nonhomologous DNA end joining: an essential repair process in multicellular eukaryotes. *Genes Cells Devoted Mol Cell Mech* **4**: 77–85.

Mali P, Yang L, Esvelt KM, *et al.* 2013. RNA-guided human genome engineering via Cas9. *Science* **339**: 823–826.

Mao Z, Jiang Y, Liu X, Seluanov A, Gorbunova V. 2009. DNA repair by homologous recombination, but not by nonhomologous end joining, is elevated in breast cancer cells. *Neoplasia N Y N* **11**: 683–691.

Maruyama T, Dougan SK, Truttmann MC, *et al.* 2015. Increasing the efficiency of precise genome editing with CRISPR-Cas9 by inhibition of nonhomologous end joining. *Nat Biotechnol* **33**: 538–542.

Matsuzaki K, Terasawa M, Iwasaki D, Higashide M, Shinohara M. 2012. Cyclin-dependent kinase-dependent phosphorylation of Lif1 and Sae2 controls imprecise nonhomologous end joining accompanied by double-strand break resection. *Genes Cells Devoted Mol Cell Mech* **17**: 473–493.

McVey M, Lee SE. 2008. MMEJ repair of double-strand breaks (director's cut): deleted sequences and alternative endings. *Trends Genet* **24**: 529–538.

Mehta A, Haber JE. 2014. Sources of DNA double-strand breaks and models of recombinational DNA repair. *Cold Spring Harb Perspect Biol* **6**: a016428.

Ménoret S, De Cian A, Tesson L, *et al.* 2015. Homology-directed repair in rodent zygotes using Cas9 and TALEN engineered proteins. *Sci Rep* **5**: 14410.

Müller M, Lee CM, Gasiunas G, *et al.* 2016. *Streptococcus thermophilus* CRISPR-Cas9 systems enable specific editing of the human genome. *Mol Ther J Am Soc Gene Ther* **24**: 636–644.

Nakagawa Y, Sakuma T, Nishimichi N, *et al.* 2016. Ultra-superovulation for the CRISPR-Cas9-mediated production of gene-knockout, single-amino-acid-substituted, and floxed mice. *Biol Open* **5**(8): 1142–1148.

Nishimasu H, Ran FA, Hsu PD, *et al.* 2014. Crystal structure of Cas9 in complex with guide RNA and target DNA. *Cell* **156**: 935–949.

Niu Y, Shen B, Cui Y, *et al.* 2014. Generation of gene-modified cynomolgus monkey via Cas9/RNA-mediated gene targeting in one-cell embryos. *Cell* **156**: 836–843.

Onuma A, Fujii W, Sugiura K, Naito K. 2016. Efficient mutagenesis by CRISPR/Cas system during meiotic maturation of porcine oocytes. *J Reprod Dev* **63**(1): 45–50.

Qin W, Dion SL, Kutny PM, *et al.* 2015. Efficient CRISPR/Cas9-mediated genome editing in mice by zygote electroporation of nuclease. *Genetics* **200**: 423–430.

Ran FA, Hsu PD, Lin C-Y, *et al.* 2013. Double nicking by RNA-guided CRISPR Cas9 for enhanced genome editing specificity. *Cell* **154**: 1380–1389.

Renaud J-B, Boix C, Charpentier M, *et al.* 2016. Improved genome editing efficiency and flexibility using modified oligonucleotides with TALEN and CRISPR-Cas9 nucleases. *Cell Rep* **14**: 2263–2272.

Roth DB, Wilson JH. 1986. Nonhomologous recombination in mammalian cells: role for short sequence homologies in the joining reaction. *Mol Cell Biol* **6**: 4295–4304.

Shen B, Zhang J, Wu H, *et al.* 2013. Generation of gene-modified mice via Cas9/RNA-mediated gene targeting. *Cell Res* **23**: 720–723.

Simsek D, Jasin M. 2010. Alternative end-joining is suppressed by the canonical NHEJ component Xrcc4-ligase IV during chromosomal translocation formation. *Nat Struct Mol Biol* **17**: 410–416.

Singh P, Schimenti JC, Bolcun-Filas E. 2015. A mouse geneticist's practical guide to CRISPR applications. *Genetics* **199**: 1–15.

Slaymaker IM, Gao L, Zetsche B, *et al.* 2016. Rationally engineered Cas9 nucleases with improved specificity. *Science* **351**: 84–88.

Sorek R, Lawrence CM, Wiedenheft B. 2013. CRISPR-mediated adaptive immune systems in bacteria and archaea. *Annu Rev Biochem* **82**: 237–266.

Sternberg SH, Redding S, Jinek M, Greene EC, Doudna JA. 2014. DNA interrogation by the CRISPR RNA-guided endonuclease Cas9. *Nature* **507**: 62–67.

Tesson L, Usal C, Ménoret S, *et al.* 2011. Knockout rats generated by embryo microinjection of TALENs. *Nat Biotechnol* **29**: 695–696.

Vartak SV, Raghavan SC. 2015. Inhibition of nonhomologous end joining to increase the specificity of CRISPR/Cas9 genome editing. *FEBS J* **282**: 4289–4294.

Wang B, Li K, Wang A, *et al.* 2015. Highly efficient CRISPR/HDR-mediated knock-in for mouse embryonic stem cells and zygotes. *BioTechniques* **59**: 201–202, 204, 206–208.

Wang H, Yang H, Shivalila CS, *et al.* 2013. One-step generation of mice carrying mutations in multiple genes by CRISPR/Cas-mediated genome engineering. *Cell* **153**: 910–918.

Wang X, Cai B, Zhou J, *et al.* 2016. Disruption of FGF5 in cashmere goats using CRISPR/Cas9 results in more secondary hair follicles and longer fibers. *PLoS One* **11**: e0164640.

Xing H-L, Dong L, Wang Z-P, *et al.* 2014. A CRISPR/Cas9 toolkit for multiplex genome editing in plants. *BMC Plant Biol* **14**: 327.

Yang H, Wang H, Jaenisch R. 2014. Generating genetically modified mice using CRISPR/Cas-mediated genome engineering. *Nat Protoc* **9**: 1956–1968.

Yang H, Wang H, Shivalila CS, *et al.* 2013. One-step generation of mice carrying reporter and conditional alleles by CRISPR/Cas mediated genome engineering. *Cell* **154**: 1370–1379.

Yasue A, Mitsui SN, Watanabe T, *et al.* 2014. Highly efficient targeted mutagenesis in one-cell mouse embryos mediated by the TALEN and CRISPR/Cas systems. *Sci Rep* **4**: 5705.

Zetsche B, Gootenberg JS, Abudayyeh OO, *et al.* 2015. Cpf1 is a single RNA-guided endonuclease of a class 2 CRISPR-Cas system. *Cell* **163**: 759–771.

Zhang L, Jia R, Palange NJ, *et al.* 2015. Large genomic fragment deletions and insertions in mouse using CRISPR/Cas9. *PLoS One* **10**: e0120396.

Zhu Z, Verma N, González F, Shi Z-D, Huangfu D. 2015. A CRISPR/Cas-mediated selection-free knockin strategy in human embryonic stem cells. *Stem Cell Rep* **4**: 1103–1111.

12 Detection of Insertion/Deletion (Indel) Events after Genome Targeting: Pros and Cons of the Available Methods

Eric P. Bennett, Ashley M. Jacobi, Garrett R. Rettig and Mark A. Behlke

12.1 Introduction

Recent advances in genome editing techniques enable modification of complex genomes at the single nucleotide level. Although genome modification can be accomplished by a variety of proven, highly efficient approaches in cells, tissues and animal models (Cong *et al.*, 2013; Fu *et al.*, 2013; Geurts *et al.*, 2010; Jinek *et al.*, 2013; Mali *et al.*, 2013; Moscou and Bogdanove, 2009; Urnov *et al.*, 2005), fast and accurate methods that identify the nature of the genetic modifications introduced remain limited. In particular, with the discovery of the CRISPR/Cas9 system and its proven utility in mammalian cells, genome editing can now be performed in large numbers using laboratory robotics. However, characterization of the resulting genetic changes can be a bottleneck. Swift and accurate methods to identify the resultant genetic modifications are few, labor-intensive and often require screening of hundreds or thousands of clones in order to obtain correctly edited bi-, tri- or multi-allelic genes. Therefore, methods are needed to screen cell populations or clones that have undergone high-throughput genome targeting and are capable of distinguishing the variety of repair products that result from these targeting events. For certain applications, such as therapeutic genome targeting, this may turn out to be of utmost importance.

In this chapter we review the available methods for precise detection of indel events in cells and discuss the practical advantages and problems related to use of these methods in genome-targeting workflows. Focus will be put on the recently described indel detection by amplicon analysis (IDAA) methodology (Yang *et al.*, 2015; Figure 12.1) that is based on amplicon fluorophore labeling and detection by DNA capillary electrophoresis and has been shown to circumvent limitations associated with previously used indel detection methods.

12.2 Indel Detection Methodologies

The development of methods for rapid editing of genomes at the single, multi-cellular and whole-animal level have revolutionized the way genome targeting is

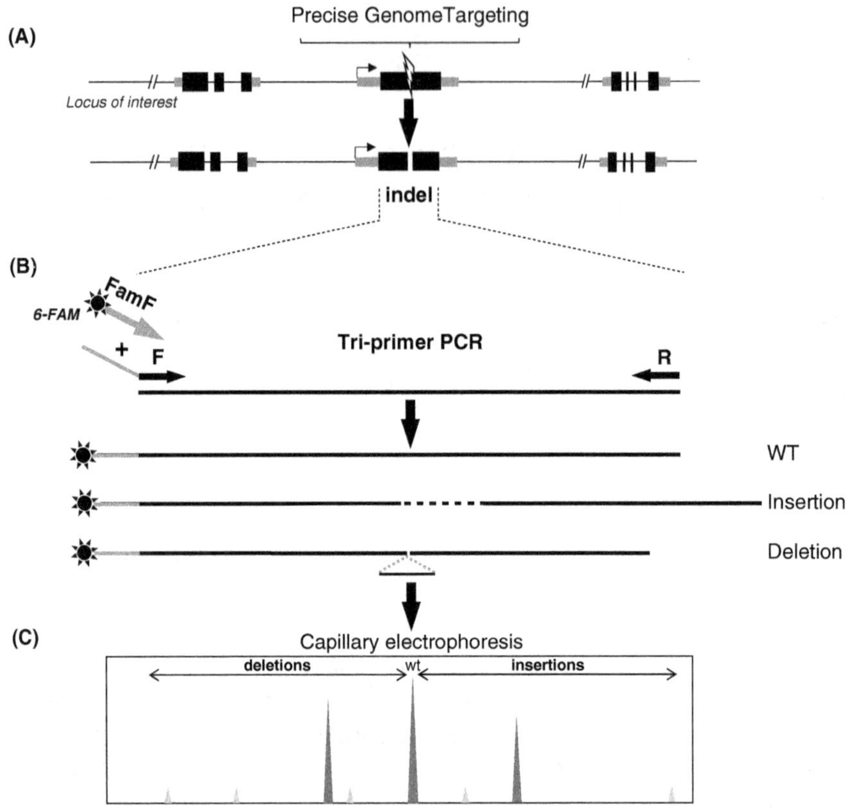

Figure 12.1 Schematic illustration of the IDAA principle. (A) Targeting of a genomic locus of interest with precise genome-targeting tools will induce indels at the target site. (B) Tri-primer amplification using the essential fluorophore-labeled (black star) universal primer (FamF), allows for homogeneous amplicon labeling, as described previously (Yang *et al.*, 2015). (C) Fluorophore-labeled amplicons (marked by black stars) are analyzed by capillary electrophoretic fragment analysis using commonly available sequencing equipment, such as ABI3130, 3500 or later instruments (Applied Biosystems/Thermo Fisher Scientific, USA). Raw fragment analysis data can be processed using freely available software packages as described recently (Lonowski *et al.*, 2017). Essential size standard (minor light gray peaks) included in the assay enables direct comparison of indels (major dark gray peaks) between run profiles.

performed. Specific genomic DNA sites can now be targeted by an arsenal of genetic tools that all lead to a double-stranded break at the target site; the break can be repaired by different enzymatic pathways that result in a variety of different classes of repair products. Repair options include classical non-homologous end joining (C-NHEJ), alternative non-homologous end joining (Alt-NHEJ), also known as microhomology-mediated end joining (MMEJ) (Bae *et al.*, 2014; Deriano and Roth, 2013; Nakade *et al.*, 2014), as well as homologous recombination (HR) (Smith, 2004; Stark *et al.*, 2004; Symington and Gautier, 2011). The effectiveness of the currently used genome-targeting tools has been shown to induce a diverse range of indels at the target site in cell pools, tissues and animals (Beumer *et al.*, 2013; Chen *et al.*, 2013; Kim *et al.*, 2013; Kuan *et al.*, 1989; Yang *et al.*, 2013). Several studies have addressed the targeting efficiency of the CRISPR/Cas9 system at different sites and sequence contexts. One such study evaluated the relative activities of 218 sgRNAs using *in vitro* Surveyor® heteroduplex mismatch cleavage assays (Liu *et al.*, 2016), while other studies assessed guide RNA cleavage activity using next-generation sequencing (NGS) analysis (Doench *et al.* 2014; 2016; Shalem *et al.*, 2013). Collectively these and other studies address the functional activity of the different gRNAs based on their ability to cause detectable indel mutations. A recent study by Van Overbeek *et al.* (2016) used NGS analysis to compare the cleavage/repair events for 223 individual gRNAs using various delivery systems and cell line model systems, and found that the repair patterns following a gRNA cleavage event were non-random, reproducible and largely determined by the C-NHEJ and MMEJ capacity of the cells. A given gRNA was associated with a unique "fingerprint" of repair products that were characteristic of that site in specific cells. These observations are consistent with recently published findings from our group (Lonowski *et al.*, 2017).

Currently there are only a few methods available to assess the full spectrum of repair products in a cell population following CRISPR/Cas9 cleavage. Choice of indel detection methodology to be used is primarily dictated by the scope of the experiment and the available analytical resources. Direct evaluation of the sequences of thousands of sites by NGS methods may be the best route to assess the full spectrum of repair products following CRISPR/Cas9 cleavage, but the time and cost constraints associated with such experiments limit widespread adoption of this method. Classic and cost-effective mutation-screening methodologies could be employed, such as denaturing high-performance liquid chromatography (DHPLC) (Eshaghpour and Crothers, 1978; Liu *et al.*, 2002), single-stranded conformational polymorphism (SSCP) analysis (Inazuka *et al.*, 1997), and denaturing gradient gel electrophoresis (DGGE) (Børresen *et al.*, 1988); however, in many cases these methodologies do not provide sufficient detail about the nature of the repair products present. In the following, we review features of the indel detection methods most commonly used for evaluation of genomic DNA following genome-targeting experiments.

12.2.1 RFLP and Fragment Analysis

The earliest indel detection methods were based on restriction fragment length polymorphism (RFLP) assays. These assays were sufficient to monitor cutting efficiencies of first-generation genome modification tools based on FOK-I fused zinc-finger nucleases (ZFNs) and TALENs in both mammalian cells and whole animals (Bibikova *et al.*, 2001; Smith *et al.*, 1999; 2000). Subsequently, RFLP assays were also used to monitor the efficacy of nuclease-induced, HR-driven target integration of single-stranded oligonucleotides (ssODNs) for simple PCR-based identification of correct targeting, with or without inclusion of diagnostic restriction enzyme cleavage sites (Chen *et al.*, 2011; Miller *et al.*, 2007; Urnov *et al.*, 2005). Although RFLP methods are cost-efficient to use, the approach requires the presence of diagnostic restriction sites at the targeted site. Furthermore, RFLP methods are laborious, lack detection sensitivity in heterogeneous cultures and are not amenable to high-throughput applications (Figure 12.2).

12.2.2 PCR-based Methods

Various PCR-based methods have also been developed for indel detection and discrimination of genome-modified cells (Yu *et al.*, 2014). One such assay, based on high-resolution melt analysis (HRMA) of fluorescently stained PCR products, is a simple, rapid, cost-efficient and sensitive method for detecting indels (Dahlem *et al.*, 2012). However, PCR applications are collectively hampered by the lack of detailed size discrimination power at the individual indel level in a given genome-targeted sample. Thus, for in-depth indel profiling of edited cell pools, tissues or organisms, PCR-based indel detection methods are inadequate (Figure 12.2).

12.2.3 Endonuclease Mismatch Cleavage (EMC) Assays

Enzymatic mutation detection methods comprise four simple steps: (1) PCR amplification of a region that includes the targeted genomic cut site, (2) formation of heteroduplex amplicons, (3) selective cleavage of heteroduplexes (but not homoduplexes) by a mismatch-sensitive endonuclease and (4) detection of the cleavage event (Yeung *et al.*, 2005). These endonuclease mismatch cleavage (EMC) assays include use of various nucleases such as plant-derived CEL-I endonuclease (Oleykowski *et al.*, 1998; Pimkin *et al.*, 2007), bacteriophage-derived T7 endonuclease-I (Mashal *et al.*, 1995), T4 endonuclease VII (T4E7) (Youil *et al.*, 1995), endonuclease V (EndoV) (Gao *et al.*, 2007) and Surveyor® nuclease (Surveyor) (Qiu *et al.*, 2004). Studies have established that the efficiency of EMC is dependent on the properties of the mismatch-cleavage endonuclease used. Recent comparative studies of the T7EI and Surveyor nucleases demonstrated that: (a) Surveyor exhibits unwanted exonucleolytic activity (Huang *et al.*, 2012), (b) T7EI outperforms Surveyor with regard to signal-to-noise ratio (Sakurai *et al.*, 2014; Vouillot *et al.*, 2015) and (c) T7EI has a preference for heteroduplex DNA formed by deletions (Huang *et al.*, 2012). One of the first descriptions of the use of EMC for genome-targeting indel detection involved Surveyor nuclease-mediated monitoring of IL2Rγ exon 5 gene-modification rates (Miller *et al.*, 2007), and subsequently EMC has become one of the most commonly used genome-targeting indel detection assays, likely due to its ease of use, low cost and fast turnaround time (Figure 12.2). However, a recent comparative study demonstrated obvious limitations in single-base mutation detection for T7EI and Surveyor enzymes (Vouillot *et al.*, 2015). This observation is in line with those made by us and others showing that T7EI poorly discriminates the presence of single-base deletion events (Yang *et al.*, 2015). In the latter case, various analytical systems were employed, aimed at improving the analytical resolution of EMC products analyzed, but in all cases failed to overcome the inherent risk of false-negative data from this approach (data not shown). In some cases where the heterogeneity of repair products is limited, mutant PCR amplicons can reanneal to form homoduplexes, which are also missed by EMC assays. Thus, while convenient, EMC assays will almost

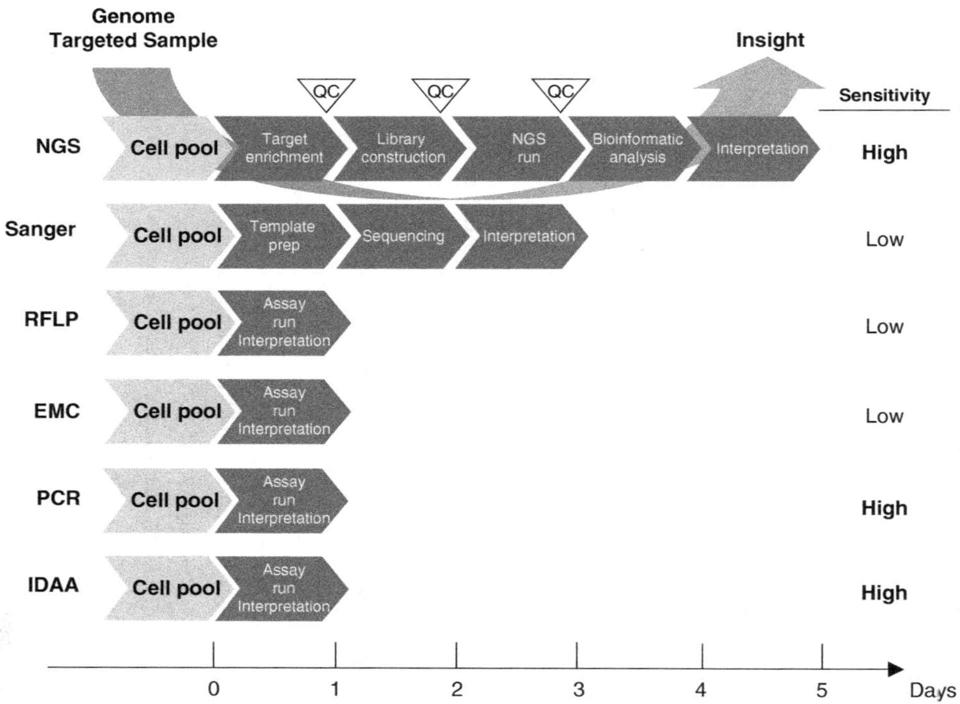

	Sensitivity	Resolution	HTRP	Cost
NGS	**High**	**High**	Low	High
Sanger	Low	Low	Low	High
RFLP	Low	Low	Low	**Low**
EMC	Low	Low	Low	**Low**
PCR	**High**	Low	**High**	**Low**
IDAA	**High**	**High**	**High**	**Low**

Figure 12.2 Schematic presentation of the assay time required and performances of the available genome-targeting indel detection methods. Approximate time estimates in days (axis shown below) for completing the required steps for each respective method (dark gray hexagons) from genome-targeted sample (light gray hexagon) to insight. QC represents necessary quality control steps required for NGS. The table on the right displays the performance (high or low) of the respective methods with regard to sensitivity, resolution, high-throughput amenability (HTRP) and cost. Preferred choices are shown in bold font. Necessary NGS quality control workflow check points are indicated by QC marked triangles.

always underestimate the degree of editing present in a cell population following a genome editing experiment.

12.2.4 DNA Sequencing

Sensitive and quantitative indel detection methods have been developed based on sequencing of amplicons covering the targeted genomic locus. By this approach, locus-derived amplicons are cloned into plasmids and individual clones are Sanger-sequenced followed by bioinformatics compilation of individual sequences and indel identification. The laborious amplicon cloning involved in these procedures has been streamlined by a method that only requires a pair of PCR reactions and Sanger-sequencing runs per sample followed by sequence trace analysis using a specially developed decomposition algorithm that identifies the major induced mutations (Brinkman *et al.*, 2014). This approach, termed Tracking of Indels by DEcomposition (TIDE), is straightforward, but indel detection is limited by indel size, lacks sensitivity for indel profiling of samples with high indel complexity (indels derived from cell pools) and is dependent on high-quality Sanger-sequencing data. The advent of next-generation sequencing (NGS) methodologies has proven to be a more powerful and widely accepted method for in-depth sequencing of amplicons covering a targeted genomic locus. The millions of NGS-generated sequence reads are bioinformatically deconvoluted and allow for accurate and highly sensitive profiling of the complete spectrum of indels induced by a particular genome-targeting tool. However, NGS-associated bioinformatic data analysis is not straightforward to those not expert in such methods. To simplify analysis, several online resources have been developed for indel quantification and characterization of NGS data, including CRISPR Genome Analyzer (CRISPR-GA) (Güell *et al.*, 2014), CRISPResso (Pinello *et al.*, 2016) and CRISP-R (Lindsay *et al.*, 2015), that all provide a report integrating NGS information. Undoubtedly, NGS-based indel profiling gives the most information and has the ability to simultaneously characterize thousands of different repair events; however, for many researchers this method is too slow, complex and costly for routine use (Figure 12.2). Thus, until recently, there was an unmet need for inexpensive, fast, sensitive and robust indel detection in the precise genome-targeting field.

12.2.5 Indel Detection by Amplicon Analysis (IDAA) Methodology

An attractive method for the discrimination of single-base differences among DNA fragments, such as amplicons, is based on capillary electrophoretic separation of DNA fragments under denaturing conditions, also known as fragment analysis (FA). FA is commonly applied to linkage and microsatellite instability mapping, loss of heterozygosity analysis, SNP genotyping, etc. In the field of genome targeting, this analytical principle has been used for analysis of amplicons that have been fluorescently labeled using fluorophore-labeled gene-specific primers followed by size discrimination using standard sequencing equipment available in most molecular biology laboratories (Beumer *et al.*, 2013; Foley *et al.*, 2009; Kim *et al.*, 2011). Although these early procedures were effective, the use of fluorophore-labeled gene-specific primers is both costly and non-flexible. For greater flexibility we introduced the tri-primer fluorophore amplicon labeling principle for IDAA (Yang *et al.*, 2015). IDAA is a single-step tri-primer amplification procedure that enables uniform fluorescence labeling of amplicons derived from a precise gene-targeted locus followed by size discrimination by FA (Figure 12.1). The resolution power of IDAA down to a single base was demonstrated in recent publications (Kosicki *et al.*, 2017; Lonowski *et al.*, 2017; Yang *et al.*, 2015) and collectively IDAA fulfills all the requirements for low cost, sensitive, high-resolution and high-throughput detection/characterization of indels, as outlined in Figures 12.1 and 12.2. Importantly, we have determined that the detection sensitivity of IDAA is similar to NGS (Lonowski *et al.*, 2017) and in the next section we exemplify the assay detection capabilities by in-depth profiling of indel formation in the short time window immediately after cellular delivery of the Cas9/gRNA complex.

12.3 CRISPR/Cas9-targeting Efficiency, Indel Profile and Dynamics

A limited number of studies have examined the in-depth indel profile and dynamics immediately after cellular delivery of CRISPR/Cas9-targeting tools. To the best of our knowledge and during the course of preparing this review, only two studies have addressed this in detail (Kosicki *et al.*, 2017; Van Overbeek *et al.*, 2016). In this regard, it is widely assumed that precise genome targeting-induced indel events occur within the first 2–3 days post-transfection of the Cas9/gRNA complex. Such studies can be performed in cells that stably express Cas9 by just transfecting the gRNA complex (sgRNA or a crRNA:tracrRNA pair) or by delivering both the Cas9 and the gRNA(s). Here, we present data from studies done in HEK293 cells with stable, low expression of *Streptococcus pyogenes* Cas9 (HEK-Cas9 cells) using Lipofectamine® RNAiMAX (Thermo Fisher Scientific, USA) and 30 nM of an Alt-R™ crRNA:tracrRNA gRNA complex (Integrated DNA Technologies, USA). The gRNA targeted a locus within the human *HPRT1* gene. The gRNA was delivered into cultured cells in triplicate biological replicates (Figure 12.3A) in 96-well plates followed by 3, 6, 12, 24, 48 or 72 h incubation before harvesting DNA using Quick Extract™ DNA Extraction Solution (Epicentre, USA). Crude lysates were analyzed by: (1) EMC assay (KAPA HiFi PCR amplification and heteroduplex formation followed by digestion with 2 units T7EI), (2) IDAA profiling as described recently (Lonowski *et al.*, 2017) or (3) NGS. T7EI-based EMC was performed on a fragment analyzer, IDAA on an ABI3500XL and NGS on a MiSeq. In addition, one sample was NGS-profiled on a MiSeq (Illumina, USA). Results are displayed in Figure 12.3B–E. Nearly identical IDAA profiles were observed across the triplicate samples included in the study (Figure 12.3A) and were in line with similar results obtained by us and others previously (Lonowski *et al.*, 2017; Van Overbeek *et al.*, 2016). The T7EI results (Figure 12.3B) show that cleavage products (7.6%) were detectable 6 h post-gRNA transfection and increased at subsequent time points. The panels indicate that a substantial fraction of amplicons (~30%) remain uncut up to 72 h. Very different

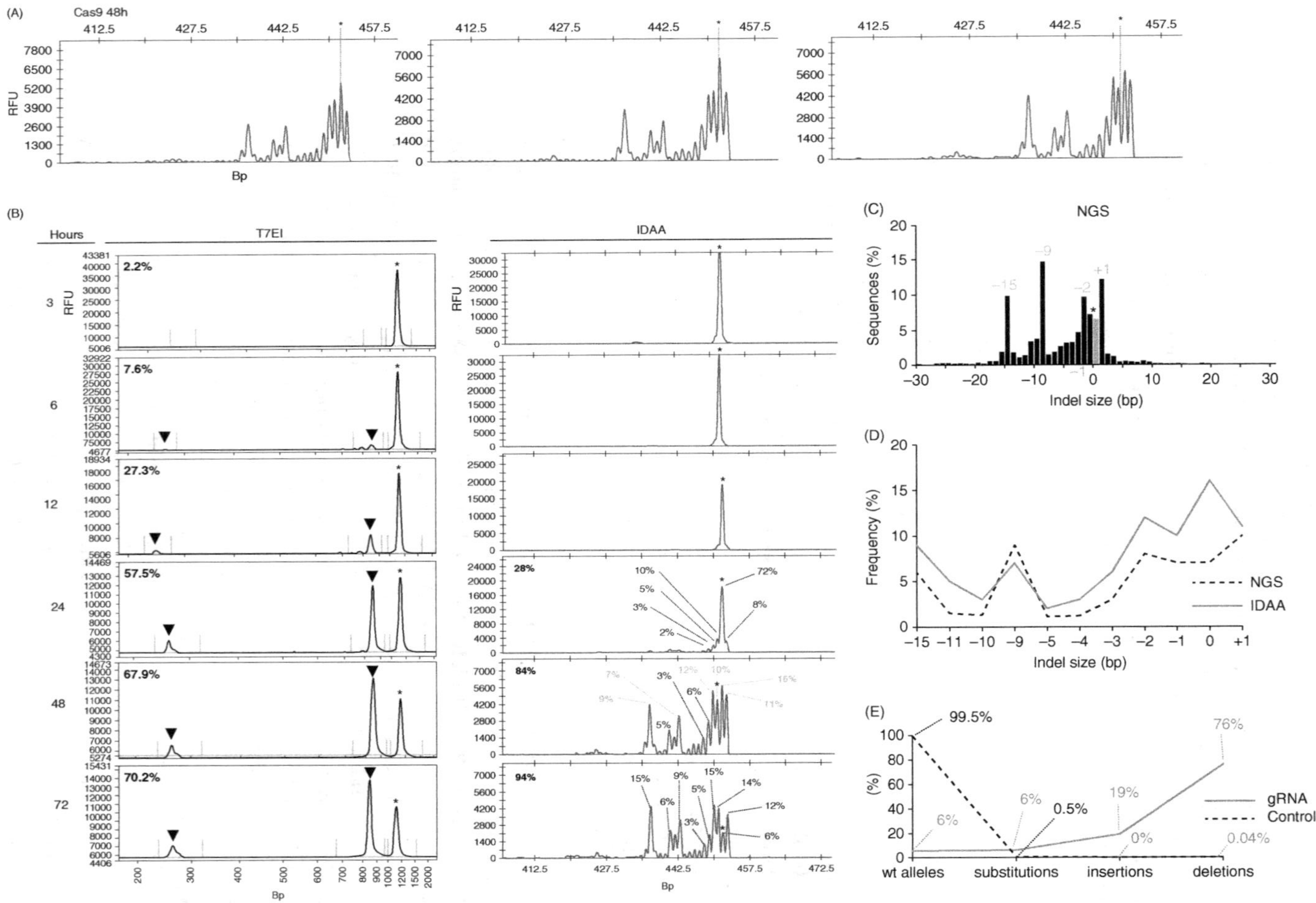

Figure 12.3 Comparative analysis of the performance of EMC, IDAA and NGS-based indel detection methods. (A) IDAA profiles for triplicate 48 h samples. (B) Representative panels from triplicate 3 h, 6 h, 12 h, 24 h, 48 h, 72 h samplings analyzed by T7EI and IDAA. For T7EI, the PCR primers used were; HPRT For primer –

dynamics were observed by IDAA profiling of the same samples, where indels only become evident after 12–24 h and accumulated rapidly thereafter, with unmodified amplicon present at only 6% for the 72 h time point. These results clearly suggest that gRNA delivery was highly efficient, that the majority of cells analyzed were exposed to the CRISPR/Cas9-targeting tools and that detectable editing events occurred in up to 94% of cells after 72 h. In addition, the 48 h time point sample was also analyzed by NGS (Figure 12.3C) and, interestingly, the IDAA and NGS profiles generated were strikingly similar (nearly identical "fingerprints"), as reported by us previously (Lonowski *et al.*, 2017; Yang *et al.*, 2015). For example, there was a high degree of agreement between the prominent NGS and IDAA indel events: +1 bp, –1 bp, –2 bp, –9 bp and –15 bp (Figure 12.2B and C). Comparative quantification of the individual indels/peaks identified is in agreement between the two methods and displays highly similar profiles (Figure 12.3D). In this regard, it should be mentioned that the MiSeq data set was not generated on the same sample used for T7EI and IDAA, but on a replicate sample that employed an NGS-specific set of amplification primers that were different from those used for IDAA (700 bp vs. 500 bp amplicon targets, respectively, details shown in the legend to

---

Caption for Figure 12.3 (cont.)

AAGAATGTTGTGATAAAAGGTGATGCT and HPRT Rev primer – -5′ACACATCCATGGGACTTCTGCCTC-3′. For IDAA tri-primer PCR primers used were; HPRTFORext : 5′- AGCTGACCGGCAGCAAAATTG ATGGGTAGTCAGGGTGCAGGT-3′ and HPRTREV : 5′- TGGCAAATGTGCCTCTCTACAA-3′ in addition to labeled FAMF universal primer (Yang *et al.*, 2015). Relative percentage of cut fragments for T7EI (filled triangles) and total indel cutting efficiency for IDAA is shown in bold in upper left corner of the respective panels. Size standards indicated by lines. Most prominent IDAA peaks were quantified and are shown, with selected highlighted peaks also observed by NGS. (C) NGS profiles for the 48 h sampling, only indels ranging from ±30 bp are shown. NGS amplicons were generated by KAPA hifi polymerase and HPRT For primer – CTTCAGGTTCCAGGTGATCA and HPRT Rev primer – CTAGACTACAGCTTTATGTGACT. Amplicons were purified by SPRI bead clean-up and Nextera™ (Illumina, USA) libraries were prepared according to manufacturer's recommendations. Sequencing was performed on an Illumina MiSeq. Data analysis was completed using a freely available computational pipeline for analyzing CRISPR genome editing experiments (Pinello *et al.*, 2016). The plain text output from the computational analysis was used to generate the graphical report for visualizing the insertion and deletion profile relative to the unedited sequences. Further data analysis was done via an in-house NGS analysis pipeline which includes a soft trim of all reads prior to comparison to a reference sequence using BLAST. From the detailed alignment data, variations relative to the reference sequence were identified and their frequencies were computed. These data are shown in Table 12.1 and represent events that occur with >1% frequency. Full sequence characterization indicates that the +1 bp, −9 bp and −15 bp events represent more than one specific editing event. In each case, there are two non-matched events at the sequencing level that would contribute to identical peaks by IDAA (panel B and Table 12.1) or the overall indel profile determined from the NGS data (Table 12.1). (D) Comparative plot of the most frequently occurring indels detected by IDAA and NGS. Of notice, the profiles are highly similar. (E) Analysis of the individual sequences identified within the differently sized amplicon fragments detected, grouped as unmodified intact amplicons (wild-type/wt alleles), intact amplicons possessing substitutions, insertions or deletions. Non-targeted control sample is shown by stippled line representing background levels for the different type of indel groups. Of notice, significant frequency of substitutions are observed (6%) among the intact/full-length amplicons detected. Wild-type/unedited peaks are indicated by an asterisk (*) in all panels shown.

Figure 12.3), thus experimental differences may account for the minor indel quantification variance observed.

In order to evaluate to what extent the respective indels identified represented homogeneous or heterogeneous events/sequences, the NGS data set was interrogated further. The panels depicted in Figure 12.3 also confirm previous observations that the predominant indel insertion event is a single-base insertion and that the nature of the deletion events are discretely heterogeneous with respect to sizes detected (Lonowski *et al.*, 2017; Yang *et al.*, 2015). The in-depth NGS data analysis also discloses that the indel profile varies in a non-random fashion. Table 12.1 displays the observed frequency of sequences represented by the major indels identified ranging from +1 to –15 bp. Examples of non-random indels having discrete sequences (fingerprints) include the identical positioning of +1bp T or A insertional events (detected at 8.6% and 1.5% respectively), the –1 bp deletion (detected at 7%, represented by a single T deletion event in 100% of cases) and –9 bp deletion events (found at frequencies of 8.2% and 1%). Fragments with a more heterogeneous sequence composition include the –2 bp deletion events found at frequencies of 4.1%, 2.7% and 1.2%. Clearly, the DNA sequences identified were found to

Table 12.1 Editing events at HPRT 38087 relative to total sequence

Event	Size	%/NGS	%/IDAA	Comparison to Reference
Ins	+1	8.6	11	CCCCTTCC<u>T</u>TAGTAATCCCCATAATT
Ins	+1	1.5		CCCCTTCC<u>A</u>TAGTAATCCCCATAATT
WT	0	11.7	16	CCCCTTCCTAGTAATCCCCATAATT
Del	–1	7.0	10	CCCCTTCC–TGTAATCCCCATAATT
Del	–2	4.1	12	CCCCTTCC––GTAATCCCCATAATT
Del	–2	2.7		CCCCTT––TAGTAATCCCCATAATT
Del	–2	1.2		CCCCTTCCT––TAATCCCCATAATT
Del	–3	1.2	6	CCCCTT–––AGTAATCCCCATAATT
Del	–3	2.1		CCCCTTCCTA–––ATCCCCATAATT
Del	–4	1.2	3	CCCCTTCC––––AATCCCCATAATT
Del	–5	1.1	2	CCCCTTCC–––––ATCCCCATAATT
Del	–9	8.2	7	CCCC–––––––––ATCCCCATAATT
Del	–9	1.0		CCCCT–––––––––TCCCCATAATT
Del	–10	1.3	3	CCCCTTCC––––––––––CATAATT
Del	–11	1.5	5	CCCCTTCC–––––––––––ATAATT
Del	–15	4.3	9	CCCC–––––––––––––––ATAATT
Del	–15	1.4		CCCCTT–––––––––––––––AATT
Sub	0	2.3		CCCCTTC*A*TAGTAATCCCCATAATT
Sub	0	1.9		CCCCTTCCTAGTAATCCCC*T*TAATT
Sub	0	1.9		CCCCTT*A*CTAGTAATCCCCATAATT
Sub	0	1.1		CCCCTTC*T*TAGTAATCCCCATAATT
Sub	0	1.0		CCCCTTCCTA*A*TAATCCCCATAATT

Ins: Insertions are shown underlined
Del: Deletions are shown by a dashed line
Sub: Substitutions are shown in grey-highlighted italics

be non-randomly positioned within the target repaired, suggesting that discrete, reproducible repair had occurred from the CRISPR/Cas9-induced double-strand breaks. These observations are in agreement with recent observations made by Van Overbeek *et al.* (2016) and us (Lonowski *et al.*, 2017).

Careful, in-depth NGS analysis revealed that a substantial number of full-length wild-type amplicons, which were previously assumed to represent unmodified wild-type alleles, in fact possessed base substitutions. While some base changes may result from PCR amplification, they may also represent previously unsuspected imperfect repair events. The frequency of these substitution events was significantly higher in Cas9/gRNA-treated cultures than was seen in control cells (Figure 12.3E). These changes are undetectable by the IDAA assay, but might be detected by the T7EI EMC assay. It would be interesting to perform NGS analysis at the 6 hour time point, where IDAA did not show editing and the T7EI EMC assay did. We speculate that some of these events may be single-base substitution repair events detected by EMC which were invisible to IDAA and leads us to wonder if such products can function as substrates for additional rounds of editing by the Cas9 RNP complex still present in the cells at this early time point (which would now be the equivalent of a single-base off-target cleavage event).

12.4 Concluding Remarks

The advent of highly efficient precise genome editing tools, such as ZFNs, TALENs and CRISPR/Cas9, call for improved methods for fast, precise and in-depth dissection of the changes in DNA sequence that use of these methods produce in cells, cell pools, tissues and animals. In this review, we presented the pros and cons of currently available methods to detect the indels induced by any of these genome editing tools. Collectively, traditional methodologies based on RFLP, PCR and EMC are adequate for basic, low-throughput genome editing purposes, but lack sufficient sensitivity, resolution and high-throughput capability for more demanding applications such as therapeutic genome editing or high-throughput studies. To this end, improved NGS methodologies and new bioinformatic analytical tools enable sufficient, sensitive and precise detection and characterization of indels. In spite of this, NGS may be prohibitively expensive and time-consuming for routine and general use in basic labs. In conclusion, the development of the IDAA methodology may meet the requirements for fast, precise, sensitive and "cost-effective" indel profiling in both basic and demanding genome editing applications such as therapeutic indel profiling.

Acknowledgements

We thank Camilla Andersen for excellent technical assistance with the IDAA experiments and Matt McNeill, Jeff Manthey and Tina Luebke for assistance with the NGS experiments. This work was financed by the University of Copenhagen Excellence Program for Interdisciplinary Research (CDO2016),

The Danish Council for Independent Research (4004-00140B) and the Danish National Research Foundation (DNRF107).

REFERENCES

Bae S, Kweon J, Kim HS, Kim J-S. 2014. Microhomology-based choice of Cas9 nuclease target sites. *Nat Methods* **11**(7): 705–706.

Beumer KJ, Trautman JK, Christian M, *et al.* 2013. Comparing zinc finger nucleases and transcription activator-like effector nucleases for gene targeting in *Drosophila. G3 (Bethesda)* **3**(10): 1717–1725.

Bibikova M, Carroll D, Segal DJ, *et al.* 2001. Stimulation of homologous recombination through targeted cleavage by chimeric nucleases. *Mol Cell Biol* **21**(1): 289–297.

Brinkman EK, Chen T, Amendola M, van Steensel B. 2014. Easy quantitative assessment of genome editing by sequence trace decomposition. *Nucleic Acids Res* **42**(22): e168.

Børresen AL, Hovig E, Brøgger A. 1988. Detection of base mutations in genomic DNA using denaturing gradient gel electrophoresis (DGGE) followed by transfer and hybridization with gene-specific probes. *Mutation Res* **202**(1): 77–83.

Chen F, Pruett-Miller SM, Huang Y, *et al.* 2011. High-frequency genome editing using ssDNA oligonucleotides with zinc-finger nucleases. *Nat Methods* **8**(9): 753–755.

Chen S, Oikonomou G, Chiu CN, *et al.* 2013. A large-scale in vivo analysis reveals that TALENs are significantly more mutagenic than ZFNs generated using context-dependent assembly. *Nucleic Acids Res* **41**(4): 2769–2778.

Cong L, Ran FA, Cox D, *et al.* 2013. Multiplex genome engineering using CRISPR/Cas systems. *Science* **339**(6121): 819–823.

Dahlem TJ, Hoshijima K, Jurynec MJ, *et al.* 2012. Simple methods for generating and detecting locus-specific mutations induced with TALENs in the zebrafish genome. *PLoS Genet* **8**(8): e1002861.

Deriano L, Roth DB. 2013. Modernizing the nonhomologous end-joining repertoire: alternative and classical NHEJ share the stage. *Annu Rev Genet* **47**: 433–455.

Doench JG, Fusi N, Sullender M, *et al.* 2016. Optimized sgRNA design to maximize activity and minimize off-target effects of CRISPR-Cas9. *Nat Biotechnol* **34**: 1–12.

Doench JG, Hartenian E, Graham DB, *et al.* 2014. Rational design of highly active sgRNAs for CRISPR-Cas9-mediated gene inactivation. *Nat Biotechnol* **32**(12): 1262–1267.

Eshaghpour H, Crothers DM. 1978. Preparative separation of the complementary strands of DNA restriction fragments by alkaline RPC-5 chromatography. *Nucleic Acids Res* **5**(5): 1627–1637.

Foley JE, Maeder ML, Pearlberg J, *et al.* 2009. Targeted mutagenesis in zebrafish using customized zinc-finger nucleases. *Nat Protocols* **4**(12): 1855–1867.

Fu Y, Foden JA, Khayter C, *et al.* 2013. High-frequency off-target mutagenesis induced by CrIsPr-Cas nucleases in human cells. *Nat Biotechnol* **31**(9): 822–826.

Gao H, Huang J, Barany F, Cao W. 2007. Switching base preferences of mismatch cleavage in endonuclease V: an improved method for scanning point mutations. *Nucleic Acids Res* **35**(1): 1–6.

Geurts AM, Cost GJ, Freyvert Y, *et al.* 2010. Knockout rats produced using designed zinc finger nucleases. *Science* **325**(5939): 2009–2011.

Güell M, Yang L, Church G. 2014. Genome editing assessment using CRISPR genome analyzer. *Bioinformatics* **30**(20): 2968–2970.

Huang MC, Cheong WC, Lim LS, Li M-H. 2012. A simple, high sensitivity mutation screening using Ampligase mediated T7 endonuclease I and Surveyor nuclease with microfluidic capillary electrophoresis. *Electrophoresis* **33**(5): 788–796.

Inazuka M, Wenz H, Sakabe M, Tahira T, Hayashi K. 1997. A streamlined mutation detection system: multicolor post-PCR fluorescence labeling and single-strand conformational polymorphism analysis by capillary electrophoresis. *Genome Res* **7**(11): 1094–1103.

Jinek M, East A, Cheng A, *et al.* 2013. RNA-programmed genome editing in human cells. *eLife* **2**(3): e00471.

Kim HH, Um E, Cho S, Jung C, Kim J. 2011. Surrogate reporters for enrichment of cells with nuclease-induced mutations. *Nat Methods* **8**(11): 941–943.

Kim Y, Kweon J, Kim J-S. 2013. TALENs and ZFNs are associated with different mutation signatures. *Nat Methods* **10**(3): 185.

Kosicki M, Rajan SS, Lorenzetti FC, *et al.* 2017. Dynamics of indel profiles induced by various CRISPR/Cas9 delivery methods. *Prog Mol Biol Trans Sci* **152**: 49–67.

Kuan SF, Byrd JC, Basbaum C, Kim YS. 1989. Inhibition of mucin glycosylation by aryl-N-acetyl-alpha-galactosaminides in human colon cancer cells. *J Biol Chem* **264**(32): 19271–19277.

Lindsay H, Burger A, Biyong B, *et al.* 2015. CrispR variants: precisely charting the mutation spectrum in genome engineering experiments. *Nat Biotechnol* **34**(7): 701–703.

Liu W, Smith DI, Rechtzigel KJ, Thibodeau SN, James CD. 2002. Denaturing high performance liquid chromatography (DHPLC) used in the detection of germline and somatic mutations. *Nucleic Acids Res* **26**(6): 1396–1400.

Liu X, Homma A, Sayadi J, *et al.* 2016. Sequence features associated with the cleavage efficiency of CRISPR/Cas9 system. *Sci Rep* **6**: 19675.

Lonowski LA, Narimatsu Y, Riaz A, *et al.* 2017. Genome editing using FACS enrichment of nuclease expressing cells and Indel Detection by Amplicon Analysis (IDAA). *Nat Protoc* **12**(3): 581–603.

Mali P, Yang L, Esvelt KM, *et al.* 2013. RNA-guided human genome engineering via Cas9. *Science* **339**(6121): 823–826.

Mashal RD, Koontz J, Sklar J. 1995. Detection of mutations by cleavage of DNA heteroduplexes with bacteriophage resolvases. *Nat Genet* **2**: 177–183.

Miller JC, Holmes MC, Wang J, *et al.* 2007. An improved zinc-finger nuclease architecture for highly specific genome editing. *Nat Biotechnol* **25**(7): 778–785.

Moscou MJ, Bogdanove AJ. 2009. A simple cipher governs DNA recognition by TAL effectors. *Science* **326**(5959): 1501.

Nakade S, Tsubota T, Sakane Y, *et al.* 2014. Microhomology-mediated end-joining-dependent integration of donor DNA in cells and animals using TALENs and CRISPR/Cas9. *Nat Comm* **5**: 5560.

Oleykowski CA, Mullins CRB, Godwin AK, Yeung AT. 1998. Mutation detection using a novel plant endonuclease. *Nucleic Acids Res* **26**(20): 4597–4602.

Pimkin M, Caretti E, Canutescu A, *et al.* 2007. Recombinant nucleases CEL I from celery and SP I from spinach for mutation detection. *BMC Biotechnol* **7**: 29.

Pinello L, Canver MC, Hoban MD, *et al.* 2016. Analyzing CRISPR genome-editing experiments with CRISPResso. *Nat Biotech* **34**(7): 695–697.

Qiu P, Shandilya H, D'Alessio JM, *et al.* 2004. Mutation detection using Surveyor nuclease. *BioTechniques* **36**(4): 702–707.

Sakurai T, Watanabe S, Kamiyoshi A, Sato M, Shindo T. 2014. A single blastocyst assay optimized for detecting CRISPR/Cas9 system-induced indel mutations in mice. *BMC Biotechnol* **14**(1): 69.

Shalem O, Sanjana NE, Hartenian E, *et al.* 2013. Genome-scale CRISPR-Cas9 knockout screening in human cells. Science **343**(6166): 84–87.

Smith GR. 2004. How homologous recombination is initiated: unexpected evidence for single-strand nicks from v(d)j site-specific recombination. *Cell* **117**(2): 146–148.

Smith J, Berg JM, Chandrasegaran S. 1999. A detailed study of the substrate specificity of a chimeric restriction enzyme. *Nucleic Acids Res* **27**(2): 674–681.

Smith J, Bibikova M, Whitby FG, *et al.* 2000. Requirements for double-strand cleavage by chimeric restriction enzymes with zinc finger DNA-recognition domains. *Nucleic Acids Res* **28**(17): 3361–3369.

Stark JM, Pierce AJ, Oh J, Pastink A, Jasin M. 2004. Genetic steps of mammalian homologous repair with distinct mutagenic consequences. *Mol Cell Biol* **24**(21): 9305–9316.

Symington LS, Gautier J. 2011. Double-strand break end resection and repair pathway choice. *Annu Rev Genet* **45**: 247–273.

Urnov FD, Miller JC, Lee YL, *et al.* 2005. Highly efficient endogenous human gene correction using designed zinc-finger nucleases. *Nature* **435**(7042): 646–651.

Van Overbeek M, Capurso D, Carter MM, *et al.* 2016. DNA repair profiling reveals nonrandom outcomes at Cas9-mediated breaks. *Molecular Cell* **63**: 1–14.

Vouillot L, Thelie A, Pollet N. 2015. Comparison of T7E1 and surveyor mismatch cleavage assays to detect mutations triggered by engineered nucleases. *G3 (Bethesda)* **5**(3): 407–415.

Yang L, Guell M, Byrne S, *et al.* 2013. Optimization of scarless human stem cell genome editing. *Nucleic Acids Res* **41**(19): 9049–9061.

Yang Z, Steentoft C, Hauge C, *et al.* 2015. Fast and sensitive detection of indels induced by precise gene targeting. *Nucleic Acids Res* **43**(9): e59.

Yeung AT, Hattangadi D, Blakesley L, Nicolas E. 2005. Enzymatic mutation detection technologies. *BioTechniques* **38**(5): 749–758.

Youil R, Kemper BW, Cotton RG. 1995. Screening for mutations by enzyme mismatch cleavage with T4 endonuclease VII. *Proc Natl Acad Sci USA* **92**(1): 87–91.

Yu C, Zhang Y, Yao S, Wei Y. 2014. A PCR based protocol for detecting indel mutations induced by TALENs and CRISPR/Cas9 in zebrafish. *PLoS One* **9**(6): e98282.

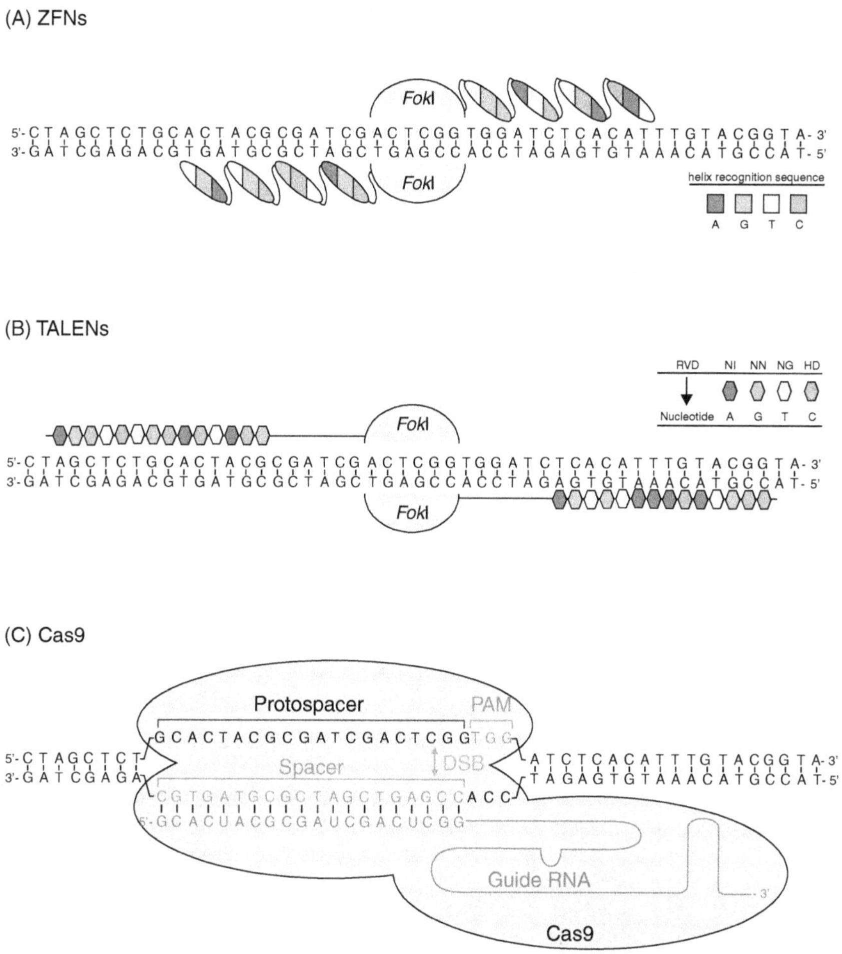

Figure 1.3 ZFN, TALEN and Cas9 DNA recognition. (A) Shown is a ZFN pair bound to its DNA target. Each monomer is composed of a DNA binding domain fused to a FokI cleavage domain. The DNA binding domain is assembled from several zinc-finger DNA recognition domains (four are shown, depicted as ovals) and each domain can recognize three unique nucleotides, noted in different colors. Strung together, the multiple domains can dictate DNA binding to unique genomic sequences with high specificity. When two monomers are brought together by binding adjacent DNA sequences, the FokI domains dimerize and the nuclease is active. (B) A TALEN pair bound to its DNA target. As with ZFNs, TALENs are composed of a DNA binding domain fused to the FokI cleavage domain. The binding specificity is dictated by the repeat variable diresidues (RVD) within each TALE repeat unit (shown are the different RVDs that bind each nucleotide). (C) Cas9 is targeted to DNA sequences by a bound sgRNA. If a protospacer adjacent motif (PAM) is present, the first 20 nucleotides of the guide RNA can form Watson–Crick base pairs with the target spacer sequence, and activate Cas9 nuclease activity.

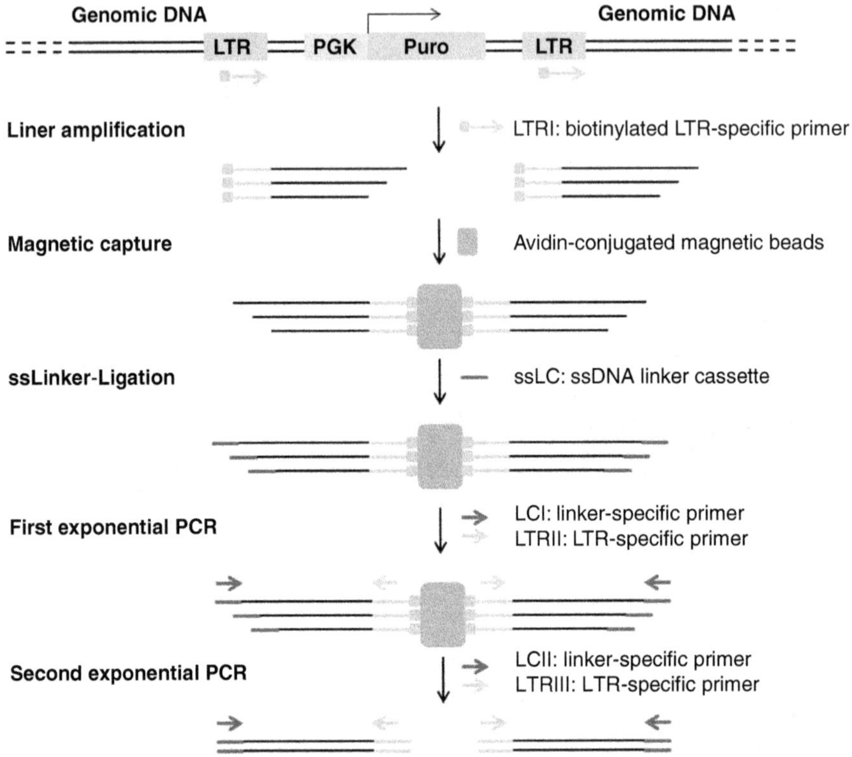

Figure 2.3 Scheme of nrLAM-PCR-mediated detection of IDLV-tagged DSBs. The process of nrLAM-PCR is initiated with linear amplification of vector-genomic DNA junctions using a 5′-biotinylated vector-specific primer (LTRI). The PCR product is captured by streptavidin-conjugated magnetic beads. The captured product is ligated to a single-strand linker cassette (ssLC) so that subsequent exponential PCR can be carried out using linker-specific primer LCI and vector-specific primer LTRII. After magnetic capture, the product is subjected to a second exponential PCR using linker- (LCII) and vector-specific primers (LTRIII).

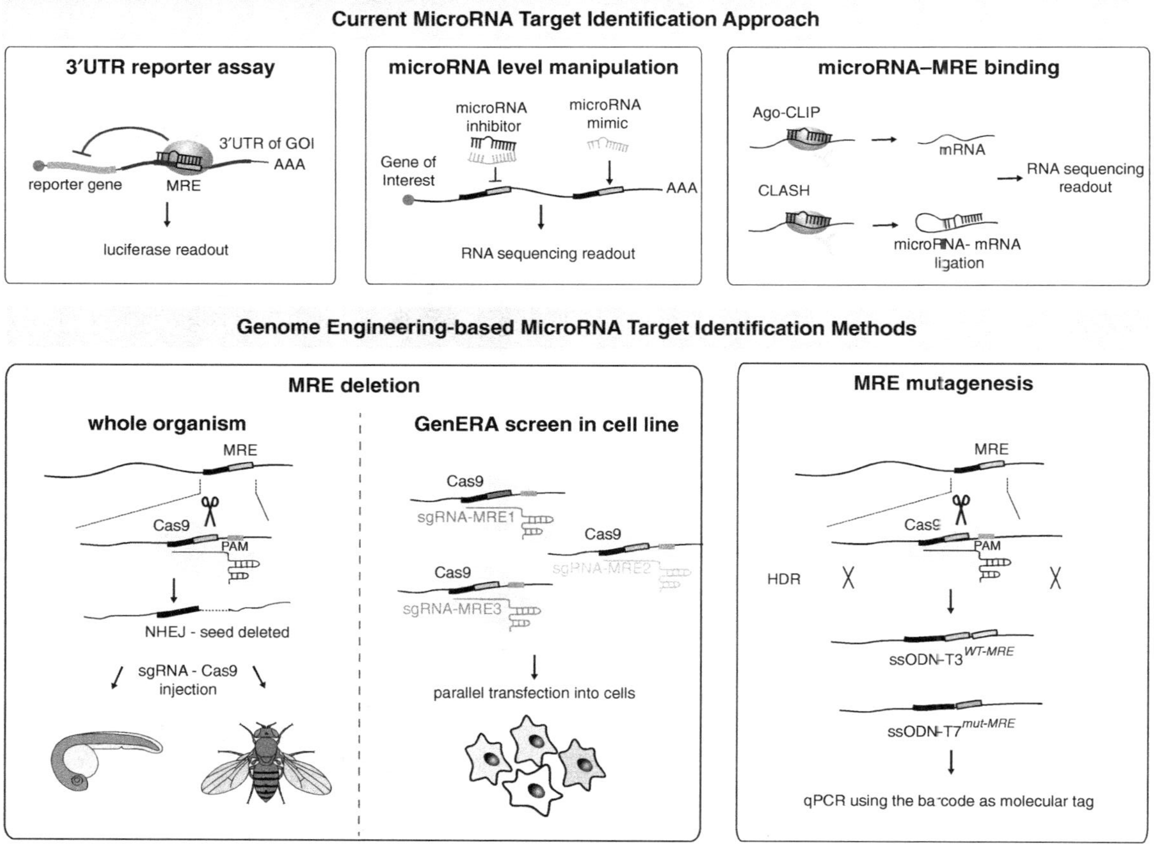

Figure 3.1 Approaches to identifying and validating miRNA targets. Current approaches include 3′UTR reporter assays where the 3′UTR of the gene of interest is coupled to a fluorescent or luminescent reporter gene. Comparison of reporter expression in the presence or absence of the MRE, or upon miRNA manipulation allows investigation of MRE activity. Targets can also be identified after manipulation of miRNA levels by miRNA inhibitors or miRNA mimics, followed by RNA-Seq to identify the resulting gene expression changes. Direct binding of miRNAs to MREs can be detected by techniques such as Ago-CLIP and CLASH that rely on crosslinking of miRNAs to their targets followed by biochemical purification of Ago-containing complexes, and subsequent analysis of bound mRNAs by RNA-Seq. Genome engineering can be employed to investigate MRE functionality by utilizing deletions resulting from inefficient NHEJ-based repair of Cas9-induced DSBs to mutate MREs in the context of a whole organism. A similar approach can be employed to screen many MREs for their functionality using a multiplexed assay in cell lines. Homology directed repair (HDR) can also be employed to make defined mutations in MREs of interest in cell lines and coupled to a molecular tag that can be used to specifically detect the effect of these changes.

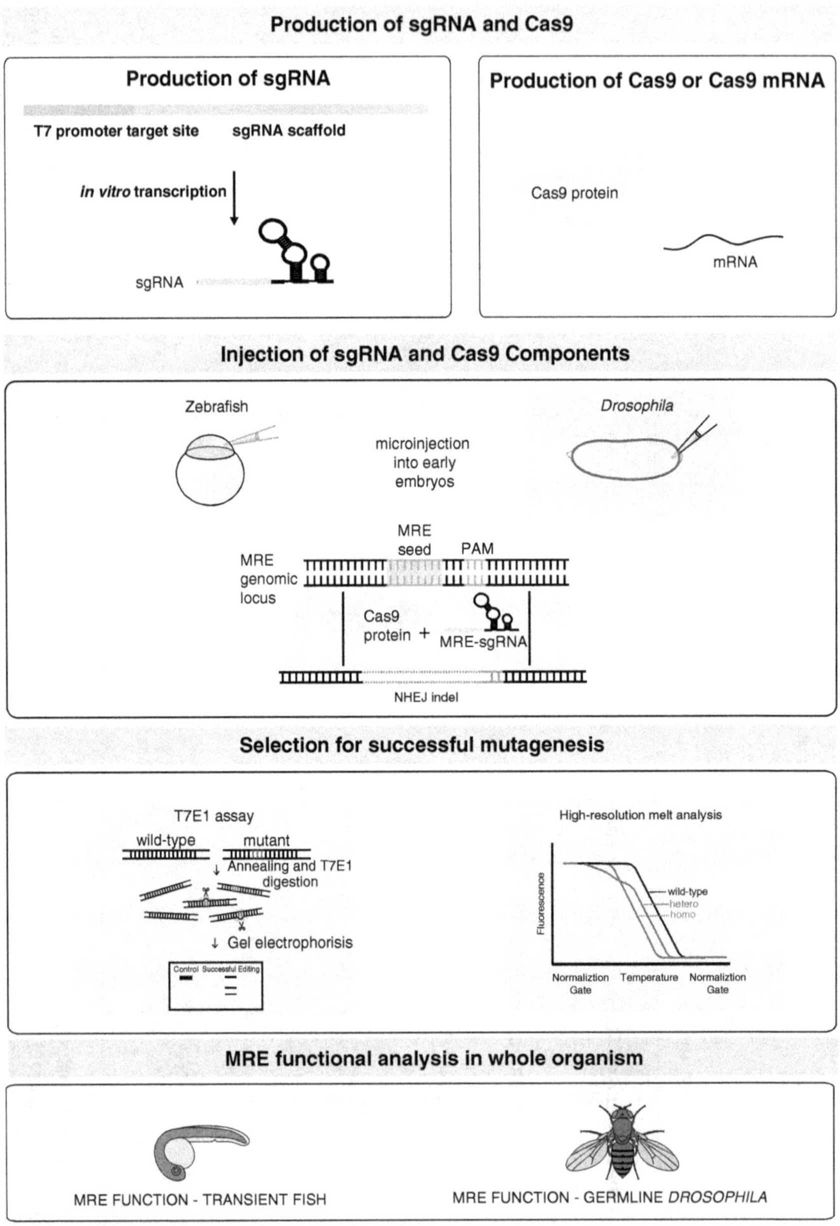

Figure 3.2 Strategy for deletion of MREs in whole organisms. sgRNAs can be produced by *in vitro* transcription using a PCR product generated from overlapping primers that contains a T7 promoter (green), followed by the specific target sequence (pink), and the sgRNA scaffold (orange). These sgRNAs are injected into early embryos of for instance zebrafish or *Drosophila* along with Cas9 mRNA or protein to induce mutagenesis at the MRE of interest. Mutations are screened by T7 endonuclease (T7E1) assay, high-resolution melt (HRM) analysis, restriction enzyme polymorphisms and DNA sequencing for those of interest, and functional analysis of the MRE deletion can be performed in the intact organism.

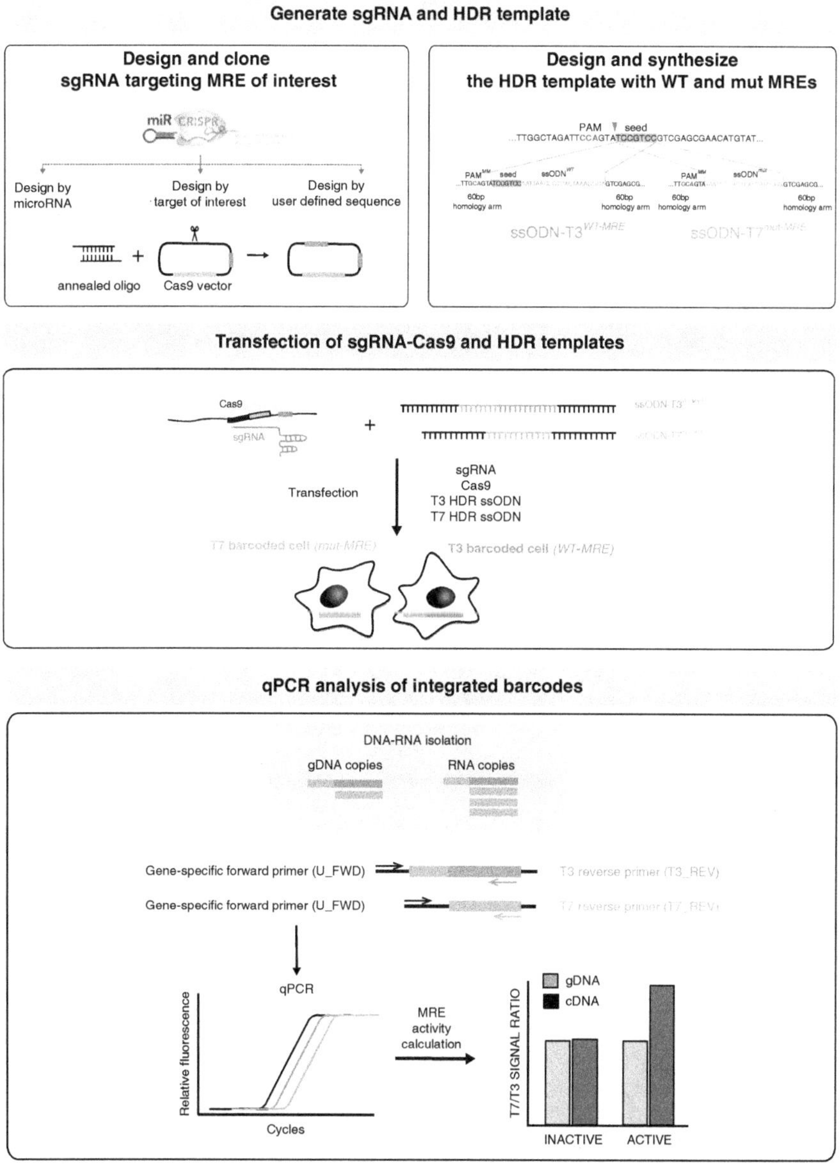

Figure 3.3 Strategy for MRE mutagenesis in cell lines. sgRNAs are designed using miR-CRISPR to target near the MRE of interest, and target sequences cloned into an appropriate expression vector (e.g. pX459). Simultaneously, two oligonucleotide HDR templates are synthesized, which contain 60 nt homology arms and either integrate a T3 sequence, whilst maintaining the MRE of interest, or a T7 sequence, whilst mutating the MRE. The CRISPR construct is co-transfected with both HDR templates into the cell line of interest, and gDNA and RNA are isolated from the pool of cells followed by reverse transcription to generate cDNA. Integration of the two HDR templates can be specifically detected by qPCR using primers specific for the T7 or T3 tag and a common gene-specific forward primer. The integration efficiency of the two HDR templates is assayed by measuring the T7/T3 ratio in gDNA and the activity of the MRE determined by an increase in this ratio when measured in the cDNA sample.

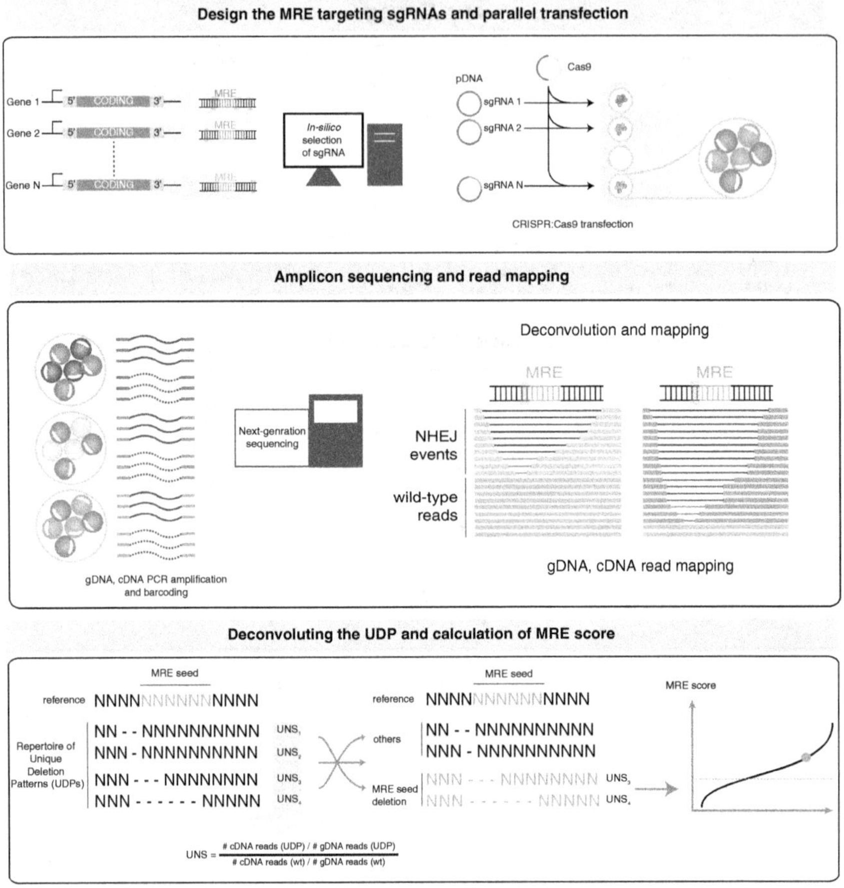

Figure 3.4 Genome engineering-based evaluation of RNA regulatory element activity (GenERA). sgRNAs are designed computationally using miR-CRISPR to target within 20 nt of MREs of interest. Oligonucleotides are cloned into sgRNA expression vectors, transfected in individual wells to generate mixed populations of mutant cells and gDNA and RNA are extracted from each well. Amplicons are designed spanning the MRE of interest, and PCR is performed on gDNA and cDNA followed by adaptor ligation, barcoding and high-throughput sequencing. Reads are mapped to each amplicon, and segregated into unique deletion profiles (UDPs). A unique normalized score (UNS) is calculated to assess the effect of each UDP on gene expression, using the ratio of cDNA to gDNA reads, normalized to WT reads. An overall MRE score can be calculated by a sum of MRE seed deleting UNSs, weighted to account for gDNA read number.

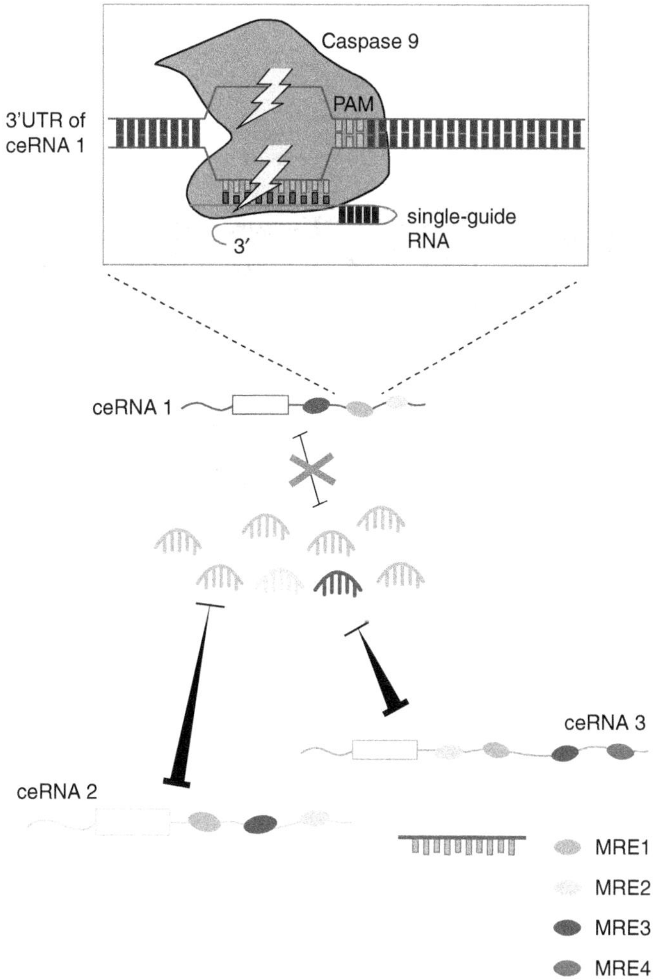

Figure 4.2 Application of the CRISPR/Cas9 system to study of microRNA:ceRNA interactions. A single-guide RNA (sgRNA) designed to target a specific genomic locus directs the Caspase 9 (Cas9) endonuclease to bind and cleave dsDNA (MRE 1 on the 3′UTR of ceRNA 1). The DSB is subsequently repaired via the NHEJ mechanism, resulting in a deletion within MRE 1, hence a reduction in miRNA binding and dysregulation of ceRNA transcript expression. Consequently, the increased pool of miRNAs can bind and alter other transcript expression (ceRNAs 2 and 3) to perturb the miRNA:ceRNA regulatory network.

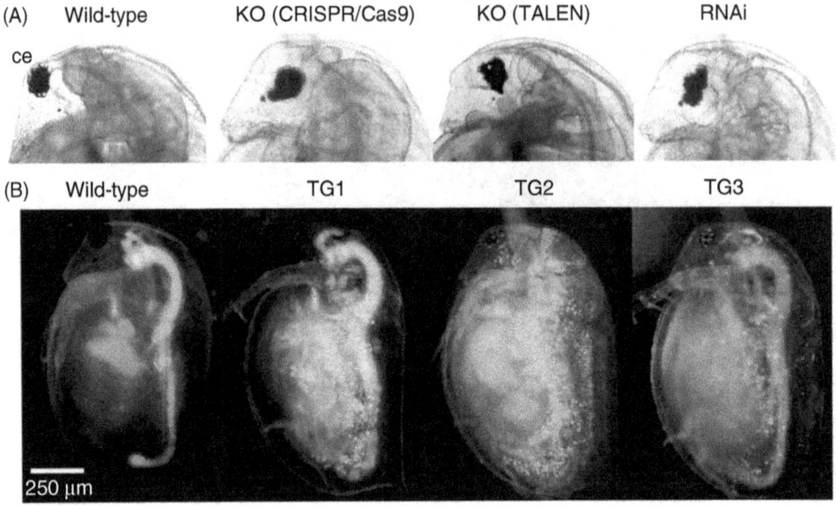

Figure 5.2 Phenotypes of knockout (A) and knockin animals (B). Each image was originally used in previous papers (Nakanishi *et al.*, 2014; 2015; 2016).

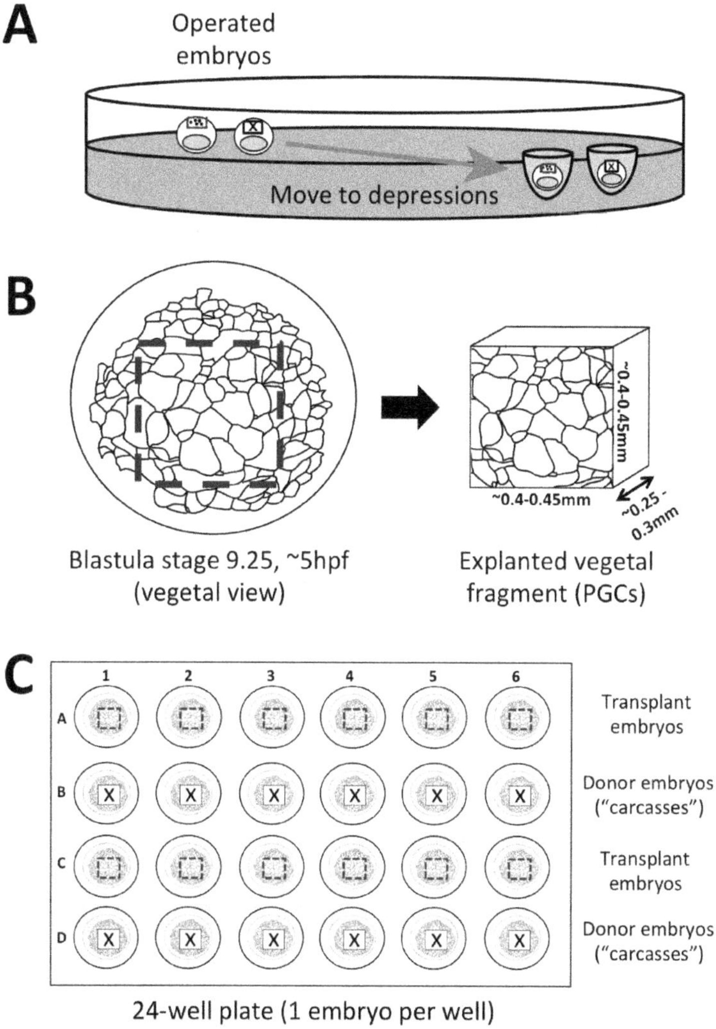

Figure 6.2 How to perform leapfrog transplants. (A) Transplantations are performed in a dish containing a thick bed of agarose and 0.3X MMR, with depressions to hold embryos while they heal post-surgery. (B) An illustration of the vegetal pole of a *Xenopus tropicalis* blastula stage embryo at 5 hpf, to show the size of vegetal blastomeres. The blue box indicates the approximate size and boundary of the region of the explanted/transplanted tissue. (C) Once transplants have healed into place, the embryos are moved to a 24-well agarose-coated dish (vegetal pole up), containing 1/9X MMR to develop overnight. Transplant donor embryos are also moved to this dish and each pair of donor and recipient embryos is tracked as the donor embryos are used as proxies to assess the mutagenesis efficiency in the transplanted PGCs carried by the "leapfrogged" embryos.

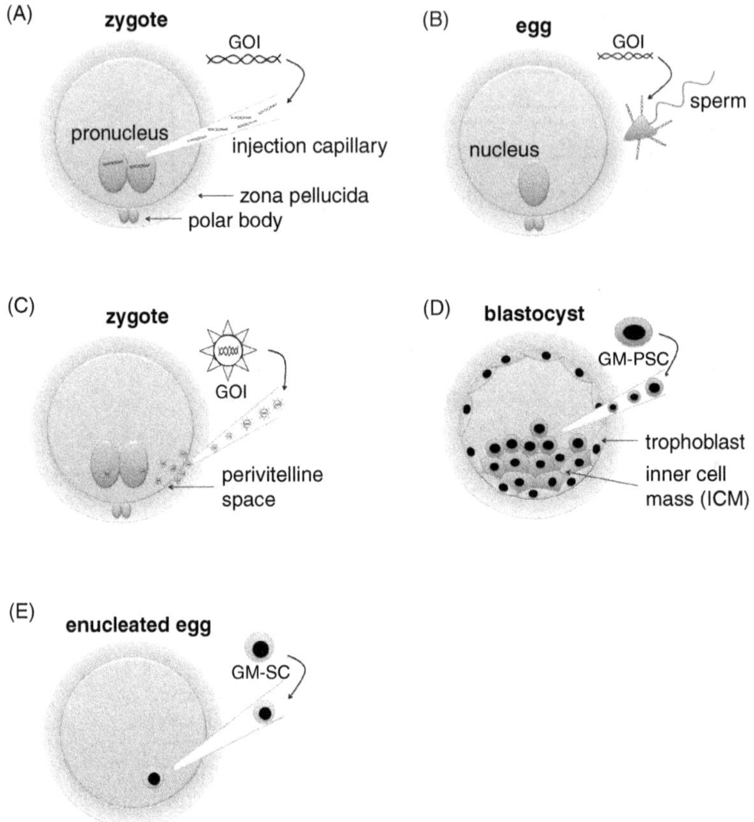

Figure 8.1 Schematic illustration of five commonly used methods to generate genetically modified animals. (A) Microinjection. GOI, gene of interest. (B) Sperm-mediated gene transfer. (C) Viral vector-mediated gene transfer. (D) Embryonic stem cell-mediated gene transfer. GM-PSC, genetically modified pluripotent stem cells. (E) Somatic cell nuclear transfer. GM-SC, genetically modified somatic cells.

Figure 9.4 *CD163* knockout pigs edited with CRISPR/Cas9 are challenged and resistant to PRRS virus. Photo credit: Melissa Samuel.

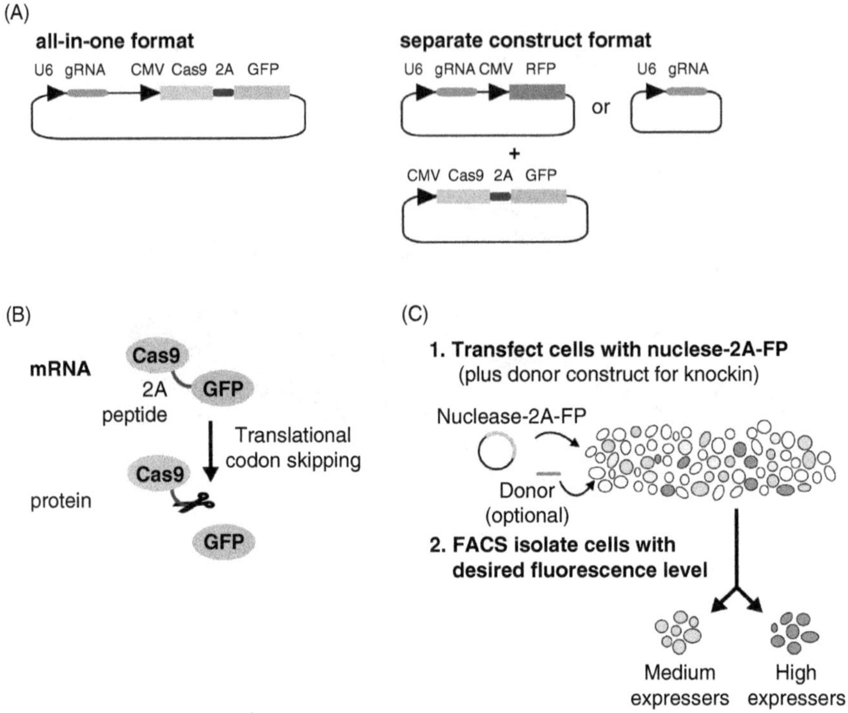

Figure 10.1 Vectors and principles of FACS-based genome editing. (A) Examples of Cas9-2A-FP vectors. (Left) All-in-one format, where Cas9-2A-FP and gRNA are expressed from the same vector. (Right) Separate construct format, where Cas9-2A-FP is expressed from one vector and gRNA from another vector that may carry a distinct FP for co-sorting or carry no marker. CMV and U6 indicate the promoters that drive expression. (B) In the Cas9-2A-FP transcript, the 2A peptide undergoes codon skipping during translation, resulting in 1:1 co-expression of nuclease and FP as separate proteins. (C) Principle of FACS-based genome editing. In the first step, cells are transfected with a nuclease-2A-FP construct. In the second step, cells with a desired level of fluorescence are isolated by FACS 1–3 days later.

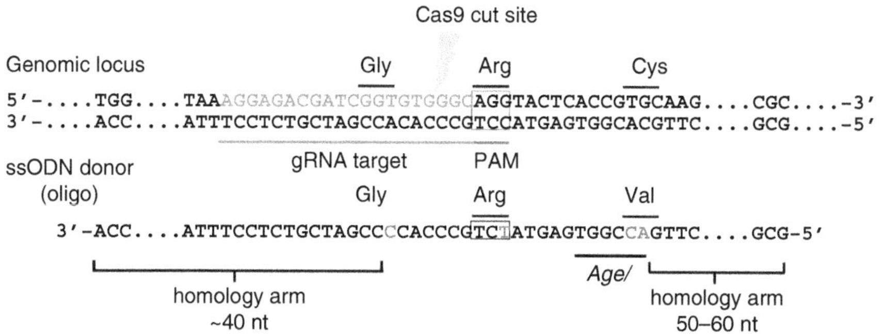

Figure 10.2 Key features of an ssODN donor for knockin editing using CRISPR/Cas9. The mutations contained in the donor are in red. The mutation-in-study is a Cys-Val codon-swap mutation, which simultaneously creates a diagnostic *AgeI* site for RFLP analysis. The gRNA target and PAM sequences are indicated, as well as the Cas9 cut site (yellow lightning). Note that the gRNA-blocking mutations are silent (Gly-Gly; Arg-Arg) and two are sufficient, since one of them mutates the PAM. Note that the effective homology arm begins at the mutation most distant to the cut site (or at the cut site itself, if there is no mutation).

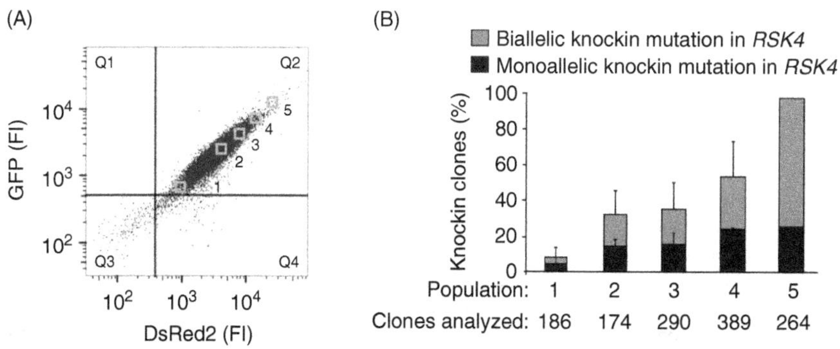

Figure 10.4 FACS-based genome editing for generation of clonal cell lines with ssODN-meditated knockin mutation. (A) FACS profile of K562 cells three days after transfection with mRNA for left ZFN-2A-GFP and right ZFN-2A-DsRed2 targeting the RSK4 locus and an ssODN donor for knockin of a Cys443Val mutation and a diagnostic *BamHI* site for RFLP analysis. The GFP and DsRed2 expression levels of individual cells (represented by dots) are measured as arbitrary fluorescence intensity (FI) units on the *y*-axis and *x*-axis, respectively. Parallel analysis of mock-transfected K562 cells was used to determine the autofluorescence level of the cells in the GFP and DsRed2 channels (not shown). Within the population with specific GFP/DsRed2 double-fluorescence (Q2), five subpopulations were defined (boxed), and single cells from each population were isolated by FACS and seeded into 96-well plates for clonal expansion. (B) After clonal expansion, each cell line was analyzed for monoallelic or biallelic knockin of the ssODN-containing mutations by RFLP analysis. The fraction of knockin modified clones is shown as a percentage on the *y*-axis and the total number of clones analyzed on the *x*-axis. The data are means +SD of three independent experiments, except for population 5, which represents one experiment.

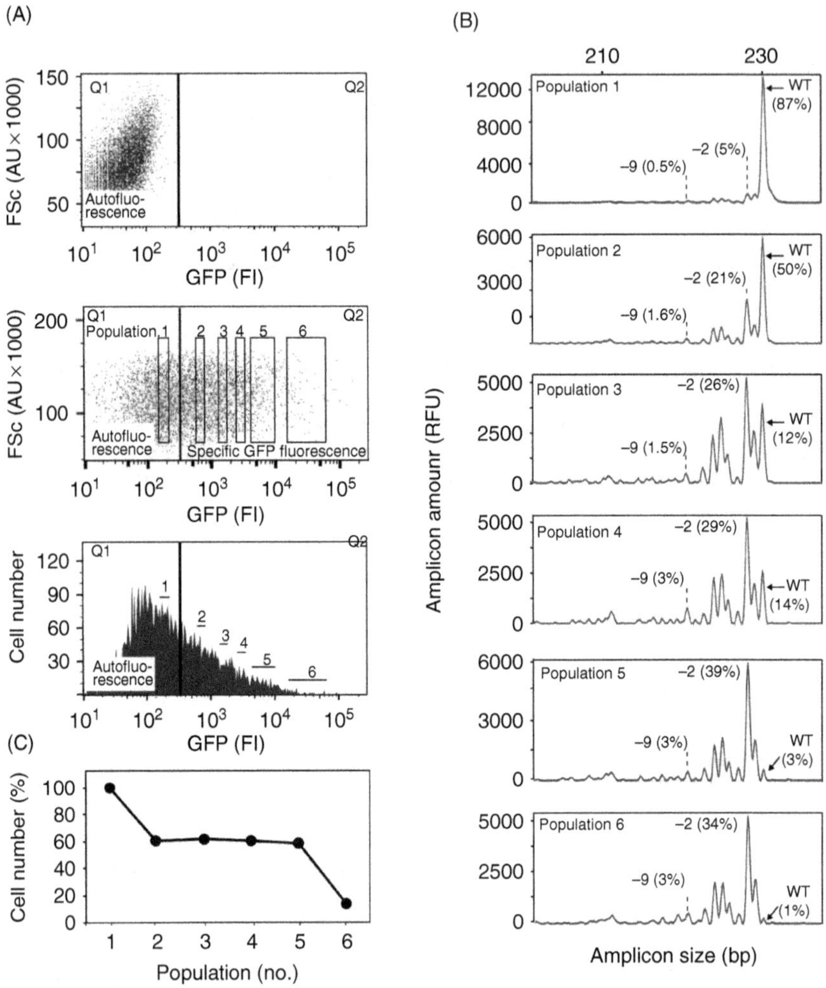

Figure 10.3 FACS-based genome editing for generation of knockout cell populations. (A) Top: FACS profile of mock-transfected K562 cells, showing the autofluorescence level of the cells in the GFP channel. Middle: FACS profile of K562 cells two days after transfection with an all-in-one pCMV-Cas9-2A-GFP construct expressing gRNA targeting the *KRAS* locus. The GFP expression level of individual cells (represented by dots) is measured as arbitrary fluorescence intensity (FI) units (*x*-axis) and cell-size as arbitrary forward scatter (FSc) units (*y*-axis). Within the pCMV-Cas9-2A-GFP-transfected cell population, six subpopulations were defined (boxed). Bottom: Another representation of the pCMV-Cas9-2A-GFP transfected cell population, showing cell number (*y*-axis) distribution at the various GFP fluorescence levels, as determined in 10 000 cells. The six distinct populations defined in the middle panel are indicated. (B) 20 000 cells from each of the six populations in (A) were bulk isolated by FACS, cultured for six days, after which an aliquot of each population was analyzed by IDAA to determine the level of indel (= knockout) mutations at the *KRAS* locus. The frequencies of wild-type and select indel alleles in the six populations are indicated. (C) Prior to IDAA, the total number of live cells in each of the six populations was determined and expressed as a percentage of population 1.

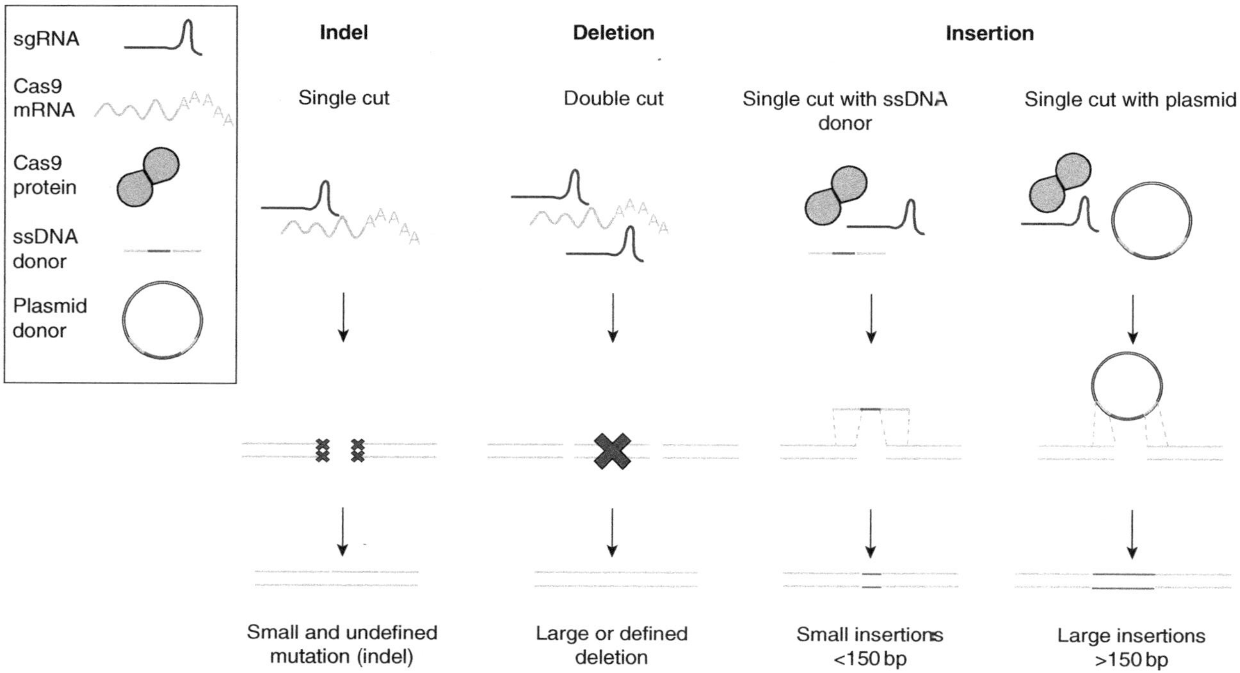

Figure 11.2 CRISPR can be used to generate a range of mutations in mice. Depending on the desired mutation, different CRISPR reagents are used to create indels and defined deletions through NHEJ or insertions through HDR.

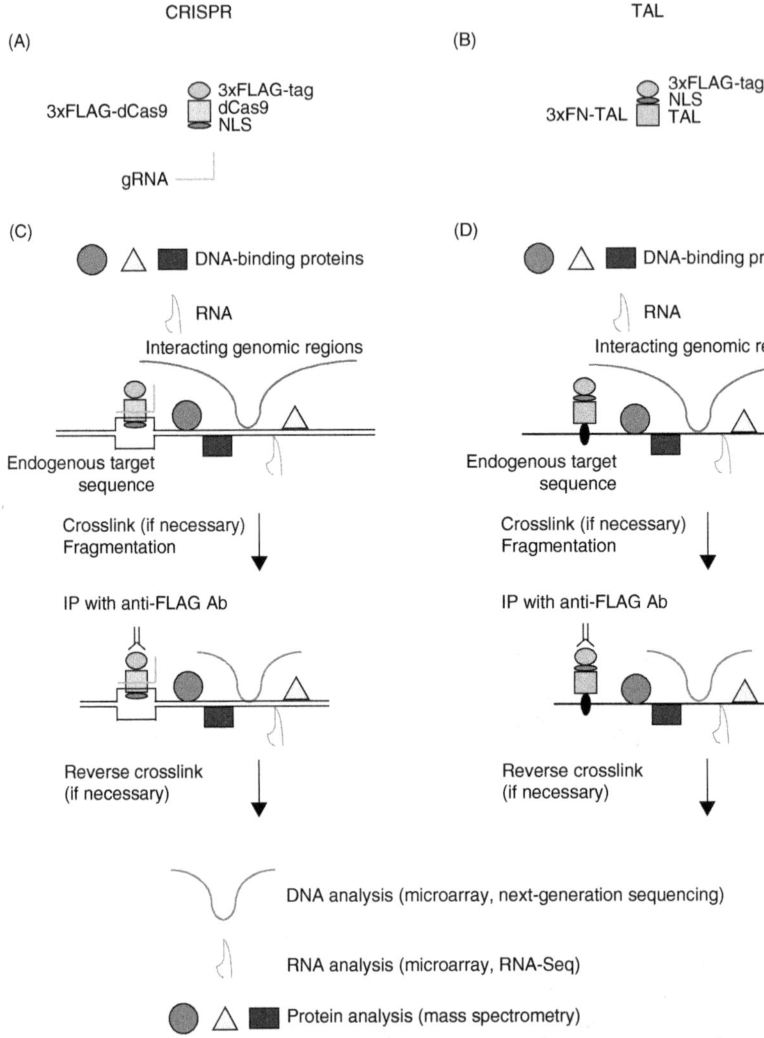

Figure 13.1 In cell enChIP. A locus of interest is tagged with an engineered DNA-binding molecule such as the CRISPR complex, which comprises a catalytically inactive form of Cas9 (dCas9) and gRNA (A), or a TAL protein (B) that recognizes the target sequence, expressed in the cells to be analyzed. (C and D) Interactions between molecules are stabilized by a crosslinker such as formaldehyde, if necessary. After fragmentation of chromatin by sonication or enzymatic digestion, the locus tagged with the engineered DNA-binding molecule is subjected to affinity purification. Crosslinking is reversed if necessary and the RNAs and other genomic regions that interact with the locus are identified by NGS or microarray analysis. Interacting proteins are identified by MS.

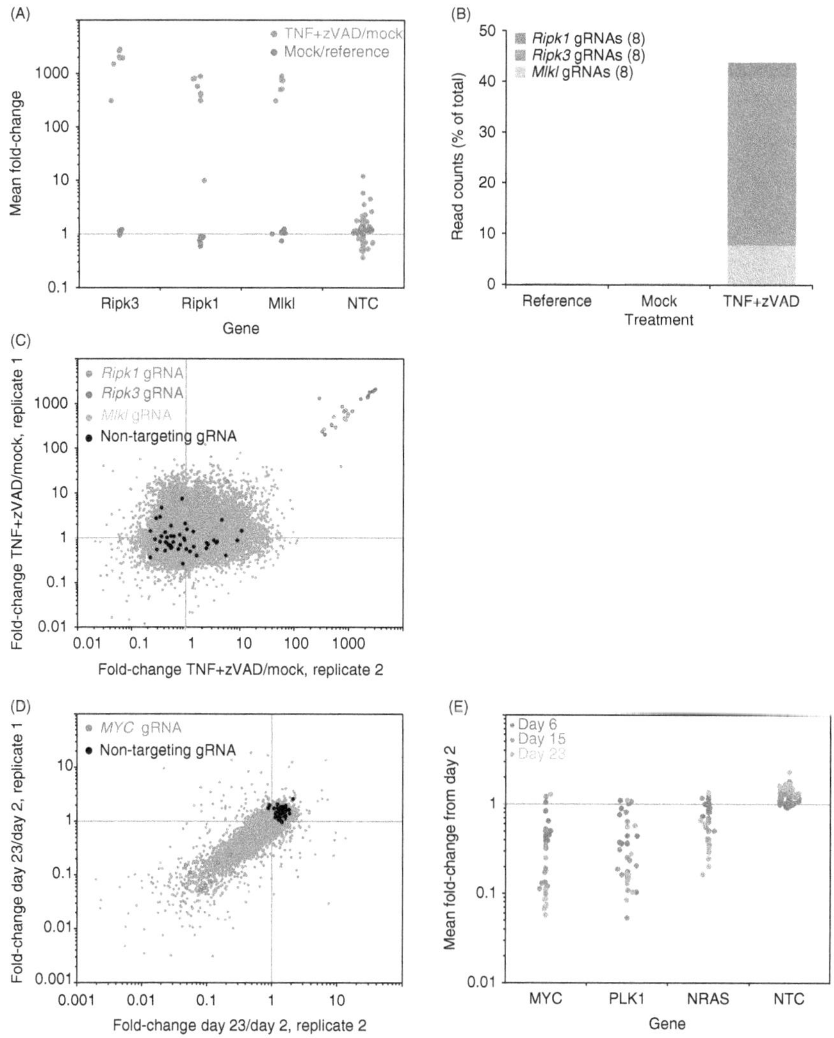

Figure 14.3 Example positive and negative screen data. (A–C) Positive selection screen. Cas9-expressing, mouse L929 cells were transduced with ~17 000 pooled sgRNAs in triplicate at 1000-fold sgRNA representation. Stable integrants were selected with antibiotic, and 15 days post-transduction, cultures were divided into treated (TNF+zVAD, to induce necroptosis) or mock populations. (A) All sgRNAs (~8/gene) targeting the core necrosome genes *Ripk3, Ripk1* and *Mlkl* show enrichment via NGS sequencing reads in TNF+zVAD-treated samples normalized to mock-treated samples (red dots; mean of replicates), but not in mock-treated samples normalized to an early reference time point (2 days post-transduction) (blue dots). (B) Enrichment of *Ripk1, Ripk3* and *Mlkl* sgRNA read counts in TNF+zVAD-treated samples compared to the early reference and mock-treated samples, shown as a percentage of the total read counts for all sgRNAs in the pool (means of replicate paired samples). (C) Correlation between two representative replicate samples for sgRNA abundance changes with TNF+zVAD treatment normalized to mock treatment. (D, E) Negative selection screen results. Cas9-expressing, human NCI-H1299 cells were transduced with ~17 000 pooled sgRNAs (~8/gene in triplicate at 1000-fold sgRNA representation). Stable integrants were selected using antibiotic, and DNA samples were collected from the cell populations at 2, 6, 15 and 23 days post-transduction. (D) Correlation of changes in sgRNA abundance between two representative replicate samples, with day 23 normalized to day 2 read counts. sgRNAs targeting MYC and 47 non-targeting control sgRNAs are highlighted. (E) Quantitation of sgRNA abundance over time for sequences targeting known oncogenes (*NRas* and *Myc*), an essential gene (*Plk1*), and non-targeting control sgRNAs (NTC). Each dot represents the mean fold-change of the three replicate sgRNA read counts normalized to the day 2 paired replicate read count.

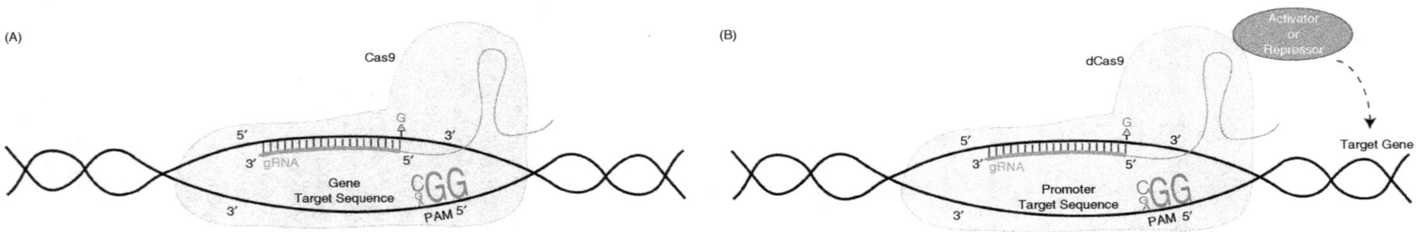

Figure 15.1 Diagram of gRNA (orange) in complex with Cas9 enzyme targeting a gene to create a knockout (A). The PAM sequence, with preferred bases emphasized is marked. For CRISPRi libraries, the modified Cas9 enzyme and associated proteins are shown acting on the targeted gene (B).

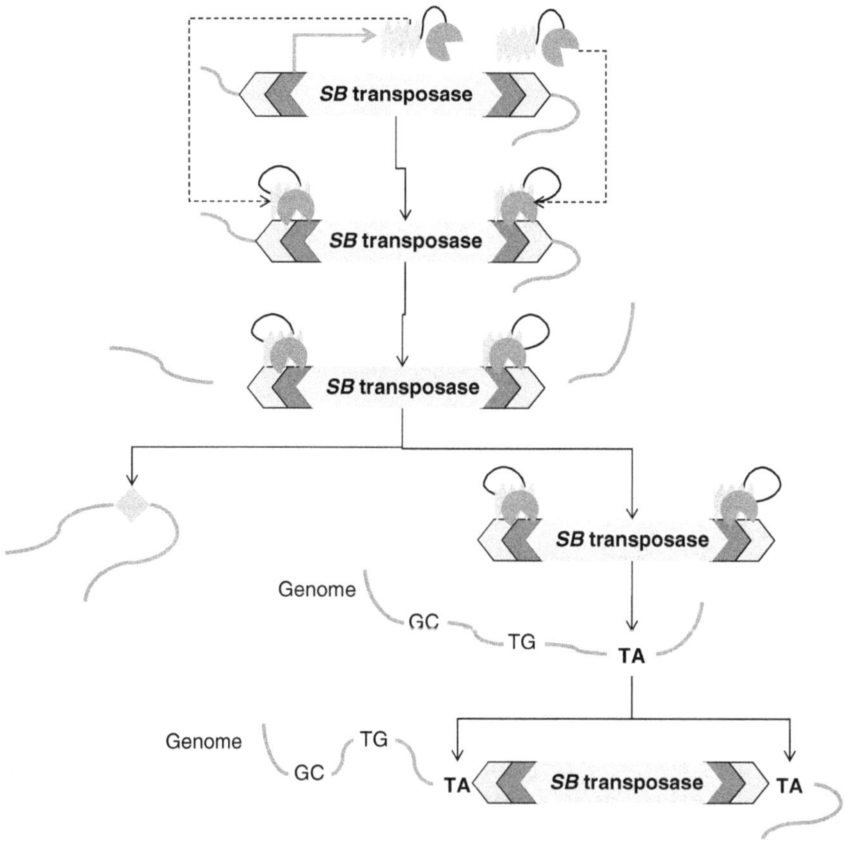

Figure 17.1 Outline of *Sleeping Beauty* (*SB*) transposition. In a natural, autonomous configuration, the transposon encodes the *SB* transposase, which is flanked by two sets of inverted repeats, IRs (darker arrow)/DRs (lighter arrow). Each of these IRs contain two binding sites for the transposase. The transposase binds to its binding sites within the IR/DR repeats. Following excision from its initial genomic locus (donor locus) the transposon integrates into a TA dinucleotide elsewhere (target locus). Transposon excision leaves a temporary break at the donor locus behind, which is repaired by the cell's repair machinery, resulting in a molecular footprint (brown rhombus). Basically the SB transposase consists of two major functional domains, the N-terminal domain (orange box with wavy outline) involved in DNA binding and a C-terminal domain (red pie) involved in catalysis. Host factors assist transposition at various steps in the reaction (not shown for simplicity).

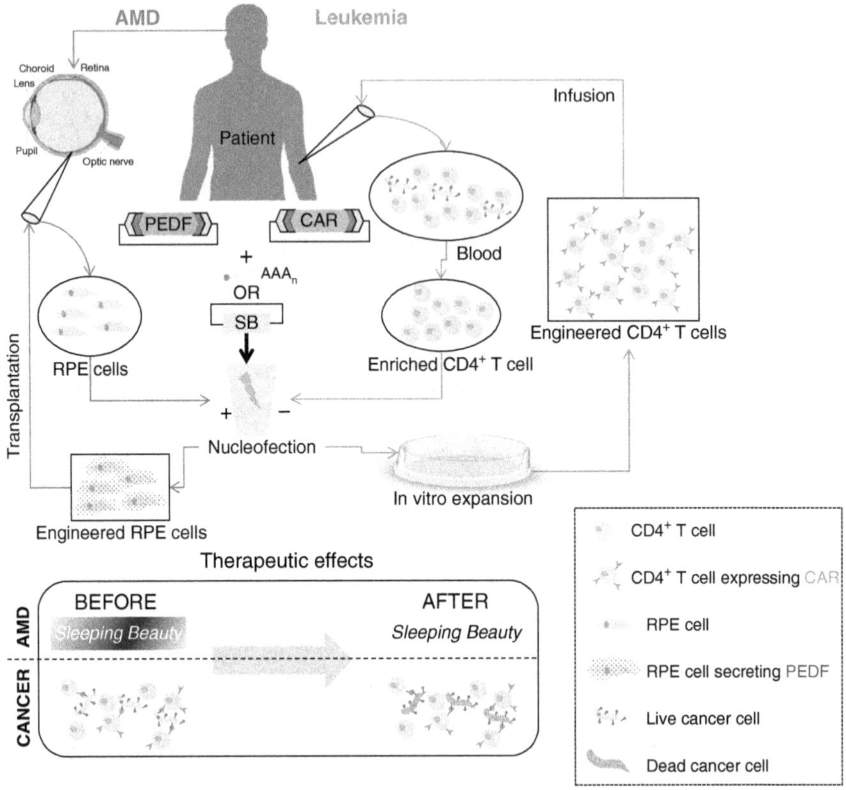

Figure 17.3 Illustration depicting the application of the *Sleeping Beauty* system for gene therapy. Schematics of two *ex vivo* strategies are shown: (1) targeting AMD (age-related macular degeneration) and (2) targeting a lymphoma. AMD treatment: First, the retinal pigment epithelial cells (RPEs) are surgically isolated from the patient (~ 10 000 cells). The cells are engineered to secrete the therapeutic gene, pigment epithelium-derived factor (PEDF). The expression vector is the *Sleeping Beauty* (non-viral system) delivered by nucleofection. The therapeutic cells are transplanted back into the patient's retina within a short span of time (~ 60 min). Note that the cells are not cultured. For details refer to Johnen *et al.* (2012; 2015; /www.targetamd.eu/). Treating B-cell malignancy: CD4$^+$ T cells are isolated from the patient's blood and are nucleofected with a chimeric antigen receptor (CAR) against the tumor antigen using the *Sleeping Beauty* system. The engineered cells are expanded, enriched and selected *in vitro* before being infused back into the respective patient. For details refer to Kebriaei *et al.* (2016); Singh *et al.* (2015). The envisioned therapeutic effects are shown in the adjoining inset.

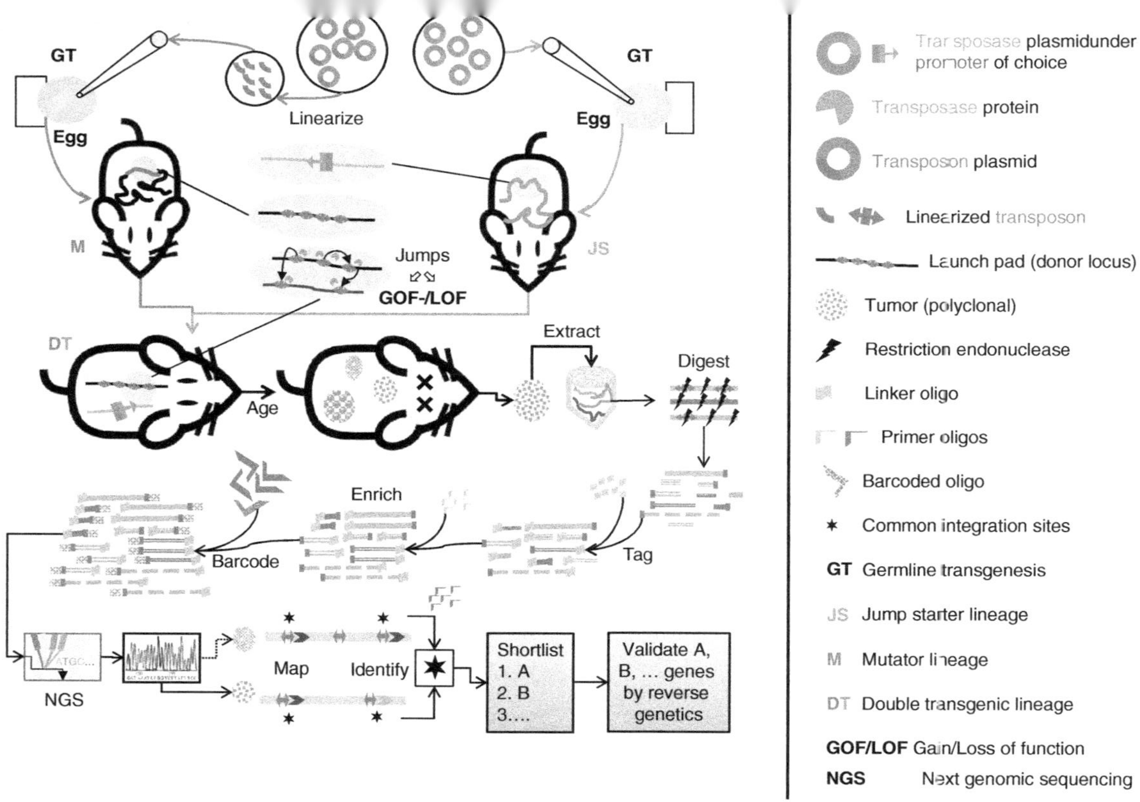

Figure 17.4 Illustration of a generalized strategy for employing the *Sleeping Beauty* transposon system for functional oncogenomics. A conditional insertional mutagenesis strategy is shown at whole organism level (illustrated as mice here). It first involves development of chosen models that are independently transgenic for both the transposon (the mutator lineage) and the transposase (the jumpstarter) lineage by germline transgenic technologies like DNA microinjections. The mutator line carries the transposon at the donor locus in a form of a concatemerized array (also referred to as the launchpad), generated by classic transgenic methodology of microinjecting linearized plasmid DNA.
The transposon is engineered to contain oncogenic promoters along with appropriate splice acceptor and donor signals. Upon insertion into genes, the mutagenic transposon is capable of altering the natural reading frames and overexpressing novel transcripts. These gain- and/or loss-of-function (GOF/LOF) events may ultimately kick-start carcinogenesis. This scenario is restricted to tissues of choice by expressing the transposase in tissue-specific manner. As the double transgenic animals "age," they develop tumors. The tumors are harvested, processed and analyzed. *De novo* insertion sites of the remobilized transposon from the launchpad are determined by an elaborate PCR method, and are mapped onto the host genome. The insertion sites are statistically filtered, and common insertion sites (CISs) to multiple harvested tumors are determined. The oncogenic likelihood of the mutated genes is validated by reverse genetic approaches. For further information, refer to Geurts *et al.* (2006); Moriarity *et al.* (2015).

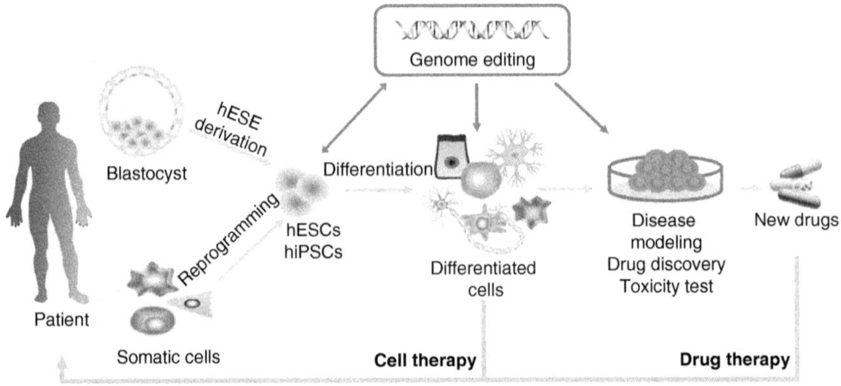

Figure 18.5 Clinic application of genome editing in hPSCs. In drug discovery, genetically modified hESCs and hiPSCs are differentiated to generate lineage-specific cells, which can be used in disease modeling to dissect the mechanisms underlying disease phenotypes, in drug screening to identify drug candidates for human diseases and in cardiac and liver toxicity testing to assess drug safety. In cell therapy, genetically modified hESCs and hiPSCs can be differentiated into functional cells to repair damaged tissues. In this context, genome editing can help to derive specialized cells from hPSCs for cell therapy with increased efficacy and safety.

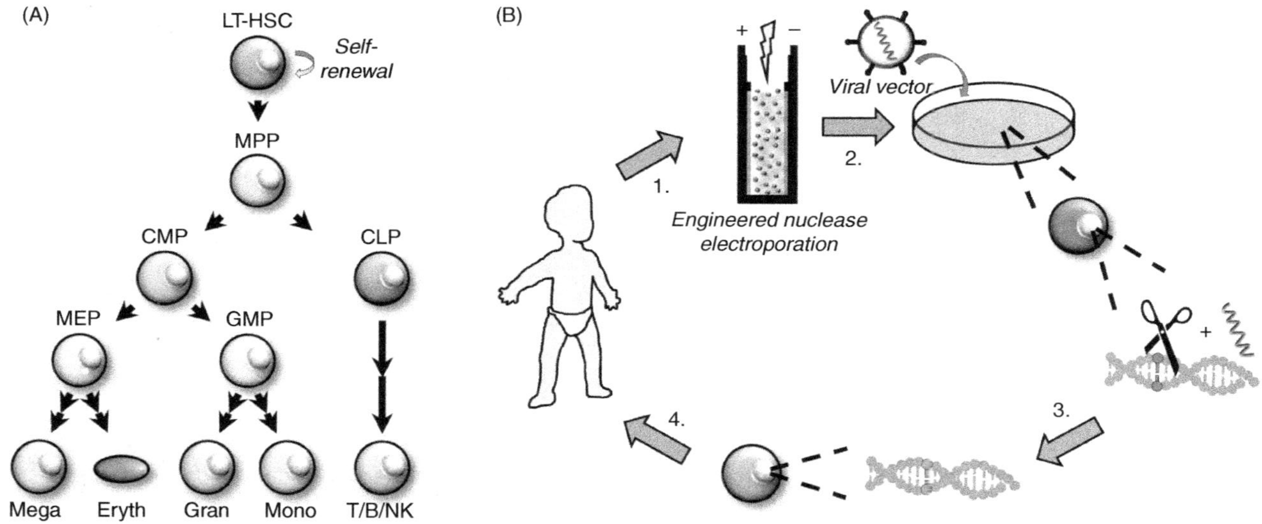

Figure 20.1 Genome editing in hematopoietic stem and progenitor cells. (A) Schematic of the hematopoietic hierarchy in which long-term repopulating hematopoietic stem cells (LT-HSCs) with self-renewal capacity give rise to multipotent progenitors (MPPs), which in turn give rise to common myeloid progenitors (CMPs) and common lymphoid progenitors (CLPs). CMPs give rise to myeloid-erythroid progenitors (MEPs) and granulocyte-monocyte progenitors (GMPs), which through several differentiation steps form fully differentiated hematopoietic cells, megakaryocytes, erythrocytes, granulocytes and monocytes. The CLPs can, through multiple differentiation steps, give rise to T cells, B cells and natural killer (NK) cells. (B) Schematic showing the putative process of autologous transplantation of genome edited patient-derived $CD34^+$ HSPCs. Step 1: $CD34^+$ HSPCs with the genetic mutation (depicted in red) can be derived from different sources: bone marrow harvested from the iliac crest, $CD34^+$ cells mobilized into the peripheral blood, or from cord blood if the patient's cord blood was cryopreserved. Cells are cultured overnight with cytokines that promote survival and proliferation of HSCs and engineered nucleases are delivered to the cells either as mRNA or protein by electroporation. Step 2: cells recover before transplantation, or for therapies entailing homologous recombination (HR), cells are added viral vectors for donor DNA delivery or alternatively, single-stranded oligodeoxynucleotides can be included in the electroporation. Step 3: in the nuclei of electroporated cells, the nuclease facilitates a specific double-strand break at the desired site (red base pair) and either induces random sequence insertion or deletions (INDELs) leading to disruption of the locus, or in the presence of a homologous donor (green DNA) induces HR leading to gene correction or transgene integration (green base pair). Step 4: cells containing the edited locus are transplanted back into the patient to give rise to hematopoiesis of corrected cells.

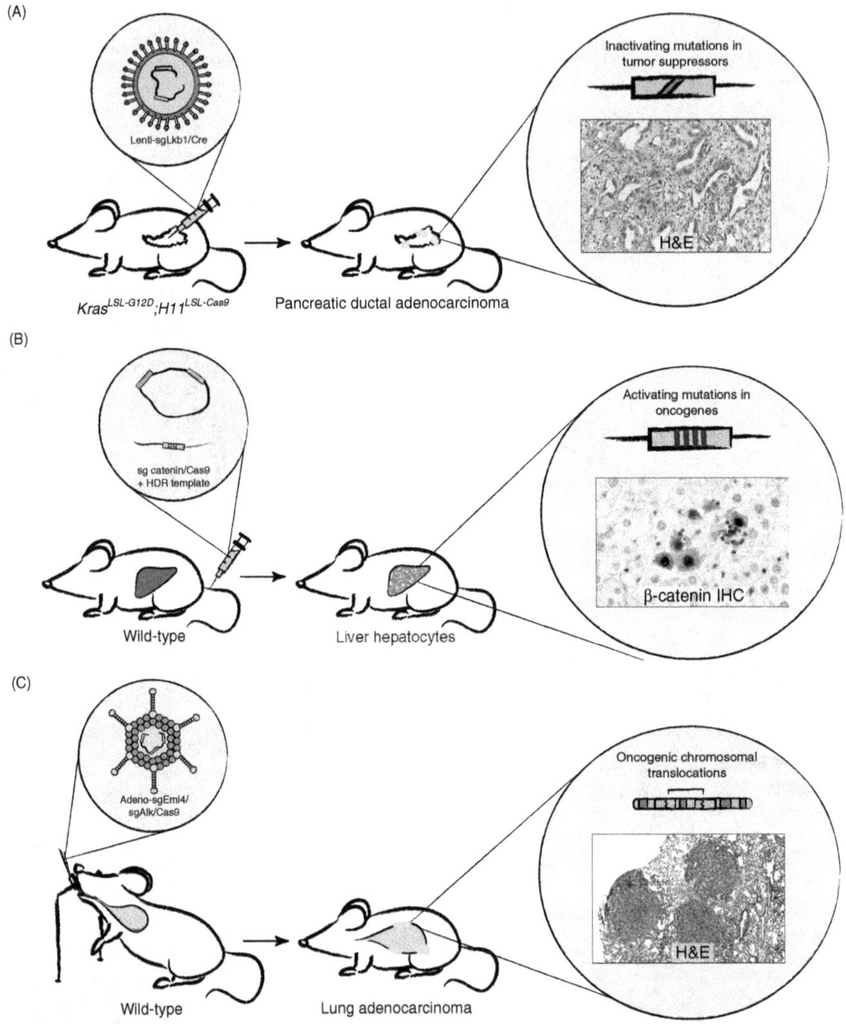

Figure 21.2 Examples of *in vivo* cancer modeling using CRISPR/Cas9 systems. (A) Tumor suppressor inactivation during carcinogenesis can be modeled using an individual sgRNA targeting the desired gene. Lentiviruses encoding both Cre and an sgRNA targeting *Lkb1* were delivered to the pancreas of a mouse with Cre-inducible *Cas9* and $Kras^{G12D}$ alleles by retrograde pancreatic ductal injection. This led to the formation of pancreatic tumors and metastases harboring *Lkb1* indels at the targeted locus (data provided by Ian Winters and Shin-Heng Chiou). (B) Modeling activating mutations in oncogenes can be achieved through the additional delivery of a template for homology-directed repair (HDR). A plasmid encoding Cas9 and an sgRNA targeting β-*catenin*, as well as a ssDNA β-*catenin* HDR template with activating mutations were delivered to the mouse liver by hydrodynamic tail vein injection. This resulted in cells with nuclear β-*catenin* expression as detected by IHC demonstrating the feasibility of this approach for modeling oncogene-induced liver cancer. (Data provided by Wen Xue and Tyler Jacks). (C) Chromosomal rearrangements in cancer can be generated in somatic cells using the CRISPR-Cas9 system. Cas9 along with two sgRNAs targeting the *Eml4* and *Alk* genes were packaged into an adenoviral vector and delivered to mouse lung by intratracheal instillation. Resulting tumors that formed in the lung harbored the desired chromosomal *Eml4-Alk* translocations (data provided by Andrea Ventura).

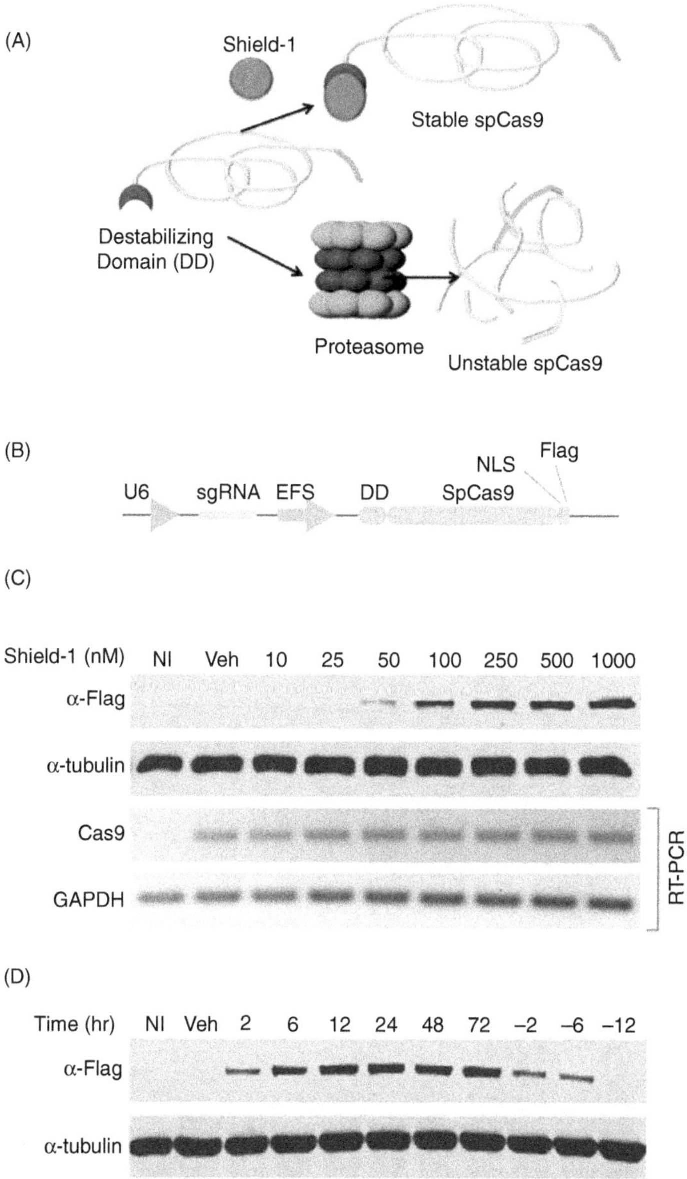

Figure 22.1 Establishment of an inducible lentiviral guide RNA expression system based on conditionally destabilized Cas9. (A) Diagram of conditionally destabilized Cas9 (DD-Cas9). Cas9 is fused to the destabilizing domain of mutant FKBP12 that triggers the rapid degradation of the fused-protein. The synthetic ligand (Shield-1) prevents this destabilizing effect and thus degradation of Cas9. Note, a Flag tag is present at the C-terminal of Cas9. (B) Schematic of the lentiviral construct utilized in our studies. (C) Ligand-dependent stabilization of DD-Cas9. Cells expressing destabilized Cas9 (DD-Cas9) and uninfected cells (NI) were either mock-treated (vehicle) or treated with increasing concentrations of Shield-1, as indicated. To measure the level of expression of Cas9, lysates were analyzed by western blot with an anti-Flag antibody. Various degrees of Shield-1-induced stabilization could be observed in a dose-dependent manner. DD-Cas9 was not detected in lysate from mock-treated cells. To verify that the destabilization of DD-Cas9 occurs post-transcriptionally, DD-Cas9 expression was also verified by RT-PCR analysis using Cas9-specific primers and GAPDH as the loading control. (D) Rapid and reversible regulation of destabilized DD-Cas9. A549 cells transduced with the DD-Cas9 lentiviral vector and uninfected cells were treated with Shield-1 (200 nM) for different time points as indicated. Samples were assayed by western blot analysis at the indicated time points (Senturk *et al.*, 2017).

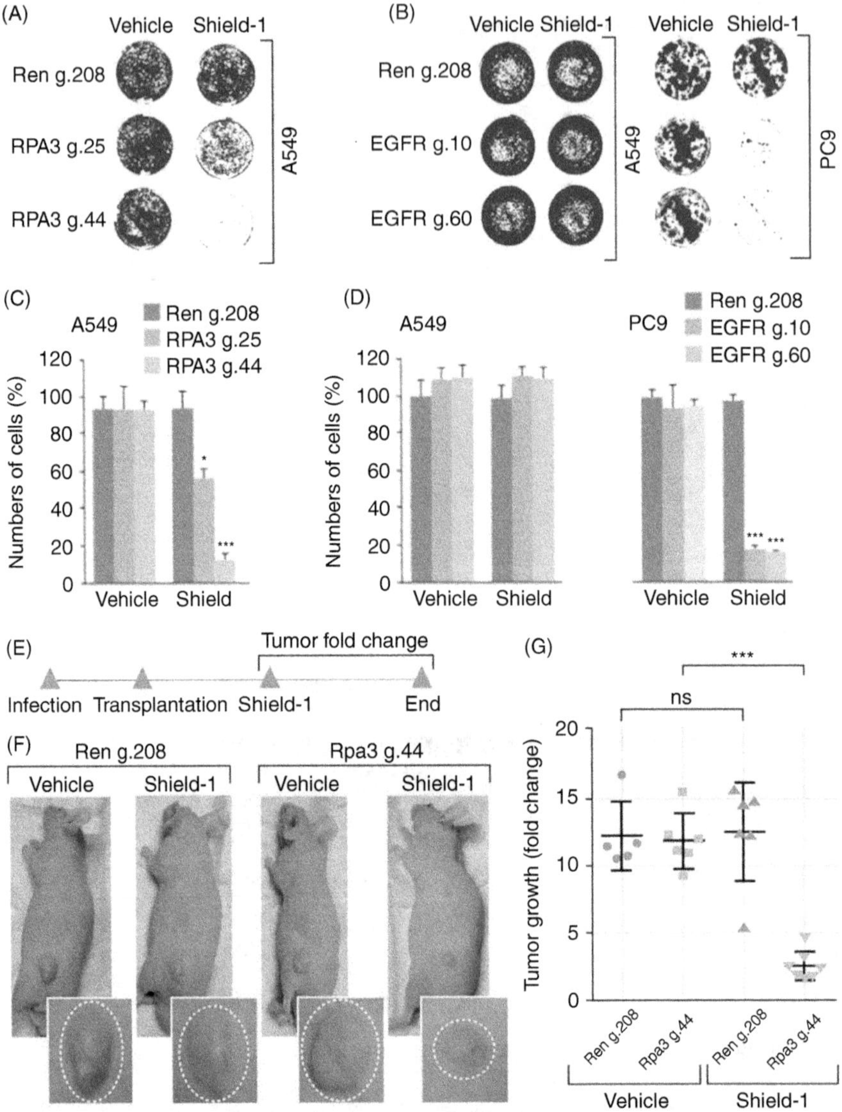

Figure 22.3 DD-Cas9 can drive robust synthetic ligand-dependent gene editing in mammalian cells and enables the identification of essential genes *in vitro* and *in vivo*. Targeting RPA3 in A549 cells and EGFR in PC9 cells resulted in a dramatic decrease in the number of viable cells following induction with Shield-1. (A) A549 cells were transduced with the DD-Cas9 vector targeting the RPA3 locus and, as a control, the Renilla gene. (B) PC9 cells and A549 cells were transduced with DD-Cas9 vector targeting the EGFR locus. In all cases cells were treated 24 hours after infection with Shield-1 (200 nM) for six days. The number of cells was assessed after six days by Crystal Violet staining. (C, D) The data in the charts represent quantification of Panels A and B, respectively, and are expressed as mean ± standard deviation upon treatment with Shield-1 and vehicle control relative to uninfected cells ($n = 4$, * $p < 0.05$, ** $p < 0.01$, *** $p < 0.005$) Student's t-test. (E) Schematic of the transplantable model system used in this study. (F) A549 cells were transduced with DD-Cas9 targeting RPA3 and as control Renilla and transplanted sub-cutaneously in immune-deficient mice. When the tumors reached an approximate size of 4–5 mm in diameter, mice were treated with Shield-1 (1 μg). Fold change in tumor growth was determined ten days after treatment. Pictures illustrate representative tumor sizes. The chart in panel (G) illustrates combined quantification of tumor volume in the indicated cohorts upon Shield-1 treatment. Data represent mean ± standard deviation and are shown relative to the negative control (Cas9 negative) *** $p < 0.0001$ (RPA3 g.44 + Shield-1 vs. – Shield-1) Student's t-test (Senturk *et al.*, 2017).

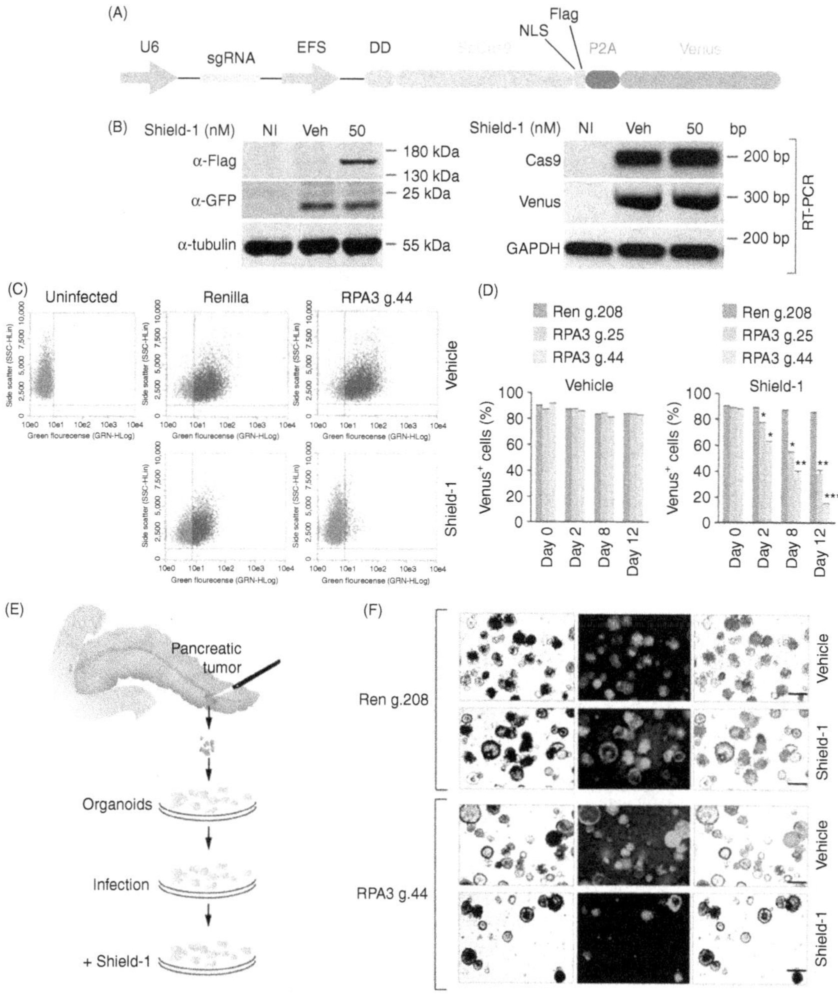

Figure 22.4 DD-Cas9 can be coupled to a modified fluorescent protein to generate a traceable system. (A) Schematic of the lentiviral DD-Cas9 plasmid containing U6 promoter-driven single-guide RNA (sgRNA) and EFS promoter-driven DD-Cas9 pieces Note, a FLAG tag is present at the Cas9 C-terminal moiety, and 2A self-cleaving peptide (P2A) separates both DD-Cas9 and the gene of interest, in this case the modified fluorescent protein Venus. (B) Shield-1-independent expression of Venus. A549 cells were transduced with the DD-Cas9/P2A/Venus lentiviral vector and treated with Shield-1 for three days. Expression of Cas9 and Venus were assayed by western blot and RT-PCR. (C) Competition assay. A549 cells were transduced at low MOI with the DD-Cas9/P2A/Venus lentiviral vector. Cells were then treated with Shield-1 or vehicle control for 12 days and quantified by FACS. The panels represent typical FACS dot plots from A549 sorted cells. (D) A chart illustratrating the percentage of Venus$^+$ cells during time in A549 cells transduced with the DD-Cas9/P2A/Venus lentiviral vector targeting the indicated genes upon treatment with Shield-1 and vehicle control, error bars are standard deviation ($n = 3$, $^* p < 0.05$, $^{**} p < 0.01$, $^{***} p < 0.005$) Student's t-test. (E) Schematic representing our test for cell vulnerabilities in pancreatic tumor organoids. (F) DD-Cas9/P2A/Venus could be used to infect primary human tumor-derived cultures and to assess their cellular vulnerabilities. Human pancreatic cancer-derived organoids were infected with the DD-Cas9/P2A/Venus lentiviral vector targeting RPA3 and as control Renilla. Organoids, cultured in 24-well plates, were treated with Shield-1 or vehicle control 72 hours post-infection. The number of Venus$^+$ organoids was determined 72 hours after infection and six days after Shield-1 treatment (Senturk *et al.*, 2017).

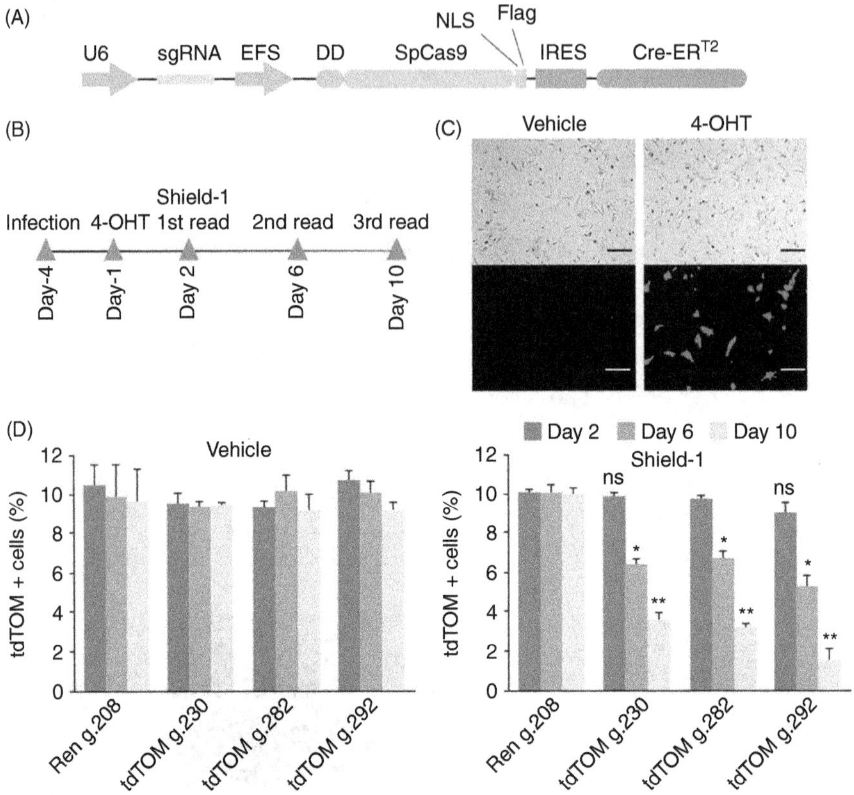

Figure 22.5 DD-Cas9 can be combined with tamoxifen-inducible Cre (Cre-ER^{T2}) to facilitate the study of genetic interactions *in vivo* in pre-existing mouse models based on the Cre-lox system. (A) Diagram of the DD-Cas9/IRES/Cre-ER^{T2} vector utilized in this study. (B) Schematic for experimental design. MEF cells obtained from CAG-LSL-tdTomato mice were transduced with the indicated vectors. Induction of Cre-mediated recombination was induced upon 4-hydroxytamoxifen treatment (5 μM) for three days. To test for efficiency of gene inactivation, cells were treated with Shield-1 (200 nM). (C) Bright-field and fluorescence images of LSL-tdTomato cells treated with 4-hydroxytamoxifen. (D) The chart represents the percentage of tdTomato-positive MEFs transduced with DD-Cas9/IRES/Cre-ER^{T2} lentiviruses carrying the indicated sgRNAs over time after exposure to Shield-1 and vehicle. Numbers of tdTomato-positive and -negative cells were determined by FACS. Each bar represents the average of three individual replicates, error bars are standard deviation ($n = 3$, * $p < 0.05$, ** $p < 0.01$) Student's t-test (Senturk *et al.*, 2017).

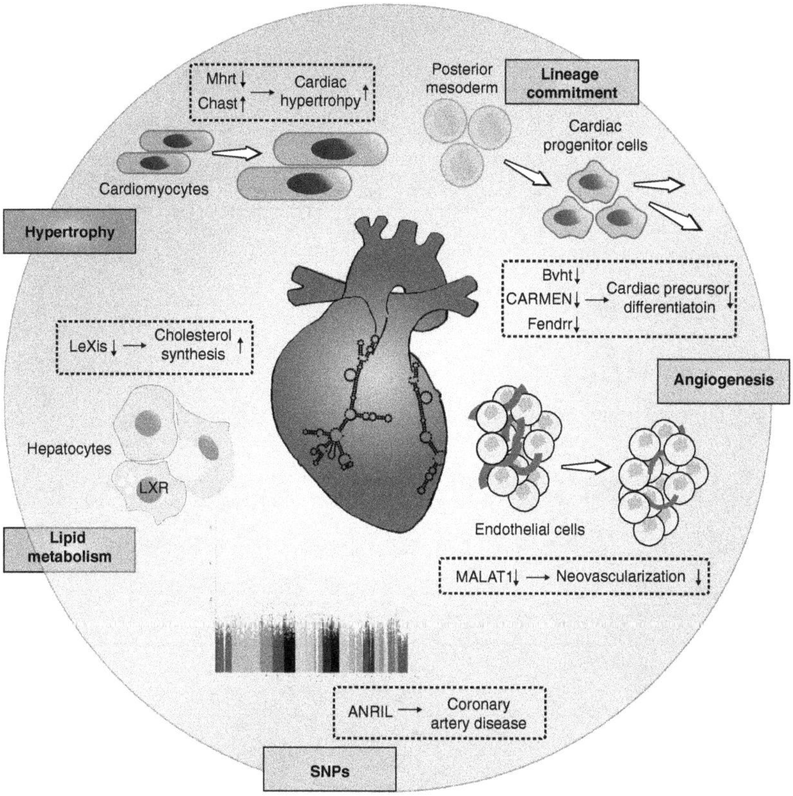

Figure 24.2 LncRNAs in cardiovascular disease. Selection of lncRNAs involved in hypertrophy, lipid metabolism, cardiac lineage development, vascular endothelium and SNPs associated with coronary artery disease.

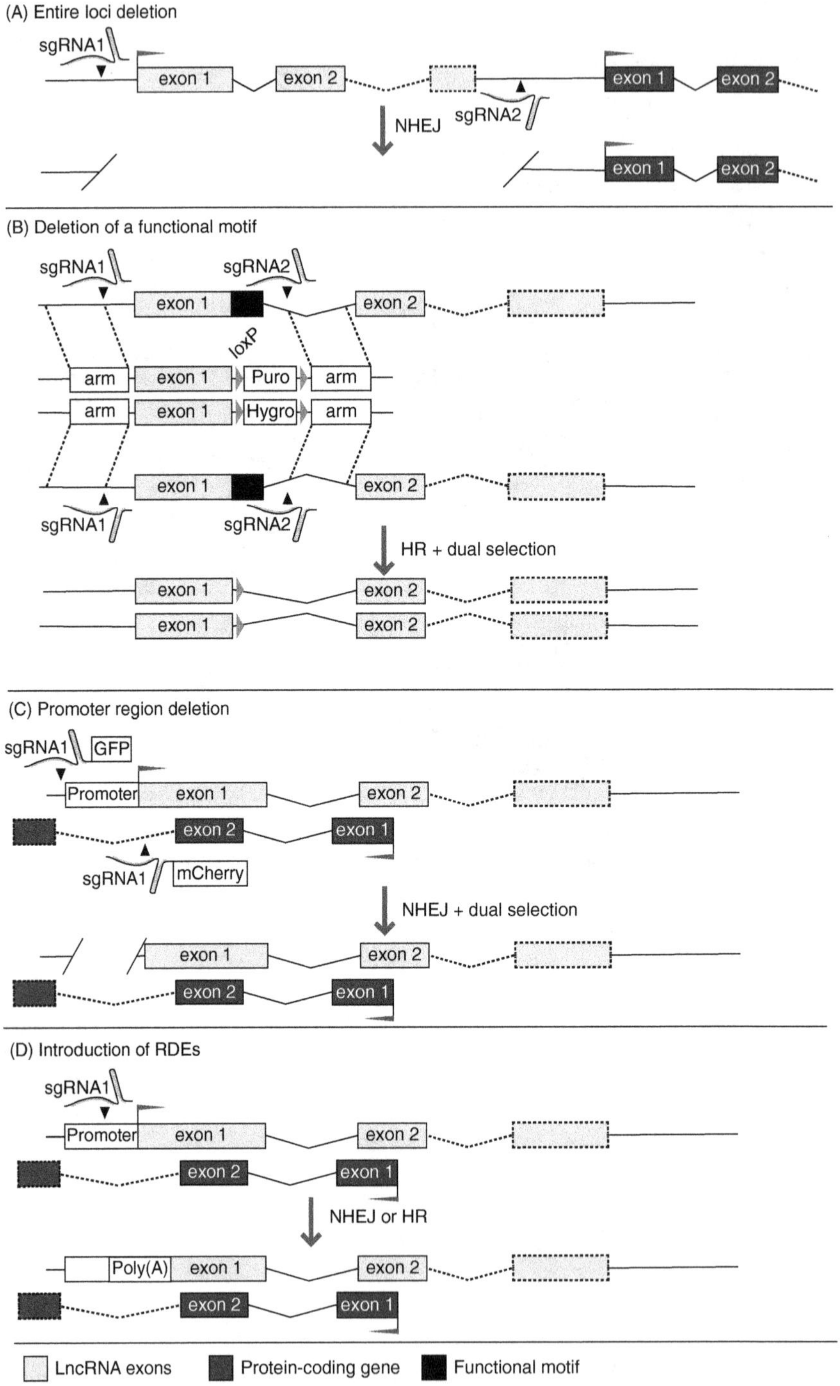

Figure 24.3 Gene editing approaches using CRISPR/Cas9. Schematic display for lncRNA silencing by (A) entire loci deletion, (B) deletion of a functional motif, (C) promoter region deletion and (D) introduction of RNA-destabilizing elements.

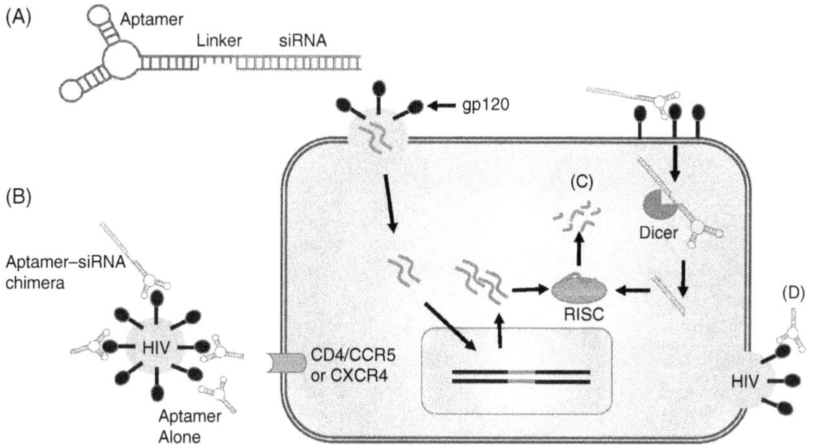

Figure 25.2 Overview of aptamer and aptamer–siRNA chimera mode of action. (A) Schematic of aptamer conjugated to siRNA via linker bridge. (B) Interaction of aptamer and aptamer–siRNA chimera for viral neutralization and entry inhibition. (C) After internalization of the aptamer–siRNA chimera, siRNA is processed by the DICER/RISC pathway, which leads to viral mRNA degradation. (D) Aptamer interaction during viral budding inhibits budding and maturation.

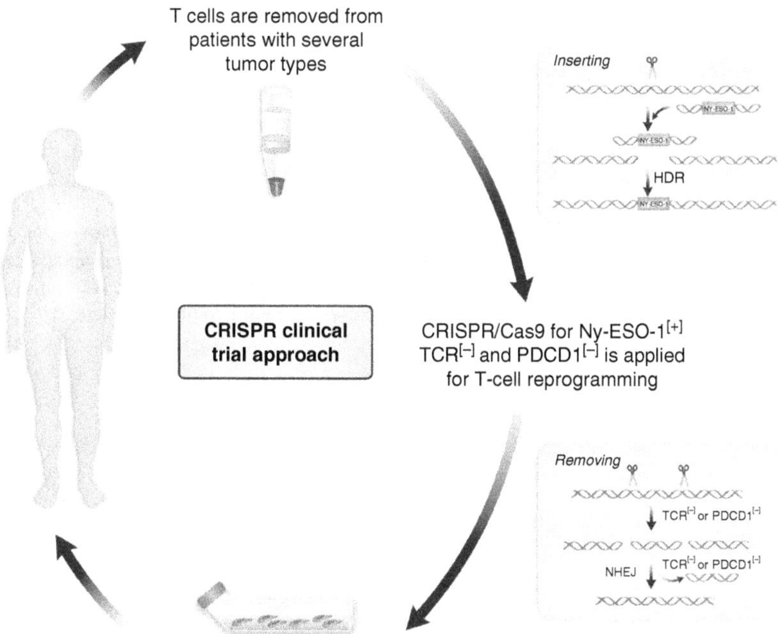

Figure 26.2 Schematic representation of the use of CRISPR/Cas9 system for cancer treatment. The protocol is based on reprograming of T cells of the immune system to destroy the tumor cells. The T cells from patients with different types of cancer are subjected to three genetic alterations: (1) inserting the receptor protein NY-ESO-1, a specific antigen of tumor cells, directing the T cells to attack the tumor, (2) removing the natural T-cell receptor, so that no interference occurs, thus ensuring the efficiency of the treatment and (3) removing the PD-1 protein gene, which negatively regulates T-cell response, enhancing the killing of tumor cells.

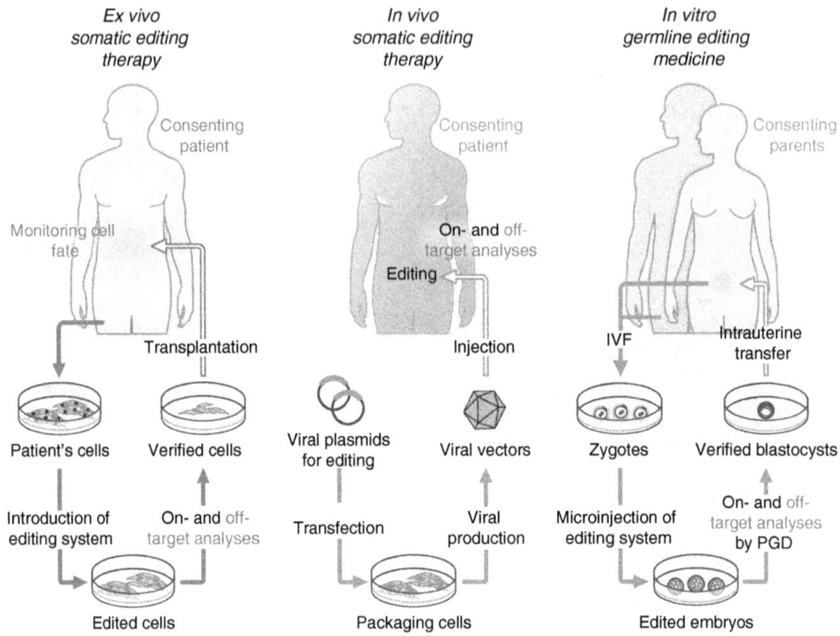

Figure 30.1 Three concepts of genome editing medicine. Autologous *ex vivo* somatic editing therapy begins by procuring somatic cells from a consenting patient. After genome editing, somatic cells are subject to on-target gene modification and off-target mutation analyses. The behavior of transplanted cells will be monitored in the patient. With regard to *in vivo* somatic editing therapy, the direct subject of genome editing is a consenting patient. A viral vector harboring the genome editing system is produced by the culturing of packaging cells. After the viral vector is delivered to the body, the patient will undergo various tests, which include the confirmation of on-target gene modification and the occurrence of off-target mutations in some tissues. *In vitro* germline editing medicine is primarily initiated by creating zygotes *in vitro*. Similarly to an assisted reproductive technology (ART) procedure that is used for male infertility, genome editing nucleases are microinjected into zygotes and are subsequently subject to preimplantation genetic diagnosis (PGD), by which no off-target mutations as well as an on-target gene modification are verified, prior to embryo transfer. IVF: *in vitro* fertilization. (Araki and Ishii, 2016. Reproduced with permission from Elsevier.)

13 Application of TAL Proteins and the CRISPR System to Purification of Specific Genomic Regions for Locus-specific Identification of Chromatin-associated Molecules

Toshitsugu Fujita and Hodaka Fujii

13.1 Introduction

Genome functions such as transcription and epigenetic regulation play central roles in biological phenomena, including development, cell differentiation and other cellular functions. Abnormal regulation of genome functions can lead to pathogenesis. Accordingly, elucidation of the molecular mechanisms underlying genome functions should facilitate understanding of the biological principles, as well as development, of effective treatments against intractable diseases such as cancers, metabolic disorders and neurological disorders. To gain mechanistic insight into the regulation of genome functions, it is essential to determine which molecules are associated with a genomic region of interest. However, methods for identifying the molecules that interact with specific genomic regions remain limited (see the review by Fujita and Fujii, 2013b).

To isolate genomic regions of interest while preserving their molecular interactions, we developed locus-specific chromatin immunoprecipitation (locus-specific ChIP) technologies consisting of insertional ChIP (iChIP) (Fujita and Fujii, 2011, 2012, 2014a; Fujita *et al.*, 2015a; Hoshino and Fujii, 2009) and engineered DNA-binding molecule-mediated ChIP (enChIP) (Fujii and Fujita, 2015; Fujita and Fujii, 2013a, 2014b, 2014c, 2015a, 2015b, 2016b; Fujita *et al.*, 2013; 2015b). In iChIP, the recognition sequences of an exogenous DNA-binding protein, such as the bacterial LexA protein (LexA-binding elements (LexA BE)), are inserted into target genomic regions. Subsequently, the exogenous DNA-binding protein, fused with a tag or tags if necessary, is expressed in cells containing the recognition sequences. After crosslinking to stabilize molecular interactions if necessary, chromatin is fragmented by sonication or other methods and subjected to affinity purification, e.g. using an antibody (Ab) against the DNA-binding protein or the fused tag(s). After reverse crosslinking when necessary, proteins, RNAs or other genomic regions interacting with the target genomic region are identified by mass spectrometry (MS), RNA sequencing (RNA-Seq), next-generation sequencing (NGS) or other methods. Recently, we also developed *in vitro* iChIP technology (Fujita and Fujii, 2014a). In this technique, the recognition sequences of an exogenous DNA-binding protein are inserted into the target genomic regions. After crosslinking (if necessary), chromatin is fragmented by sonication or enzymatic digestion. The resultant fragmented chromatin is incubated with recombinant or synthetic exogenous DNA-binding molecules and subjected to affinity purification. The associated molecules are identified as described above.

Although iChIP has been successfully applied in a wide range of contexts (see review by Fujita and Fujii, 2016a), knockin of the recognition sequences of an exogenous DNA-binding protein is time-consuming and requires significant effort. However, recent advances in genome editing technologies have made knockin procedures much more straightforward and less tedious (see relevant chapters of this book). Nevertheless, methods that do not require knockin procedures are still desired, and we developed enChIP to meet this need. In this chapter, we describe recent advances in enChIP technology, as well as its applications.

13.2 Methodology

enChIP consists of conventional (in cell) enChIP and the more recently developed *in vitro* enChIP. In conventional enChIP, engineered DNA-binding molecules used for locus-tagging are expressed in the cells to be analyzed. By contrast, in *in vitro* enChIP, loci are tagged by mixing recombinant or synthetic engineered DNA-binding molecules with fragmented chromatin in a test tube.

13.2.1 Scheme for Conventional (In Cell) enChIP

Conventional enChIP is depicted in Figure 13.1 (Fujita and Fujii, 2013a, 2014b; Fujita *et al.*, 2013; 2015b). The method proceeds as follows:

1, Generate an engineered DNA-binding molecule recognizing the target sequence. Suitable molecules include zinc-finger proteins (ZFPs) (Pabo *et al.*, 2001), transcription activator-like proteins (TALs) (Bogdanove and Voytas, 2011), and the clustered regularly interspaced short palindromic repeats (CRISPR) system (Doudna and Charpentier, 2014; Harrison *et al.*, 2014). The CRISPR system used in enChIP analysis consists of a mutant dCas9 protein lacking nuclease activity, and a guide RNA (gRNA). If necessary, one or more nuclear localization signals (NLSs) or tags for affinity purification can be fused to the engineered DNA-binding molecule.

2, Express the engineered DNA-binding molecule in the cells to be analyzed. When transfection efficiency is high, a transient expression strategy can be used. Otherwise, stable transformants should be generated using retroviral or lentiviral expression systems, or other conventional approaches. We deposited retroviral and lentiviral expression vectors for enChIP in Addgene (www.addgene.org/crispr/purify/, www.addgene.org/Hodaka_Fujii/).

3, Crosslink the cells using formaldehyde or another appropriate reagent, if necessary. Cells should be stimulated as required prior to the crosslinking step. For stable isotope labeling by amino acids in cell culture (SILAC) analysis, cells should be cultured in SILAC media (see Section 13.3.1.1).

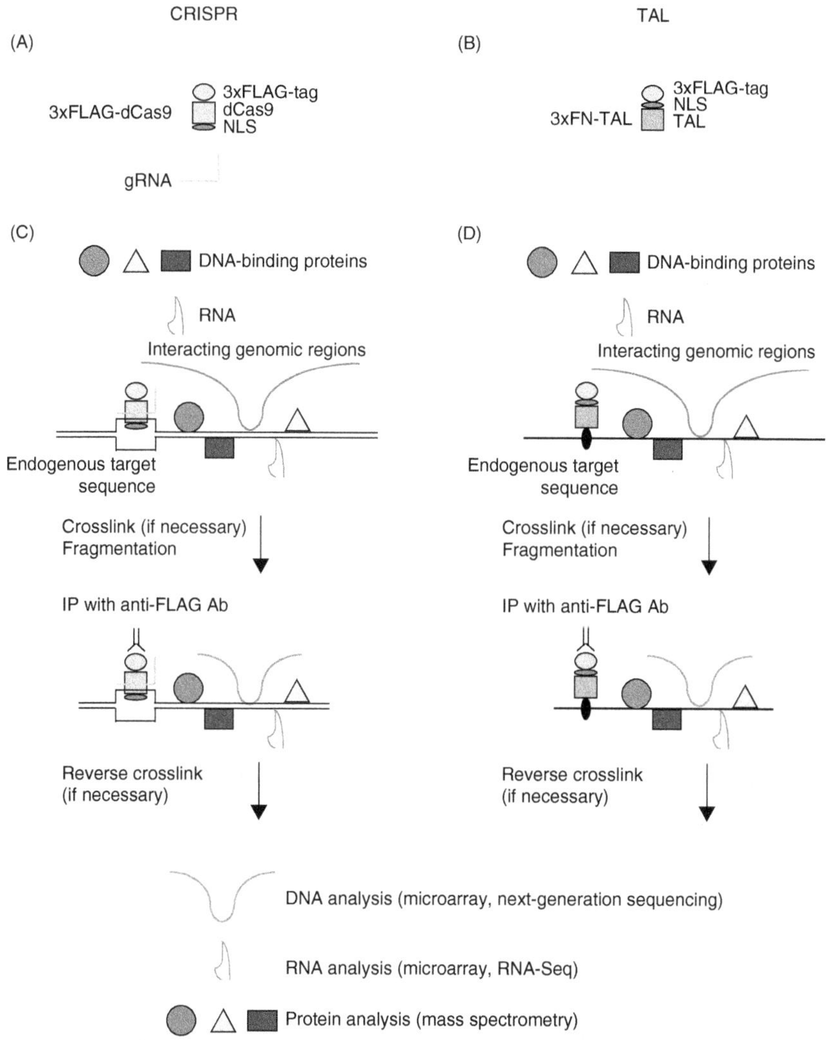

Figure 13.1 In cell enChIP. A locus of interest is tagged with an engineered DNA-binding molecule such as the CRISPR complex, which comprises a catalytically inactive form of Cas9 (dCas9) and gRNA (A), or a TAL protein (B) that recognizes the target sequence, expressed in the cells to be analyzed. (C and D) Interactions between molecules are stabilized by a crosslinker such as formaldehyde, if necessary. After fragmentation of chromatin by sonication or enzymatic digestion, the locus tagged with the engineered DNA-binding molecule is subjected to affinity purification. Crosslinking is reversed if necessary and the RNAs and other genomic regions that interact with the locus are identified by NGS or microarray analysis. Interacting proteins are identified by MS. (A black-and-white version of this figure will appear in some formats. For the color version, please refer to the plate section.)

4. Prepare the chromatin fraction and fragment it by sonication or digestion with one or more endonucleases.
5. Purify the chromatin complexes tagged with the engineered DNA-binding molecule using an Ab or other affinity reagents against the fused tag(s) or the engineered DNA-binding molecule itself.

6, Reverse-crosslink the isolated chromatin, if a crosslinker was used in step 3.
7, Identify proteins associated with the target genomic region by MS. Identify RNAs and other genomic regions by NGS or microarray analysis.

13.2.2 Scheme for *In Vitro* enChIP

The scheme for *in vitro* enChIP is as follows (Figure 13.2) (Fujita and Fujii, 2014a; Fujita *et al.*, 2016):

1, Crosslink the cells using formaldehyde or another appropriate reagent, if necessary. Cells should be stimulated as required prior to the crosslinking step. For SILAC analysis, cells should be cultured in SILAC media (see Section 13.3.1.1).
2, Prepare the chromatin fraction and fragment it by sonication or digestion with one or more endonucleases.
3, Incubate fragmented chromatin with recombinant or synthetic engineered DNA-binding molecules. ZFPs and TALs can be generated as recombinant proteins. When using the CRISPR complex, dCas9 can be generated as a recombinant protein, whereas the gRNA can be chemically synthesized or generated by *in vitro* transcription.
4, Purify the chromatin complexes tagged with the engineered DNA-binding molecule using an Ab or other affinity reagents against the fused tag(s) or the engineered DNA-binding molecule itself.
5, Reverse-crosslink the isolated chromatin, if a crosslinker was used in step 3.
6, Identify proteins associated with the target genomic region by MS. Identify RNAs and other genomic regions by NGS or microarray analysis.

13.2.3 Choosing Between In Cell and *In Vitro* enChIP

In vitro enChIP has several advantages over in cell enChIP. First, in the former technique, it is not necessary to express engineered DNA-binding molecules in the cells to be analyzed; instead, chromatin prepared from wild-type/non-transgenic cells can be used. Establishment of cells for *in vivo* enChIP analysis takes time and effort, especially when enChIP analysis is attempted in primary cells. Second, it might be difficult to express engineered DNA-binding molecules in cells of non-model organisms or pathogens. In such cases, *in vitro* enChIP is the method of choice. Third, expression of engineered DNA-binding molecules in the cells might have aberrant effects on expression of target genes due to CRISPR interference (CRISPRi) (Bikard *et al.*, 2013; Qi *et al.*, 2013) or other mechanisms.

On the other hand, the yields of *in vitro* enChIP are significantly lower than those of in cell enChIP (Fujita *et al.*, 2016). Consequently, to obtain comparable amounts of purified specific genomic regions, significantly more cells must be used for *in vitro* enChIP. This issue has a particularly serious impact on the identification of proteins by MS. However, it is still feasible to use *in vitro* enChIP to identify RNAs and genomic regions associated with a target genomic region. Thus, the choice between in cell or *in vitro* enChIP should be made based

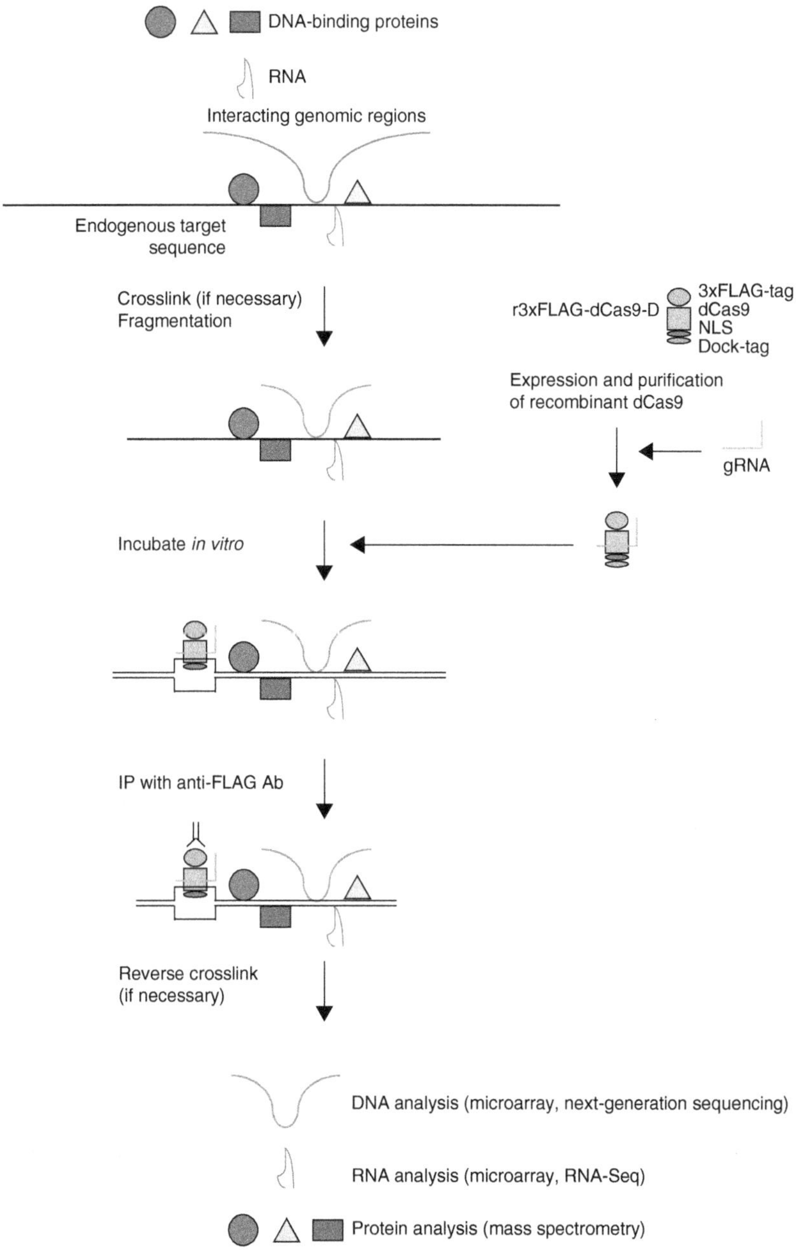

Figure 13.2 *In vitro* enChIP. An engineered DNA-binding molecule such as the CRISPR complex comprising recombinant dCas9 protein plus synthesized gRNA, or a recombinant TAL protein, is prepared *in vitro*. Fragmented chromatin from the cells to be analyzed is incubated with the engineered DNA-binding molecule and subjected to affinity purification. Molecules associated with the target genomic region are analyzed as described for in cell enChIP. This figure illustrates *in vitro* enChIP using CRISPR.

on the feasibility of establishing cells for in cell enChIP, the nature of the target molecules and other factors.

13.3 Results and Discussion

13.3.1 Identification of Proteins Associated with a Specific Genomic Region

13.3.1.1 enChIP-MS

We used enChIP using CRISPR combined with MS (enChIP-MS) to identify proteins associated with the promoter region of the interferon (IFN)-inducible *IFN regulatory factor-1 (IRF-1)* gene (Fujita and Fujii, 2013a, 2014b). In the latter of the two studies, we employed SILAC, a quantitative form of MS (Ong *et al.*, 2002). The results revealed proteins involved in transcriptional regulation, acetyltransferases, DNA topoisomerases, histone variants, histone deacetylation/co-repressor components and other proteins as being specifically associated with the promoter in an IFNγ-specific manner (Fujita and Fujii, 2014b).

We also performed enChIP-MS analysis using a tagged TAL protein to identify proteins associated with mouse telomeres (Fujita *et al.*, 2013). We detected known telomere-binding proteins as well as novel ones.

In both attempts described above, we did not detect all proteins already known to interact with the target genomic regions. For example, the Stat1 protein plays a central role in IFNγ-induced transcription of the *IRF-1* gene (Durbin *et al.*, 1996; Meraz *et al.*, 1996). However, we failed to detect Stat1 in our enChIP-MS analysis. Likewise, the shelterin proteins (e.g. TRF-1 and TRF-2) are associated with telomeres (de Lange, 2005), but we could not detect them in our enChIP-MS analysis. There are several potential explanations for these results. First, these proteins might not be optimal targets of MS analysis. In other words, they might be pulled down in the purified complexes, but not detected by MS due to some intrinsic properties such as the positions of trypsin cleavage sites or peptide length. Consistent with this idea, we detected shelterin-binding proteins (Fujita *et al.*, 2013), suggesting that shelterins did in fact exist in the purified complexes. Alternatively, the sensitivities of the MS analyses used in previous studies may have been insufficient to detect some of the proteins associated with the target loci. Notably in this regard, the stoichiometry of binding of transcription factors such as Stat1 is usually quite low, e.g. 1–4 molecules per locus. Therefore, it is more difficult to detect transcription factors than epigenetic regulators, which bind to target loci more broadly with higher stoichiometry. Accordingly, it may be necessary to significantly increase cell number in order to detect transcription factors. Future improvement in the sensitivity of MS analyses should enable direct identification of transcription factors by enChIP-MS.

13.3.1.2 Another Method for Identifying Proteins Associated with Specific Genomic Regions

Another technique for purifying specific genomic regions is proteomics of isolated chromatin (PICh) (Déjardin and Kingston, 2009). PICh utilizes nucleic acid probes with a tag such as biotin for affinity purification. After fragmentation of crosslinked chromatin, the DNA is partially denatured and hybridized with the probe for affinity purification. PICh has been used to identify proteins interacting with telomeres (Antão *et al.*, 2012; Déjardin and Kingston, 2009) and ribosomal RNA gene promoters (Ide and Dejardin, 2015), both of which are multi-copy loci. It would be of interest to apply PICh to low-copy loci.

13.3.1.3 Methods for Detecting Localization of Proteins on the Genome

Several techniques have been used to detect localization of proteins on the genome:

1. *Biochemical approaches*: ChIP has been used to map proteins of interest, especially in combination with NGS (ChIP-Seq) (Collas, 2010). This approach is very powerful when good antibodies are available.
2. *Imaging approaches*: Localization of a specific protein in the nucleus can be determined by imaging approaches using fluorescent Abs against the target protein or target proteins fused to fluorescent proteins, such as green fluorescent protein, in combination with fluorescent *in situ* hybridization (FISH) or other techniques to visualize specific genomic regions (see below) (Fisher and Merkenshlager, 2002; Fraser and Bickmore, 2007). However, these imaging techniques are not non-biased methods and they require both candidate proteins and genomic loci. In addition, their resolution is usually quite low.

The combination of locus-specific approaches such as enChIP-MS and PICh with ChIP-Seq and imaging analyses should facilitate a comprehensive understanding of the roles of proteins in regulating genome function.

13.3.2 Identification of RNAs Associated with a Specific Genomic Region

13.3.2.1 enChIP-RT-PCR and enChIP-RNA-Seq

We also analyzed RNAs associated with specific genomic regions using enChIP in combination with RT-PCR or RNA-Seq. When we purified mouse telomere regions by enChIP using a TAL protein that recognizes a telomere repeat, followed by RT-PCR, we were able to successfully detect the RNA component of telomerase (*Terc*) in the telomere complexes (Fujita *et al.*, 2013). Furthermore, enChIP combined with RNA-Seq (enChIP-RNA-Seq) detected known telomere-binding RNAs such as the RNA component of mitochondrial RNA processing endoribonuclease (*Rmrp*) and telomeric repeat-containing RNAs (TERRAs), as well as RNAs whose interactions with telomeres were not previously reported (Fujita *et al.*, 2015b). To our knowledge, locus-specific ChIP combined with RNA-Seq or microarray analysis such as enChIP-RNA-Seq is a unique technology that enables non-biased identification of RNAs associated with specific genomic regions.

13.3.2.2 Methods for Detecting Localization of RNAs in the Genome

Several techniques have been developed for determining the localization of RNA species on the genome:

1. *Biochemical approaches*: Oligonucleotide-mediated affinity purification approaches have been used to detect localization of a specific RNA of interest (Chu *et al.*, 2011; Engreitz *et al.*, 2013; Mariner *et al.*, 2008; Simon *et al.*, 2011). These approaches are a counterpart of ChIP, in that they detect the binding sites of RNA species rather than proteins of interest.
2. *Imaging approach*: RNA-FISH depends on hybridization of fluorescent probes with RNAs and genomic regions (see below) (Kwon, 2013). This is not a non-biased approach, and requires selection of candidate RNAs. On the other hand, the ability to localize specific RNAs within the whole nucleus, as well as in other regions of the cell, would facilitate understanding of their biological functions.

enChIP-RNA-Seq is complementary to these techniques and represents a valuable tool for analyzing RNA functions in the nucleus.

13.3.3 Identification of Genomic Regions Associated with a Specific Genomic Region

13.3.3.1 enChIP-Seq

Recently, the dynamics of nuclear organization and its roles in nuclear functions have attracted intense attention from those researching genome functions. Among the important factors that influence nuclear organization are physical interactions between genomic regions. We showed that enChIP combined with NGS (enChIP-Seq) enables identification of interactions between genomic regions (Fujita *et al.*, 2017). enChIP-Seq is thus a useful method for identifying potential regulatory genomic regions, such as enhancers and silencers, as well as for detecting transcription factories and other types of nuclear organization.

13.3.3.2 Methods for Detecting Interactions Between Genomic Regions

Several other techniques have been used to detect interactions between genomic regions.

1. *Biochemical approaches*:
    a. Chromosome conformation capture (3C) (de Wit and de Laat, 2012; Dekker *et al.*, 2002) and its derivatives have been used to detect interactions between genomic regions. In 3C and its derivatives, molecules are crosslinked with formaldehyde or another crosslinker(s). Subsequently, chromatin is digested with a restriction enzyme(s), and the DNA ends within a chromatin complex are ligated. After purification of DNA, genomic loci interacting with the target genomic region are detected by PCR using locus-specific primers or, in a more non-biased fashion, by NGS. Some derivatives of 3C, such as Hi-C, can be used to detect genome-wide interactions (de Wit and de Laat, 2012). On the

other hand, 3C and its derivatives depend on intra-molecular ligation reactions of DNA ends in a chromatin complex for detection of interactions between genomic regions. Because it is very difficult to distinguish intra-molecular ligation reactions from inter-molecular ligation reactions, which cause non-specific background signals, signals detected by 3C-based methods must be confirmed by other methods. In addition, digestion of DNA using restriction enzymes under non-optimal conditions (e.g. in chromatin crosslinked prior to fragmentation) may be incomplete, leading to high background signals, especially for detection of short-range interactions (e.g. several kb) within a gene locus. Several studies have attempted to overcome these problems (Kalhor *et al.*, 2011; Nagano *et al.*, 2015).

b. The in-gel replication of interacting DNA segments (INGRID) assay is a recently reported method for detecting interactions between genomic regions (Gavrilov *et al.*, 2014). The initial steps of INGRID are identical to those of 3C, i.e. crosslinking, fragmentation and solubilization. However, the chromatin fragments are distributed randomly in a layer of polyacrylamide gel, and DNA fragments of interest are subjected to in-gel PCR amplification. Subsequently, target genomic regions are detected with molecular beacons, each consisting of a fluorophore and a quencher, and they emit fluorescence after hybridization with target genomic regions. The INGRID method is ligation-free, thereby avoiding problems caused by the ligation reaction. On the other hand, the method requires selection of candidate interacting genomic regions, and so is not useful for non-biased screening. However, INGRID provides an attractive approach for confirming interactions between genomic regions identified by non-biased methods.

2. *Imaging approaches*:

   a. FISH (Fraser and Bickmore, 2007) involves labeling of genomic regions with nucleic acid probes, each of which is often conjugated to a distinct fluorescent probe. The labeled cells are observed by microscopy to determine whether the signals of the fluorescent probes are in proximity to one another. One advantage of FISH is that it is a single-cell analysis, facilitating comparative analysis of different cell states such as phases of the cell cycle. However, due to this characteristic, many cells must be individually analyzed in order to calculate statistical significance. In addition, the resolution of FISH analysis is low due to the optical limitations of microscopy, even when using super-resolution microscopy (Egner and Hell, 2005; Ji *et al.*, 2008; Sahl and Moerner, 2013). Furthermore, because cells are fixed, live cell imaging is not feasible. Finally, FISH is not a non-biased approach.

   b. Live imaging of specific genomic loci has been reported (Chen *et al.*, 2013; Lindhout *et al.*, 2007; Miyanari *et al.*, 2013). In particular, the system using CRISPR has been used successfully to detect low-copy loci (Chen *et al.*, 2013). Tagging of genomic regions with distinct fluorescent proteins would enable detection of their interactions in living cells, making it possible to analyze dynamic interactions over the course of the cell cycle or under other conditions. However, these systems still suffer from background fluorescence from the fluorescent proteins fused with the engineered DNA-binding molecules, and consequently require multiple tagging of a single locus. In addition, as with FISH, the resolution is low and the method is not non-biased.

Compared with the methods described above, enChIP-Seq has several unique characteristics:

1. enChIP-Seq enables non-biased identification of genomic regions interacting with a target genomic locus. In other words, this method detects "one-to-all" interactions.
2. When chromatin is fragmented by sonication, enChIP-Seq does not require any enzymatic reactions (such as ligation), and thus avoids the associated problems described above.
3. enChIP-Seq can achieve high-resolution analysis. Resolution is flexibly determined by size-fractionation prior to NGS. One of the factors determining resolution is background signals from long DNA fragments containing the target genomic regions. Size-fractionation can remove those DNA fragments containing the target genomic regions longer than a threshold, enabling reduction of background signals. For example, if DNA fragments smaller than 1 kbp are collected after enChIP and reverse-crosslinking, interacting genomic regions more than 1 kbp away from the target locus can be readily detected. Thus, the combination of enChIP-Seq with other complementary analyses described above should enable high-confidence detection of interactions between genomic regions. A comparison chart of non-biased methods for analysis of specific genomic regions is shown in Table 13.1.

13.4 Concluding Remarks

In this review, we outlined enChIP technology, an essential component of locus-specific ChIP technologies. Combination of enChIP with downstream analyses such as MS and NGS enables comprehensive identification of molecules that interact with target genomic regions. Thus, the enChIP approach represents an "all-in-one" method for elucidation of the molecular mechanisms underlying genome functions, as well as an essential tool for analysis of the "4D-Nucleome," i.e. the dynamic three-dimensional structures of genomic DNA in the nucleus.

Table 13.1 Comparison of non-biased methods for analyzing specific genomic regions

Method	DNA analysis	Protein analysis	RNA analysis	Include enzyme reactions	Low-copy number genes	Need transgenic	Allele-specific analysis
in cell iChIP	Feasible	Feasible	Feasible	No	Feasible	Yes (LexA BE, tagged LexA)	Feasible
in vitro iChIP	Feasible	Feasible	Feasible	No	Feasible	Yes (LexA BE)	Feasible
in cell enChIP	**Feasible**	**Feasible**	**Feasible**	**No**	**Feasible**	**Yes (CRISPR complex, TAL)**	**Feasible**
***in vitro* enChIP**	**Feasible**	**Feasible**	**Feasible**	**No**	**Feasible**	**No**	**Feasible**
3C and its derivatives	Feasible	Not feasible	Not feasible	Yes	Feasible	No	Not feasible
PICh	Not reported	Feasible (telomeres, rDNA)	Not reported	No	Not reported	No	Not feasible

REFERENCES

Antão JM, Mason JM, Déjardin J, Kingston RE. 2012. Protein landscape at *Drosophila melanogaster* telomere-associated sequence repeats. *Mol Cell Biol* **32**: 2170–2182.

Bikard D, Jiang W, Samai P, *et al.* 2013. Programmable repression and activation of bacterial gene expression using an engineered CRISPR-Cas system. *Nucleic Acids Res* **41**: 7429–7437.

Bogdanove AJ, Voytas DF. 2011. TAL effectors: customizable proteins for DNA targeting. *Science* **333**: 1843–1846.

Chen B, Gilbert LA, Cimini BA, *et al.* 2013. Dynamic imaging of genomic loci in living human cells by an optimized CRISPR/Cas system. *Cell* **155**: 1479–1491.

Chu C, Qu K, Zhong FL, Artandi SE, Chang HY. 2011. Genomic maps of long noncoding RNA occupancy reveal principles of RNA-chromatin interactions. *Mol Cell* **44**: 667–678.

Collas P. 2010. The current state of chromatin immunoprecipitation. *Mol Biotechnol* **45**: 87–100.

de Lange T. 2005. Shelterin: the protein complex that shapes and safeguards human telomeres. *Genes Dev* **19**: 2100–2110.

de Wit E., de Laat W. 2012. A decade of 3 C technologies: insights into nuclear organization. *Genes Dev* **26**: 11–24.

Déjardin J, Kingston RE. 2009. Purification of proteins associated with specific genomic loci. *Cell* **136**: 175–186.

Dekker J, Rippe K, Dekker M, Kleckner N. 2002. Capturing chromosome conformation. *Science* **295**: 1306–1311.

Doudna JA, Charpentier E. 2014. Genome editing: the new frontier of genome engineering with CRISPR-Cas9. *Science* **346**: 1258096.

Durbin JE, Hackenmiller R, Simon MC, Levy DE. 1996. Targeted disruption of the mouse Stat1 gene results in compromised innate immunity to viral diseases. *Cell* **84**: 443–450.

Egner A, Hell SW. 2005. Fluorescence microscopy with super-resolved optical sections. *Trends Cell Biol* **15**: 207–215.

Engreitz JM, Pandya-Jones A, McDonel P, *et al.* 2013. The Xist lncRNA exploits three-dimensional genome architecture to spread across the X chromosome. *Science* **341**: 1237973.

Fisher AG, Merkenschlager M. 2002. Gene silencing, cell fate and nuclear organisation. *Curr Opin Genet Dev* **12**: 193–197.

Fraser P, Bickmore W. 2007. Nuclear organization of the genome and the potential for gene regulation. *Nature* **447**: 413–417.

Fujii H, Fujita T. 2015. Isolation of specific genomic regions and identification of their associated molecules by engineered DNA-binding molecule-mediated chromatin immunoprecipitation (enChIP) using the CRISPR system and TAL proteins. *Int J Mol Sci* **16**: 21802–21812.

Fujita T, Asano Y, Ohtsuka J, *et al.* 2013. Identification of telomere-associated molecules by engineered DNA-binding molecule-mediated chromatin immunoprecipitation (enChIP). *Sci Rep* **3**: 3171.

Fujita T, Fujii H. 2011. Direct identification of insulator components by insertional chromatin immunoprecipitation. *PLoS One* **6**: e26109.

Fujita T, Fujii H. 2012. Efficient isolation of specific genomic regions by insertional chromatin immunoprecipitation (iChIP) with a second-generation tagged LexA DNA-binding domain. *Adv Biosci Biotechnol* **3**: 626–629.

Fujita T, Fujii H. 2013a. Efficient isolation of specific genomic regions and identification of associated proteins by engineered DNA-binding molecule-mediated chromatin immunoprecipitation (enChIP) using CRISPR. *Biochem Biophys Res Commun* **439**: 132–136.

Fujita T, Fujii H. 2013b. Locus-specific biochemical epigenetics/chromatin biochemistry by insertional chromatin immunoprecipitation. *ISRN Biochem* **2013**: 913273.

Fujita T, Fujii H. 2014a. Efficient isolation of specific genomic regions retaining molecular interactions by the iChIP system using recombinant exogenous DNA-binding proteins. *BMC Mol Biol* **15**: 26.

Fujita T, Fujii H. 2014b. Identification of proteins associated with an IFNγ-responsive promoter by a retroviral expression system for enChIP using CRISPR. *PLoS One* **9**: e103084.

Fujita T, Fujii H. 2014c. Identification of proteins interacting with genomic regions of interest in vivo using engineered DNA-binding molecule-mediated chromatin immunoprecipitation (enChIP). *Bio Protoc* **4**: e1124.

Fujita T, Fujii H. 2015a. Applications of engineered DNA-binding molecules such as TAL proteins and the CRISPR/Cas system in biology research. *Int J Mol Sci* **16**: 23143–23164.

Fujita T, Fujii H. 2015b. Isolation of specific genomic regions and identification of associated molecules by engineered DNA-binding molecule-mediated chromatin immunoprecipitation (enChIP) using CRISPR. *Methods Mol Biol* **1288**: 43–52.

Fujita T, Fujii H. 2016a. Biochemical analysis of genome functions using locus-specific chromatin immunoprecipitation technologies. *Gene Regul Syst Bio* **10**: 1–9.

Fujita T, Fujii H. 2016b. Isolation of specific genomic regions and identification of associated molecules by enChIP. *J Vis Exp* **107**: e53478.

Fujita T, Kitaura F, Fujii H. 2015a. A critical role of the Thy28-MYH9 axis in B cell-specific expression of the Pax5 gene in chicken B cells. *PLoS One* **10**: e0116579.

Fujita T, Yuno M, Fujii H. 2016. Efficient sequence-specific isolation of DNA fragments and chromatin by in vitro enChIP technology using recombinant CRISPR ribonucleoproteins. *Genes Cells* **21**: 370–377.

Fujita T, Yuno M, Okuzaki D, Ohki R, Fujii H. 2015b. Identification of non-coding RNAs associated with telomeres using a combination of enChIP and RNA sequencing. *PLoS One* **10**: e0123387.

Fujita T, Yuno M, Suzuki Y, Sugano S, Fujii H. 2017. Identification of physical interactions between genomic regions by enChIP-Seq. *Genes Cells* **22**: 506–520.

Gavrilov AA, Chetverina HV, Chermnykh ES, Razin SV, Chetverin AB. 2014. Quantitative analysis of genomic element interactions by molecular colony technique. *Nucleic Acids Res* **42**: e36.

Harrison MM, Jenkins BV, O'Connor-Giles KM, Wildonger J. 2014. A CRISPR view of development. *Genes Dev* **28**: 1859–1872.

Hoshino A, Fujii H. 2009. Insertional chromatin immunoprecipitation: a method for isolating specific genomic regions. *J Biosci Bioeng* **108**: 446–449.

Ide S, Dejardin J. 2015. End-targeting proteomics of isolated chromatin segments of a mammalian ribosomal RNA gene promoter. *Nat Commun* **6**: 6674.

Ji N, Shroff H, Zhong H, Betzig E. 2008. Advances in the speed and resolution of light microscopy. *Curr Opin Neurobiol* **18**: 605–616.

Kalhor R, Tjong H, Jayathilaka N, Alber F, Chen L. 2011. Genome architectures revealed by tethered chromosome conformation capture and population-based modeling. *Nat Biotechnol* **30**: 90–98.

Kwon S. 2013. Single-molecule fluorescence in situ hybridization: quantitative imaging of single RNA molecules. *BMB Rep* **46**: 65–72.

Lindhout BI, Fransz P, Tessadori F, *et al.* 2007. Live cell imaging of repetitive DNA sequences via GFP-tagged polydactyl zinc finger proteins. *Nucleic Acids Res* **35**: e107.

Mariner PD, Walters RD, Espinoza CA, *et al.* 2008. Human Alu RNA is a modular transacting repressor of mRNA transcription during heat shock. *Mol Cell* **29**: 499–509.

Meraz MA, White JM, Sheehan KC, *et al.* 1996. Targeted disruption of the Stat1 gene in mice reveals unexpected physiologic specificity in the JAK-STAT signaling pathway. *Cell* **84**: 431–442.

Miyanari Y, Ziegler-Birling C, Torres-Padilla ME. 2013. Live visualization of chromatin dynamics with fluorescent TALEs. *Nat Struct Mol Biol* **20**: 1321–1324.

Nagano T, Várnai C, Schoenfelder S, *et al.* 2015. Comparison of Hi-C results using in-solution versus in-nucleus ligation. *Genome Biol* **16**: 175.

Ong SE, Biagoev B, Kratchmarova I, *et al.* 2002. Stable isotope labeling by amino acids in cell culture, SILAC, as a simple and accurate approach to expression proteomics. *Mol Cell Proteomics* **1**: 376–386.

Pabo CO, Peisach E, Grant RA. 2001. Design and selection of novel Cys2His2 zinc finger proteins. *Annu Rev Biochem* **70**: 313–340.

Qi LS, Larson MH, Gilbert LA, *et al.* 2013. Repurposing CRISPR as an RNA-guided platform for sequence-specific control of gene expression. *Cell* **152**: 1173–1183.

Sahl SJ, Moerner WE. 2013. Super-resolution fluorescence imaging with single molecules. *Curr Opin Struct Biol* **23**: 778–787.

Simon MD, Wang CI, Kharchenko, PV, *et al.* 2011. The genomic binding sites of a noncoding RNA. *Proc Natl Acad Sci USA* **108**: 20497–20502.

14 Application of CRISPR for Pooled, Vector-based Functional Genomic Screening in Mammalian Cell Lines

Trinna L. Cuellar, Marinella Callow, Ted Lau, Colin Watanabe, Russell Bainer, Mike Costa and Benjamin Haley

14.1 Development of a Molecular Genetics Toolkit

The advent of whole-genome sequencing ushered in a new era of biological experimentation, focused on reverse engineering of cellular systems to better understand disease and development. With the gene sequences in hand, their direct and indirect relationships could be dissected via functional genomics, a field enlivened by the discovery of RNA interference (RNAi) (Fire *et al.*, 1998). This technology enabled targeted, large-scale assessment of gene activity via loss-of-function (LOF) screening, an approach that has proven to be particularly valuable for studies utilizing mammalian cell-based systems (Dorsett and Tuschl, 2004). Subsequent advances in multiplex construction of vector-based RNAi reagents (specifically short hairpin RNAs/shRNAs) and next-generation sequencing (NGS) technologies facilitated LOF screens targeting hundreds or thousands of genes within a single tissue culture flask (Boettcher and Hoheisel, 2010). However, incomplete gene suppression and off-targeting potential can cloud the interpretation of RNAi screens (Mohr *et al.*, 2010; Sigoillot and King, 2011). Recently, the applied use of CRISPR (clustered regularly interspaced short palindromic repeats), a bacterial genome defense mechanism repurposed to facilitate gene knockout and modification in eukaryotic cells, has shifted the functional genomics paradigm, resulting in more robust target discovery and validation efforts than had previously been imagined (Cong *et al.*, 2013; Jinek *et al.*, 2013; Mali *et al.*, 2013).

Delivery of the CRISPR-associated DNA endonuclease, Cas9, as well as a small, partially structured RNA (single-guide RNA or sgRNA) into a cell triggers double-stranded DNA (dsDNA) breaks at user-defined positions (Sander and Joung, 2014). These breaks often result in complete gene disruption by way of error-prone DNA repair mechanisms. sgRNAs, which impart specificity to this process, can be expressed from retro or lentiviral contexts and cloned *en masse* into large pools, akin to the shRNAs that comprise vector-based RNAi libraries (Sanjana, 2016). Working within a variety of cell-based systems, separate groups have found stable expression of both Cas9 and a library of sgRNAs to be tolerated and effective for LOF screening (Koike-Yusa *et al.*, 2014; Shalem *et al.*, 2014; Wang *et al.*, 2014).

Major advantages of CRISPR-based screening include high efficacy per targeting reagent, permitting libraries with lower complexity relative to those utilizing shRNAs, and greater phenotypic penetrance due to functional distinction between knockout (CRISPR) vs. knockdown (RNAi) (Sanjana, 2016; Shalem *et al.*, 2014).

14.2 sgRNA Design and Library Development

This chapter is intended to provide a series of practical considerations and expected results related to the development and implementation of lentivirus-expressed sgRNA libraries for *in vitro* CRISPR-based LOF screening. We will focus on libraries containing sgRNAs designed to associate with *Streptococcus pyogenes* Cas9, although similar libraries may be developed for alternate sgRNA/Cas9 variants, such as those derived from *Streptococcus aureus* (Ran *et al.*, 2015). Cas9 is programmed with a ~100 nucleotide (nt) RNA, which is a chimera of the two small RNAs (the crRNA, which contains the ~20 nt targeting segment as well as a short sequence that will base pair with the tracrRNA, a semi-structured RNA that associates with Cas9) that imbue the microbial Type II CRISPR system with its ability to target and cleave DNA in a sequence-specific manner (Jinek *et al.*, 2012). The sgRNA will direct Cas9 to genomic loci that contain two core elements: (1) similarity to a ~20 nt, user-defined targeting sequence within the sgRNA and (2) an NGG motif, known as a protospacer adjacent motif (PAM), immediately 3′ to the targeting sequence (Figure 14.1). Upon association with the target site, a pair of molecular scissors will reorient within Cas9, positioning the Cas9 nuclease domains such that they will induce a site-specific dsDNA break (Sternberg *et al.*, 2015). These dsDNA breaks are typically repaired by either non-homologous end joining (NHEJ) or, possibly, microhomology-mediated end joining (MMEJ), in

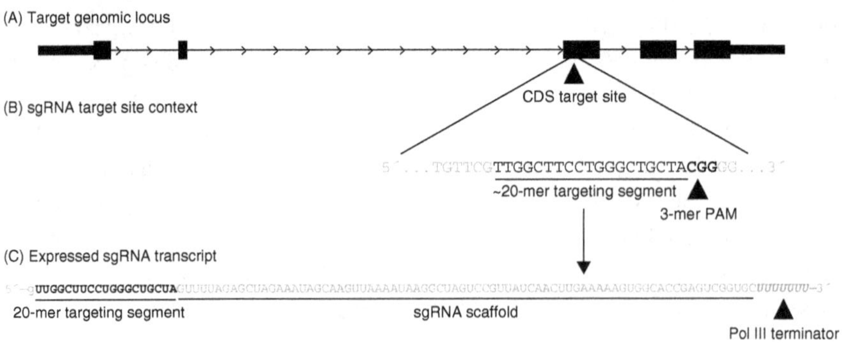

Figure 14.1 Basic sgRNA design. A theoretical locus is depicted, including introns (thin lines), untranslated (thin blocks) and translated (thick blocks) regions of the expressed mRNA. An sgRNA target site is identified by the ~20 nt targeting segment, followed by an NGG protospacer adjacent motif (PAM). The ~20 nt targeting segment is incorporated into expressed sgRNA transcripts (comprising linked crRNA and tracrRNA elements), which will direct Cas9 to the user-defined genomic location.

such a way that frame-shifting insertions or deletions (indels) can be introduced, leading to gene disruption (van Overbeek *et al.*, 2016).

14.2.1 sgRNA Library Composition

Vector-based RNAi screening benefited from the inclusion of >25 shRNAs per gene in each pool, a feature necessitated by the promiscuity of RNAi-mediated gene silencing and the relatively low percentage (<20%) of highly active shRNA sequences (Bassik *et al.*, 2009). However, screens that employ high-coverage libraries can be challenging due to the large numbers of cells that need to be cultured to reach the experimental gold standard; 1000-fold representation and triplicate assay sampling. Lentivirus-mediated expression of CRISPR reagents has been shown to trigger more consistent, potent and specific gene inactivation relative to shRNAs, reducing the need for such high complexity within a comparable sgRNA library (Evers *et al.*, 2016; Hart *et al.*, 2014; 2015; Morgens *et al.*, 2016; Shalem *et al.*, 2014). In our experience, mouse- or human-specific sgRNA libraries made up of 6–8 targeting sequences per gene, formatted as either genome-scale or pathway/gene family-specific sublibraries, provide a reasonable compromise between gene coverage and pool size. Several sgRNA libraries specific for human or mouse cell-based studies are available from public resources (Addgene) or for purchase from a host of vendors. Custom library construction is feasible, and detailed protocols are provided elsewhere (Wang *et al.*, 2016b).

Ideally, sgRNA libraries will contain a set of non-targeting controls (NTCs) to assess and/or normalize technical and biological variation across conditions or screen replicates, and, where possible, positive controls to ensure Cas9/sgRNA function within the screened cells. Where assay-specific positive controls are not available, sgRNAs targeting essential genes or tumor suppressors can be used to confirm library efficacy. Common NTCs include sgRNAs designed so that they target DNA lacking an identifiable PAM, or sgRNAs designed to target and cleave untranscribed portions of the genome. The latter NTC format could be beneficial for distinguishing generic dsDNA break-induced vs. gene-specific phenotypes within particular screening contexts.

14.2.2 sgRNA Targeting Sequence Selection

The size and complexity of eukaryotic genomes makes the design of genome-wide sgRNA libraries a substantial task. Considerations for library design include the genes of interest, the annotations used to define these genes, and which elements (i.e. exons) of these genes should be selected for targeting. Proper delineation of the sequence shared across the transcribed or predicted isoforms of each gene can be difficult and is compounded by differences among genome annotations, alignment tools and cell line to cell line genetic variation (Mohr *et al.*, 2016).

It has been suggested that targeting sgRNAs to annotated functional domains can improve the frequency of gene disruptive indels, while others found effective sgRNAs across all but the last 10% of targeted protein coding sequences (CDSs) (Doench *et al.*, 2016; Munoz *et al.*, 2016; Shi *et al.*, 2015). We have observed guides targeting the most conserved and shared (between both transcribed and annotated isoforms) regions in the first half of the CDS to be generally effective. These sequences may be enriched for functional domains, as evidenced by their evolutionary persistence. Local DNA accessibility could interfere with Cas9-mediated DNA cutting, an issue that may be minimized by spacing sgRNA target sites across the specified CDS (Hart *et al.*, 2015; Hinz *et al.*, 2015; Horlbeck *et al.*, 2016; Isaac *et al.*, 2016; Smith *et al.*, 2016). While the criteria for designing highly efficient sgRNAs is a field of active study, several sgRNA-specific off-target prediction tools have been developed and detailed reviews on the subject of Cas9-targeting specificity can be found elsewhere (Tycko *et al.*, 2016).

Furthermore, the accuracy of sgRNA efficiency and specificity prediction is dependent upon experimental conditions and the genetic makeup of the cell lines being studied. This is exemplified by separate studies that have observed target-independent cytotoxicity triggered by sgRNAs that induce dsDNA breaks within amplified regions of the genome, and proper interpretation of cell-based CRISPR screens that make use of viability as an end-point should take gene copy number variation of each cellular context into account (Aguirre *et al.*, 2016; Munoz *et al.*, 2016; Wang *et al.*, 2015). Non-overlapping screening and validation approaches, such as performing a CRISPR screen followed by validation of hits using RNAi, or vice versa, may limit technology-dependent liabilities (Morgens *et al.*, 2016).

14.2.3 sgRNA Expression Context

To avoid the confounding effects of multiple guide RNAs being present and active in a given cell, pooled vector-based screens require single-copy integration of the unique targeting elements across the cell population. Therefore, the choice between an "all-in-one" sgRNA + Cas9 virus-based expression system or a two-part system that splits Cas9 and the sgRNA expression between separate vectors will impact both screen set-up and efficacy. In general, it is advisable to use the split vector format, since this permits both higher titer viral particle generation, due to the reduced packaging requirements relative to an all-in-one format, and pre-screen validation of Cas9-stable cell lines with either project-specific or generic sgRNA controls (Sanjana *et al.*, 2014). CRISPR activity in a given cell line is likely to benefit from moderate-to-high levels of Cas9, which is more achievable using the two-vector system, although we have observed that extreme levels of Cas9 may impact cellular homeostasis (data not shown). In contrast, all-in-one vectors allow for the simplified application of CRISPR screening across cell lines by virtue of a single transduction and selection step. However, this convenience comes at the cost of efficiency, since both Cas9 and each sgRNA will be expressed from a single copy per cell. For primary cells or lines with limited passaging potential, the all-in-one vector system may be the only feasible approach.

Once the library plasmids are in hand, production of the sgRNA library lentiviral particles is performed by large-scale co-transfection of immortalized human embryonic kidney 293T (HEK293T) cells with the sgRNA plasmid pool along with lentiviral packaging element-expressing plasmids compatible with the sgRNA library vector system. After 2–3 days of cell culture and viral particle generation, the virus-enriched supernatant is collected, concentrated and cryopreserved for later use. At this point, it is useful to assess virus library titers and to perform NGS analysis of the genome-embedded sgRNAs from a Cas9-negative target cell line transduced with the lentiviral particles. Without Cas9, sgRNA expression will be inert, and NGS reads from the transduced cell line can be used to identify the distribution of sgRNAs in active viral particles across the pool.

14.3 Screen Set-up and Execution

14.3.1 General Considerations for Pooled sgRNA/Cas9 Screening

CRISPR-based, pooled vector screens can be applied for the identification of factors governing positively or negatively selectable phenotypes, scored by sgRNA enrichment or depletion, respectively, throughout the course of the assay (Figure 14.2). Such screens have been used to assess the effects of gene knockout on cell growth and viability, drug resistance, host–pathogen interactions and selection of cells expressing particular marker genes (Hart *et al.*, 2015; Marceau *et al.*, 2016; Savidis *et al.*, 2016; Shalem *et al.*, 2014; Wang *et al.*, 2015; 2016a; Zhang *et al.*, 2016). Regardless of the CRISPR screen being undertaken, measures must be put in place to maximize Cas9 potency, minimize off-targeting potential, optimize the dynamic range of the phenotype being assessed in the screen and reduce variability across sampled replicates. Here, we will focus on

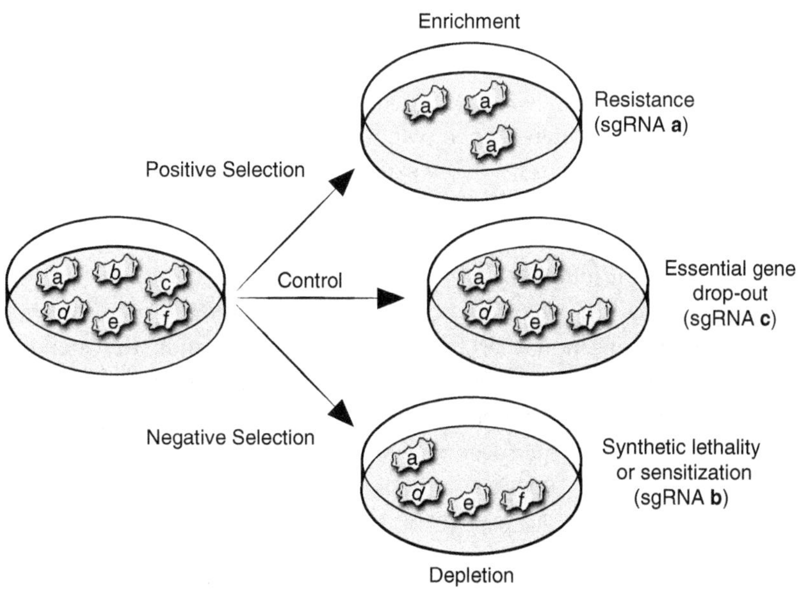

Figure 14.2 Overview of CRISPR screening concepts and expectations. For cells infected with a library of sgRNA expression constructs targeting a diversity of genes, positive selection results in preferential growth of cells carrying sgRNAs that confer resistance to a treatment and enrichment of these sgRNAs in the cell population (upper right). In a cell growth assay, negative selection will result in depletion of sgRNAs within the pool for general, essential genes (center right) and genes selectively required in combination with particular genotypes or drug treatments (lower right).

screens using a split-plasmid system, in which Cas9 and the sgRNA pool are expressed from separate vectors.

Prior to any screen, the target cell lines must be developed to ensure that: (1) neither expression of Cas9 and sgRNAs nor antibiotic or other selection methodology impacts the desired phenotype and (2) Cas9/sgRNA expression results in measurable gene disruption. This process is initiated via stable integration of a Cas9 expression cassette into the desired cell line(s), resulting in a polyclonal population of cells with variable Cas9 levels. Subsequently, the mixed population or, optionally, single-cell clones derived from that population, should be evaluated for Cas9 function, using low (< 0.5) multiplicity of infection (MOI) lentiviral transduction of assay-specific or generic positive (e.g. sgRNA targeting essential genes such as *Plk1*) and negative control sgRNAs.

The next step is to establish the growth characteristics of the Cas9-expressing cell line(s) in a manner that mimics the screen itself. This includes determining the optimal seeding density, growth rate and days to confluence in large-scale culture systems. Separately, sgRNA library titrations should be performed on a small scale, which will enable transduction at an MOI of < 0.5 (to reduce the likelihood that two different sgRNAs integrate into a cell) and selection of sufficient cell integrant numbers under the optimized conditions at the time of the screen. While positive and negative selection screens may have distinct considerations that should be evaluated before initiating the project (described in more detail below), sgRNA library size, the number of assay replicates and time points, and the desired sgRNA fold representation are common parameters that dictate the quantity of cells and viral particles needed to properly execute any pooled vector screen. The timing of both Cas9 DNA cutting activity and cellular mechanisms of DNA repair compounded by target gene product turnover and phenotype expression should be considered when planning the duration of the screen and frequency of sampled time points. In our experience, the process of DNA damage and functional gene disruption is nearly complete ~10 days after sgRNA library transduction, although some phenotypes may not reach maximal expressivity until ≥18 days post-transduction.

Throughout the screen, care should be taken to utilize the cell numbers and, accordingly, amount of genomic DNA sampled in order to maintain the intended representation of the library pool. A contraction at any single step may decrease the power to quantify sgRNA depletion or enrichment with statistical significance (Strezoska *et al.*, 2012). For perspective, a single replicate for a screen using a 20 000 sgRNA library at 1000-fold representation will require 2×10^7 cells (i.e. 20 000 sgRNAs × 1000 cells per sgRNA) following selection for stable sgRNA expression and at each subsequent passage.

After transduction and selection for cells with stable sgRNA expression, genomic DNA should be collected 2–3 days following selection for stable sgRNA expression. This will serve as a reference for both transduction quality and library representation before significant knockout phenotypes are induced. DNA samples should then be collected at progressive time points, which will be evaluated relative to the early reference or time-matched, mock control samples in order to assess knockout phenotype penetrance. Several reagents are available for efficient, high-quality genomic DNA preparation. Solid-phase DNA capture technologies may be prone to recovery artifacts due to column overloading and poor elution. Therefore, we recommend the use of column-free approaches, such as the Gentra Puregene system (Qiagen), that employs precipitation to concentrate and purify the DNA in an easily scalable format.

After collecting the DNA samples, NGS-ready libraries can be generated via PCR-based amplification of genome-integrated sgRNAs, with careful consideration taken to avoid sampling bias. For accurate determination of DNA quantity, we suggest using fluorimetry-based systems, such as the Qubit (ThermoFisher), for both genomic DNA and amplicon assessment. The amount of template used for library amplicon generation and, thus, proper sampling of the sgRNA pool in any given population, requires one to factor sgRNA pool size, library representation and genomes assessed. Using the theoretical screen mentioned above as an example, 6.6 pg (mass of DNA in diploid human cell) × 20 000 (# of sgRNAs in the pool) × 1000 (fold representation) = 132 μg of genomic DNA template required per replicate in order to maintain full library representation at the library amplicon generation phase. DNA samples taken prior to or in the absence of selection for stable sgRNA integration should be oversampled to account for the presence of sgRNA-negative cells within the population.

Preparation of the screen-derived NGS libraries is performed using sgRNA vector-specific primers and high-fidelity DNA polymerases with non-saturating PCR conditions to reduce both PCR-introduced sequence errors and amplification bias for overenriched or efficiently amplified sequences. In order to avoid substrate or contaminant-mediated inhibition of the reaction, it is advisable to run, and subsequently pool, multiple small-scale PCRs for each screen replicate. PCR products must be purified prior to NGS, and we recommend using SPRI paramagnetic bead technology (Beckman Coulter), which is fast, robust and automation-friendly.

14.3.2 Positive Selection Screens

Perhaps the most robust use of pooled vector screening technology is in positive selection contexts, whereby the knockout phenotype for a given gene results in an enrichment of the sgRNAs targeting the gene relative to that of the general sgRNA population. The strength of positive selection screening comes from the induction of very specific phenotypes, wherein those sgRNAs that do not trigger the desired effect are excluded or depleted from the treated cell population. This, in turn, leads to modest background, permitting the use of lower sgRNA representation relative to negative selection screens and, likely, fewer off-target effects.

An example data set from a positive selection screen is provided in Figure 14.3A–C. Here, we investigated the genetic components of necroptosis, a form of lytic, caspase-independent cell death induced upon TNF-α (TNF) binding to tumor necrosis factor receptor TNFR1 (Newton and Manning, 2016). Downstream of TNFR1, a signaling cascade mediated by RIPK1, RIPK3 and MLKL leads to activation of this process. Cas9-expressing murine fibroblast L929 cells, which are susceptible to necroptosis upon stimulation with TNF and zVAD (a peptide

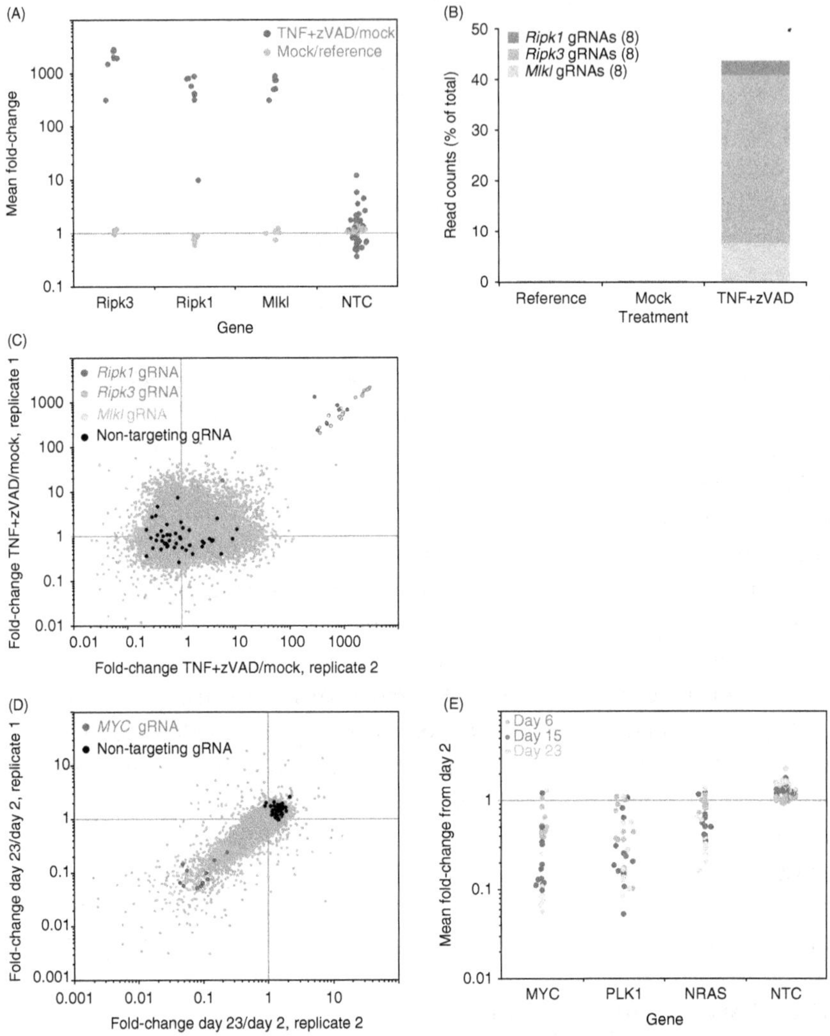

Figure 14.3 Example positive and negative screen data. (A–C) Positive selection screen. Cas9-expressing, mouse L929 cells were transduced with ~17 000 pooled sgRNAs in triplicate at 1000-fold sgRNA representation. Stable integrants were selected with antibiotic, and 15 days post-transduction, cultures were divided into treated (TNF+zVAD, to induce necroptosis) or mock populations. (A) All sgRNAs (~8/gene) targeting the core necrosome genes *Ripk3, Ripk1* and *Mlkl* show enrichment via NGS sequencing reads in TNF+zVAD-treated samples normalized to mock-treated samples (red dots; mean of replicates), but not in mock-treated samples normalized to an early reference time point (2 days post-transduction) (blue dots). (B) Enrichment of *Ripk1, Ripk3* and *Mlkl* sgRNA read counts in TNF+zVAD-treated samples compared to the early reference and mock-treated samples, shown as a percentage of the total read counts for all sgRNAs in the pool (means of replicate paired samples). (C) Correlation between two representative replicate samples for sgRNA abundance changes with TNF+zVAD treatment normalized to mock treatment. (D, E) Negative selection screen results. Cas9-expressing, human NCI-H1299 cells were transduced with ~17 000 pooled sgRNAs (~8/gene in triplicate at 1000-fold sgRNA representation). Stable integrants were selected using antibiotic, and DNA samples were collected from the cell populations at 2, 6, 15 and 23 days post-transduction. (D) Correlation of changes in sgRNA abundance between two representative replicate samples, with day 23 normalized to day 2 read counts. sgRNAs targeting MYC and 47 non-targeting control sgRNAs are highlighted. (E) Quantitation of sgRNA abundance

competitive caspase inhibitor), hereafter referred to as TNF+zVAD, were infected with a mouse-specific sgRNA library targeting ~2000 genes (~8 sgRNA/gene), including guides specific for the core necrosome pathway, at 1000-fold sgRNA representation per replicate. NGS sequencing was scaled to achieve the sgRNA fold representation maintained throughout cell infection and passaging, producing a median sgRNA read count of ~1000. Time-matched mock-treated cell samples were required for sgRNA read count normalization of TNF+zVAD-treated samples to control for sgRNAs that increased in abundance by inducing cell proliferation rather than conferring necroptosis resistance. Comparison of time-matched mock and TNF+zVAD treated samples, performed in triplicate, revealed a strong enrichment for individual sgRNAs targeting genes encoding the known regulators, *Ripk1, Ripk3* and *Mlkl* (Figure 14.3A, C), and these sgRNAs represented a significant proportion of the read counts for the treated samples (Figure 14.3B). At the concentrations of TNF+zVAD used in this screen, ~0.5% of the cells survive treatment independent of sgRNA or Cas9 expression. Though limited, the sgRNAs present in this residual population is expected to contribute to variability between replicate samples.

14.3.3 Negative Selection Screens

Negative selection screens are used for identifying genes essential to particular processes such as cell viability, proliferation or differentiation, by virtue of target sgRNA depletion compared to a reference population (Sanjana, 2016). Importantly, off-target effects are more likely to decrease cell viability or proliferation and induce sgRNA drop-out rather than conferring specific resistance phenotypes and sgRNA enrichment. Thus, it is critical that negative selection screen assay conditions (developed during pre-screen cell line validation) enable maximal signal-to-noise ratios and sgRNA representation (we recommend ≥500-fold) throughout the duration of the screen.

The expected results of a negative selection screen are depicted in Figure 14.3D–E. Here, a Cas9-stable, human NCI-H1299 non-small-cell lung cancer line has been transduced with a human-specific sgRNA library targeting ~2000 genes with a coverage of ~8 sgRNAs/gene and 1000-fold sgRNA representation per replicate, as above for the positive selection example. Progressive depletion can be observed for the bulk of sgRNAs targeting both essential and cell cycle factors (e.g. *PLK1*), as well as cell line-specific oncogenes and growth regulators, like *MYC* and *NRAS*, two genes known to harbor activating mutations in NCI-H1299 cells, while the NTCs show little change over time. As discussed earlier, the kinetics and

---

Caption for Figure 14.3 (cont.)

over time for sequences targeting known oncogenes (*NRAS* and *MYC*), an essential gene (*PLK1*), and non-targeting control sgRNAs (NTC). Each dot represents the mean fold-change of the three replicate sgRNA read counts normalized to the day 2 paired replicate read count. (A black-and-white version of this figure will appear in some formats. For the color version, please refer to the plate section.)

penetrance of a knockout phenotype may differ, target-by-target, and assaying the population at multiple time points ensures detection of a comprehensive set of screen hits.

14.3.4 Data Assessment and Hit Selection

After sequencing amplicon libraries derived from the screened cell populations, a number of quality control measures should be undertaken to ensure the data are usable for target prioritization. Read coverage should be assessed to confirm that each library has been sequenced at sufficient and, ideally, similar depth. For example, in order to avoid undersampling, we suggest sequencing the library amplicons to a depth that scales in proportion to the fold representation of the screen and number of sgRNAs within the pool, for each replicate (i.e. 20 000 unique sgRNAs in the pool × 1000 fold representation = 2×10^7 sequencing reads). Variation between samples in terms of total sequencing depth may be due to poor quality of amplicon DNA or improper flow cell loading of the sequencing instrument. Furthermore, using the sequencing conditions described above, the vast majority of sgRNAs included in the pool should be detected in amplicon libraries derived from cell populations collected at an early reference point (immediately before or after selection of stable sgRNA integration).

Correlation heatmaps and non-parametric principal component analysis (PCA) are used to verify that replicates are well correlated (Spearman's pairwise rho ~0.8) and that the majority of the observed variance is consistent with differences between experimental treatments of samples. In positive selection screens, a random cohort of sgRNAs that do not contribute to the phenotype will be present in the small number of cells that survive selection. The stochastic differences among this fraction will contribute to a weaker correlation between replicates, when comparing the total sgRNA population.

sgRNA libraries are designed to incorporate multiple targeting reagents per gene so as to minimize the influence of off-target or inactivate sgRNAs during screen hit prioritization. Several software packages exist to estimate and integrate the read counts for each sgRNA to identify genes with robust enrichment or depletion within a screen, such as gCrisprTools, MAGeCK or various geneset enrichment tests (https://bioconductor.org/packages/release/bioc/html/gCrisprTools.html; Li *et al.*, 2014; Yu *et al.*, 2016). All of these tools identify differences in sgRNA abundance using an analytical framework similar to those used to identify differentially expressed genes in RNA-Seq experiments; they then use the sgRNA estimates to derive robust gene-level scores on the basis of the estimated significance of the change in sgRNA levels (using *P*-values from a parametric model) or the ranks of the sgRNA signals relative to other guides included in the screen.

While the ranking of sgRNAs is typically very stable across replicates, the selective dynamics in competitive screens introduce significant normalization and estimation challenges, and in practice the measured behavior of individual sgRNAs may not be accurately estimated. Consequently, summarization methods that depend on the relative rankings of sgRNA signals, such as robust rank aggregation or Wilcoxon tests, are more likely to yield accurate results than those that depend on parametric tests such as direct linear modeling (Kolde *et al.*, 2012; Wilcoxon, 1946). At the same time, rank-based tests may be considerably more intensive computationally and can lack resolution for ranking top hits. Accordingly, there are practical advantages and disadvantages to the various aggregation approaches. Highly ranked targets with few of the individual sgRNAs driving the majority of the signal should thus be viewed with caution, regardless of their associated gene-level scores.

14.3.5 Screen Validation

Hit validation often represents the most time-intensive portion of a CRISPR screen, and the breadth and depth of target characterization will depend on any number of variables. A common first step of this process is technical validation of the reagents, which entails arraying and assaying select sgRNAs from the library that scored as hits, all within the original Cas9-expressing screening cell line. This can either be done with individually cloned lentiviral constructs or transiently transfected, synthetic/*in vitro* transcribed sgRNA or crRNA/tracrRNA reagents (Doudna and Charpentier, 2014). Note that false negative results may be caused by inefficient knockout or by outgrowth of clones with wild-type or heterozygous genotypes. A "mini pooled screen" with a custom-generated library comprising select sgRNA sequences from the screen, positive control sgRNAs and an equal portion of non-targeting sequences may also be considered for technical validation; this better enables long-term cell culture and recapitulates the competitive environment of the primary screen. Subsequently, cells exposed to specific sgRNAs should be evaluated by directly sequencing genomic DNA at the sgRNA target sites and/or western blotting (or other protein analyses) for the target gene product to ensure robust, on-target gene disruption has occurred. Rescue of the knockout phenotype may be possible by introduction of an sgRNA-resistant cDNA, although careful attention should be paid to maintain transgene expression at or near physiological levels.

Beyond technical validation of the screening reagents, the screen hits and screening system itself can be validated through the use of sgRNA sequences that differ from those in the screening library, an expanded set of target cell lines (where available and feasible), or alternative gene disruption and silencing technologies (Boettcher and McManus, 2015). With respect to the latter, RNAi represents an obvious choice for cross-validation, as this permits gene silencing in the absence of DNA damage, which, as described above, can itself induce non-specific cellular phenotypes. Once satisfied with the technical reproducibility of the screen hits, biological validation and mechanistic studies should be carried out based on the specifics of the model system and functional or pathway-specific characteristics of the gene(s) chosen for detailed follow-up.

14.4 Conclusions

In a short time, CRISPR has revolutionized the functional genomics field by enabling simple-to-design, inexpensive and large-scale gene disruption capabilities, across a variety of model systems. Application of this technology for targeted or genome-scale LOF screening holds particularly great promise, as it circumvents the issues of reagent potency and specificity that has plagued the previous generation of screening tools. However, the technology in not infallible, and careful attention should be paid towards reagent design, the model system under study and robustness of the phenotypic readout, as well as technical nuances of the sgRNA detection and quantitation process when considering or analyzing a CRISPR-mediated genetic screen.

While DNA damage induced by Cas9 is effective for gene disruption, off-targeting or non-specific, DNA break-triggered effects cannot be fully avoided. For this reason, alternative screening approaches, such as RNAi or CRISPR interference, where Cas9 has been modified so as to enable targeted transcriptional suppression in the absence of DNA cutting, should be considered, particularly in cell-based models with significant genome amplification or in situations where incomplete gene inactivation is desirable (Gilbert *et al.*, 2014; Morgens *et al.*, 2016).

Looking ahead, continued refinement of genome annotations, enhancements of sgRNA expression contexts, characterization of Cas9 variants (or similar proteins), as well as improved NGS technologies and their requisite analytical toolkits will enable larger, more effective CRISPR-based screens.

REFERENCES

Aguirre AJ, Meyers RM, Weir BA, *et al.* 2016. Genomic copy number dictates a gene-independent cell response to CRISPR/Cas9 targeting. *Cancer Discov* **6**: 914–929.

Bassik MC, Lebbink RJ, Churchman LS, *et al.* 2009. Rapid creation and quantitative monitoring of high coverage shRNA libraries. *Nat Methods* **6**: 443–445.

Boettcher M, Hoheisel JD. 2010. Pooled RNAi screens: technical and biological aspects. *Curr Genomics* **11**: 162–167.

Boettcher M, Mcmanus MT. 2015. Choosing the right tool for the job: RNAi, TALEN, or CRISPR. *Mol Cell* **58**: 575–585.

Cong L, Ran FA, Cox D, *et al.* 2013. Multiplex genome engineering using CRISPR/Cas systems. *Science* **339**: 819–823.

Doench JG, Fusi N, Sullender M, *et al.* 2016. Optimized sgRNA design to maximize activity and minimize off-target effects of CRISPR-Cas9. *Nat Biotechnol* **34**: 184–191.

Dorsett Y, Tuschl T. 2004. siRNAs: applications in functional genomics and potential as therapeutics. *Nat Rev Drug Discov* **3**: 318–329.

Doudna JA, Charpentier E. 2014. Genome editing: the new frontier of genome engineering with CRISPR-Cas9. *Science* **346**: 1258096.

Evers B, Jastrzebski K, Heijmans JP, *et al.* 2016. CRISPR knockout screening outperforms shRNA and CRISPRi in identifying essential genes. *Nat Biotechnol* **34**: 631–633.

Fire A, Xu S, Montgomery MK, *et al.* 1998. Potent and specific genetic interference by double-stranded RNA in *Caenorhabditis elegans*. *Nature* **391**: 806–811.

Gilbert LA, Horlbeck MA, Adamson B, *et al.* 2014. Genome-scale CRISPR-mediated control of gene repression and activation. *Cell* **159**: 647–661.

Hart T, Brown KR, Sircoulomb F, Ottapel R, Moffat J. 2014. Measuring error rates in genomic perturbation screens: gold standards for human functional genomics. *Mol Syst Biol* **10**: 733.

Hart T, Chandrashekhar M, Aregger M, *et al.* 2015. High-resolution CRISPR screens reveal fitness genes and genotype-specific cancer liabilities. *Cell* **163**: 1515–1526.

Hinz JM, Laughery MF, Wyrick JJ. 2015. Nucleosomes inhibit Cas9 endonuclease activity in vitro. *Biochemistry* **54**: 7063–7066.

Horlbeck MA, Witkowsky LB, Guglielmi B, *et al.* 2016. Nucleosomes impede Cas9 access to DNA in vivo and in vitro. *Elife* **5**: e12677.

Isaac RS, Jiang F, Doudna JA, *et al.* 2016. Nucleosome breathing and remodeling constrain CRISPR-Cas9 function. *Elife* **5**: e13450.

Jinek M, Chylinski K, Fonfara I., *et al.* 2012. A programmable dual-RNA-guided DNA endonuclease in adaptive bacterial immunity. *Science* **337**: 816–821.

Jinek M, East A, Cheng A, *et al.* 2013. RNA-programmed genome editing in human cells. *Elife* **2**: e00471.

Koike-Yusa H, Li Y, Tan EP, Velasco-Herrera Mdel C, Yusa K. 2014. Genome-wide recessive genetic screening in mammalian cells with a lentiviral CRISPR-guide RNA library. *Nat Biotechnol* **32**: 267–273.

Kolde R, Laur S, Adler P, Vilo J. 2012. Robust rank aggregation for gene list integration and meta-analysis. *Bioinformatics* **28**: 573–580.

Li W, Xu H, Xiao T, *et al.* 2014. MAGeCK enables robust identification of essential genes from genome-scale CRISPR/Cas9 knockout screens. *Genome Biol* **15**: 554.

Mali P, Yang L, Esvelt KM, *et al.* 2013. RNA-guided human genome engineering via Cas9. *Science* **339**: 823–826.

Marceau CD, Puschnik AS, Majzoub K, *et al.* 2016. Genetic dissection of Flaviviridae host factors through genome-scale CRISPR screens. *Nature* **535**: 159–163.

Mohr S, Bakal C, Perrimon N. 2010. Genomic screening with RNAi: results and challenges. *Annu Rev Biochem* **79**: 37–64.

Mohr SE, Hu Y, Ewen-Campen B, *et al.* 2016. CRISPR guide RNA design for research applications. *FEBS J* **283**(17): 3232–3238.

Morgens DW, Deans RM, Li A, Bassik MC. 2016. Systematic comparison of CRISPR/Cas9 and RNAi screens for essential genes. *Nat Biotechnol* **34**: 634–636.

Munoz DM, Cassiani PJ, Li L, *et al.* 2016. CRISPR screens provide a comprehensive assessment of cancer vulnerabilities but generate false-positive hits for highly amplified genomic regions. *Cancer Discov* **6**(8): 900–913.

Newton K, Manning G. 2016. Necroptosis and inflammation. *Annu Rev Biochem* **85**: 743–763.

Ran FA, Cong L, Yan WX, *et al.* 2015. In vivo genome editing using *Staphylococcus aureus* Cas9. *Nature* **520**: 186–191.

Sander JD, Joung JK. 2014. CRISPR-Cas systems for editing, regulating and targeting genomes. *Nat Biotechnol* **32**: 347–355.

Sanjana NE. 2016. Genome-scale CRISPR pooled screens. *Anal Biochem* **532**: 95–99.

Sanjana NE, Shalem O, Zhang F. 2014. Improved vectors and genome-wide libraries for CRISPR screening. *Nat Methods* **11**: 783–784.

Savidis G, McDougall WM, Meraner P, *et al.* 2016. Identification of Zika virus and dengue virus dependency factors using functional genomics. *Cell Rep* **16**: 232–246.

Shalem O, Sanjana NE, Hartenian E, *et al.* 2014. Genome-scale CRISPR-Cas9 knockout screening in human cells. *Science* **343**: 84–87.

Shi J, Wang E, Milazzo JP, *et al.* 2015. Discovery of cancer drug targets by CRISPR-Cas9 screening of protein domains. *Nat Biotechnol* **33**: 661–667.

Sigoillot FD, King RW. 2011. Vigilance and validation: keys to success in RNAi screening. *ACS Chem Biol* **6**: 47–60.

Smith JD, Suresh S, Schlecht U, *et al.* 2016. Quantitative CRISPR interference screens in yeast identify chemical-genetic interactions and new rules for guide RNA design. *Genome Biol* **17**: 45.

Sternberg SH, Lafrance B, Kaplan M, Doudna JA. 2015. Conformational control of DNA target cleavage by CRISPR-Cas9. *Nature* **527**: 110–113.

Strezoska, Z., Licon, A., Haimes, J., *et al.* 2012. Optimized PCR conditions and increased shRNA fold representation improve reproducibility of pooled shRNA screens. *PLoS One* **7**: e42341.

Tycko J, Myer VE, Hsu PD. 2016. Methods for optimizing CRISPR-Cas9 genome editing specificity. *Mol Cell* **63**: 355–370.

Van Overbeek M, Capurso D, Carter MM, *et al.* 2016. DNA repair profiling reveals nonrandom outcomes at Cas9-mediated breaks. *Mol Cell* **63**(4): 633–646.

Wang H, Lu B, Castillo J, *et al.* 2016a. Tankyrase inhibitor sensitizes lung cancer cells to endothelial growth factor receptor (EGFR) inhibition via stabilizing angiomotins and inhibiting YAP signaling. *J Biol Chem* **291**: 15256–15266.

Wang T, Birsoy K, Hughes NW, *et al.* 2015. Identification and characterization of essential genes in the human genome. *Science* **350**: 1096–1101.

Wang T, Lander ES, Sabatini DM. 2016b. Large-scale single guide RNA library construction and use for CRISPR-Cas9-based genetic screens. *Cold Spring Harb Protoc* **2016**: pdb top086892.

Wang T, Wei JJ, Sabatini DM, Lander ES. 2014. Genetic screens in human cells using the CRISPR-Cas9 system. *Science* **343**: 80–84.

Wilcoxon, F. 1946. Individual comparisons of grouped data by ranking methods. *J Econ Entomol* **39**: 269.

Yu J, Silva J, Califano A. 2016. ScreenBEAM: a novel meta-analysis algorithm for functional genomics screens via Bayesian hierarchical modeling. *Bioinformatics* **32**: 260–267.

Zhang R, Miner JJ, Gorman MJ, *et al.* 2016. A CRISPR screen defines a signal peptide processing pathway required by flaviviruses. *Nature* **535**: 164–168.

15 Generation and Utilization of CRISPR/Cas9 Screening Libraries in Mammalian Cells

Megan L. Ludwig, Nicole M. Michmerhuizen,* Rebecca C. Hoesli,* Jaqueline E. Mann,* Samantha N. Devenport,* Aditi S. Kulkarni,* Andrew C. Birkeland* and J. Chad Brenner

15.1 Introduction

The adaptation of the clustered regularly interspaced short palindromic repeats (CRISPR)/Cas9 system to generate knockouts in mammalian cells has resulted in significant scientific advances. By choosing any 20 base pair sequence guide RNA (gRNA), as long as it is next to a protospacer adjacent motif (PAM) of 5′-NGG-3′, to direct the Cas9 endonuclease and establish knockouts has resulted in unmatched speed and flexibility for creating these genetic models compared to other gene-editing methods. This power and efficiency has allowed researchers to apply this technology on a genome-wide scale that was previously infeasible using TALENs or zinc-finger nucleases. Since late 2013, several CRISPR screening libraries have been developed, targeting anywhere from several hundred to thousands of genes and discovering genes essential to cell proliferation, resistance to clinical treatments and involved in processes of toxicity (Shalem *et al.*, 2014; Wang *et al.*, 2014; Zhou *et al.*, 2014).

To date, the generation of genome-wide knockout libraries is the predominant application for CRISPR/Cas9 in large-scale screening, but other methods are being actively developed. CRISPR interference, or CRISPRi, has also been used to transcriptionally repress target genes. While inactive Cas9 (dCas9) alone can sterically inhibit transcription of targets, robust silencing can be achieved by fusing dCas9 with a known transcriptional repressor, such as the repressor domain of KRAB (Gilbert *et al.*, 2013; 2014). dCas9 fusion proteins have also been applied in large-scale gene activation screens (Gilbert *et al.*, 2014; Konermann *et al.*, 2015). Konermann *et al.* optimized a CRISPR-Cas9-mediated activation method (CRISPRa) utilizing dCas9 fusions with multiple activation domains to maximize transcriptional activation (Konermann *et al.*, 2015). This strategy, combined with a pool of 70 290 gRNA, was used to identify genes for which gain-of-function perturbations could confer resistance to BRAF inhibition. As predicted, known BRAF resistance pathway members, such as EGFR, were enriched following

* Authors contributed equally.

pharmacological BRAF inhibition (Konermann *et al.*, 2015). As CRISPRi/CRISPRa technology continues to develop, broader applications are envisioned, including manipulation of multiple genes in single cells in order to unveil novel interaction networks (Zalatan *et al.*, 2015).

This chapter will focus on the creation and use of a CRISPR/Cas9 screening library for large-scale genetic studies. Here, we discuss important considerations in design and methodology to these screens.

15.2 Methodology

15.2.1 Deciding Library Coverage

The first consideration in designing or using a CRISPR screening library is the number of genetic targets, whether genome-wide or a targeted panel. The size of the library is critical to establishing optimal parameters in future steps, such as how many cells to screen, sequencing depth and calling statistical significance. Many recent CRISPR screens in human cells have utilized whole-genome gRNA libraries targeting 7000 to 25 000 genes (Gilbert *et al.*, 2014; Konermann *et al.*, 2015; Shalem *et al.*, 2014; 2015; Wang *et al.*, 2014; Zhou *et al.*, 2014). However, it is also necessary to have multiple gRNAs targeting each gene. While the whole-genome gRNAs target up to 25 000 genes, the actual library size is around 300 000 gRNAs, which results in up to 10× coverage (that is, 10 unique gRNAs per target gene). Coverage of 3–6×, or 3–6 unique gRNAs for each gene, is currently most common.

15.2.2 Designing Guide RNAs

Effective gRNA design hinges first on targeting an appropriate site within the gene of interest that has as few off-target binding sites as possible. In general, a large portion of the coding region of a given gene is targetable. While the design of each gRNA/CRISPR complex can be dependent on the specific experimental needs, for effective knockouts, most gRNAs are designed to target early exons such that they induce insertions or deletions (indels) after the start codon. Targeting sequences near the 3′ end of the gene is less favorable since frameshift mutations that occur near the end of a protein are less likely to alter expression (Doench *et al.*, 2014; Wang *et al.*, 2014).

To improve target cleavage, the efficacy and/or stability of the gRNA sequence must be optimized (Hsu *et al.*, 2013). The protospacer adjacent motif (PAM) is a critical element to consider. The PAM sequence is directly adjacent to the 20 base pair gRNA, and generally given as a motif of 5′-NGG-3′. However, it has been noted that cytosine is favored and thymine is disfavored as the variable nucleotide in the PAM for effective gRNA design. The preference for cytosine may be a result of RNA polymerase III termination at U-rich regions (since the downstream transcript is U-rich). The gRNA nucleotide located directly adjacent to the PAM is also very important; in this site, guanine is preferred and cytosine is undesirable (Doench *et al.*, 2014; Figure 15.1). Various other changes to specific nucleotides within the gRNA sequence have also been shown to significantly affect efficacy,

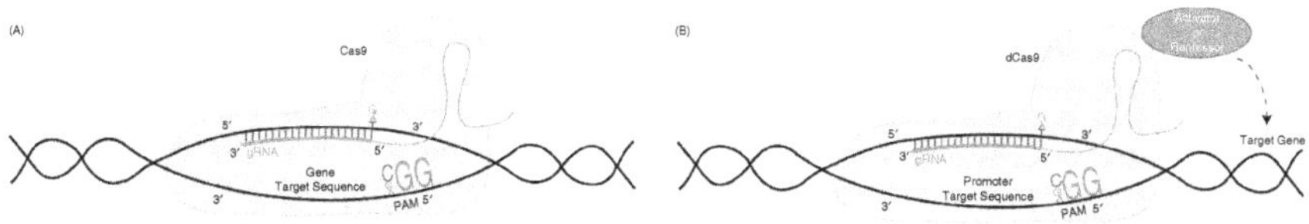

Figure 15.1 Diagram of gRNA (orange) in complex with Cas9 enzyme targeting a gene to create a knockout (A). The PAM sequence, with preferred bases emphasized is marked. For CRISPRi libraries, the modified Cas9 enzyme and associated proteins are shown acting on the targeted gene (B). (A black-and-white version of this figure will appear in some formats. For the color version, please refer to the plate section.)

although these requirements are not necessarily generalizable across gRNAs, gene targets and cell types (Mohr *et al.*, 2016).

Other characteristics of the gRNA sequence are also important to consider when designing gRNAs for experiments. For example, gRNAs with a G/C content that is too high or too low are often less effective (Doench *et al.*, 2014; Wang *et al.*, 2014). Additionally, if the target sequence displays high nucleosome occupancy or is typically in the coiled (or closed) chromatin state, gRNA efficacy will suffer (Smith *et al.*, 2016). While the gRNA features described here apply in most CRISPR applications, other more specialized characteristics differ between individual cell lines, organisms and techniques. Ongoing work seeks to better understand these differences in order to improve the selectivity of gRNAs at their intended target sites.

15.2.3 Minimizing Off-target Effects of gRNAs

Because the gRNA does not have to match perfectly with the target DNA sequence to effectively induce genetic alterations, there is a possibility for off-target effects when using CRISPR. Even though gRNAs are often most effective when perfectly matched to the target sequence, some degree of mismatch is tolerated by the CRISPR enzyme. In fact, the distance of the mismatch from the PAM site may be one variable in determining off-target efficiency, as it has been suggested that nucleotide differences generally do not decrease efficacy as substantially if they are found at sites more distant from the PAM. As might be expected, multiple mismatches decrease efficacy further than single mismatches, particularly if they are close to each other (Hsu *et al.*, 2013). In fact, gRNAs have been shown to bind to and cleave DNA regions with as many as five mismatches and in some cases the mutation frequency is as high at an off-target site as it is at the intended target site (Fu *et al.*, 2013). In other genomic "hotspots," located at some centromeres and telomeres, off-target effects are especially common (Tsai *et al.*, 2015); however, attributes of these hotspot regions have not yet been fully described, and it is still relatively unclear as to how often gRNAs cause genomic alterations in off-target regions.

Consequently, when utilizing CRISPR for screening, increasing the number of guide RNAs per target should mitigate undesirable off-target effects and reduce bias, since each gRNA has a different sequence and cleavage profile (Doench *et al.*, 2014). In order to decrease non-specific effects further, other efforts are necessary and are an active area of investigation. Shorter gRNAs, with 17 or 18 nucleotides rather than 20, have shown improved specificity, at least in human cells (Fu *et al.*, 2014). While nickases have also been proposed to improve efficacy, this technology is currently not useful in a knockout screening format (Cho *et al.*, 2014; Mali *et al.*, 2013; Ran *et al.*, 2013). Furthermore, although the ratio of Cas9 to gRNA is not a direct feature of gRNA design, this parameter can also be modulated to improve specificity (Hsu *et al.*, 2013). Consequently, as the field continues to evolve, gRNA libraries may be modified to alter specificity and design parameters that can be adapted to meet the needs of diverse experimental questions. In order to predict which gRNAs might be functional, recent work has developed gRNA and CRISPR library design software. Many of these programs are publicly available and their early evaluation has been promising (Heigwer *et al.*, 2016; Xie *et al.*, 2014).

15.2.4 Choosing the gRNA and Cas9 Expression System

There are multiple methods for introducing the gRNA and Cas9 endonuclease into cells, with transient or stable expression. Currently, the most common method of gRNA library expression is using lentiviral vectors. Lentiviral constructs are effective across a wide variety of cell types, and give stable long-term expression of the gRNA and/or Cas9 (Agrotis and Ketteler, 2015). Additionally, as lentiviral constructs integrate into the genome, the gRNA can be used as a unique identification of the knockout that was established in the cell.

Lentiviral constructs expressing a gRNA and a Cas9 together are common, but a two-vector system where Cas9 and a library of gRNAs are introduced separately has also been established (Figure 15.2). The two-vector system allows for an initial

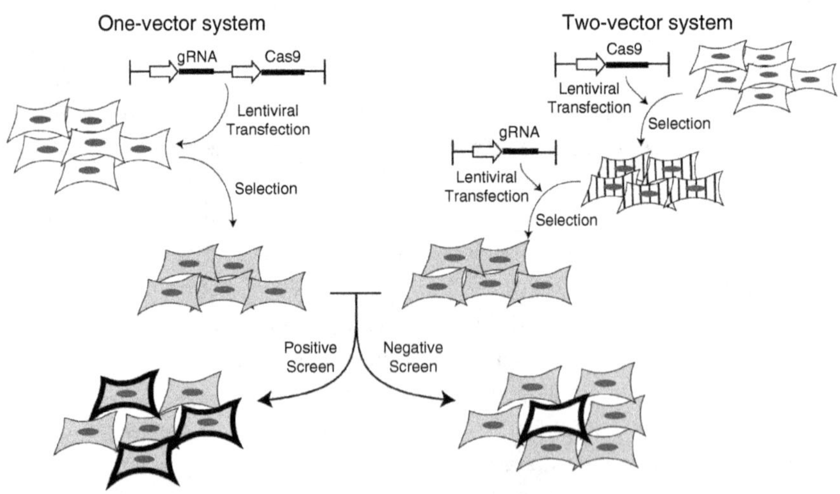

Figure 15.2 Overview of steps needed to introduce lentiviral CRISPR library into cells, using either a one- or a two-vector system. After selection, the pooled library can then be screened by positive selection, where the hits are observed at high frequency (outlines cells), or by negative selection, where the hits are observed at low frequency or lost from the surviving population (empty cell).

selection of a stable pool of cells expressing Cas9, and then subsequent introduction of the gRNA library with a secondary means of selection (either a different antibiotic or a fluorescent marker) (Koike-Yusa *et al.*, 2014; Wang *et al.*, 2014). It is possible that a one-vector system that contains the gRNA, Cas9 and all the needed expression machinery creates a large lentiviral construct that might be difficult to virally package. The two-vector system also has experimental flexibility, with the ability to have a stable pool of Cas9-expressing cells against a variety of smaller, targeted gRNA libraries (Zhou *et al.*, 2014). However, the one-vector system has been successfully used in a variety of cell types for a variety of screens (Shalem *et al.*, 2014; Wallace *et al.*, 2016; Wu *et al.*, 2016). The disadvantage of using lentiviral vectors is that the long-term expression of the gRNA and Cas9 has unknown effects on off-target regions of the DNA or on other cell functions. While short-term expression has been shown to be enough to induce knockouts, it has been suggested that continual long-term expression increases the chances of off-target effects (Agrotis and Ketteler, 2015).

Other vector-based systems, such as adeno-associated vectors or plasmid expression vectors allow for Cas9 and gRNA expression, but do not integrate. This makes it difficult to determine successful hits from the screen, as there is currently no method for identifying the gRNA or target region without a depth of next-generation sequencing that is cost prohibitive and analytically infeasible. Other transient methods of expressing gRNAs or Cas9, while potentially successful in establishing knockout, activation or suppression cell phenotypes, result in the inability to screen for successful events and testing of a pooled library.

15.2.5 Introducing the Library into Mammalian Cells

Introducing the CRISPR screening library into cells is as simple as a lentiviral transduction. However, an important concern is introducing the library into the cells at a low multiplicity of infection (MOI) so that there is a low chance of multiple constructs integrating into one cell. Generally, small-scale tests are run to determine an optimal MOI (5–30%) before the conditions are repeated for a large-scale transduction. Coverage is again a concern. For example, a library of 65 000 gRNAs can be covered at >400× if a population of 30 million cells are maintained. If the MOI is 30%, then 100 million cells need to be transduced. After lentiviral transduction, there is a selection process using either antibiotics or sorting by fluorescent marker depending on the vector choice. While the kinetics of CRISPR/Cas9 activity are largely unknown and most likely vary per cell line, generally several days are given between selections for the library pool and experimental treatment.

15.2.6 Experimental Design

Pooled CRISPR screens can be classified as either positive or negative selection. Positive selection screens use a strong selective pressure, such as a toxin, to eliminate the majority of cells from the population, enriching for cells with genetic perturbations conferring protection from the toxin. For example, to identify genes functioning in DNA mismatch repair (MMR), Wang *et al.* cultured cells in the presence of 6-thioguanine (6-TG), which induces cell cycle arrest via the MMR pathway. Cas9-inducible KBM7 cells were transduced with a pool of 73 000 gRNAs targeting 7031 genes, and incubated with 6-TG. Sequencing of the surviving population revealed that gRNAs targeting the four known MMR pathway components, MSH2, MSH6, MLH1 and PMS2, were greatly enriched when compared to the original library (Wang *et al.*, 2014). This type of screen can be powerful in identifying genes that convey resistance to drugs or other death-inducing treatments.

CRISPR/Cas9-mediated knockout libraries can also be used in negative selection screens, aimed at identifying genes essential for survival under the chosen experimental condition by monitoring the population for depletion of certain gRNAs. To verify that their methods are also appropriate for negative selection screening, Wang *et al.* assessed representation of gRNAs in library-infected Cas9-KBM7 cells at initial seeding and after 12 cell doublings. As expected, gRNAs targeting ribosomal protein genes were dramatically depleted in the final population, demonstrating the efficacy of this strategy in identifying essential genes (Wang *et al.*, 2014). Negative selection screens can also be used to identify targets that convey sensitivity to a low selection process, which could identify possible combinations of inhibitors for future clinical use. It is important to note that negative selection screens have additional technical concerns. For successful identification of a loss of gRNAs, then the CRISPR pool must have each gRNA represented in many cells (>300) with many gRNAs targeting the same gene. Deep sequencing coverage of the treated library is also needed to be able to successfully identify the low-frequency events or drop-outs.

15.2.7 Experimental Conclusion

At the conclusion of the experiment, each gRNA can be PCR-amplified and sequenced to identify the surviving population because of the integrated lentiviral gRNA construct. However, there are an unknown number of gRNAs and targeted genes, making it difficult to determine proper sequencing coverage for the treatment. Library size, as well as choosing positive or negative selection screens are the main considerations. Additionally, for either screen type, the gRNA coverage needs to be compared to an initial library or a non-treated pool for which the general assumption is that full coverage has been established.

15.3 Results and Discussion

15.3.1 Identification of Significant gRNAs and Genes

In most CRISPR/Cas9 screens, the general idea is to compare two or more populations of cells being subjected to a particular condition or treatment. The effects of the condition or treatment are measured by quantifying levels of gRNAs in the different cell populations. An increase in gRNA quantity after treatment indicates a disruption of the function of a target gene that provides some benefit to the cells. On the other hand, a decrease of gRNA numbers indicates that the disruption of the target gene function is not beneficial to the cell for survival.

Many software analysis pipelines, modules and packages are available for computational analyses of CRISPR/Cas9 screening data sets. Although these software packages use different algorithms, their common output is a list of significant gRNAs and genes. The first step of these analysis pipelines is to obtain read counts from each of the cell population samples in the study. These read counts are then normalized to account for library size, read count distribution, read depth and other factors. For example, the study design may need to decide if reads with one or more mismatches from gRNAs in the reference library will be used to quantify total read numbers. If there are biological or technical replicates in the experiment, variance estimation is performed to capture the variance across replicates. By incorporating the normalized read counts, variance estimation and other parameters, the statistical significance of each gRNA is calculated. These gRNAs are then ranked and the top gRNAs chosen for further investigation of pathways and genes of interest.

Model-based Analysis of Genome-wide CRISPR/Cas9 Knockout (MAGeCK) is one of the most popular tools for analyzing data from CRISPR/Cas9 screens. MAGeCK uses the median normalization method for read count normalization followed by mean-variance estimation to compare replicates in an experiment. Using this mean-variance estimation model, each gRNA is assigned a statistical significance. Genes are considered essential if their gRNAs are repeatedly on the top of the ranking list after applying the robust rank aggregation (RRA) algorithm. Based on RRA results, both positively and negatively selected essential genes can be identified. To identify pathways that are enriched, the same RRA algorithm is then applied to the list of top-ranked essential genes (Li *et al.*, 2014).

MaGeCK-VISPR is an improvised version of MAGeCK with many advanced capabilities. Quality control measures are incorporated to help in assessing the quality of the experiment. Another important feature is the ability to test if genes are essential under multiple conditions at the same time. To estimate gene significance, an expectation maximization (EM) algorithm is used instead of the RRA. This EM-based method can also estimate the efficiency of the gRNA knockout along with estimating gene ranking and significance. Another very beneficial feature is the ability to visualize, and interactively explore, the QC as well as the analysis results (Li *et al.*, 2015).

Other data analysis pipelines include the shALIGN (Sims *et al.*, 2011) and BiNGS!SL-seq (Kim and Tan, 2012), both of which use different R packages for the statistical data analysis. Empirical Analysis of Digital Gene Expression Data in R (edgeR) is a popular R package commonly used for various types of gene expression data such as RNA-Seq, microarray, etc. A customized workflow in edgeR can be used for analyzing data from CRISPR/Cas9 screens (Dai *et al.*, 2014).

15.3.2 Comparing CRISPR Libraries to siRNA and shRNA Screens

The development of the CRISPR/Cas system for precise genome editing has significantly advanced functional genomics screening; however, it is important to understand the utility of traditional functional genetics screening approaches. Previously, functional genomics studies relied upon small interfering RNA (siRNA) and short hairpin RNA (shRNA) technologies to explore gene function through loss-of-function studies. Although the use of siRNA and shRNA revolutionized the field at their introduction, the addition of CRISPR to the field of functional genomics has provided several advantages over siRNA and shRNA techniques, but also some limitations where siRNA and/or shRNA approaches may still be advantageous.

In one study, Feng *et al.* showed that CRISPR techniques resulted in improved gene inactivation efficiencies as compared to shRNA (Koike-Yusa *et al.*, 2014). This translates to improved sensitivity in evaluating gene function, as a complete knockout of a gene can reveal gene effects that are not adequately demonstrated by the knockdown of expression achieved by shRNA or siRNA screens (Wade, 2015). CRISPR additionally has the advantage of being able to create permanent cell line pools, which can be used to perform repeated experiments to further delineate gene function (Wade, 2015). Likewise, high-density gene saturation editing has also enabled functional genomics screens that assess biochemical activity of proteins, whereas shRNA-based approaches require complex multistep rescue techniques.

Like shRNA or siRNA, CRISPR also has the ability to reduce gene expression by recruitment of transcriptional repressors or direct transcriptional interference, as well as increase gene expression by recruiting transcription activators (Wade, 2015). CRISPR also has the benefit of being able to explore the effects of non-transcribed genes, as it directly acts on a DNA and does not rely upon transcription or translation like siRNA and shRNA screens (Wade, 2015). Finally, pooled CRISPR screens are cheaper than siRNA due to the transfection agents required to carry out siRNA screens (Wade, 2015), but may have similar cost points to shRNA pooled screens. Functional genomic approaches leveraging plate formats and transfection assays are beginning to emerge for CRISPR technology, but these are still in the early phases of development.

Despite the advantages of CRISPR, siRNA and shRNA screens will likely remain an important tool in functional genomics screening, as siRNA and shRNA screens often provide complementary information to CRISPR, and do maintain some advantages. For example, siRNA and shRNA screens are still ideal for targeting essential genes, as complete knockdown of essential functioning genes will result in cell death (Taylor and Woodcock, 2015). siRNA and shRNA screens also have a larger target pool than CRISPR, as they can essentially target any existing RNA. CRISPR, on the other hand, requires particular sequences in addition to the gRNA site to recruit the necessary machinery to cause double-stranded breaks (Taylor and Woodcock, 2015). This limits the number of cut sites in the genome; however, modified Cas9 enzymes are being rapidly developed to overcome this issue, regulate differential cut site specificity and enzyme efficiency. In fact, in the near future, we may perform pooled CRISPR screens with high-density gRNA libraries and multiple different Cas9 enzymes to analyze different effects and compare off-target and on-target activities of multiple different Cas9 enzymes.

Researchers have previously gained significant experience with siRNA and shRNA screens, allowing ample time for protocol refinement and an improvement in the rate of off-target effects (OTEs) in siRNA and shRNA screens (Wade, 2015), but one current limitation of CRISPR-based screens is the unknown consequences of false negatives due to variations in gRNA accuracy, and due to improved efficiency of NHEJ, resulting in accurate repair of the double-strand breaks (Wade, 2015). Other technical limitations that will need to be overcome include the fact that CRISPR screens require lysis of the host cell populations in order to identify the genes of interest and thus do not allow for any image-based phenotypic screen (Wade, 2015). Cas9 is also constitutively active, and while this has not been shown to harm the cell, it can result in increasing OTEs (Taylor and Woodcock, 2015). Finally, as mentioned above, the CRISPR screens require complete knockout of the gene of interest, and thus if there are many copies of the gene, as in cancer cells, CRISPR may not efficiently demonstrate loss of function in that gene (Taylor and Woodcock, 2015). As such, knockouts of genes, such as lncRNAs, without open reading frames or genes with high copy number in the cell population remain challenging for CRISPR, but not for si/shRNA-mediated approaches.

Consequently, while we have entered an era where knockout-based functional genomics screens are viable in almost any cell population, si/shRNA screening will continue to complement CRISPR-based technology. These alternative approaches can be easily leveraged for performing secondary or tertiary screens on nominated targets advancing from CRISPR screens or simply provide companion methodology that may enhance the chances of a successful screen under particular experimental conditions (e.g. assessing the effect of genes with high copy number in cancer cell lines).

Collectively, CRISPR technology is rapidly advancing and quickly becoming a standard tool in cell and molecular biology labs across the world. Our early experience with pooled CRISPR screening and creation of individual gene knockout lines has complemented the rapidly evolving knowledge base building around the technology. It is clear that knockout screening technology will lead to new medical and translational advances that were simply not possible before Cas9 was advanced into human models, and represents a clear story of how truly basic research can make a significant impact on the greater scientific community.

Grant Support

Dr. J. Chad Brenner received funding from NIH Grants T32 DC005356, U01-DE 025184, P30-CA046592-S1, R01-CA194536. M. Ludwig was supported by T32-GM 007315.

REFERENCES

Agrotis A, Ketteler R. 2015. A new age in functional genomics using CRISPR/Cas9 in arrayed library screening. *Frontiers in Genetics* **6**: 300.

Cho SW, Kim S, Kim Y, *et al.* 2014. Analysis of off-target effects of CRISPR/Cas-derived RNA-guided endonucleases and nickases. *Genome Res* **24**(1): 132–141.

Dai Z, Sheridan JM, Gearing LJ, *et al.* 2014. edgeR: a versatile tool for the analysis of shRNA-seq and CRISPR-Cas9 genetic screens. *F1000Res* **3**: 95.

Doench JG, Hartenian E, Graham DB, *et al.* 2014. Rational design of highly active sgRNAs for CRISPR-Cas9-mediated gene inactivation. *Nat Biotechnol* **32**(12): 1262–1267.

Fu Y, Foden JA, Khayter C, *et al.* 2013. High-frequency off-target mutagenesis induced by CRISPR-Cas nucleases in human cells. *Nat Biotechnol* **31**(9): 822–826.

Fu Y, Sander JD, Reyon D, *et al.* 2014. Improving CRISPR-Cas nuclease specificity using truncated guide RNAs. *Nat Biotechnol* **32**(3): 279–284.

Gilbert LA, Horlbeck MA, Adamson B, *et al.* 2014. Genome-scale CRISPR-mediated control of gene repression and activation. *Cell* **159**(3): 647–661.

Gilbert LA, Larson MH, Morsut L, *et al.* 2013. CRISPR-mediated modular RNA-guided regulation of transcription in eukaryotes. *Cell* **154**(2): 442–451.

Heigwer F, Zhan T, Breinig M, *et al.* 2016. CRISPR library designer (CLD): software for multispecies design of single guide RNA libraries. *Genome Biol* **17**: 55.

Hsu PD, Scott DA, Weinstein JA, *et al.* 2013. DNA targeting specificity of RNA-guided Cas9 nucleases. *Nat Biotechnol* **31**(9): 827–832.

Kim J, Tan AC. 2012. BiNGS!SL-seq: a bioinformatics pipeline for the analysis and interpretation of deep sequencing genome-wide synthetic lethal screen. *Methods Mol Biol* **802**: 389–398.

Koike-Yusa H, Li Y, Tan EP, *et al.* 2014. Genome-wide recessive genetic screening in mammalian cells with a lentiviral CRISPR-guide RNA library. *Nat Biotechnol* **32**(3): 267–273.

Konermann S, Brigham MD, Trevino AE, *et al.* 2015. Genome-scale transcriptional activation by an engineered CRISPR-Cas9 complex. *Nature* **517**(7536): 583–588.

Li W, Köster J, Xu H, *et al.* 2015. Quality control, modeling, and visualization of CRISPR screens with MAGeCK-VISPR. *Genome Biol* **16**(1): 1–13.

Li W, Xu H, Xiao T, *et al.* 2014. MAGeCK enables robust identification of essential genes from genome-scale CRISPR/Cas9 knockout screens. *Genome Biol* **15**(12): 554.

Mali P, Aach J, Stranges PB, *et al.* 2013. CAS9 transcriptional activators for target specificity screening and paired nickases for cooperative genome engineering. *Nat Biotechnol* **31** (9): 833–838.

Mohr SE, Hu Y, Ewen-Campen B, *et al.* 2016. CRISPR guide RNA design for research applications. *FEBS J* **283**(17): 3232–3238.

Ran FA, Hsu PD, Lin CY, *et al.* 2013. Double nicking by RNA-guided CRISPR Cas9 for enhanced genome editing specificity. *Cell* **154**(6): 1380–1389.

Shalem O, Sanjana NE, Hartenian E, *et al.* 2014. Genome-scale CRISPR-Cas9 knockout screening in human cells. *Science* **343**(6166): 84–87.

Shalem O, Sanjana NE, Zhang F. 2015. High-throughput functional genomics using CRISPR-Cas9. *Nat Rev Genet* **16**(5): 299–311.

Sims D, Mendes-Pereira AM, Frankim J, *et al.* 2011. High-throughput RNA interference screening using pooled shRNA libraries and next generation sequencing. *Genome Biol* **12**(10): R104.

Smith JD, Suresh S, Schlecht U, *et al.* 2016. Quantitative CRISPR interference screens in yeast identify chemical-genetic interactions and new rules for guide RNA design. *Genome Biol* **17**: 45.

Taylor J, Woodcock S. 2015. A perspective on the future of high-throughput RNAi screening: will CRISPR cut out the competition or can RNAi help guide the way? *J Biomol Screen* **20**(8): 1040–1051.

Tsai SQ, Zheng Z, Nguyen NT, *et al.* 2015. GUIDE-seq enables genome-wide profiling of off-target cleavage by CRISPR-Cas nucleases. *Nat Biotechnol* **33**(2): 187–197.

Wade M. 2015. High-throughput silencing using the CRISPR-Cas9 system: a review of the benefits and challenges. *J Biomol Screen* **20**(8): 1027–1039.

Wallace J, Hu R, Mosbruger TL, *et al.* 2016. Genome-wide CRISPR-Cas9 screen identifies microRNAs that regulate myeloid leukemia cell growth. *PLoS One* **11**(4): e0153689.

Wang T, Wei JJ, Sabatini DM, Lander ES. 2014. Genetic screens in human cells using the CRISPR-Cas9 system. *Science* **343**(6166): 80–84.

Wu Y, Zhou L, Wang X, *et al.* 2016. A genome-scale CRISPR-Cas9 screening method for protein stability reveals novel regulators of Cdc25A. *Cell Discov* **2**: 16014.

Xie S, Shen B, Zhang C, Huang X, Zhang Y. 2014. sgRNAcas9: a software package for designing CRISPR sgRNA and evaluating potential off-target cleavage sites. *PLoS One* **9**(6): e100448.

Zalatan JG, Lee ME, Almeida R, *et al.* 2015. Engineering complex synthetic transcriptional programs with CRISPR RNA scaffolds. *Cell* **160**(1–2): 339–350.

Zhou Y, Zhu S, Cai C, *et al.* 2014. High-throughput screening of a CRISPR/Cas9 library for functional genomics in human cells. *Nature* **509**(7501): 487–491.

Part IV

Genome Editing in Stem Cells and Regenerative Biology

16 Targeted Genome Editing Using Nuclease-assisted Vector Integration

Michael Gapinske,* Laurent Grant*, Alexander Brown and Pablo Perez-Pinera

16.1 Introduction

Gene editing technologies use engineered nucleases to introduce targeted modifications in the genomes of living cells. Three different generations of artificial DNA-binding domains have been developed for gene editing applications, including zinc-finger nucleases (ZFNs) (Gonzalez *et al.*, 2010; Mandell and Barbas, 2006; Pabo *et al.*, 2001), transcription activator-like effector nucleases (TALENs) (Boch *et al.*, 2009; Moscou and Bogdanove, 2009) and the clustered regularly interspaced short palindromic repeats (CRISPR)/Cas9 RNA-guided nuclease (RGN) system (Cong *et al.*, 2013; Doudna and Charpentier, 2014; Hsu *et al.*, 2014; Jinek *et al.*, 2013; Mali *et al.*, 2013; Wright *et al.*, 2016). All gene editing tools function by introducing a targeted double-strand break (DSB) within the genomic DNA. Although all ZFNs, TALENs and CRISPRs have been used successfully in this manner, it is the CRISPR system that has revolutionized the gene editing field, providing an efficient means to modify target genomic loci that does not require protein engineering.

In *S. pyogenes*, naturally expressed CRISPR RNAs (crRNAs) and trans-activating-crRNA (tracrRNA) form a complex that serves to guide the Cas9 nuclease to specific DNA sequences (Doudna and Charpentier, 2014). Acting as an immune response towards mobile genetic elements, such as viruses, the CRISPR/Cas9 system selectively destroys foreign DNA based on complementarity between the endogenously expressed crRNA and the target sequence. For recognition and cleavage by *S. pyogenes* CRISPR/Cas9, each target must be ~17 to 21 bases in length and immediately precede a suitable protospacer adjacent motif (PAM), of the form NGG. The *S. pyogenes* CRISPR system has been adapted for use in mammalian cells by addition of nuclear localization sequences (NLS) to Cas9 and utilization of a single-guide RNA (sgRNA) molecule, which, for convenience, replaces the combined functionality of both crRNA and tracrRNA.

* These authors contributed equally to this work.

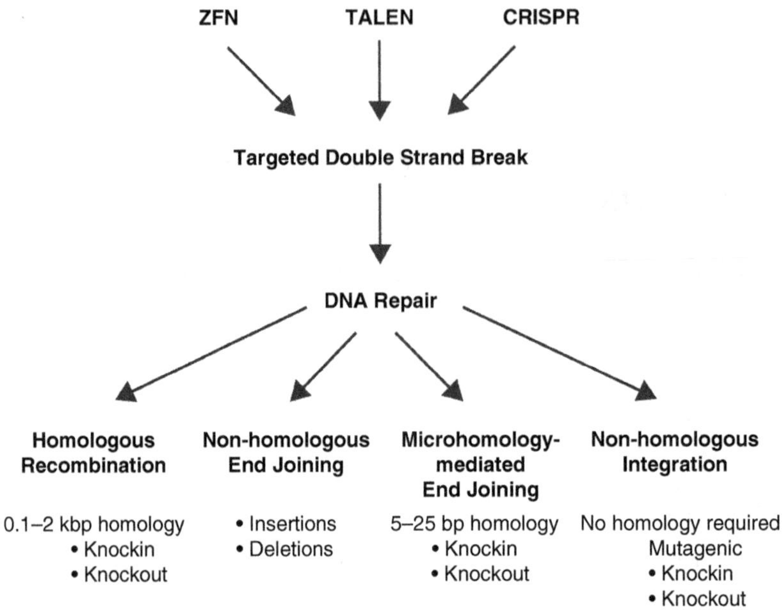

Figure 16.1 DNA repair pathways. ZFN, TALEN and CRISPR gene editing tools induce a DSB at target sites in the genome that activate DNA repair pathways in the cells including homologous recombination, non-homologous end joining, microhomology-mediated end joining and non-homologous integration. These pathways can be hijacked to achieve targeted integration of heterologous DNA at target sites.

Naturally occurring or induced DSBs within the genomic DNA of mammalian cells trigger a DNA damage response that activates DNA repair mechanisms. Three different cellular DNA repair mechanisms have been successfully utilized to introduce desired mutations at target sites in genomic DNA, including non-homologous end joining (NHEJ), homology-directed repair (HDR) and microhomology-mediated end joining (MMEJ; Figure 16.1).

Non-homologous end joining is a repair pathway that involves the direct ligation of DSB ends, without need for a homologous template (Jackson and Bartek, 2009; Lieber, 2010). This is the preferred repair mechanism used by most mammalian cells. Because this process does not require template DNA it can occur during any phase of the cell cycle. Once the DSB occurs, the free DNA ends are bound by Ku heterodimers, which align the two ends and protect against degradation. Ku then serves as a scaffold for the binding of several other NHEJ factors, forming a stable complex at the site of damage. This complex recruits DNA processing enzymes, which prepare the DSB ends for ligation, either by cleaving 5′ and 3′ overhangs or adding bases to achieve the necessary blunt ends. Finally, the NHEJ complex ligates the ends of the DSB, even if they are not perfectly compatible. During this error-prone repair process, single or multiple random bases can be gained or lost, thus leading to insertion/deletion mutations (indels). Most NHEJ-based gene editing technologies exploit the introduction of indels to create missense, nonsense and frameshift mutations at functionally relevant loci to achieve either partial or total knockout of gene function.

Homology-directed repair is a DSB repair mechanism that uses templates that possess an inherent homology to the DNA flanking the DSB. Unlike NHEJ, when the cells utilize HDR, the 5′ ends of the broken DNA are digested to create 3′ overhangs. In subsequent steps, a repair complex forms and the 3′ strand invades and progressively displaces the complementary strand of the selected repair template, or donor, DNA (Jackson and Bartek, 2009; Moynahan and Jasin, 2010). The final step is the resolution of the resultant Holliday junction that is formed. In mammalian cells, HDR repair is used when there is a homologous piece of DNA present, mostly in the S and G2 phases of the cell cycle. However, the use of this repair pathway can be stimulated by providing a DNA repair template that the cells can use in the place of a sister chromatid, or other homologous alleles. This repair template is typically a vector with homology arms of lengths oscillating between a few hundred base pairs to thousands of base pairs (Byrne *et al.*, 2014). The desired DNA modification is positioned between the homology arms, and, upon DNA repair, this change will be introduced into the genome (Figure 16.1).

Microhomology-mediated end joining is a DNA repair pathway that uses 5–25 base pair microhomologous sequences at both ends of the DSB during the alignment of broken ends before joining (Figure 16.1; McVey and Lee, 2008; Nussenzweig and Nussenzweig, 2007). MMEJ can be stimulated through the use of repair templates with short homology arms. The PITCh (Precise Integration into Target Chromosome) system employs MMEJ for both TALEN- and CRISPR/Cas9-mediated gene editing, termed TAL-PITCh (Sakuma *et al.*, 2015) and CRIS-PITCh (Nakade *et al.*, 2014; Sakuma *et al.*, 2016), respectively. TAL-PITCh requires the construction of a vector with a region of ~33 bases homologous to the genomic target site, such that a DSB can be induced in both the genome and the vector with one pair of TALENs. The fragments (~10–20 bp each) of DNA on either side of the cut site serve as the short homology arms for MMEJ, which lead to integration of the entire vector. CRIS-PITCh uses a similar method of vector integration, in that it integrates a vector using short arms that are complementary to the genomic DNA on either side of the Cas9 cut site. MMEJ DSB repair integrates the plasmid into the genome using these short arms. This repair pathway is poorly understood and construction of donor vectors unique to each target site is still required. Additionally, like HDR, this repair pathway is used at low frequency and for many purposes it is too inefficient.

Nuclease-assisted vector integration. Despite the high rates of DSB that can be achieved with RGNs, introduction of heterologous DNA at target sites remains challenging since the generation of recombination vectors remains a laborious and costly process and the efficiency of integration that can be accomplished with current techniques is low. While much research has been focused on increasing the efficiency of each DNA repair mechanism to facilitate integration, none of these strategies has provided a robust and consistent solution to this problem.

Alternatively, strain development can be streamlined exploiting nuclease-assisted vector integration (NAVI). By simply delivering gene editing tools to induce a targeted DSB simultaneously in the genome and in a vector, the free ends of the genomic DNA can ligate the free ends of the heterologous DNA thus leading to vector integration at the target site. Since NHEJ is likely the preferred DNA repair pathway, integration mediated by NHEJ often occurs at higher rates than integration mediated by HDR or MMEJ. Two important caveats of NAVI is the lack of directionality of the insert, as well as the introduction of mutations at the junctions of the genomic DNA with the vector. However, unlike HDR, NAVI facilitates integration of DNA fragments in excess of 45 kb in short time frames at low cost (Brown *et al.*, 2016).

In this chapter we describe the procedures and considerations for creation of vectors suitable for NAVI. We demonstrate the utility of this method through the implementation of a strategy for knocking out the *ETFA* gene, a protein that catalyzes the initial step of the mitochondrial fatty acid β-oxidation by shuttling electrons between primary flavoprotein dehydrogenases and the membrane-bound electron transfer flavoprotein ubiquinone oxidoreductase (Finocchiaro *et al.*, 1988).

16.2 Methods

The first step in the workflow of the experiments consists of identifying suitable target sites for high on-target activity, while minimizing off-target effects (Figure 16.2). These sequences are used to prepare vectors to target simultaneously the specific site in genomic DNA, as well as the transfer vector. These tools are delivered to mammalian cells, which can be selected to identify clonal populations with the desired change in the genome. Finally, each clone is individually validated using methods appropriate to detect the molecular change introduced.

16.2.1 Single-guide RNA Design

While multiple free online tools are available for designing sgRNA, we prefer Benchling (https://benchling.com/crispr). This free online DNA analysis suite allows importing genomic sequences from multiple species, as well as identifying all potential targets within a user-defined window of DNA. In addition, Benchling

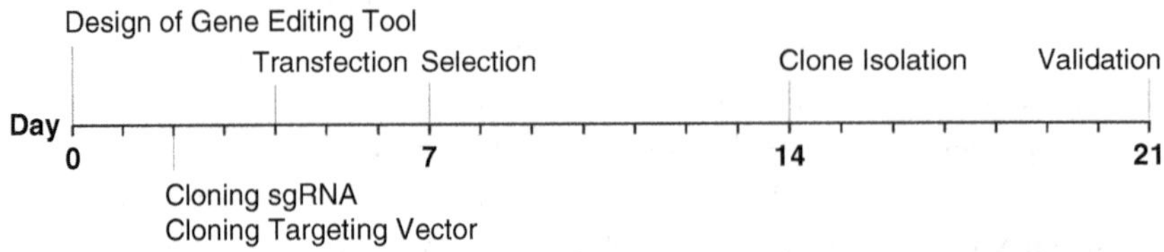

Figure 16.2 Overall experimental workflow and timeline to generate cell lines with targeted gene knockout.

provides on-target (Doench *et al.*, 2014) and off-target scores (Hsu *et al.*, 2013) for each sgRNA to easily predict activity and specificity.

16.2.2 Construction of Vectors

The introduction of off-target mutations is an important concern when using RGNs (Fu *et al.*, 2013; 2014; Pattanayak *et al.*, 2013). While it has been shown that delivery of Cas9-sgRNA ribonucleic acid complexes is an effective technique to introduce DSBs with high efficiency and reduced off-target effects (Liang *et al.*, 2015; Yu *et al.*, 2016), in our experience the use of plasmid DNA is more versatile. Consequently, this chapter describes protocols to accomplish targeted integration utilizing Cas9 and sgRNA expression plasmids.

16.2.2.1 Vectors for Expression of sgRNA and Cas9

These experiments have been performed using plasmid pSPgRNA (Addgene #47108), which has two BbsI/BpiI sites between a human U6 promoter and the sgRNA loop, for cloning the protospacer element of the expressed sgRNA. The plasmid encoding spCas9 was also obtained from Addgene (Plasmids #41815). The transfer vectors are derived from a pCDNA3.1 backbone, into which a puromycin or hygromycin expression cassette has been cloned.

16.2.2.2 Oligonucleotides for sgRNA Construction

i. *Target selection*: Typically, any coding region within the gene of interest can be used as a target for knockout since the indels introduced will often lead to a premature stop codon, resulting in RNA nonsense-mediated decay (Popp and Maquat, 2016).
ii. *Oligonucleotide design*: For each target sequence, two oligonucleotides are needed:
   a. A sense oligonucleotide with the form 5′-CACC G-X20-3′, where X20 represents the 20 bases of the protospacer in genomic DNA preceded by four bases that are complementary to the sgRNA vector overhangs and a fifth base, which is G in order to more efficiently initiate transcription of RNA from the upstream U6 promoter.
   b. An antisense oligonucleotide with the format: 5′-AAAC Y20-C-3′ where Y20 represents the reverse complement of the target sequence, and AAAC and the last C are vector-complementing overhangs.
   The sequences of the oligonucleotides for assembly of sgRNAs that target *ETFA* are:
      i. 5′-CACCGTGCATGCTCAGCTATTACCA-3′
      ii. 5′-AAACTGGTAATAGCTGAGCATGCAC-3′.

16.2.2.3 Oligonucleotides for Targeting the Transfer Vector

The sgRNA introduces a DSB at the target site but, for vector integration, it is also necessary to introduce a DSB in the vector that will be integrated. The chosen vector contains an expression cassette encoding a drug-resistance gene, which enables the selection of cell populations that harbor the integrated plasmid. We will refer to this vector as the transfer vector. While it is possible to clone the genomic target sequence into the transfer vector, we find it more convenient to utilize a separate sgRNA that targets an existing sequence within the transfer vector. The method for cloning sgRNA plasmids that target the transfer vector is identical to the method for cloning other sgRNAs. The target sequences to build an sgRNA that introduces a DSB in the transfer vectors are:

iii. 5′-CACCGACCGGGTCTTCGAGAAGACC-3′
iv. 5′-AAACGGTCTTCTCGAAGACCCGGTC-3′

16.2.2.4 Cloning of the sgRNA Expression Plasmids

The first step consists of preparing stocks of each oligonucleotide at 100 μM in nuclease-free molecular-biology-grade water. These oligonucleotides are combined in a 100 μL reaction containing 1 μL of each sense and antisense oligonucleotide and 98 μL of tris-buffered saline (TBS) and annealed by incubation at 95 °C for 5 minutes, followed by 25 °C for 3 minutes. Next, 1 μL of annealed oligonucleotides is mixed with 170 ng sgRNA vector, 2 μL 10× T4 ligase buffer, 1 μL of T4 ligase, 1 μL BpiI, 1 μL T4 polynucleotide kinase (PNK), and molecular-biology-grade water to a final reaction volume of 20 μL. The oligonucleotides are phosphorylated and incorporated into the vector through an optimized, simultaneous digestion/ligation cycling reaction: 37 °C for 5 minutes, 16 °C for 10 minutes, repeat for 3–5 cycles. The product of the ligation is transformed by mixing 1.5 μL of the reaction product with 30 μL of chemically competent *E. coli*, spread onto prewarmed LB agar containing 100 μg/mL carbenicillin and incubated at 37 °C overnight.

Typically, greater than 90% of transformants contain the desired construct, however, vector identity can be validated by colony PCR with the standard M13 forward primer and the antisense cloning oligonucleotide used as a reverse primer prior to confirmation by sequencing, if desired. The correct PCR products are ~330 base pairs long when visualized using agarose gel electrophoresis. One validated colony is grown overnight in 5 mL of LB broth with 100 μg/mL carbenicillin and the plasmid DNA is isolated using the Qiagen Spin Miniprep Kit. Each construct is verified by DNA sequencing, performed using the M13 forward primer.

16.2.3 Cell Culture and Transfection

HCT116 cells are colorectal carcinoma cells with a diploid genome. We typically knock out both alleles of a single gene of interest by co-transfecting two vectors that contain antibiotic resistance expression cassettes (each conferring resistance to either puromycin or hygromycin), an RGN targeting the genomic DNA and an RGN that induces a DSB simultaneously in both transfer vectors. The use of media supplemented with both puromycin and hygromycin selects cell populations that incorporate one expression cassette conferring resistance into each allele, thus ensuring biallelic modification.

Transfection is performed using Lipofectamine 2000™ in 24-well plates. The use of low-passage HCT116 cells in logarithmic growth is critical for high transfection efficiency. The DNA-Lipofectamine complexes are prepared by mixing the DNA with 50 μL of Opti-MEM in a microfuge tube and, in a separate tube, 2 μL of Lipofectamine 2000™ with 50 μL of Opti-MEM. After 5 minutes, the contents of both tubes are combined and incubated for 20 minutes. During this incubation, HCT116 cells are trypsinized, harvested and resuspended at 10^6 cells/mL in complete DMEM (10% FBS, 1% penicillin/streptomycin). Finally, the 100 μL DNA-Lipofectamine reagent mixture is pipetted into one well of a 24-well culture dish and mixed with 400 μL of cell suspension. Three days after transfection, the cells are passed to a 10 cm plate and grown in DMEM containing 10% FBS, 1% penicillin/streptomycin, 0.5 μg/mL puromycin and 100 μg/mL hygromycin. After seven days, clonal cell populations are readily visible, picked via pipette under the microscope, and transferred to a single well in a 24-well plate.

16.2.4 Knockout Validation

The cells are trypsinized when they become 50% confluent in the 24-well plate. Half of the cells are plated back into a 24-well plate and half are harvested for analysis of integration (Figure 16.3). Genomic DNA is isolated using the EarthOx Animal Genomic DNA Purification Mini Kit and PCR is performed using KAPA2G Robust PCR kits (KAPA Biosystems). A typical 25 μL reaction uses 20–100 ng of genomic DNA, Buffer A (5 μL), Enhancer (5 μL), dNTPs (0.5 μL), 10 μM forward primer (1.25 μL), 10 μM reverse primer (1.25 μL), KAPA2G Robust DNA Polymerase (0.5 U) and water (up to 25 μL). The DNA sequence of the primers for the *ETFA* target were:

WT FW: 5′-TAGCTAGATTTTCGCTATAGAACAAG-3′
WT REV: 5′-GAGCCCAGTTTGGGTTACAA-3′
TV REV: 5′-TGCCCTTGTCTTGTAGTTTCC-3′

The cycling parameters for the PCR reaction are 1 min at 95 °C followed by 30 cycles of 10 s at 95 °C, 10 s at 60 °C, and 10 s at 72 °C. The PCR products are visualized in 2% agarose gels and images captured using a ChemiDoc-It² (UVP). Clones possessing readily detectable amplicon levels of the expected sizes for genomic integration, without detectable wt amplification, are chosen for expansion.

Lack of target protein expression can be analyzed by western blot to ensure gene knockout (Figure 16.3). Briefly, cells are lysed with loading buffer, boiled for 5 min, and loaded in NuPAGE® Novex 4–12% Bis-Tris Gel polyacrylamide gels, which are then transferred to nitrocellulose membranes. Non-specific antibody binding is blocked with 50 mM Tris/150 mM NaCl/0.1% Tween-20 (TBS-T) with 5% non-fat milk for 30 min. The membranes are then incubated with primary antibodies anti-GAPDH (Cell Signaling Technology) or anti-ETFA (Santa Cruz

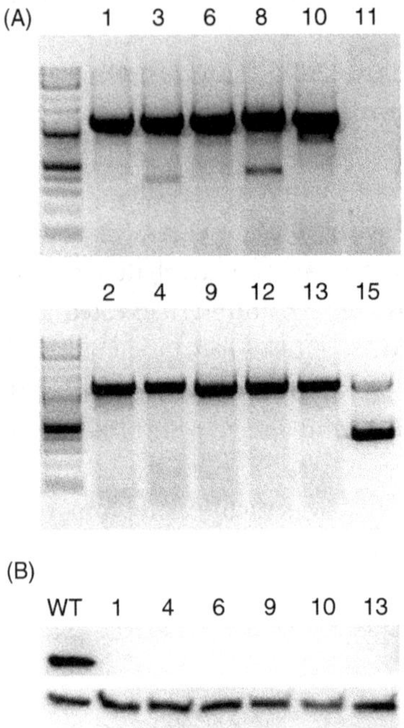

Figure 16.3 Effective generation of cell lines with disruption of ETFA. A CRISPR system was designed to introduce two vectors containing expression cassettes that confer resistance to either puromycin or hygromycin. Selection with both antibiotics ensures that the cell lines generated contain biallelic modifications at the target locus, which disrupted the ETFA gene, as determined by PCR to detect integration in their genomic DNA (A) or western blot (B).

Biotechnology) in 5% BSA or 5% non-fat milk in TBS-T diluted 1:1000 for 60 min, and then membranes are washed with TBS-T for 30 min. After labeling with primary antibodies, the membranes are incubated with antirabbit HRP-conjugated antibody (Sigma-Aldrich) diluted 1:10 000 for 30 min, and washed with TBS-T for 30 min. Finally, membranes are visualized using the Clarity™ ECL Western Blotting Substrate (Bio-Rad) and images are captured using a ChemiDoc-It^2 (UVP).

16.3 Results and Discussion

16.3.1 Selection of Gene Editing Tools

Gene editing tools are known to introduce off-target effects, which may lead to confounding results when generating cell models, since the target vector may integrate at unintended sites and confer resistance to the drugs used for selection. This problem is particularly severe with RGNs, which rely on individual sgRNAs to provide the sequence specificity needed to modify a single locus in complex genomes that often contain billions of base pairs. Several methods have been developed to minimize off-target effects introduced by Cas9. Some of the most common include the use of truncated sgRNAs (Fu *et al.*, 2014), inactivated Cas9 with FokI domains (Guilingeret *al.*, 2014; Tsai *et al.*, 2014), novel versions of Cas9 with reduced off-target effects (Kleinstiver *et al.*, 2016; Slaymaker *et al.*, 2016) and methods for the reduction of the residence time of Cas9 within the cells, which include delivery of mRNA or ribonucleic acid complexes (Liang *et al.*, 2015). Based on our experience, the cost of using plasmid DNA is significantly lower than other methods, and, while we have observed off-target effects frequently in populations of pooled cells, when analyzing predicted off-target sites in single clones, only in rare instances do we detect off-target integration events. More importantly, this method for rapid generation of gene knockouts is highly efficient, regardless of the method chosen to introduce DSBs. Therefore methods that increase the specificity of the DSB are likely to also increase the specificity of integration.

16.3.2 Cloning

The effectiveness of traditional T4 cloning for introduction of oligonucleotides in the sgRNA vectors can be variable, depending on the user. The protocol we describe in this chapter is very effective and reduces the number of steps that may introduce errors. In this single-step reaction, the restriction enzyme digests the plasmids, while ligation of the oligonucleotides by the T4 ligase destroys the restriction sites. At the end of the 15 minute reaction the only circular plasmid in the reaction contains the properly cloned sequences. The combination of this single-step reaction with the use of 5 minute transformations significantly reduces the time needed to prepare sgRNA vectors.

16.3.3 Integration Efficiency

We have measured integration efficiency by counting the number of colonies that survived selection after transfection in relation to the number of cells that used NHEJ to repair the DSB (Brown *et al.*, 2016). Using this quantification method, we have determined that integration efficiency can be as high as 30% of the NHEJ events that occur during repair of the DSB. More importantly, when introducing monoallelic or biallelic modifications, the integration efficiency was over 80% of the lines screened. In the case of *EFTA*, we screened 12 colonies and 75% contained homozygous integrations (Figure 16.3). Overall, screening of two colonies is sufficient to ensure that at least one will contain the mutation of interest. This minimizes the resources needed for this type of work and limits the time frame for successful generation of transgenic cell lines to less than three weeks (Brown *et al.*, 2016).

16.3.4 Selection

Typically, screening of clones generated by NHEJ without selection requires performing several rounds of serial dilutions of cells followed by PCR analysis and/or sequencing of their genomic DNA. Typically, the higher the efficiency of the gene-editing tools at inducing DSBs, the lower the number of colonies that must be screened. When the efficiency of the RGN is low or when multiplexed allelic modifications are desired, the screening becomes time- and resource-consuming. Unfortunately, there is also a direct correlation between high activity of Cas9 at on-target sites and introduction of off-target effects. The use of drug selection overcomes this problem, since even low-efficiency RGNs can generate hundreds to thousands of clonal cell lines. Importantly, the use of cancer cell lines in research is very common and, because these cells are frequently polyploid, there is a high demand for methods that can efficiently knock out multiple alleles. Since our method relies on the integration of a vector that confers resistance to one drug in each allele, modification of more than two alleles can be accomplished by transfecting transfer vectors that confer resistance to three, four, five or six drugs. However, as the number of target alleles increases, the rate of allele modification decreases – being as low as 10% for simultaneous disruption of four alleles (Brown *et al.*, 2016). Despite this decreased rate, 10% exceeds by many orders of magnitude the number of clones that need to be screened for detection of modification in four alleles without selection.

16.3.5 NHEJ Mutations

This genetic engineering strategy relies on the use of NHEJ by the cells to link genomic DNA to the linearized vector. Because NHEJ is a mutagenic pathway, the junction between the genomic DNA and the vector contains mutations similar to those introduced during NHEJ repair of DSBs in genomic DNA in the absence of a vector. Furthermore, we have reported a lack of directionality during linear vector integration in genomic DNA (Brown *et al.*, 2016). These random mutagenic events add complexity to the structure of the genome; however, the end result is identical to that observed without using transfer vectors. Finally, it is important to note that cell lines that are heterozygous for integration of the transfer vector can still be knockouts, since Cas9 often introduces indels in the allele that does not integrate the transfer vector. In these lines, PCR to detect integration will generate products that are apparently wt, however analysis of gene expression or sequencing will often confirm a true gene knockout. This consideration is important in experiments that intend to modify a single allele to obtain heterozygous cells.

16.4 Concluding Remarks

We describe here a method for rapid generation of clonal populations of cells with gene knockout that requires minimal work and can be completed by users with little expertise in molecular biology techniques in less than three weeks. This system is capable of introducing large fragments of heterologous DNA at target sites in the genome of cells *in vivo*. Despite some limitations, such as the lack of single base resolution, this system has broad applicability in genetic engineering.

REFERENCES

Boch J, Scholze H, Schornack S, *et al.* 2009. Breaking the code of DNA binding specificity of TAL-type III effectors. *Science* **326**(5959): 1509–1512.

Brown A, Woods WS, Perez-Pinera P. 2016. Multiplexed targeted genome engineering using a universal nuclease-assisted vector integration system. *ACS Synth Biol* **5**(7): 582–588.

Byrne SM, Ortiz L, Mali P, Aach J, Church GM. 2014. Multi-kilobase homozygous targeted gene replacement in human induced pluripotent stem cells. *Nucleic Acids Res* **43**(3): e21.

Cong L, Ran FA, Cox D, *et al.* 2013. Multiplex genome engineering using CRISPR/Cas systems. *Science* **339**(6121): 819–823.

Doench JG, Hartenian E, Graham DB, *et al.* 2014. Rational design of highly active sgRNAs for CRISPR-Cas9–mediated gene inactivation. *Nat Biotechnol* **32**(12): 1262–1267.

Doudna JA, Charpentier E. 2014. Genome editing: the new frontier of genome engineering with CRISPR-Cas9. *Science* **346**(6213): 1258096.

Finocchiaro G, Ito M, Ikeda Y, Tanaka K. 1988. Molecular cloning and nucleotide sequence of cDNAs encoding the alpha-subunit of human electron transfer flavoprotein. *J Biol Chem* **263**(30): 15773–15780.

Fu Y, Foden JA, Khayter C, *et al.* 2013. High-frequency off-target mutagenesis induced by CRISPR-Cas nucleases in human cells. *Nat Biotechnol* **31**(9): 822–826.

Fu Y, Sander JD, Reyon D, Cascio VM, Joung JK. 2014. Improving CRISPR-Cas nuclease specificity using truncated guide RNAs. *Nat Biotechnol* **32**(3): 279–284.

Gonzalez B, Schwimmer LJ, Fuller RP, *et al.* 2010. Modular system for the construction of zinc-finger libraries and proteins. *Nat Protoc* **5**(4): 791–810.

Guilinger JP, Thompson DB, Liu DR. 2014. Fusion of catalytically inactive Cas9 to FokI nuclease improves the specificity of genome modification. *Nat Biotechnol* **32**(6): 577–582.

Hsu PD, Lander ES, Zhang F. 2014. Development and applications of CRISPR-Cas9 for genome engineering. *Cell* **157**(6): 1262–1278.

Hsu PD, Scott DA, Weinstein JA, *et al.* 2013. DNA targeting specificity of RNA-guided Cas9 nucleases. *Nat Biotechnol* **31**(9): 827–832.

Jackson SP, Bartek J. 2009. The DNA-damage response in human biology and disease. *Nature* **461**(7267): 1071–1078.

Jinek M, East A, Cheng A, *et al.* 2013. RNA-programmed genome editing in human cells. *Elife* **2**: e00471.

Kleinstiver BP, Pattanayak V, Prew MS, *et al.* 2016. High-fidelity CRISPR-Cas9 nucleases with no detectable genome-wide off-target effects. *Nature* **529**(7587): 490–495.

Liang X, Potter J, Kumar S, *et al.* 2015. Rapid and highly efficient mammalian cell engineering via Cas9 protein transfection. *J Biotechnol* **208**: 44–53.

Lieber MR. 2010. The mechanism of double-strand DNA break repair by the nonhomologous DNA end-joining pathway. *Annu Rev Biochem* **79**(1): 181–211.

Mali P, Yang L, Esvelt KM, *et al.* 2013. RNA-guided human genome engineering via Cas9. *Science* **339**(6121): 823–826.

Mandell JG, Barbas CF. 2006. Zinc finger tools: custom DNA-binding domains for transcription factors and nucleases. *Nucleic Acids Res* **34**(Web Server): W516–W523.

McVey M, Lee SE. 2008. MMEJ repair of double-strand breaks (director's cut): deleted sequences and alternative endings. *Trends Genet* **24**(11): 529–538.

Moscou MJ, Bogdanove AJ. 2009. A simple cipher governs DNA recognition by TAL effectors. *Science* **326**(5959): 1501.

Moynahan ME, Jasin M. 2010. Mitotic homologous recombination maintains genomic stability and suppresses tumorigenesis. *Nat Rev Mol Cell Biol* **11**(3): 196–207.

Nakade S, Tsubota T, Sakane Y, *et al.* 2014. Microhomology-mediated end-joining-dependent integration of donor DNA in cells and animals using TALENs and CRISPR/Cas9. *Nat Commun* **5**: 5560.

Nussenzweig A, Nussenzweig MC. 2007. A backup DNA repair pathway moves to the forefront. *Cell* **131**(2): 223–225.

Pabo CO, Peisach E, Grant RA. 2001. Design and selection of novel Cys2His2 zinc finger proteins. *Annu Rev Biochem* **70**(1): 313–340.

Pattanayak V, Lin S, Guilinger JP, *et al.* 2013. High-throughput profiling of off-target DNA cleavage reveals RNA-programmed Cas9 nuclease specificity. *Nat Biotechnol* **31**(9): 839–843.

Popp MW, Maquat LE. 2016. Leveraging rules of nonsense-mediated mRNA decay for genome engineering and personalized medicine. *Cell* **165**(6): 1319–1322.

Sakuma T, Nakade S, Sakane Y, Suzuki KT, Yamamoto T. 2016. MMEJ-assisted gene knock-in using TALENs and CRISPR-Cas9 with the PITCh systems. *Nat Protoc* **11**(1): 118–133.

Sakuma T, Takenaga M, Kawabe Y, *et al.* 2015. Homologous recombination: independent large gene cassette knock-in in CHO cells using TALEN and MMEJ-directed donor plasmids. *Int J Mol Sci* **16**(10): 23849–23866.

Slaymaker IM, Gao L, Zetsche B, *et al.* 2016. Rationally engineered Cas9 nucleases with improved specificity. *Science* **351**(6268): 84–88.

Tsai SQ, Wyvekens N, Khayter C, *et al.* 2014. Dimeric CRISPR RNA-guided FokI nucleases for highly specific genome editing. *Nat Biotechnol* **32**(6): 569–576.

Wright AV, Nunez JK, Doudna JA. 2016. Biology and applications of CRISPR systems: harnessing nature's toolbox for genome engineering. *Cell* **164**(1–2): 29–44.

Yu X, Liang X, Xie H, *et al.* 2016. Improved delivery of Cas9 protein/gRNA complexes using lipofectamine CRISPRMAX. *Biotechnol Lett* **38**(6): 919–929.

17 Genome Engineering Using *Sleeping Beauty* Transposition in Vertebrates

Shreevathsa S. Chilkunda, Vaishnavi Raghunathan, Suneel A. Narayanavari, Zoltán Ivics and Zsuzsanna Izsvák

17.1 Introduction

A multitude of cutting-edge technologies have emerged, changing our perspective of genome engineering and its utility for human and veterinary medicine. Among them is a powerful class of tools, developed from transposable elements (TEs). TEs are stretches of DNA sequences that are mobile in the genomes of virtually all organisms. They occupy a high proportion of a genome, ranging from 10% in several fish species, 45% in the human genome and up to more than 80% of the maize genome (Mandal *et al.*, 2008; Munoz-Lopez *et al.*, 2010; Padeken *et al.*, 2015).

By virtue of their activity, TEs exist in a constant evolutionary arms race, with their hosts overcoming their defensive mechanisms until they succumb to mutational damage, which eventually restricts their activity to a bare minimum or shuts it down altogether. TEs and their hosts can co-exist as long as their mobilization is strictly regulated to keep the mutagenic load caused by their mobilization to a minimum. Basic research on TEs continuously fosters applications and the development of novel technologies. These efforts begin to pay dividends, as we witness the success of translating TE research for different applications, including manipulation of cells in culture, animal biotechnology, gene function annotation to connect genetics and physiology in model organisms, cancer research to identify genes and regulatory networks relevant for cancer, and gene therapeutic applications in humans. Indeed, technologies based on transposon-mediated gene transfer seem to greatly contribute to genomic manipulations in various species.

One popular non-viral approach to gene delivery is based on the *Sleeping Beauty* (*SB*) transposon system (Ivics *et al.*, 1997). *SB* belongs to the *Tc1/mariner* family of DNA transposons that has been adapted as a versatile tool for genome manipulation and engineering (Ivics *et al.*, 2009; 2015; Izsvak *et al.*, 2010b; Narayanavari *et al.*, 2016). The following sections will discuss the mechanism of gene transfer using *SB*, the main features and characteristics of *SB*-based vectors, and its advantages for use in gene therapy.

17.2 The *Sleeping Beauty* Transposon System

SB was the first active DNA transposon that was reconstructed from bits and pieces of ancient transposons from *Salmonid* fish genomes by *in vitro* evolution (Ivics *et al.*, 1997). The resurrected *SB* transposon has a simple structure: it is 1639 bp in length with 230 bp terminal inverted repeats (IRs) at each end (Ivics *et al.*, 1997). The IRs contain the 30 bp recognition sequences for the transposase in a direct repeat orientation (DRs; Izsvák *et al.*, 2000). In Nature, it exists as an autonomous entity with the transposon encoding the transposase open reading frame (ORF), which is 1023 bp in length.

Since the transposase has the ability to mobilize the transposon *in trans*, it can be separated from the transposon, enabling other DNA sequences to replace it (Ivics *et al.*, 1997). While the N-terminal region of the transposase is mediating substrate recognition, the C-terminus contains the catalytic domain, performing both cleavage and reintegration of the transposon (Craig, 1995; Plasterk *et al.*, 1999).

The transposition of *SB* is a cut-and-paste process whereby the excision of the transposon from its initial genomic location is followed by insertion into another locus. The process itself can be divided into four major steps: (1) the transposase binds to the IRs of the transposon, (2) the transposase subunits pair the two ends of the transposon and hold them to form a synaptic complex, (3) excision of the transposon from the donor site and (4) reintegration at the target site (Figure 17.1).

17.3 Characteristics of the *Sleeping Beauty* Transposon Vector

A binary system: The transposon and the transposase can be supplied as two separate components. In this set-up the transposase is replaced by a sequence or a gene of interest (GOI) that is flanked by the IRs, and the transposase is supplied *in trans* in the form of DNA or mRNA (Figure 17.2).

Efficacy: A hyperactive variant of *SB*, termed SB100X ("Molecule of the Year, 2009," http://iscspm.webnode.com/), yields efficient, stable gene transfer in various cell types, including therapeutically relevant primary cells (Belay *et al.*, 2010; Jin *et al.*, 2011; Mátés *et al.*, 2009). The activity of the transposon is further improved by optimizing the transposon (Cui *et al.*, 2002; Wang, 2016). SB transposition supports high and persistent transgene expression *in vitro* (Grabundzija *et al.*, 2010) as well as *in vivo* (Mátés *et al.*, 2009).

Precise integration: During transposition, the cargo flanked by the two IRs is liberated from the donor molecule. The left-behind donor molecule (usually a plasmid DNA) does not get integrated via a transposase-mediated mechanism. The substrate recognition of *SB* is a highly precise process, leading to genomic integration of DNA sequences whose boundaries are defined by the transposon end sequences at the IRs (Wang *et al.*, 2014). This property of *SB* is not a general feature of all the transposon systems. As an example, the *piggyBac* transposon (*PB*) system results in aberrant, pseudo-transposition events more frequently

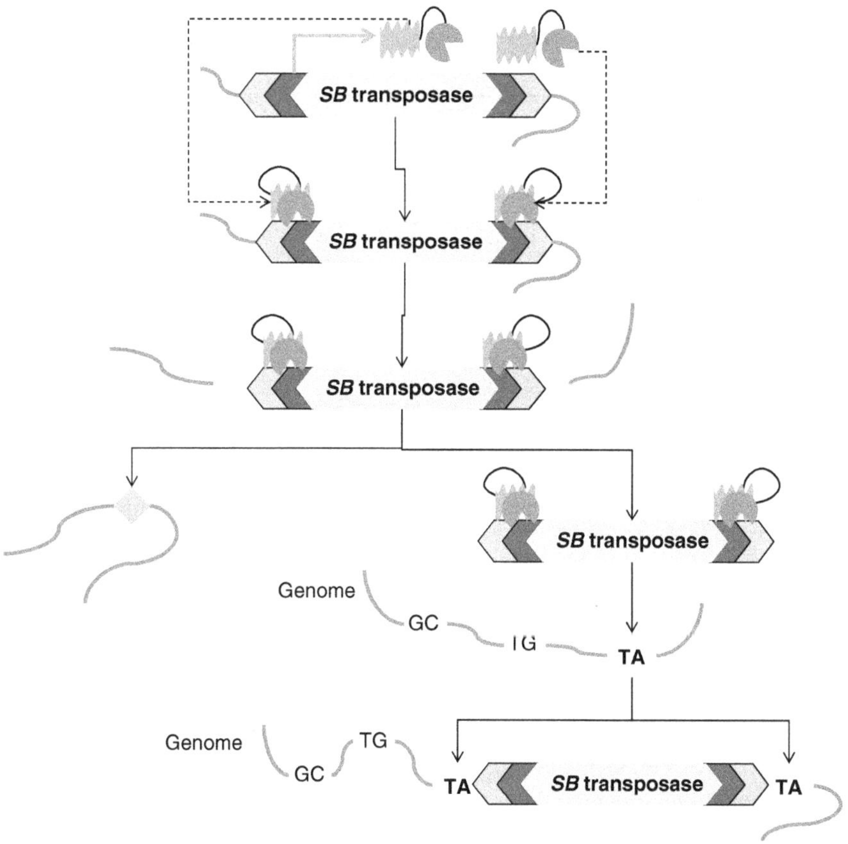

Figure 17.1 Outline of *Sleeping Beauty* (*SB*) transposition. In a natural, autonomous configuration, the transposon encodes the *SB* transposase, which is flanked by two sets of inverted repeats, IRs (darker arrow)/DRs (lighter arrow). Each of these IRs contain two binding sites for the transposase. The transposase binds to its binding sites within the IR/DR repeats. Following excision from its initial genomic locus (donor locus) the transposon integrates into a TA dinucleotide elsewhere (target locus). Transposon excision leaves a temporary break at the donor locus behind, which is repaired by the cell's repair machinery, resulting in a molecular footprint (brown rhombus). Basically the SB transposase consists of two major functional domains, the N-terminal domain (orange box with wavy outline) involved in DNA binding and a C-terminal domain (red pie) involved in catalysis. Host factors assist transposition at various steps in the reaction (not shown for simplicity). (A black-and-white version of this figure will appear in some formats. For the color version, please refer to the plate section.)

(Wang *et al.*, 2013). Importantly, the transposase – even in excess – does not appear to generate non-specific double-strand breaks in the genome (Galla *et al.*, 2011). The transposition process is not highly specific for target sequences; the transposase directs the integration of its cargo into a TA dinucleotide.

Molecular footprint: Tc1/mariner-like transposons leave behind a unique molecular footprint at the excision site. These footprints are generated by the host's DNA repair machinery for sealing the staggered double-strand break created by the transposase (Izsvak *et al.*, 2004). *SB* leaves a footprint of 3 bp flanked by TA repeats representing the target site duplications (Luo *et al.*, 1998).

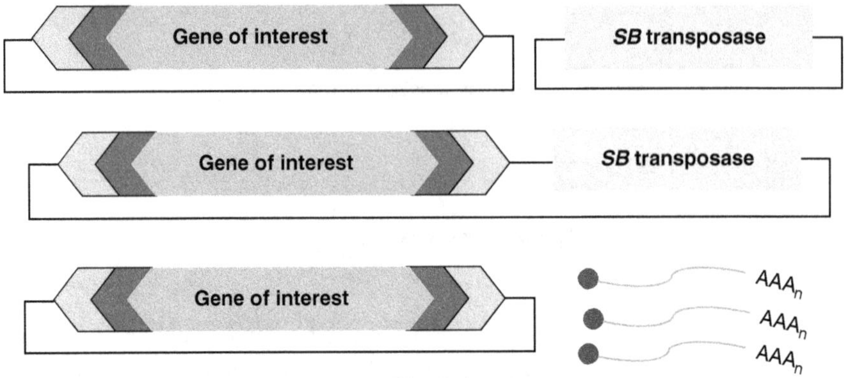

Figure 17.2 Schematics of the binary *Sleeping Beauty* system. The transposase gene can be moved out from the integrating transposon, and replaced by the gene of interest. Only the gene of interest, flanked by the recognition sequences of the transposase (see Figure 17.1) would be cleaved from the donor molecule and integrated into the target molecule by the transposase-mediated recombination. The integrating unit can be supplied as a simple plasmid vector, with the transposase supplied *in trans*, either on the same plasmid backbone, on a separate plasmid or even in the form of ready-to-translate mRNA synthesized *in vitro*. Note that the transposon system can be also combined with viral delivery (Narayanavari *et al.*, 2016; not shown).

Integration profile: SB has demonstrated a nearly random integration profile in the human genome (Ammar *et al.*, 2012; Gogol-Döring *et al.*, 2016; Voigt *et al.*, 2012). Notably, the conformation of the DNA flanking the TA-dinucleotide influences the integration site preference (Vigdal *et al.*, 2002). In contrast to the close-to-random integration pattern of *SB*, other transposons (e.g. *piggyBac*; Gogol-Doring *et al.*, 2016) or viral vectors (e.g. HIV/lenti) integrate preferentially into transcriptionally active regions (Ciuffi *et al.*, 2005; Schröder *et al.*, 2002). Compared to other integrating delivery vectors, *SB* appears to have the highest chance to integrate into a "safe harbor" locus (Gogol-Doring *et al.*, 2016).

Targeting: SB does not have the intrinsic ability to selectively target certain sites of regions in the genome. Nevertheless, a proof of concept has emerged that it could be engineered for targeted genomic manipulation (10–30%) (Ammar *et al.*, 2012; Ivics *et al.*, 2007; Voigt *et al.*, 2012; Yant *et al.*, 2007). Thus, in principle, the insertion could be fine-tuned to avoid detrimental phenotypes from undesired insertional mutagenesis. Transposons, unlike other genome targeting tools such as CRISPR/Cas9, and TALEN or zinc-finger nucleases, promote stable and long-term expression by directly inserting their genetic cargo into genomes. Importantly, this knockin feature of transposition is not dependent on the homology-dependent repair activity of the cell.

17.3.1 No Requirement for Active Cell Cycling

SB was demonstrated to be capable of transposing in primary, non-dividing T cells (Frommolt *et al.*, 2006; Singh *et al.*, 2014; 2015), suggesting that it does not require active cell division. This is an important feature, as many of the therapeutically relevant cells are non-dividing.

17.3.2 Post-integration Stability in the Human Genome

There are no *SB*-like transposons endogenously present in mammalian genomes. By contrast, and quite surprisingly, a domesticated *piggyBac* transposon-derived gene (piggyBac transposable element derived 5, PGBD5) can serve as an endogenous transposase source. Curiously, PGBD5 is capable of mobilizing insect-derived *piggyBac* transposon vectors in the human genome (Henssen *et al.*, 2015; Ivics, 2016).

High cargo capacity: Although the size of the cargo has been shown to inversely affect transposition efficiency (Izsvak *et al.*, 2000; Swierczek *et al.*, 2012; Wang *et al.*, 2014), there is no strict size limitation of the cargo for transposon-mediated gene transfer. When combined with bacterial artificial chromosomes (BACs), SB could deliver sequences up to 100 kb in human embryonic stem (ES) cells (Rostovskaya *et al.*, 2012). Notably, a "sandwich" transposon, in which the cargo to be mobilized is flanked by two complete *SB* elements arranged in an inverted orientation is suitable to efficiently deliver transgenes >10 kb (Zayed *et al.*, 2004).

Non-viral: Although DNA transposons like *SB* are mobile genetic elements, just like integrating viruses (including retroviruses), they are non-infectious and require no additional biosafety than Biological Safety Level 1 (BSL1).

No reverse transcription: Being a DNA transposon, *SB* does not require reverse transcription. This feature is advantageous, especially when the cargo is "complex," and whose processing would be limited by the intrinsic capacity of a reverse transcriptase, ultimately resulting in rearrangements in the cargo (Holkers *et al.*, 2013).

Low immunogenicity: Immunogenicity is an issue for several viral vectors that require repeated administration (e.g. adenoviral) to maintain the desired expression level of the therapeutic gene (Daya *et al.*, 2008). In contrast, non-viral, plasmid-based delivery systems, including *SB*, trigger no robust immune response, and therefore are safer than their viral counterparts (Hackett *et al.*, 2010). Immunogenicity could be mitigated by removing sequences of bacterial origin (e.g. minicircle technology; Monjezi *et al.*, 2016).

Inherently low promoter/enhancer activity: SB vectors have inherently low enhancer/promoter activity and therefore elicit low risk of trans-activating undesired gene expression at their integration sites (Bell *et al.*, 2007; Moldt *et al.*, 2007; Walisko *et al.*, 2008). Still, negligible promoter/enhancer activity is not an inherent feature of transposons, as the *piggyBac* transposon (*PB*) exhibits significant promoter/enhancer activity (Wilson *et al.*, 2007). Notably, insulator (barrier or enhancer-blocking) sequences flanking the transposon can be combined with *SB*, and could prevent accidental trans-activation of endogenous promoters by transgene cassettes (Bell *et al.*, 2007; Moldt *et al.*, 2007; Walisko *et al.*, 2008).

Stable expression in stem cells: SB is capable of providing efficient gene delivery (knockin) into various somatic and embryonic stem cells, where it supports stable and faithful gene expression (Belay *et al.*, 2010; Voigt *et al.*, 2016; Wang *et al.*, 2014; 2016).

17.4 Applications in Non-viral Gene Delivery

Molecular reconstruction of the *SB* transposon was a milestone in non-viral gene delivery methods, making genome manipulation possible in vertebrate species (Ivics *et al.*, 1997; 2015; Narayanavari *et al.*, 2016).

17.4.1 *Sleeping Beauty*, a Powerful Tool in Vertebrate Transgenesis

There is a constant need for simple and efficient transgenic tools, which can be exploited in a variety of model systems. Transposon-based transgenesis has significantly contributed to germline transgenesis in biomedically important species (Ivics *et al.* 2009; 2014b and c; Katter *et al.*, 2013).

Notably, *SB*-mediated transgenesis is less prone to genetic mosaicism and gene silencing in comparison to the classical method of direct pronuclear injection or lentivirus-mediated transgenesis (Katter *et al.*, 2013). Combined with specific promoter elements, the transposon vector can direct transgene expression in transgenic lines in a tissue-specific manner (Chandrashekran *et al.*, 2014; Szebenyi *et al.*, 2015).

There are established protocols for using *SB* to achieve high-efficiency germline transgenesis, while promoting sustained transgene expression in zebrafish (Davidson *et al.*, 2003), mouse and rat (Ivics *et al.*, 2014c; Katter *et al.*, 2013a), rabbits (Ivics *et al.*, 2014b), pigs (Ivics *et al.*, 2014a) and cattle (Garrels *et al.*, 2016). Remarkably, a method for generating multi-transgenic animals using *SB* was successful even in a large animal model of cattle. The established protocol is capable of generating complex transgenic genotypes within a feasible time frame (Garrels *et al.*, 2016).

17.4.2 In Gene Therapy

Advantages of non-viral vectors, including the *SB* system, include the relatively low cost of producing sufficient amounts for a clinical treatment, the ease of engineering and modifying the therapeutic construct, its stability during storage and its valuable safety features.

17.4.2.1 Integrative Approaches

17.4.2.1.1 Combining *Sleeping Beauty* and Non-viral Delivery

Gene therapy approaches combine *SB* with non-viral gene delivery technologies, including *in vivo* hydrodynamic injection (Bell *et al.*, 2007), liposome-mediated tranfection (Ma *et al.*, 2013), electroporation (Park *et al.*, 2013) and nanodelivery (Kren *et al.*, 2009). The hyaluronan- and asialoorosomucoid-coated nanoparticles encapsulating the *SB* transposon system that expresses the B-domain-deleted canine FVIII was a successful *in vivo* approach to target hepatocytes (Kren *et al.*, 2009). This resulted in hemophilia A mice exhibiting stable improved phenotype by using a simple intravenous injection for delivering the cassettes. Furthermore, long-term amelioration of a metabolic disorder of the liver was also achieved using *SB* packaged into proteoliposomes in a rat model of Crigler–Najjar syndrome type 1 following intravenous injection (Wang *et al.*, 2009). Importantly, neither of these methods elicited an immunogenic response.

17.4.2.1.2 Combining Non-viral *Sleeping Beauty* and Viral Delivery

The *SB* system poses a great potential to promote genome integration when coupled to viral particles that do not normally integrate as a part of their replicative cycle. Viral-SB transposon hybrids exist with adenovirus (Yant *et al.*, 2002), adeno-associated virus (Zhang *et al.*, 2013), herpesrvirus (Bowers *et al.*, 2006; Peterson *et al.*, 2007; Silva *et al.*, 2009; 2010) and baculovirus (Luo *et al.*, 2011; Turunen *et al.*, 2014). Alternatively, a combined vector that delivers SB with IDLV (integrase defective lentiviral vector; Moldt *et al.*, 2011; Staunstrup *et al.*, 2009; Vink *et al.*, 2009) integrates in a fairly random manner, following the safer integration profile of *SB*. These hybrid vectors couple the efficient delivery properties with the relatively safe and reliable features of an integrating expression vector (Narayanavari *et al.*, 2016), highlighting the feasibility and significance of novel approaches for gene therapy, in both *ex vivo* and *in vivo* applications.

17.4.3 Selected Examples

The ability of *SB* to mediate stable, long-term expression in a therapeutic mouse model was initially reported by Yant *et al.* (2000), in which extended expression of the human clotting factor IX (FIX) in FIX-deficient mice was demonstrated as a therapeutic procedure. Since then, several diseases have been targeted in both preclinical and clinical studies by employing *SB* (Table 17.1). Here we highlight some examples.

SB-based somatic cell gene therapy has been modeled for a variety of human hematological disorders such as haemophilia A, haemophilia B, sickle cell disease and Fanconi anaemia type C (Table 17.1). *SB* has also been successfully used to correct β-globin deficiency in mouse and human erythro-leukemia cell lines *in vitro* (Zhu *et al.*, 2007) and to stably deliver the FA-C gene in human lymphoblastoid cells derived from patients (Hyland *et al.*, 2011).

The transposase-encoding messenger RNA (mRNA) of *SB* was tested for its feasibility to mediate stable gene expression (Wilber *et al.*, 2006). A full phenotypic correction for treating a murine model of FAH (fumarylacetoacetate hydrolase) deficiency, an inherited tyrosinemia, was demonstrated (Wilber *et al.*, 2007).

The *SB* system is capable of handling and delivering large cargos like dystrophin and dysferlin coding sequences, which poses a challenge in other gene transfer approaches. *SB*-mediated delivery of the dystrophin gene in muscle cell lines gave rise to normal myogenic properties (Muses *et al.*, 2011). Furthermore, an attempt to deliver the full-length dysferlin (DYSF) cDNA into dysferlin-deficient H2K A/J myoblasts and further engraftment into muscles of DYSF mutant mice showed

Table 17.1 Applications of *Sleeping Beauty* in clinical trials, gene therapy and genetic screens

Disease models	Type of study	Model organism	Reference
Clinical Trials			
Alzheimer's disease	Clinical trial	Human	Eyjolfsdottir *et al.*, 2016
B Lymphoma, CAR-T	Clinical trial	Human	Kebriaei *et al.*, 2016
AMD	Clinical trial	Human	Johnen *et al.*, 2012; Johnen *et al.*, 2015; www.targetamd.eu/
Gene Therapies			
Transgenic mice	*in vivo*	Mouse	Richter *et al.*, 2016
Duchenne muscular dystrophy	*in vivo*	Mouse	Escobar *et al.*, 2016
Hypercholesterolemia	*in vivo*	Mouse	Turunen *et al.*, 2016
Melanoma	*in vivo*	Mouse	Perna *et al.*, 2015
Diabetes type I	*in vivo*	Mouse	Elso *et al.*, 2015
Alzheimer's disease	*In vivo*	Rat	Fjord-Larsen *et al.*, 2012
Duchenne muscular dystrophy	*in vivo*	Mouse	Muses *et al.*, 2011
Fanconi anemia	Clinical trial	Human	Hyland *et al.*, 2011
Hemophilia, FIX	*in vivo*	Dog	Hausl *et al.*, 2010
Mucopolysaccharidosis	*in vivo*	Mouse	Aronovich *et al.*, 2009
Sickle cell disease	*in vivo*	Mouse	Belcher *et al.*, 2010
Jaundice	*in vivo*	Rat	Wang *et al.*, 2009
Lymphoid Malignancies	*in vitro*	Human	Huang *et al.*, 2008
Glioblastoma	*in vivo*	Mouse	Ohlfest, 2005
Pulmonary hypertension	*in vivo*	Rat	Liu *et al.*, 2006a
Hemophilia, FVIII	*in vivo*	Mouse	Liu *et al.*, 2006b
Junctional epidermolysis bullosa	*in vivo*	Mouse	Ortiz-Urda *et al.*, 2003
Huntington disease	*in vitro*	Human	Chen *et al.*, 2005
α1-antitrypsin	*in vivo*	Mouse	Mikkelsen *et al.*, 2003
Tyrosinemia I	*in vivo*	Mouse	Montini, 2002
Mutagenesis Screens			
Osteosarcoma	*in vivo*	Mouse	Moriarity *et al.*, 2015
Non–small cell lung carcinoma	*in vivo*	Mouse	Dorr *et al.*, 2015
Histiocytic sarcoma	*in vivo*	Mouse	Been *et al.*, 2014
Non–melanoma skin cancer	*in vivo*	Mouse	Quintana *et al.*, 2013
Prostate adenocarcinoma	*in vivo*	Mouse	Rahrmann *et al.*, 2009
Schwann cell sarcoma	*in vivo*	Mouse	Rahrmann *et al.*, 2013
Hepatitis B virus-induced hepatocellular carcinoma	*in vivo*	Mouse	Bard-Chapeau *et al.*, 2014
B-cell precursor acute lymphoblastic leukemia (BCP-ALL)	*in vivo*	Mouse	van der Weyden *et al.*, 2011
High-grade astrocytoma	*in vivo*	Mouse	Bender *et al.*, 2010
Colorectal carcinoma	*in vivo*	Mouse	Starr *et al.*, 2009
Non-viral hepatocellular carcinoma	*in vivo*	Mouse	Keng *et al.*, 2009
Leukemia and lymphoma	*in vivo*	Mouse	Collier *et al.*, 2009

promising results in the restoration of gene expression and muscle regeneration (Escobar *et al.*, 2016).

SB has also been used to deliver RNAi molecules against specific exons of Htt (Huntington) resulting in up to 90% repression of the mutant gene (Chen *et al.*, 2005). Furthermore, *SB*-mediated delivery of miR-29 prevented inflammatory macrophage infiltration, and could improve the clinical prognosis of pulmonary fibrosis in a preclinical mouse model (Xiao *et al.*, 2012).

A recently developed approach includes a hybrid *in vivo* hematopoietic stem/progenitor cell (HSPC) transduction. The HSPC engineering strategy does not require myeloablation and transplantation (Richter *et al.*, 2016). The simple protocol consists of a subcutaneous injection of granulocyte colony-stimulating factor/G-CSF/AMD3100 to mobilize HSPCs from the bone marrow into the peripheral blood stream, followed by an intravenous injection of an integrating, *SB*-based helper-dependent adenovirus vector system. The *in vivo* HSPC transduction approach allows for a stable expression of a gene of interest in primitive HSPCs. The approach is well tolerated, is specific to hematopoietic tissues, is not associated with genotoxicity and thus has a high potential in therapy (Ren *et al.*, 2016).

The hybrid SB/adeno-system-mediated gene transfer has also been demonstrated in large mammals, wherein a high-capacity adenovirus was employed to deliver a hyperactive *SB* system (Yant *et al.*, 2002) expressing the canine factor IX (cFIX) cassette in canine liver. This strategy resulted in persistent phenotypic correction in a canine model for hemophilia B using a single intravenous injection (Hausl *et al.*, 2010).

The first clinical trial with encapsulated cell biodelivery (ECB) implants that deliver nerve growth factor (NGF) has been reported (Eyjolfsdottir *et al.*, 2016). The technology aims at halting the degeneration of cholinergic neurons and the associated cognitive decline in patients with Alzheimer's disease (AD). The implant consists of an NGF-producing genetically engineered human cell line encapsulated behind a semipermeable hollow fiber membrane that allows the influx of nutrients and the efflux of NGF. The cells are engineered using the SB100X system to stably produce the therapeutic gene product. The device, together with the engineered cells is removable. The strategy can be also used to release antiepileptic compounds, such as galanin, providing seizure-suppressant effects (Nikitidou *et al.*, 2014).

For patients with B-lymphoid malignancies, limited treatment options are available following relapse after autologous/allogeneic hematopoietic stem cell transplantation (HSCT). In these cases, engineered T-cell-based immunotherapy can specifically target a disease by employing antitumor mechanisms that are non-overlapping with the conditioning regimen used for HSCT. Remarkably, *SB* followed by years of fine-tuning has reached the clinic, and the first clinical trial exploiting *SB* has taken place (Kebriaei *et al.*, 2012). The first set of clinical trials used the concept of adoptive T-cell immunotherapy for cancer (Williams, 2008). The patients are treated with an autologous transplant of modified tumor-specific T cells expressing a CD19-directed chimeric antigen receptor (CAR; Kebriaei *et al.*,

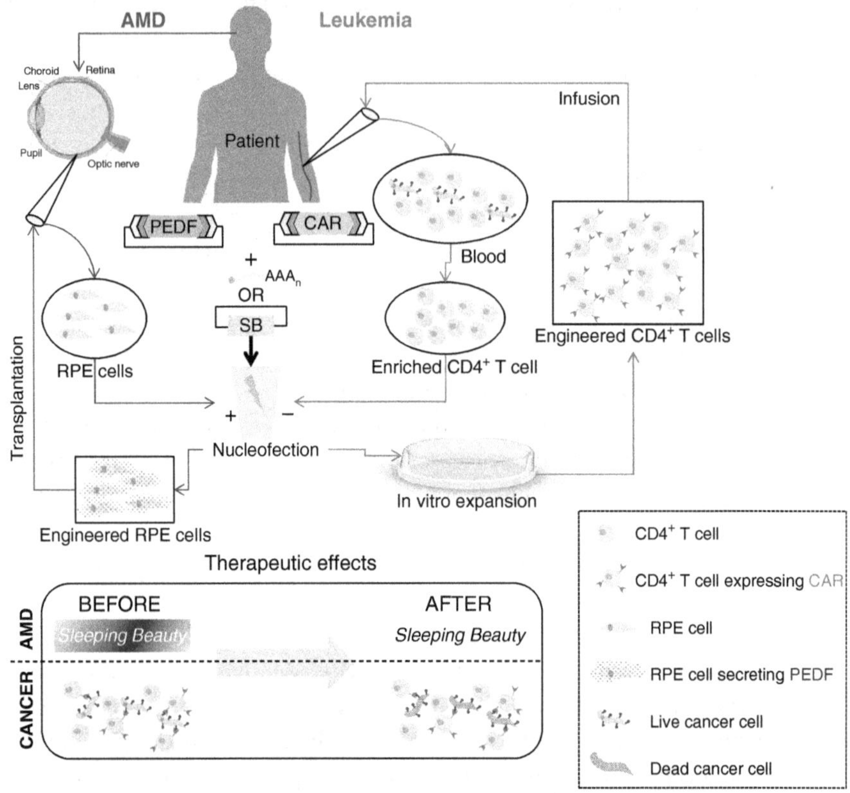

Figure 17.3 Illustration depicting the application of the *Sleeping Beauty* system for gene therapy. Schematics of two *ex vivo* strategies are shown: (1) targeting AMD (age-related macular degeneration) and (2) targeting a lymphoma. AMD treatment: First, the retinal pigment epithelial cells (RPEs) are surgically isolated from the patient (~ 10 000 cells). The cells are engineered to secrete the therapeutic gene, pigment epithelium-derived factor (PEDF). The expression vector is the *Sleeping Beauty* (non-viral system) delivered by nucleofection. The therapeutic cells are transplanted back into the patient's retina within a short span of time (~ 60 min). Note that the cells are not cultured. For details refer to Johnen *et al.* (2012; 2015; /www.targetamd.eu/). Treating B-cell malignancy: CD4$^+$ T cells are isolated from the patient's blood and are nucleofected with a chimeric antigen receptor (CAR) against the tumor antigen using the *Sleeping Beauty* system. The engineered cells are expanded, enriched and selected *in vitro* before being infused back into the respective patient. For details refer to Kebriaei *et al.* (2016); Singh *et al.* (2015). The envisioned therapeutic effects are shown in the adjoining inset. (A black-and-white version of this figure will appear in some formats. For the color version, please refer to the plate section.)

2012; Figure 17.3). Results from the Phase 1 clinical trial revealed that 99.9% of *SB* integrations occurred precisely in TA dinucleotides, without integration hotspots. The report demonstrates a successful expansion of *SB*-engineered modified T cells, with 84% CAR expression. The protocol was evaluated as being safe and recommended for further clinical development (Kebriaei *et al.*, 2016).

In a different attempt, *SB*-mediated transfer of pigment epithelium-derived factor (PEDF) *ex vivo* into retinal and iris pigment epithelial cells enabled long-term expression and secretion of PEDF (Johnen *et al.*, 2012). This successful preclinical study fuels a gene therapy clinical trial attempt coordinated by the

European consortium TargetAMD (www.targetamd.eu/). TargetAMD aims to deliver the therapeutic gene PEDF by *SB*-mediated gene transfer *ex vivo* to treat the exudative form of age-related macular degeneration (AMD) (Figure 17.3).

17.4.4 In Functional Oncogenomics

The *SB* system offers a flexibility to design for effective insertional mutagenesis. Several forward genetic screens exploit the *SB* system for insertional mutagenesis in mouse models to decipher the underlying causative factors in various cancer models (Collier *et al.*, 2005; Moriarity *et al.*, 2015; reviewed in Moriarity and Largaespada, 2015). The continuously increasing list includes non-melanoma skin cancer (Quintana *et al.*, 2013), non-viral hepato-cellular carcinoma (Keng *et al.*, 2009), hepatitis B virus-induced hepato-cellular carcinoma (Bard-Chapeau *et al.*, 2014), colorectal carcinoma (Starr *et al.*, 2009), B-cell precursor acute lymphoblastic leukemia (BCP-ALL) (van der Weyden *et al.*, 2011), lymphoma (Collier *et al.*, 2009), high-grade astrocytoma (Bender *et al.*, 2010), osteosarcoma (Moriarity *et al.*, 2015), prostate adenocarcinoma (Rahrmann *et al.*, 2009), histiocytic sarcoma (Been *et al.*, 2014) and non-small-cell lung carcinoma (Dorr *et al.*, 2015). Recently, oncogenomic screens have been adapted for human cells (Chen *et al.*, 2016; Molyneux *et al.*, 2014). These forward genetic screens also facilitate the identification of novel therapeutic targets (Figure 17.4).

17.4.5 In Stem Cell Research

Reprogramming of somatic cells provides a new opportunity to create and study patient-derived or disease-specific pluripotent stem cells that can be further exploited for mechanistic studies, treatment and therapy. Viral vectors used in the generation of induced pluripotent stem cells (iPSCs) have certain safety concerns, including insertional mutagenesis, reactivation of endogenous oncogenes or repression of tumor suppressor genes.

In consideration of safety and efficiency, efforts have been aimed at employing the *SB* system to generate iPSCs (Grabundzija *et al.*, 2013; Molina-Estevez *et al.*, 2013; Muenthaisong *et al.*, 2012; Sebe *et al.*, 2016; Talluri *et al.*, 2013; 2015). The reprogramming efficiency achieved with *SB* was comparable to that of the retroviral vectors (Grabundzija *et al.*, 2013).

Aiming at personalized gene or cell therapy, patient-derived iPSCs need to undergo genetic modification with the therapeutic transgene, followed by differentiation and transplantation. To achieve this, a strategy could be to engineer the *SB* transposon with the reprogramming cassette flanked by heterospecific loxP sites. After the generation of iPSCs, the reprogramming cassette could be exchanged using Cre-recombinase-mediated exchange with a therapeutic gene construct flanked by the recombination sites (Grabundzija *et al.*, 2013). This strategy could achieve both reprogramming and phenotype correction.

SB can provide stable gene expression in various somatic and embryonic stem cells (Belay *et al.*, 2010; Davis *et al.*, 2013; Fatima *et al.*, 2016; Grabundzija *et al.*, 2013; Kues *et al.*, 2013; Mátés *et al.*, 2009; Muenthaisong *et al.*, 2012; Talluri *et al.*, 2013; 2015). Compared to CRISPR/Cas9, TALENs and zinc-finger nucleases

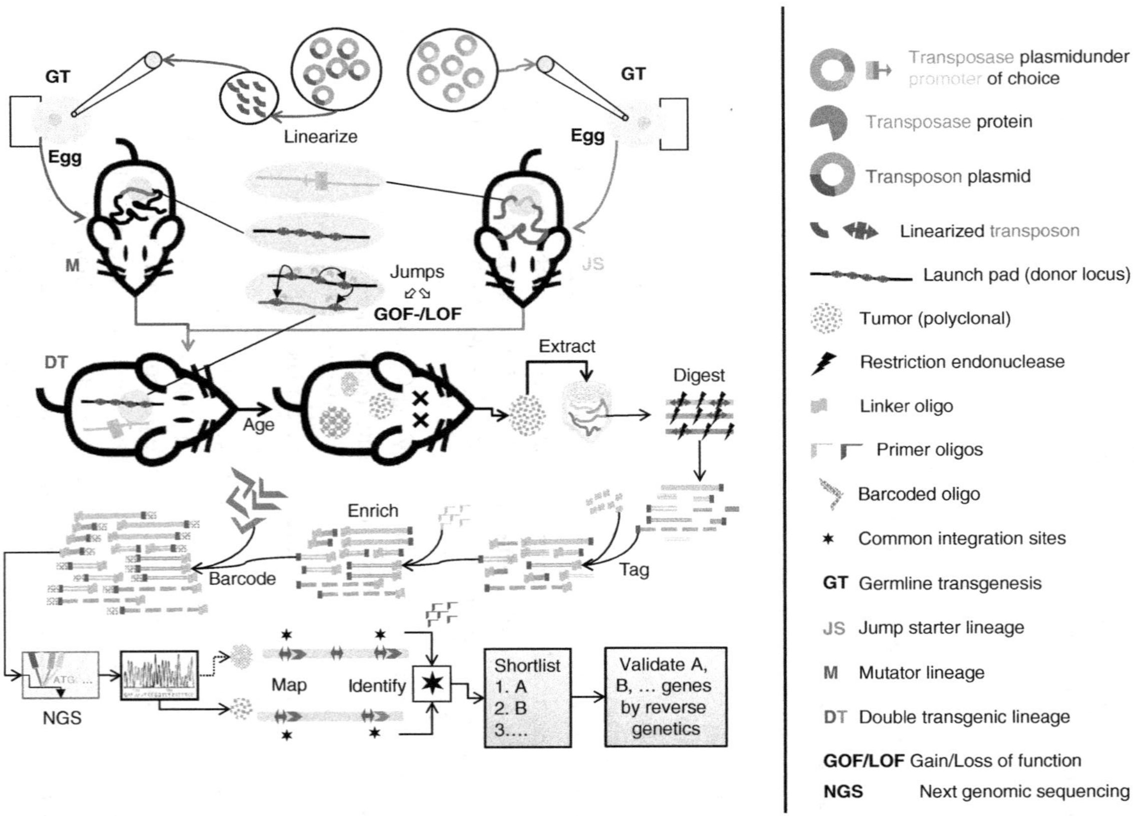

Figure 17.4 Illustration of a generalized strategy for employing the *Sleeping Beauty* transposon system for functional oncogenomics. A conditional insertional mutagenesis strategy is shown at whole organism level (illustrated as mice here). It first involves development of chosen models that are independently transgenic for both the transposon (the mutator lineage) and the transposase (the jump starter) lineage by germline transgenic technologies like DNA microinjections. The mutator line carries the transposon at the donor locus in a form of a concatemerized array (also referred to as the launchpad), generated by classic transgenic methodology of microinjecting linearized plasmid DNA.

(ZNFs), *SB* is more efficient in generating stable gene insertions in stem cells. In combination with a LTR7/HERVH reporter, the *SB* system was recently used to enrich and maintain naïve-like human embryonic stem cells (Wang *et al.*, 2014; 2016).

17.4.6 For Germline Mutagenesis

Germline mutagenesis is an important approach in functional genomics aiming at deciphering genotype–phenotype relationships in animal models. *SB* has been used for germline transgenesis in a variety of vertebrates like the amphibian *Xenopus laevis* (Sinzelle *et al.*, 2006), fishes like zebrafish (*Danio rerio*) (Balciunas *et al.*, 2004; Davidson *et al.*, 2003), Nile tilapia (*Oreochromis niloticus*) (He *et al.*, 2013) and Medaka fish (*Oryzias latipes*) (Grabher *et al.*, 2003), in rodents (Carlson *et al.*, 2003; Elso *et al.*, 2015; Horie *et al.*, 2003; Ivics *et al.*, 2011; Izsvak *et al.*, 2010a; Staunstrup *et al.*, 2011). Moreover, it has also been used for stable gene delivery and generating insertional mutants in a non-vertebrate chordate *Ciona intestinalis* (a Tunicate). In these strategies *SB* is used as an insertional mutagenic agent. The library of mutants is subjected to a phenotype-based screening. The mutagenesis can be conducted *in vitro* or *in vivo*. The *SB* system can be also used to express a library of small hairpin RNAs to identify gene targets in a phenotypic screen (e.g. genes regulating liver regeneration; Wuestefeld *et al.*, 2013). In contrast to a candidate-based strategy, the *SB*-transposon-based unbiased mutagenesis approach is suitable to identify genes with unexpected phenotype.

17.5 Summary and Outlook

In this revolutionary era it has become possible to manipulate the genome of various organisms with unprecedented ease and flexibility. Researchers have an increasing number of sophisticated tools to modify the genome. Importantly, the tools can be also combined with each other or with other recombination systems. Depending on the aim, careful evaluation of the pros and cons of a particular engineering tool helps to identify the most appropriate strategy. During the last

Caption for Figure 17.4 (cont.)

The transposon is engineered to contain oncogenic promoters along with appropriate splice acceptor and donor signals. Upon insertion into genes, the mutagenic transposon is capable of altering the natural reading frames and overexpressing novel transcripts. These gain- and/or loss-of-function (GOF/LOF) events may ultimately kick-start carcinogenesis. This scenario is restricted to tissues of choice by expressing the transposase in tissue-specific manner. As the double transgenic animals "age," they develop tumors. The tumors are harvested, processed and analyzed. *De novo* insertion sites of the remobilized transposon from the launchpad are determined by an elaborate PCR method, and are mapped onto the host genome. The insertion sites are statistically filtered, and common insertion sites (CISs) to multiple harvested tumors are determined. The oncogenic likelihood of the mutated genes is validated by reverse genetic approaches. For further information, refer to Geurts *et al.* (2006); Moriarity *et al.* (2015). (A black-and-white version of this figure will appear in some formats. For the color version, please refer to the plate section.)

decade, *SB*-transposon-based strategies have become invaluable in gene annotation, cancer research, transgenesis, stem cell research and gene therapy.

Success of a gene therapy technology relies on three aspects of the system, delivery, efficacy and, last but not least, safety. Several improvements made in each of these sectors support the implementation of the *SB* system in clinical trials. Further enhancements in targeted and efficient gene delivery into various cell types could pave the way for further clinical applications of *SB*. Coupling *SB* to cutting-edge non-viral (flow-through electroporation, nanotechnology) delivery technologies would further improve *ex vivo* applications. Viral delivery combined with stable expression of the therapeutic gene, provided by *SB*, would be suitable for *in vivo* delivery. Due to its simplicity, flexibility and affordability, *SB* offers a powerful alternative to retro/lentiviral vectors. Efficiently targeting *SB* integrations to safe harbors remains an ambitious challenge. In fact, nuclease-coupled targeting systems (e.g. ZNF/TALEN-nucleases, but especially CRISPR/Cas9) appear to be more suitable for precise editing of short genomic sequences. Indeed, the targeted nuclease systems are excellent choices for knockout (even simultaneously at multiple loci) strategies (Cong *et al.*, 2013; Mali *et al.*, 2013). For knockin transgenic applications, by contrast, the *SB* transposon might be a preferable choice in a wide variety of cell types, especially when the task is to deliver large cargos. Transposable elements, such as *SB*, are able to conduct transgene integration autonomously, and are not dependent on the homology-dependent repair pathways of the host, which is often a bottleneck for site-specific gene insertion systems in certain cell types (e.g. in stem cells). Reliable transgene expression in stem cells might be a unique feature of *SB* (Belay *et al.*, 2010; Mátés *et al.*, 2009; Richter *et al.*, 2016; Wang *et al.*, 2014). Microinjection of a plasmid DNA carrying the transgene DNA combined with mRNA provides early integration of the transgene without mosaic genotypes in various vertebrate species. Copy number-dependent transgene expression suggests that the *SB* vector does not trigger significant transgene silencing.

In the last decade, *SB* has been an extensively used tool to generate mutant libraries, in both somatic and germ cells. As a payoff, *SB*-based genetic screens annotated several gene networks and pathways, associated with various cellular mechanisms (not restricted to cancer). Unlike CRISPR/Cas9-mediated screens, the transposon-based mutagenic screens are capable of both knocking out and overexpressing a transcript, which has proved to be a powerful feature of transposon-based oncogenic screens. Besides *in vivo* strategies, conducted in animal models, *SB*-based screens could be also run in various human tissues. Due to its different integration pattern, *piggyBac* (*PB*) (a transposon of insect origin) has been shown to be an alternative genomic tool in many applications (Woodard *et al.*, 2015). The promise of *PB*-based strategies is primarily in mutagenesis and transgenesis, while safety concerns associated with PB transposition would limit its application in therapy.

REFERENCES

Ammar I, Gogol-Doring A, Miskey C, *et al.* 2012. Retargeting transposon insertions by the adeno-associated virus Rep protein. *Nucleic Acids Res* **40**(14): 6693–6712.

Aronovich EL, Bell JB, Khan SA, *et al.* 2009. Systemic correction of storage disease in MPS I NOD/SCID mice using the sleeping beauty transposon system. *Mol Ther* **17**(7): 1136–1144.

Balciunas D, Davidson AE, Sivasubbu S, *et al.* 2004. Enhancer trapping in zebrafish using the Sleeping Beauty transposon. *BMC Genomics* **5**(1): 62.

Bard-Chapeau EA, Nguyen AT, Rust AG, *et al.* 2014. Transposon mutagenesis identifies genes driving hepatocellular carcinoma in a chronic hepatitis B mouse model. *Nat Genet* **46**(1): 24–32.

Been RA, Linden MA, Hager CJ, *et al.* 2014. Genetic signature of histiocytic sarcoma revealed by a sleeping beauty transposon genetic screen in mice. *PLoS One* **9**(5): e97280.

Belay E, Matrai J, Acosta-Sanchez A, *et al.* 2010. Novel hyperactive transposons for genetic modification of induced pluripotent and adult stem cells: a nonviral paradigm for coaxed differentiation. *Stem Cells* **28**(10): 1760–1771.

Belcher JD, Vineyard JV, Bruzzone CM, *et al.* 2010. Heme oxygenase-1 gene delivery by Sleeping Beauty inhibits vascular stasis in a murine model of sickle cell disease. *J Mol Med (Berl)* **88**(7): 665–675.

Bell JB, Podetz-Pedersen KM, Aronovich EL, *et al.* 2007. Preferential delivery of the Sleeping Beauty transposon system to livers of mice by hydrodynamic injection. *Nat Protoc* **2**(12): 3153–3165.

Bender AM, Collier LS, Rodriguez FJ, *et al.* 2010. Sleeping beauty-mediated somatic mutagenesis implicates CSF1 in the formation of high-grade astrocytomas. *Cancer Res* **70**(9): 3557–3565.

Bowers W, Mastrangelo M, Howard D, *et al.* 2006. Neuronal precursor-restricted transduction via in utero CNS gene delivery of a novel bipartite HSV amplicon/transposase hybrid vector. *Mol Ther* **13**(3): 580–588.

Carlson CM, Dupuy AJ, Fritz S, *et al.* 2003. Transposon mutagenesis of the mouse germline. *Genetics* **165**(1): 243–256.

Chandrashekran A, Sarkar, R, Thrasher, A, *et al.* 2014. Efficient generation of transgenic mice by lentivirus-mediated modification of spermatozoa. *FASEB J* **28**(2): 569–576.

Chen HJ, Wei Z, Sun J, *et al.* 2016. A recellularized human colon model identifies cancer driver genes. *Nat Biotechnol* **34**(8): 845–851.

Chen ZJ, Kren BT, Wong PY, Low WC, Steer CJ. 2005. Sleeping Beauty-mediated down-regulation of huntingtin expression by RNA interference. *Biochem Biophys Res Commun* **329**(2): 646–652.

Ciuffi A, Llano M, Poeschla E, *et al.* 2005. A role for LEDGF/p75 in targeting HIV DNA integration. *Nat Med* **11**(12): 1287–1289.

Collier LS, Adams DJ, Hackett CS, *et al.* 2009. Whole-body sleeping beauty mutagenesis can cause penetrant leukemia/lymphoma and rare high-grade glioma without associated embryonic lethality. *Cancer Res* **69**(21): 8429–8437.

Collier LS, Carlson CM, Ravimohan S, Dupuy AJ, Largaespada DA. 2005. Cancer gene discovery in solid tumours using transposon-based somatic mutagenesis in the mouse. *Nature* **436**(7048): 272–276.

Cong L, Ran FA, Cox D, *et al.* 2013. Multiplex genome engineering using CRISPR/Cas systems. *Science* **339**(6121): 819–823.

Craig NL. 1995. Unity in transposition reactions. *Science* **270**(5234): 253–254.

Cui Z, Geurts AM, Liu G, Kaufman CD, Hackett PB. 2002. Structure-function analysis of the inverted terminal repeats of the sleeping beauty transposon. *J Mol Biol* **318**(5): 1221–1235.

Davidson AE, Balciunas D, Mohn D, *et al.* 2003. Efficient gene delivery and gene expression in zebrafish using the Sleeping Beauty transposon. *Dev Biol* **263**(2): 191–202.

Davis RP, Nemes C, Varga E, *et al.* 2013. Generation of induced pluripotent stem cells from human foetal fibroblasts using the Sleeping Beauty transposon gene delivery system. *Differentiation* **86**(1–2): 30–37.

Daya S, Berns KI. 2008. Gene therapy using adeno-associated virus vectors. *Clin Microbiol Rev* **21**(4): 583–593.

Dorr C, Janik C, Weg M, *et al.* 2015. Transposon mutagenesis screen identifies potential lung cancer drivers and CUL3 as a tumor suppressor. *Mol Cancer Res* **13**(8): 1238–1247.

Elso CM, Chu EP, Alsayb MA, *et al.* 2015. Sleeping Beauty transposon mutagenesis as a tool for gene discovery in the NOD mouse model of type 1 diabetes. *G3 (Bethesda)* **5**(12): 2903–2911.

Escobar H, Schöwel V, Spuler S, Marg A, Izsvák Z. 2016. Full-length dysferlin transfer by the hyperactive Sleeping Beauty transposase restores dysferlin-deficient muscle. *Mol Ther Nucleic Acids* **5**: e277.

Eyjolfsdottir H, Eriksdotter M, Linderoth B, *et al.* 2016. Targeted delivery of nerve growth factor to the cholinergic basal forebrain of Alzheimer's disease patients: application of a second-generation encapsulated cell biodelivery device. *Alzheimers Res Ther* **8**(1): 30.

Fatima A, Ivanyuk D, Herms S, *et al.* 2016. Generation of human induced pluripotent stem cell line from a patient with a long QT syndrome type 2. *Stem Cell Res* **16**(2): 304–307.

Fjord-Larsen L, Kusk P, Emerich DF, *et al.* 2012. Increased encapsulated cell biodelivery of nerve growth factor in the brain by transposon-mediated gene transfer. *Gene Ther* **19**(10): 1010–1017.

Frommolt R, Rohrbach F and Theobald M. 2006. Sleeping Beauty transposon system – future trend in T-cell-based gene therapies? *Future Oncol (London, England)* **2**(3): 345–349.

Galla M, Schambach A, Falk CS, *et al.* 2011. Avoiding cytotoxicity of transposases by dose-controlled mRNA delivery. *Nucleic Acids Res* **39**(16): 7147–7160.

Garrels W, Talluri TR, Apfelbaum R, *et al.* 2016. One-step multiplex transgenesis via Sleeping Beauty transposition in cattle. *Sci Rep* **6**: 21953.

Geurts AM, Collier LS, Geurts JL, *et al.* 2006. Gene mutations and genomic rearrangements in the mouse as a result of transposon mobilization from chromosomal concatemers. *PLoS Genet* **2**(9): e156.

Gogol-Döring A, Ammar I, Gupta S, *et al.* 2016. Genome-wide profiling reveals remarkable parallels between insertion site selection properties of the MLV retrovirus and the piggyBac transposon in primary human CD4(+) T cells. *Mol Ther* **24**(3): 592–606.

Grabher C, Henrich T, Sasado T, *et al.* 2003. Transposon-mediated enhancer trapping in medaka. *Gene* **322**: 57–66.

Grabundzija I, Irgang M, Mátés L, *et al.* 2010. Comparative analysis of transposable element vector systems in human cells. *Mol Ther* **18**(6): 1200–1209.

Grabundzija I, Wang J, Sebe A, *et al.* 2013. Sleeping Beauty transposon-based system for cellular reprogramming and targeted gene insertion in induced pluripotent stem cells. *Nucleic Acids Res* **41**(3): 1829–1847.

Hackett PB, Largaespada DA, Cooper LJ. 2010. A transposon and transposase system for human application. *Mol Ther* **18**(4): 674–683.

Hausl MA, Zhang W, Muther N, *et al.* 2010. Hyperactive sleeping beauty transposase enables persistent phenotypic correction in mice and a canine model for hemophilia B. *Mol Ther* **18**(11): 1896–1906.

He X, Li J, Long Y, *et al.* 2013. Gene transfer and mutagenesis mediated by Sleeping Beauty transposon in Nile tilapia (*Oreochromis niloticus*). *Transgenic Res* **22**(5): 913–924.

Henssen AG, Henaff E, Jiang E, *et al.* 2015. Genomic DNA transposition induced by human PGBD5. *Elife* **4**: e10565.

Holkers M, Maggio I, Liu J, *et al.* 2013. Differential integrity of TALE nuclease genes following adenoviral and lentiviral vector gene transfer into human cells. *Nucleic Acids Res* **41**(5): e63.

Horie K, Yusa K, Yae K, *et al.* 2003. Characterization of Sleeping Beauty transposition and its application to genetic screening in mice. *Mol Cell Biol* **23**(24): 9189–9207.

Huang X, Guo H, Kang J, *et al.* 2008. Sleeping Beauty transposon-mediated engineering of human primary T cells for therapy of CD19+ lymphoid malignancies. *Mol Ther* **16**(3): 580–589.

Hyland KA, Olson ER, Clark KJ, *et al.* 2011. Sleeping Beauty-mediated correction of Fanconi anemia type C. *J Gene Med* **13**(9): 462–469.

Ivics Z. 2016. Endogenous transposase source in human cells mobilizes piggyBac transposons. *Mol Ther* **24**(5): 851–854.

Ivics Z, Garrels W, Mátés L, *et al.* 2014a. Germline transgenesis in pigs by cytoplasmic microinjection of Sleeping Beauty transposons. *Nat Protoc* **9**(4): 810–827.

Ivics Z, Hackett PB, Plasterk RH, Izsvak Z. 1997. Molecular reconstruction of Sleeping Beauty, a Tc1-like transposon from fish, and its transposition in human cells. *Cell* **91**(4): 501–510.

Ivics, Z, Hiripi L, Hoffmann OI, *et al.* 2014b. Germline transgenesis in rabbits by pronuclear microinjection of Sleeping Beauty transposons. *Nat Protoc* **9**(4): 794–809.

Ivics Z, Izsvak Z. 2015. Sleeping Beauty transposition. *Microbiol Spectr* **3**(2): MDNA3-0042-2014.

Ivics Z, Izsvak Z, Medrano G, Chapman KM, Hamra FK. 2011. Sleeping Beauty transposon mutagenesis in rat spermatogonial stem cells. *Nat Protoc* **6**(10): 1521–1535.

Ivics Z, Katzer A, Stuwe EE, *et al.* 2007. Targeted Sleeping Beauty transposition in human cells. *Mol Ther* **15**(6): 1137–1144.

Ivics Z, Li MA, Mates L, *et al.* 2009. Transposon-mediated genome manipulation in vertebrates. *Nat Methods* **6**(6): 415–422.

Ivics, Z, Mátés L, Yau TY, *et al.* 2014c. Germline transgenesis in rodents by pronuclear microinjection of Sleeping Beauty transposons. *Nat Protoc* **9**(4): 773–793.

Izsvak, Z, Frohlich J, Grabundzija I, *et al.* 2010a. Generating knockout rats by transposon mutagenesis in spermatogonial stem cells. *Nat Methods* **7**(6): 443–445.

Izsvak Z, Hackett PB, Cooper LJ, Ivics Z. 2010b. Translating Sleeping Beauty transposition into cellular therapies: victories and challenges. *Bioessays* **32**(9): 756–767.

Izsvak Z, Ivics Z, Plasterk RH. 2000. Sleeping Beauty, a wide host-range transposon vector for genetic transformation in vertebrates. *J Mol Biol* **302**(1): 93–102.

Izsvak Z, Stuwe EE, Fiedler D, *et al.* 2004. Healing the wounds inflicted by Sleeping Beauty transposition by double-strand break repair in mammalian somatic cells. *Mol Cell* **13**(2): 279–290.

Jin Z, Maiti S, Huls H, *et al.* 2011 The hyperactive Sleeping Beauty transposase SB100X improves the genetic modification of T cells to express a chimeric antigen receptor. *Gene Ther* **18**(9): 849–856.

Johnen S, Djalali-Talab Y, Kazanskaya O, *et al.* 2015. Antiangiogenic and neurogenic activities of sleeping beauty-mediated PEDF-transfected RPE cells in vitro and in vivo. *BioMed Res Int* **2015**: 863845.

Johnen S, Izsvák Z, Stöcker M, *et al.* 2012. Sleeping Beauty transposon-mediated transfection of retinal and iris pigment epithelial cells. *Invest Ophthalmol Vis Sci* **53**(8): 4787–4796.

Katter K, Geurts AM, Hoffmann O, *et al.* 2013. Transposon-mediated transgenesis, transgenic rescue, and tissue-specific gene expression in rodents and rabbits. *FASEB J* **27**(3): 930–941.

Kebriaei P, Huls H, Jena B, *et al.* 2012. Infusing CD19-directed T cells to augment disease control in patients undergoing autologous hematopoietic stem-cell transplantation for advanced B-lymphoid malignancies. *Hum Gene Ther* **23**(5): 444–450.

Kebriaei P, Singh H, Huls MH, *et al.* 2016. Phase I trials using Sleeping Beauty to generate CD19-specific CAR T cells. *J Clin Invest* **126**(9): 3363–3376.

Keng VW, Villanueva A, Chiang DY, *et al.* 2009. A conditional transposon-based insertional mutagenesis screen for genes associated with mouse hepatocellular carcinoma. *Nat Biotechnol* **27**(3): 264–274.

Kren BT, Unger GM, Sjeklocha L, *et al.* 2009. Nanocapsule-delivered Sleeping Beauty mediates therapeutic Factor VIII expression in liver sinusoidal endothelial cells of hemophilia A mice. *J Clin Invest* **119**(7): 2086–2099.

Kues WA, Herrmann D, Barg-Kues B, *et al.* 2013. Derivation and characterization of Sleeping Beauty transposon-mediated porcine induced pluripotent stem cells. *Stem Cells Dev* **22**(1): 124–135.

Liu, L, Liu, H, Visner G, Fletcher BS. 2006a. Sleeping Beauty-mediated eNOS gene therapy attenuates monocrotaline-induced pulmonary hypertension in rats. *FASEB Journal* **20** (14): 2594–2596.

Liu L, Mah C, Fletcher B. 2006b. Sustained FVIII expression and phenotypic correction of hemophilia A in neonatal mice using an endothelial-targeted Sleeping Beauty transposon. *Mol Ther* **13**(5): 1006–1015.

Luo G, Ivics Z, Izsvák Z, Bradley A. 1998. Chromosomal transposition of a Tc1/mariner-like element in mouse embryonic stem cells. *Proc Natl Acad Sci USA* **95**(18): 10769–10773.

Luo W-Y, Shih Y-S, Hung C-L, *et al.* 2011. Development of the hybrid Sleeping Beauty-baculovirus vector for sustained gene expression and cancer therapy. *Gene Ther* **19**(8): 844–851.

Ma K, Wang DD, Lin Y, *et al.* 2013. Synergetic targeted delivery of Sleeping-Beauty transposon system to mesenchymal stem cells using LPD nanoparticles modified with a phage-displayed targeting peptide. *Adv Funct Mater* **23**(9): 1172–1181.

Mali P, Yang L, Esvelt KM, *et al.* 2013. RNA-guided human genome engineering via Cas9. *Science* **339**(6121): 823–826.

Mandal PK, Kazazian HH. 2008. SnapShot: vertebrate transposons. *Cell* **135**(1): 192–192.

Mátés L, Chuah MK, Belay E, *et al.* 2009. Molecular evolution of a novel hyperactive Sleeping Beauty transposase enables robust stable gene transfer in vertebrates. *Nat Genet* **41**(6): 753–761.

Mikkelsen JG, Yant SR, Meuse L, *et al.* 2003. Helper-independent Sleeping Beauty transposon-transposase vectors for efficient nonviral gene delivery and persistent gene expression in vivo. *Mol Ther* **8**(4): 654–665.

Moldt B, Miskey C, Staunstrup NH, *et al.* 2011. Comparative genomic integration profiling of Sleeping Beauty transposons mobilized with high efficacy from integrase-defective lentiviral vectors in primary human cells. *Mol Ther* **19**(8): 1499–1510.

Moldt B, Yant SR, Andersen PR, Kay MA, Mikkelsen JG. 2007. Cis-acting gene regulatory activities in the terminal regions of Sleeping Beauty DNA transposon-based vectors. *Hum Gene Ther* **18**(12): 1193–1204.

Molina-Estevez FJ, Lozano ML, Navarro S, *et al.* 2013. Brief report: impaired cell reprogramming in nonhomologous end joining deficient cells. *Stem Cells* **31**(8): 1726–1730.

Molyneux SD, Waterhouse PD, Shelton D, *et al.* 2014. Human somatic cell mutagenesis creates genetically tractable sarcomas. *Nat Genet* **46**(9): 964–972.

Monjezi R, Miskey C, Gogishvili T, *et al.* 2016. Enhanced CAR T-cell engineering using non-viral Sleeping Beauty transposition from minicircle vectors. *Leukemia* **31**(1): 186–194.

Montini E. 2002. In vivo correction of murine tyrosinemia type I by DNA-mediated transposition. *Mol Ther* **6**(6): 759–769.

Moriarity, BS, Largaespada, DA. 2015. Sleeping Beauty transposon insertional mutagenesis based mouse models for cancer gene discovery. *Curr Opin Genet Dev* **30**: 66–72.

Moriarity BS, Otto GM, Rahrmann EP, *et al.* 2015. A Sleeping Beauty forward genetic screen identifies new genes and pathways driving osteosarcoma development and metastasis. *Nat Genet* **47**(6): 615–624.

Muenthaisong S, Ujhelly O, Polgar Z, *et al.* 2012. Generation of mouse induced pluripotent stem cells from different genetic backgrounds using Sleeping beauty transposon mediated gene transfer. *Exp Cell Res* **318**(19): 2482–2489.

Munoz-Lopez M, Garcia-Perez JL. 2010. DNA transposons: nature and applications in genomics. *Curr Genom* **11**(2): 115–128.

Muses S, Morgan JE, Wells DJ. 2011. Restoration of dystrophin expression using the Sleeping Beauty transposon. *PLoS Curr* **3**: RRN1296.

Narayanavari SA, Chilkunda SS, Ivics Z, Izsvak Z. 2016. Sleeping Beauty transposition: from biology to applications. *Crit Rev Biochem Mol Bio* **52**(1): 18–44.

Nikitidou L, Torp M, Fjord-Larsen L, *et al.* 2014. Encapsulated galanin-producing cells attenuate focal epileptic seizures in the hippocampus. *Epilepsia* **55**(1): 167–174.

Ohlfest JR. 2005. Phenotypic correction and long-term expression of factor VIII in hemophilic mice by immunotolerization and nonviral gene transfer using the Sleeping Beauty transposon system. *Blood* **105**(7): 2691–2698.

Ortiz-Urda S, Lin Q, Yant SR, *et al.* 2003. Sustainable correction of junctional epidermolysis bullosa via transposon-mediated nonviral gene transfer. *Gene Ther* **10**(13): 1099–1104.

Padeken J, Zeller P, Gasser SM. 2015. Repeat DNA in genome organization and stability. *Curr Opin Genet Dev* **31**: 12–19.

Park J-S, Kim B-H, Park SG, *et al.* 2013. Induction of rat liver tumor using the Sleeping Beauty transposon and electroporation. *Biochem Biophys Res Commun* **434**(3): 589–593.

Perna D, Karreth FA, Rust AG, *et al.* 2015. BRAF inhibitor resistance mediated by the AKT pathway in an oncogenic BRAF mouse melanoma model. *Proc Natl Acad Sci USA* **112**(6): E536–E545.

Peterson EB, Mastrangelo MA, Federoff HJ, Bowers WJ. 2007. Neuronal specificity of HSV/Sleeping Beauty amplicon transduction in utero is driven primarily by tropism and cell type composition. *Mol Ther* **15**(10): 1848–1855.

Plasterk RH, Izsvak Z, Ivics Z. 1999. Resident aliens: the Tc1/mariner superfamily of transposable elements. *Trends Genet* **15**(8): 326–332.

Quintana RM, Dupuy AJ, Bravo A, *et al.* 2013. A transposon-based analysis of gene mutations related to skin cancer development. *J Invest Dermatol* **133**(1): 239–248.

Rahrmann EP, Collier LS, Knutson TP, *et al.* 2009. Identification of PDE4D as a proliferation promoting factor in prostate cancer using a Sleeping Beauty transposon-based somatic mutagenesis screen. *Cancer Res* **69**(10): 4388–4397.

Rahrmann EP, Watson AL, Keng VW, *et al.* 2013. Forward genetic screen for malignant peripheral nerve sheath tumor formation identifies new genes and pathways driving tumorigenesis. *Nat Genet* **45**(7): 756–766.

Ren J, Stroncek DF. 2016. Gene therapy simplified. *Blood* **128**(18): 2194–2195.

Richter M, Saydaminova K, Yumul R, *et al.* 2016. In vivo transduction of primitive hematopoietic stem cells after mobilization and intravenous injection of integrating adenovirus vectors. *Blood* **128**(18): 2206–2217.

Rostovskaya M, Fu J, Obst M, *et al.* 2012. Transposon-mediated BAC transgenesis in human ES cells. *Nucleic Acids Res* **40**(19): e150–e150.

Schröder ARW, Shinn P, Chen H, *et al.* 2002. HIV-1 integration in the human genome favors active genes and local hotspots. *Cell* **110**(4): 521–529.

Sebe A, Ivics Z. 2016. Reprogramming of human fibroblasts to induced pluripotent stem cells with Sleeping Beauty transposon-based stable gene delivery. *Methods Mol Biol* **1400**: 419–427.

Silva SD, Mastrangelo MA, Lotta LT, *et al.* 2009. Extending the transposable payload limit of Sleeping Beauty (SB) using the Herpes Simplex Virus (HSV)/SB amplicon-vector platform. *Gene Ther* **17**(3): 424–431.

Silva SD, Mastrangelo MA, Lotta LT, *et al.* 2010. Herpes simplex virus/Sleeping Beauty vector-based embryonic gene transfer using the HSB5 mutant: loss of apparent transposition hyperactivity in vivo. *Hum Gene Ther* **21**(11): 1603–1613.

Singh H, Huls H, Kebriaei P, Cooper LJN. 2014. A new approach to gene therapy using Sleeping Beauty to genetically modify clinical-grade T cells to target CD19. *Immunol Rev* **257**(1): 181–190.

Singh H, Moyes JS, Huls MH, Cooper LJ. 2015. Manufacture of T cells using the Sleeping Beauty system to enforce expression of a CD19-specific chimeric antigen receptor. *Cancer Gene Ther* **22**(2): 95–100.

Sinzelle L, Vallin J, Coen L, *et al.* 2006. Generation of trangenic *Xenopus laevis* using the Sleeping Beauty transposon system. *Transgenic Res* **15**(6): 751–760.

Starr TK, Allaei R, Silverstein KA, *et al.* 2009. A transposon-based genetic screen in mice identifies genes altered in colorectal cancer. *Science* **323**(5922): 1747–1750.

Staunstrup NH, Moldt B, Mátés L, *et al.* 2009. Hybrid lentivirus-transposon vectors with a random integration profile in human cells. *Mol Ther* **17**(7): 1205–1214.

Staunstrup NH, Sharma N, Bak RO, *et al.* 2011. A Sleeping Beauty DNA transposon-based genetic sensor for functional screening of vitamin D3 analogues. *BMC Biotechnol* **11**: 33.

Swierczek M, Izsvak Z, Ivics Z. 2012. The Sleeping Beauty transposon system for clinical applications. *Exp Opin Biol Ther* **12**(2): 139–153.

Szebenyi K, Furedi A, Kolacsek O, *et al.* 2015. Visualization of calcium dynamics in kidney proximal tubules. *J Am Soc Nephrol* **26**(11): 2731–2740.

Talluri TR, Herrmann D, Barg-Kues B, *et al.* 2013. Transposon-mediated reprogramming of livestock somatic cells to induced pluripotent stem cells. *Reprod Biol* **13**: 21.

Talluri TR, Kumar D, Glage S, *et al.* 2015. Derivation and characterization of bovine induced pluripotent stem cells by transposon-mediated reprogramming. *Cell Reprog* **17**(2): 131–140.

Turunen TA, Kurkipuro J, Heikura T, *et al.* 2016. Sleeping Beauty transposon vectors in liver-directed gene delivery of LDLR and VLDLR for gene therapy of familial hypercholesterolemia. *Mol Ther* **24**(3): 620–635.

Turunen TAK, Laakkonen J, Alasaarela L, Airenne KJ, Ylä-Herttuala S. 2014. Sleeping Beauty–baculovirus hybrid vectors for long-term gene expression in the eye. *J Gene Med* **16**(1–2): 40–53.

van der Weyden L, Giotopoulos G, Rust AG, *et al.* 2011. Modeling the evolution of ETV6-RUNX1-induced B-cell precursor acute lymphoblastic leukemia in mice. *Blood* **118**(4): 1041–1051.

Vigdal TJ, Kaufman CD, Izsvak Z, Voytas DF, Ivics Z. 2002. Common physical properties of DNA affecting target site selection of sleeping beauty and other Tc1/mariner transposable elements. *J Mol Biol* **323**(3): 441–452.

Vink CA, Gaspar HB, Gabriel R, *et al.* 2009. Sleeping Beauty transposition from nonintegrating lentivirus. *Mol Ther* **17**(7): 1197–1204.

Voigt F, Wiedemann L, Zuliani C, *et al.* 2016. Sleeping Beauty transposase structure allows rational design of hyperactive variants for genetic engineering. *Nat Commun* **7**: 11126.

Voigt K, Gogol-Doring A, Miskey C, *et al.* 2012. Retargeting sleeping beauty transposon insertions by engineered zinc finger DNA-binding domains. *Mol Ther* **20**(10): 1852–1862.

Walisko O, Schorn A, Rolfs F, *et al.* 2008. Transcriptional activities of the Sleeping Beauty transposon and shielding its genetic cargo with insulators. *Mol Ther* **16**(2): 359–369.

Wang J, Singh M, Sun C, *et al.* 2016. Isolation and cultivation of naive-like human pluripotent stem cells based on HERVH expression. *Nat Protoc* **11**(2): 327–346.

Wang, J, Xie G, Singh M, *et al.*, 2014. Primate-specific endogenous retrovirus-driven transcription defines naive-like stem cells. *Nature* **516**(7531): 405–409.

Wang X, Sarkar DP, Mani P, *et al.* 2009. Long-term reduction of jaundice in Gunn rats by nonviral liver-targeted delivery of Sleeping Beauty transposon. *Hepatology* **50**(3): 815–824.

Wang Y. 2016. Regulated complex assembly safeguards the fidelity of Sleeping Beauty transposition. *Nucleic Acids Res* **45**(1): 311–326.

Wang Y, Wang J, Devaraj A, *et al.* 2014. Suicidal autointegration of sleeping beauty and piggyBac transposons in eukaryotic cells. *PLoS Genet* **10**(3): e1004103.

Wilber A, Wangensteen KJ, Chen Y, *et al.* 2006. Correction of the murine model of hereditary tyrosinemia type I using messenger RNA as a source of transposase for Sleeping Beauty mediated integration of the FAH gene. *Mol Ther* **13**: S155–S156.

Wilber A, Wangensteen KJ, Chen Y, *et al.* 2007. Messenger RNA as a source of transposase for Sleeping Beauty transposon-mediated correction of hereditary tyrosinemia type I. *Mol Ther* **15**(7): 1280–1287.

Williams DA. 2008. Sleeping beauty vector system moves toward human trials in the United States. *Mol Ther* **16**(9): 1515–1516.

Wilson MH, Coates CJ, George AL. 2007. PiggyBac transposon-mediated gene transfer in human cells. *Mol Ther* **15**(1): 139–145.

Woodard LE, Wilson MH. 2015. piggyBac-ing models and new therapeutic strategies. *Trends Biotechnol* **33**(9): 525–533.

Wuestefeld T, Pesic M, Rudalska R, *et al.* 2013. A direct in vivo RNAi screen identifies MKK4 as a key regulator of liver regeneration. *Cell* **153**(2): 389–401.

Xiao J, Meng X-M, Huang XR, *et al.* 2012. miR-29 inhibits bleomycin-induced pulmonary fibrosis in mice. *Mol Ther* **20**(6): 1251–1260.

Yant SR, Ehrhardt A, Mikkelsen JG, *et al.* 2002. Transposition from a gutless adeno-transposon vector stabilizes transgene expression in vivo. *Nat Biotechnol* **20**(10): 999–1005.

Yant SR, Huang Y, Akache B, Kay MA. 2007. Site-directed transposon integration in human cells. *Nucleic Acids Res* **35**(7): e50.

Yant SR, Meuse L, Chiu W, *et al.* 2000. Somatic integration and long-term transgene expression in normal and haemophilic mice using a DNA transposon system. *Nat Genet* **25**(1): 35–41.

Zhang W, Solanlu M, Müther N, *et al.* 2013. Hybrid adeno-associated viral vectors utilizing transpose-mediated somatic integration for stable transgene expression in human cells. *PLoS One* **8**(10): e76771.

Zayed H, Izsvak Z, Walisko O, Ivics Z. 2004. Development of hyperactive sleeping beauty transposon vectors by mutational analysis. *Mol Ther* **9**(2): 292–304.

18 Genome Editing of Pluripotent Stem Cells

Zhili Rong and Yang Xu

18.1 Introduction

Human pluripotent stem cells (hPSCs) hold great promise for regenerative medicine due to their ability to self-renew indefinitely and to differentiate into any cell type in the body (De Los Angeles *et al.*, 2015). They can be classified into two major cell types, human embryonic stem cells (hESCs) and human induced pluripotent stem cells (hiPSCs). The former are derived from human blastocysts, and the latter are reprogrammed from somatic cells by introducing transcription factors and/or small chemical molecules (De Los Angeles *et al.*, 2015). With the unique properties of self-renewal and pluripotency, hPSCs could be broadly used in biomedical research and human therapy, such as mimicking human embryonic development, disease modeling, drug discovery and cell therapy (Avior *et al.*, 2016; Trounson and DeWitt, 2016).

To fully realize this potential, it is important to develop technologies for precise genome editing of hPSCs. Genome editing refers to "the process of making targeted modifications to the genome, its contexts (e.g. epigenetic marks), or its outputs (e.g. transcripts)"(Hsu *et al.*, 2014). The DNA sequence of the genome can be deleted, inserted or replaced, and, therefore, the function of specific genes and regulatory elements can be dissected (Figure 18.1). In addition, the epigenetic marks of chromatin could be targeted and modified without any changes in DNA sequence itself, and thus the function of specific post-transcriptional modifications could be analyzed, such as acetylation of histones and methylation of DNA (Figure 18.1). Genome editing technologies have advanced dramatically over recent years. The breakthrough of CRISPR (clustered regularly interspaced short palindromic repeats) technology has triggered a revolution in the field of genome editing and fundamentally improved our approach to biomedical research (Dominguez *et al.*, 2016; Doudna and Charpentier, 2014; Hsu *et al.*, 2014). In this chapter, we discuss the genome editing technologies used in hPSCs and their contribution to dissecting the basic biological properties and realizing the potential of hPSCs in regenerative therapy.

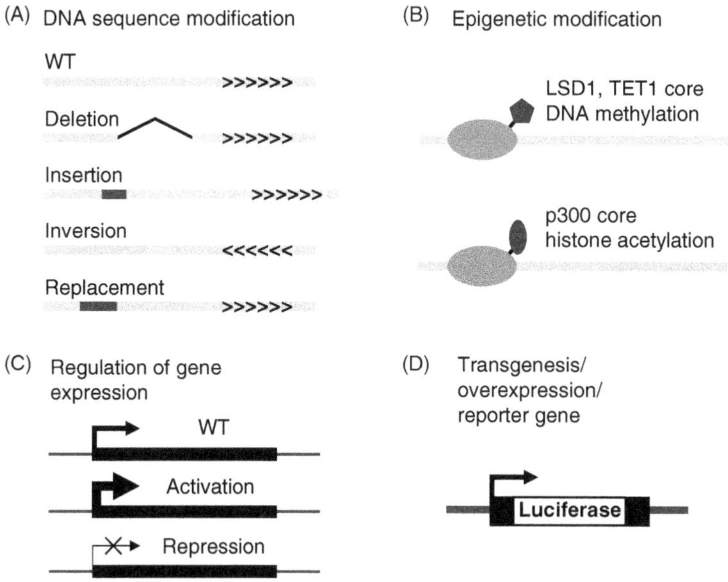

Figure 18.1 Genome editing can change DNA sequence and epigenetic marks. (A) Genome editing can modify DNA primary sequence and lead to deletion, insertion, inversion and replacement. (B) dCas9 and the DNA-binding domain of ZFN and TALEN can be fused with LSD1 or the catalytic domain of TET1 to change DNA methylation status, or with the core catalytic domain of p300 to change histone acetylation. (C) Modification of both DNA sequence and epigenetic marks can be harnessed to regulate the expression of endogenous genes, such as activating gene expression, repressing gene expression or completely disrupting a gene. (D) Gene expression cassette can be targeted into a specific genomic locus for the expression of the transgenic gene.

18.2 Genome Editing Technology in Pluripotent Stem Cells

18.2.1 Double-stranded DNA Break and Genome Editing

How does genome editing work in cells? Double-stranded DNA break (DSB) repair pathways are the basis of current genome editing methods. DSBs spontaneously occur in cells and are generally repaired by two pathways, the non-homologous end joining (NHEJ) pathway in an error-prone fashion and the homologous recombination (HR) pathway in a high-fidelity fashion (Symington and Gautier, 2011). In NHEJ, the DSB ends are occupied by the Ku70-Ku80 heterodimer and directly ligated, frequently resulting in insertions, deletions, substitutions and translocations (when two ends of different genomic locus are ligated) at the join (Figure 18.2). In contrast, HR is initiated with DSB end resection by nucleases and helicases, generating 3′ single-stranded DNA overhangs onto which the RAD51 recombinase assembles as a nucleoprotein filament. The filament can invade a homologous DNA template to form intermediates, which are resolved to produce accurately repaired DNA fragments. NHEJ is active throughout the cell cycle and is favored in the G1 phase, while HR is more prevalent in the S and G2 phases due to the presence of an identical sister chromatid as a repair template.

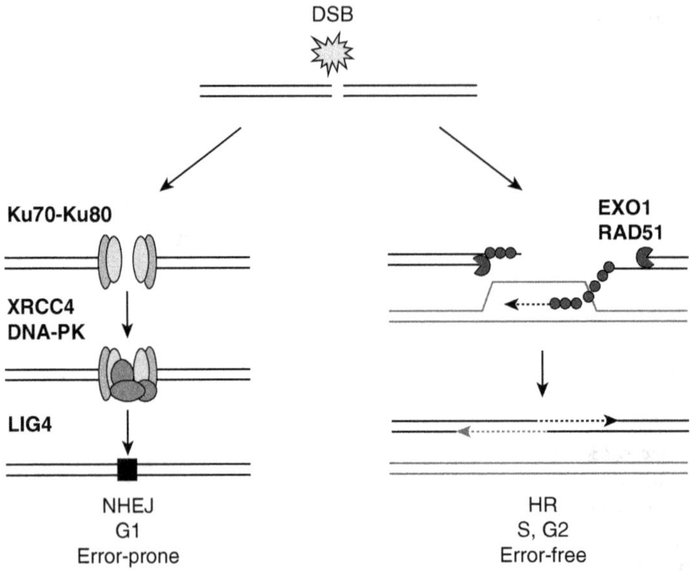

Figure 18.2 DNA double-strand break (DSB) repair pathways. DNA DSB damage in mammalian cells is primarily repaired by non-homologous end joining (NHEJ) and homologous recombination (HR). NHEJ is initiated by the Ku70-Ku80 heterodimer binding to the DNA ends, processed with XRCC4 and DNA-PK(Ku70/Ku80/DNA-PKcs) and finalized with the ligation of DNA ends by ligase 4 (LIG4). HR is initiated by extensive resection of DNA ends by EXO1, processed with RAD51 filament formation and other events, and finalized with ligation and resolution of the heteroduplex by a crossover or a non-crossover event. NHEJ is used primarily in the G1 phase of the cell cycle in an error-prone fashion, and HR is used primarily in the S and G2 phase and error-free, following replication of the genome.

18.2.2 Genome Editing Approaches

Five genome editing approaches have been successfully applied in hPSCs with different levels of efficiency, including conventional gene targeting, BAC (bacterial artificial chromosome)-based targeting, ZFNs (zinc-finger nucleases), TALENs (transcription activator-like effector nucleases) and the CRISPR/Cas9 system (Figure 18.3).

18.2.2.1 Conventional Gene Targeting

Distinct from most other mammalian cells, mouse embryonic stem cells (mESCs) display high activity of the HR DNA repair pathway and therefore allow genetic manipulation by homologous recombination (Capecchi, 1989; Koller and Smithies, 1992). Conventional gene targeting is the initial method capable of providing conclusive information on gene function. However, the use of this approach has been limited by several factors, especially the extremely low efficiency (1 in 10^6–10^9cells) in gene targeting, and thus requires time-consuming and labor-intensive screening steps to identify the correctly modified cells (Capecchi, 1989). This targeting approach in hPSCs has proven to be even less efficient than in mESCs, and thus is not adaptable for routine application. Therefore, only a few successful examples of conventional gene targeting in

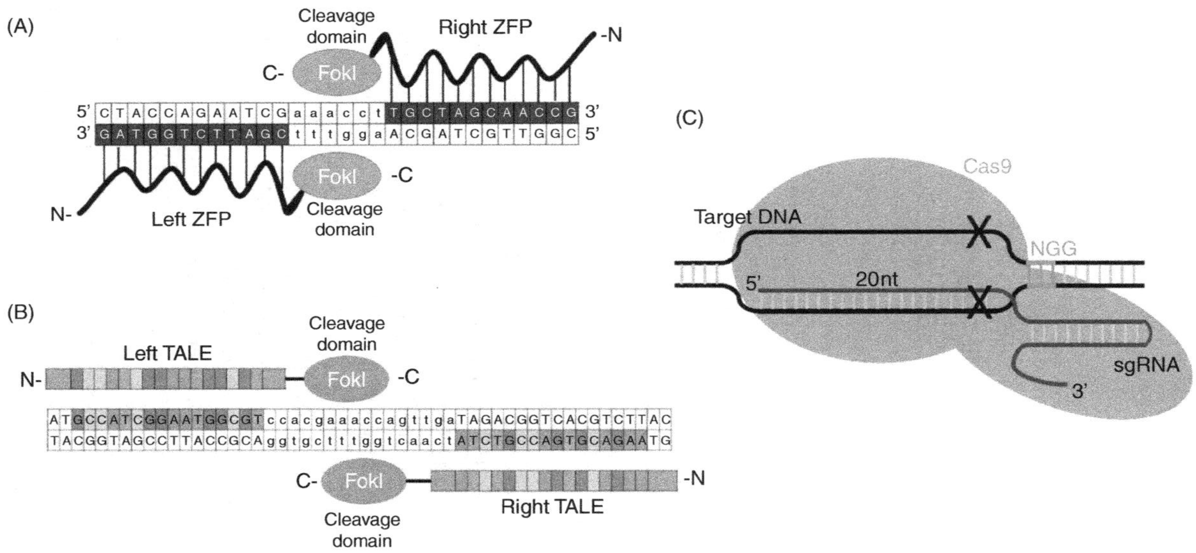

Figure 18.3 Recognition and cleavage of DNA by ZFNs, TALENs and CRISPR/Cas9 systems. (A) A schematic representation of a ZFN pair bound to a DNA target. To increase the specificity, a pair of ZFNs are typically designed to recognize the flanking arms of the cutting site. Each ZFN contains a zinc-finger domain at the N-terminus and a FokI cleavage domain at the C-terminus. Only when the pair of ZFNs simultaneously bind to the target site, do the FokI cleavage domains form a dimer to cut the 5–7 bp spacer sequence between the two binding sites. An individual zinc-finger recognizes a 3 bp DNA sequence and a pair of ZFNs are generally designed to recognize an 18–36 bp DNA fragment, excluding the spacer. (B) A schematic representation of a TALEN pair bound to a DNA target. Similar to ZFNs, a TALEN pair is required to ensure recognition with high specificity, and the FokI cleavage domain is fused to a TALE DNA-binding domain. Different from ZFNs, an individual TAL repeat recognizes a single base pair and a pair of TALENs are typically designed to recognize a 30–40 bp DNA sequence, excluding the 12–20 bp spacer. (C) A schematic representation of the Cas9/gRNA complex bound to a DNA target. The 20 nt sequence at the 5′ end of the gRNA directs the Cas9/gRNA complex to the DNA target site by base paring. An "NGG" protospacer-adjacent motif (PAM) sequence at the 3′ end of the 20 nt DNA target site is required for recognition. Once tightly bound, Cas9 can cut both strands of DNA 3 bp upstream of the PAM.

hPSCs have been reported, and sometimes require more robust screening strategies (Davis *et al.*, 2008;, Xue *et al.*, 2009; Zwaka and Thomson, 2003).

To increase targeting efficiency, adeno-associated virus (AAV), which can be modified to be non-integrating and maintained as episomal DNA in the cell, is used to replace the conventional plasmid delivery approach to deliver the DNA template for HR. With this approach, *HPRT* and *HMGA1* (high-mobility-group AT-hook) genes can be targeted in hESCs and hiPSCs with over 1% targeting efficiency (Khan *et al.*, 2010; 2011). Helper-dependent adenoviral vectors (HDAdVs) have also been developed to deliver large DNA templates up to 37 kb to correct laminopathy-associated LMNA mutations in patient-specific iPSCs (Liu *et al.*, 2011). Compared with the plasmid-based delivery approach, the viral vector-based approach appears to increase the targeting efficiency. However, virus preparation and handling is a time-consuming and complicated process, and the use of this approach is restricted to a few labs.

18.2.2.2 BAC (Bacterial Artificial Chromosome)-based Gene Targeting

In the conventional targeting approach, the DSBs at the editing site occur spontaneously. If more DSBs occur, the targeting efficiency could reduce, in principle. To this end, a BAC is used as a DNA repair template, as it can contain long homologous arms at both sides (the size of a BAC can be up to ~200kb) and therefore incidence of DSBs on the arms could increase as well. In support of this notion, the BAC-based approach can achieve high frequency of gene targeting via homologous recombination in hESCs and hiPSCs to dissect the function of genes, establish reporter cell lines and ectopically express genes at specific loci. This approach led to the first reported homozygous knockout hESCs. In this context, by sequentially disrupting both alleles of ATM (ataxia telangiectasia mutated) with BAC-targeting vectors, we established ATM(–/–) hESCs as cell models for the major human genetic instability syndrome ataxia telangiectasia (Song *et al.*, 2010). By generating p53(–/–) hESCs, we revealed that p53 plays a critical role in maintaining the genome stability of hESCs (Song *et al.*, 2010). By generating p53 acetylation site knockin mutant hESCs, we demonstrated that acetylation of human p53 is critical for activating p53-dependent DNA damage response (Chung *et al.*, 2014). By knocking in the *mCherry* gene into the *NESTIN* locus of hESCs, we could track differentiation of hESCs towards neural progenitor cells (Fu *et al.*, 2014a). By inserting a suicide gene thymidine kinase (TK) into the *NANOG* locus, we could eliminate hESCs remaining among differentiated cells and thus decrease teratoma formation risk (Rong *et al.*, 2012). By knocking in *CTLA4-Ig* and *PD-L1* genes into the *HPRT* locus, we demonstrated that the genetically modified hESCs and their derivatives could evade immune rejection when transplanted into an immune system humanized mouse model (Rong *et al.*, 2014). Compared with the conventional targeting approach, the BAC strategy could increase the targeting efficiency up to 15% with an average around 5%, which guarantees feasibility for the complex genomic editing in hPSCs. However, the construction of the BAC targeting vector is even more complicated and time-consuming than the conventional gene-targeting vector.

18.2.2.3 Zinc-finger Nucleases

Zinc-finger nucleases (ZFNs) contain two domains, a programmable, sequence-specific DNA-binding domain and a non-specific DNA cleavage domain (Bibikova *et al.*, 2003; Porteus and Baltimore, 2003; Urnov *et al.*, 2010). The Cys2-His2 zinc-finger domain is one of the most common types of DNA-binding motifs found in eukaryotes. An individual zinc-finger consists of approximately 30 amino acids and binds a 3 bp DNA sequence with varying levels of selectivity. Individual zinc-fingers can be arranged in tandem to form the DNA-binding domain, binding to an extended DNA sequence of around 18 bps, which is unique within a genome. The nuclease domain of the FokI restriction enzyme is generally used to cleave DNA (Urnov *et al.*, 2010). Since the FokI nuclease domain must dimerize to activate the catalytic activity, ZFNs are typically designed in pairs that recognize sequences flanking the target site of interest. Instead of utilizing spontaneous DSBs for homologous recombination, as in the conventional approach, ZFNs can introduce DSBs at the target site, thus increasing targeting efficiency, and also permit selection of cells repaired through both the NHEJ and HR pathways. NHEJ can introduce frameshifts into the coding sequence of a gene, resulting in premature truncations that effectively knock out the gene. As NHEJ is the major repair pathway, ZFNs greatly improve the knockout efficiency. For the HR repair pathway, a double-stranded DNA fragment or a single-stranded DNA oligonucleotide can serve as the exogenous repair template. In the case of the latter, homology arms of as short as 20 nucleotides in length are sufficient to knock in DNA fragments into hPSCs, which dramatically simplifies the cloning process for the targeting DNA template (Soldner *et al.*, 2011). Specific recognition of a unique site throughout the whole genome is the obvious advantage of ZFNs over previous approaches, and most importantly, expands the field of genome editing to epigenetics and gene expression manipulations. ZFNs can be used to change genomic structure and function at the site of interest when the DNA-binding domain is fused with different effector domains, such as transcriptional activators and repressors, recombinases, transposases, DNA and histone methyltransferases, and histone acetyltransferases (Beerli *et al.*, 1998; Gersbach *et al.*, 2011; Thakore *et al.*, 2016; Yant *et al.*, 2007). In addition, the successful development of ZFNs makes it possible to edit genomes in cells other than pluripotent stem cells and organisms, which could not be achieved with previous approaches (Carroll, 2014). Despite the advantages described above, ZFNs have proven to be technically difficult for investigators to engineer from scratch because it is labor-intensive to design and validate zinc-finger motifs that can bind a DNA sequence of interest with high affinity and specificity (Bhakta *et al.*, 2013; Gupta *et al.*, 2012; Maeder *et al.*, 2008; Ramirez *et al.*, 2008; Sander *et al.*, 2011).

18.2.2.4 Transcription Activator-like Effector Nucleases

Similar to ZFNs, transcription activator-like effector nucleases (TALENs) contain a reprogrammable DNA-binding domain and a nuclease effector domain (Carroll, 2014). The difference between ZFNs and TALENs is the nature of the DNA-binding domain. In TALENs, the basic DNA-binding module is a TAL repeat with a 1-repeat to 1 bp correspondence. Each TAL repeat has 33–35 amino acids, with two adjacent amino acids (RVD, the repeat-variable di-residue) conferring specificity for a single base pair (Boch *et al.*, 2009; Moscou and Bogdanove, 2009). A tandem array with 15–25 repeats can bind corresponding contiguous DNA sequences 15–25 bp in length. As TALENs and ZFNs share identical component structure, a DNA-binding domain plus an effector domain, they function in a similar way to generate knockout/knockin mutations and to change local genome structure and functions (Carroll, 2014; Thakore *et al.*, 2016). Compared to ZFNs, TALENs are far easier to engineer. TAL repeats can be designed to bind DNA sequences of interest with high success rates, and thus TALENs can be designed and constructed in a short time. However, the TAL repeats are almost identical to each other except for the RVD, which makes them hard to construct.

18.2.2.5 CRISPR/Cas9 Systems

The CRISPR/Cas9 system is adapted from archaea and bacterial adaptive immune systems known as clustered regularly interspaced short palindromic repeats (CRISPRs) and CRISPR-associated (Cas) systems (Doudna and Charpentier, 2014). In the type II CRISPR/Cas system, CRISPR locus encodes acquired "spacers" (a short fragment of foreign DNA incorporated from invading phage and plasmid DNA) and repeat sequences following each spacer. Transcription of the whole CRISPR locus yields a pre-crRNA, which is processed to yield crRNAs consisting of spacer-repeat fragments, several Cas proteins (including Cas9 protein) and a *trans*-activating crRNA (tracrRNA). After processing, a mature complex consisting of Cas9/tracrRNA/crRNA can locate a target dsDNA sequence and cut both strands. Target recognition is achieved by sequence complementation between the spacer in crRNA and the target "protospacer" sequence, as well as the presence of a protospacer-adjacent motif (PAM) sequence at the 3′ end of the protospacer sequence. DNA cleavage is achieved by Cas9, which is a DNA nuclease and cuts the target sequence 3 bp upstream of the PAM.

Currently, the CRISPR/Cas9 system has become the most popular approach to edit genomes. A typical engineered CRISPR/Cas9 system contains two components, a Cas9 protein and a guide RNA (an RNA engineered by joining tracrRNA and crRNA), which can form a complex to recognize and cleave a DNA sequence (Hsu *et al.*, 2014). Distinct from ZFNs and TALENs, which utilize DNA-binding motifs to achieve DNA target specificity, CRISPR/Cas9 system utilizes RNA to recognize the DNA sequence of interest. Thus, the CRISPR/Cas9 system can be easily adapted to target a genomic sequence by simply changing the guide RNA (gRNA). This feature makes the CRISPR/Cas9 system much easier to engineer than ZFNs and TALENs. It also makes it possible to simultaneously target multiple genomic sites in the same cell and even to generate a library targeting numerous sites in the genome to screen drugs, and to decipher gene functions and pathways (Shalem *et al.*, 2015). Similar to ZFNs and TALENs, when the Cas9 protein is mutated to a nuclease-dead version, dCAs9, the CRISPR/Cas9 system can be used to change genome local structure and function by fusing dCas9 with various enzymes (Dominguez *et al.*, 2016; Thakore *et al.*, 2016). When the Cas9 protein is mutated to a nickase version, it can introduce single-stranded DNA break at the genomic site of interest (Ran *et al.*, 2013). More importantly, besides the Cas9 protein, the gRNA could be engineered to carry RNA aptamers or fused to long non-coding RNAs, thus greatly expanding the utility of the CRISPR/Cas9 system in biological research (Konermann *et al.*, 2015; Shechner *et al.*, 2015; Zalatan *et al.*, 2015).

With the rapid development of CRISPR/Cas9 technology, more and more Cas9 proteins have been identified from natural species, such as *Streptococcus pyogenes, Streptococcus thermophiles, Neisseria meningitides* and *Treponema denticola*. The variant Cas9 proteins are of different sizes and require different PAMs, and thus could be more convenient to handle and cover more regions in a given genome (Barrangou and Doudna, 2016). In addition, with the engineering of Cas9 protein, novel artificial Cas9 proteins are endowed with new properties, such as different PAM requirements, inducible activity and higher fidelity (Dominguez *et al.*, 2016). Furthermore, proteins with similar functions to Cas9 have been identified, such as Cpf1, C2c1 and C2c2; the first two can recognize and cut DNA, while the last can recognize and cut RNA (Abudayyeh *et al.*, 2016; East-Seletsky *et al.*, 2016; Shmakov *et al.*, 2015; Zetsche *et al.*, 2015).

When the CRSIPR/Cas9 system was initially reported, the specificity of this system was challenged. Several strategies have been developed to improve the specificity, including Cas9n nickase to double nick DNA, generating double-stranded breaks (Ran *et al.*, 2013), a pair of dCas9/FokI fusion nucleases (Guilinger *et al.*, 2014; Tsai *et al.*, 2014), truncated guide RNAs (Fu *et al.*, 2014b), a mutated Cas9 with weakened DNA-binding affinity fused to a zinc-finger motif-based or a TALE repeat-based DNA-binding domain (Bolukbasi *et al.*, 2015) and mutated Cas9 proteins with higher fidelity (Kleinstiver *et al.*, 2016; Slaymaker *et al.*, 2016).

In summary, the CRISPR/Cas system is currently the most simple and powerful genome editing and regulation tool with high efficacy, specificity and versatility. However, a bottleneck that remains for all the genome editing tools is the low efficiency of precise editing in hPSCs.

18.3 Application of Genome Editing Technology in Pluripotent Stem Cells

18.3.1 Disease Modeling

Despite the extensive effort to develop animal models (e.g. zebrafish and mouse) to mimic human diseases, these models cannot fully recapitulate the pathology of human diseases. As hPSCs could be differentiated into cell types that are difficult to obtain from patients as primary cells (e.g. neurons), they have been widely used to model human diseases to reveal fundamental mechanisms underlining the pathogenesis of the diseases (Avior *et al.*, 2016). In early reports, human iPSCs derived from patients and healthy donors were differentiated into disease-related cell types, and the different phenotypes observed were claimed to be relevant to diseases. These studies are potentially flawed because they fail to account for possible confounders that might be responsible for the phenotypic differences, such as genetic background, age and gender. The most rigorous comparison would be between cell lines that differ only in the disease mutations and are otherwise isogenic (Figure 18.4). For example, correction of the G2019S mutation of the *LRRK2* gene in iPSCs derived from a Parkinson's disease (PD) patient could rescue the nuclear aberration phenotype in iPSC-derived neural stem cells, while introduction of the same mutation into wild-type hESCs could induce the same phenotype as the patient's iPSCs, establishing the necessity and sufficiency of the G2019S mutation for the disease phenotype (Liu *et al.*, 2012). Therefore, an ideal disease for modeling is one that arises from a single gene mutation (monogenic disorder), shows high penetrance and is associated with a cell-autonomous phenotype (Avior *et al.*, 2016). These characteristics are generally sufficient to predict an *in vitro* observable phenotype in PSC-derived cells. For diseases resulting from mutations of multiple genes (polygenic disorders, e.g. autism spectrum disorder), it has been technically difficult to correct all the genes in patient-derived iPSCs or to introduce all the mutations in wild-type PSCs, making it impossible to study the necessity and sufficiency of the mutations for the disease phenotype. Fortunately, recent developments in the CRISPR/Cas9 technology have enabled the simultaneous genetic modification of several genes in a single cell and thus could mitigate this hurdle in using hPSCs in disease modeling (Hsu *et al.*, 2014).

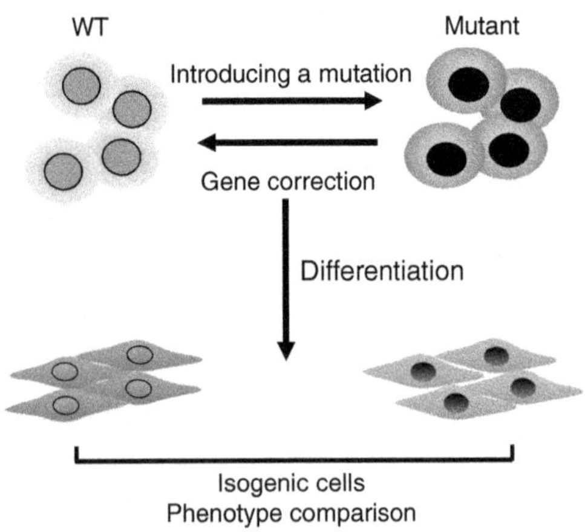

Figure 18.4 Isogenic cells created by genome editing are critical for disease modeling and drug discovery. The mutations in iPSCs derived from the patient can be corrected via genome editing, while the same mutation could be introduced into WT iPSCs derived from healthy donors or WT hESCs. The resulting genetically modified hPSCs differ from their parental cells exclusively at the edited locus and are otherwise isogenic. Disease-specific cellular pathological phenotypes found in the cell types differentiated from these isogenic cell lines can be directly attributed to the mutation. If the correction of the mutation in patient iPSCs rescues disease-specific phenotypes, it demonstrates the necessity of the mutation for the disease. If the introduction of the mutation in WT hiPSCs or hESCs leads to disease-specific phenotypes, it indicates the sufficiency of the mutation for the disease.

An obvious disadvantage of disease modeling with hPSCs (disease-in-a-dish) is its inability to reproduce the complex physiological conditions in an individual. To overcome this hurdle, co-culture systems consisting of various cell types and organoid cultures (mini-organs) have been developed to investigate cell–cell interactions and cell non-autonomous functions (Lancaster and Knoblich, 2014). Organoid cultures are small functional tissue units consisting of distinct cell types, and to some extent are organized and function similarly to the normal organ, and thus could model some features of normal tissues. An alternative strategy would be to orthotopically incorporate hPSC-derived cells into animals to construct "humanized models," such as transplanting hPSC-derived hepatocytes into damaged mouse liver and interrogating the effects of human genetic variation in the context of mouse physiological conditions (Yusa *et al.*, 2011).

18.3.2 Drug Discovery

In addition to dissecting the mechanisms underlining the pathogenesis of diseases, hPSC-based disease models could be used to develop therapies, e.g. drug discovery (Figure 18.5). To this end, a reliable readout should be defined first, which should be the disease-relevant phenotypes to be rescued in a specific cell type derived from hPSCs. There are two main strategies to identify potential drugs: the candidate drug approach and the high-throughput screening (HTS) approach (Avior *et al.*, 2016). In the former approach, a small group of defined molecules and compounds are tested on disease-affected cells and isogenic control cells. These drug candidates could be selected based on the acquired knowledge of mechanisms underlying the treated phenotypes, or ones that have been used to treat similar diseases in other models. In principle, this approach could be applied in personalized or precision medicine. Through screening existing drugs, those most effective for the particular patient could be identified using the patient's iPSC-based cell model.

In contrast, HTS can test a large number of compounds, using many more cells. In the HTS approach, the treated phenotype must be automatically measured and quantified, which is generally limited to simple and straightforward phenotypes, such as cell viability or protein expression. Additionally, the requirement of a large number of hPSC-derived cells could be an obstacle for certain cell types, due to the lack of a robust differentiation processes.

Recently, some prescription drugs have been withdrawn from clinical use or trial because of severe side effects of cardiotoxicity or hepatic toxicity, highlighting the need for more reliable assays to test cardiotoxicity in drug discovery. In addition to their use in testing drug efficacy, hPSC-derived cells such as cardiomyocytes and hepatocytes can be used to test drug toxicity. For example, hPSC-derived cardiomyocytes could be used to test drug cardiotoxicity and hPSC-derived cardiomyocytes also have the potential to fulfill this need (Braam *et al.*, 2010). One of the major reasons for the failure to discover drugs using cancer cell lines is the lack of a proper and vigorous control cell line, and thus the drug candidates generally show low specificity, killing both affected cells and healthy cells. In principle, isogenic

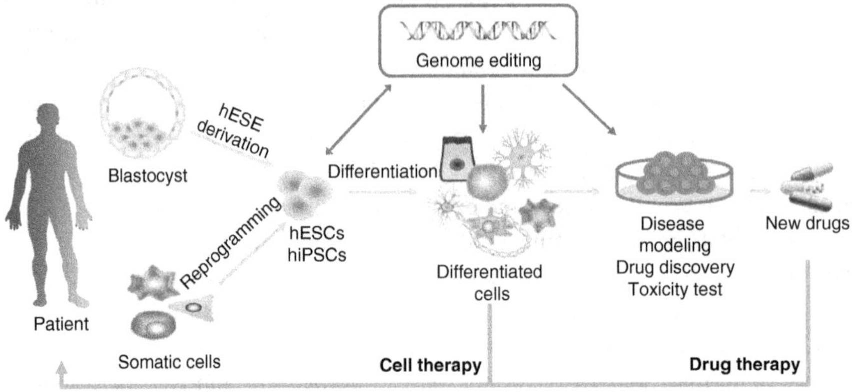

Figure 18.5 Clinic application of genome editing in hPSCs. In drug discovery, genetically modified hESCs and hiPSCs are differentiated to generate lineage-specific cells, which can be used in disease modeling to dissect the mechanisms underlying disease phenotypes, in drug screening to identify drug candidates for human diseases and in cardiac and liver toxicity testing to assess drug safety. In cell therapy, genetically modified hESCs and hiPSCs can be differentiated into functional cells to repair damaged tissues. In this context, genome editing can help to derive specialized cells from hPSCs for cell therapy with increased efficacy and safety. (A black-and-white version of this figure will appear in some formats. For the color version, please refer to the plate section.)

cells created with genome editing in hPSCs would help to eliminate those candidates with low specificity and thus facilitate the drug discovery process.

18.3.3 Cell Therapy

In addition to disease modeling and drug discovery, genome editing in hPSCs has great potential in cell therapy (Figure 18.5). For genetic disorders, particularly monogenic disorders, genome editing could permanently eliminate the symptoms of the diseases. For example, the CRISPR/Cas9 system has been used to correct a metabolic liver disease in mice (Yang *et al.*, 2016) and ZFN-mediated CCR5 gene disruption in autologous T cells to treat HIV patients has been advanced to clinic trial (Tebas *et al.*, 2014). Currently, several hPSC-based cell therapies have entered clinic trials, exemplified by hESC-derived retinal pigment epithelium to treat macular degeneration and Stargardt's macular dystrophy in Phase 1/2 clinic trials (Schwartz *et al.*, 2015). Therefore, genome editing could correct disease mutations in the iPSCs derived from patients, and the modified iPSCs could be differentiated into the cell types that are affected in patients in order to treat the disease.

18.3.4 Other Applications for Genome Editing in hPSCs

Genome editing not only advances the clinic application of hPSCs through creating novel and powerful human disease models and drug screening platforms, but it also provides an unprecedented opportunity to study the fundamental principles of cell biology and human early embryonic development. The basic principles of cell biology are generally first discovered in experimental animals and later confirmed in human cell lines. However, for studies with human cell lines, aberrant cancer cell lines are mostly used. One of the obvious disadvantages of cancer cell lines is that they have unstable and poorly defined genomes, which could have dramatic effects on cellular behavior. Another is that cancer cell line is pathologic and unable to reproduce phenotypes in normal tissue cells. Therefore, normal cells or tissues derived from hPSCs could serve as more physiologically relevant platforms for human biology research. Previously, overexpression and knockdown with shRNAs have been the most popular methods to investigate the functions of particular proteins. Overexpression experiments can lead to artificial phenotypes, and knockdown experiments might result in incomplete elimination of gene expression and repression of off-target genes. Genome editing in hPSCs is able to insert a tag to a protein in the genome and therefore investigate the protein functions at physiological expression levels. Disruption of genes in the genome could eliminate their expression to reach more reliable conclusions. For the areas of research where biological processes (e.g. neural biology) diverge between human and other species, the hPSC-based platforms for cell biology research are particularly important. For studies of human early embryonic development, there were few models before hPSCs, and hPSC-based cell models can be successfully established and used to explore fundamental mechanisms and pivotal genes that govern the early development of human embryos. For example, several genes have been proven to be essential for human pancreas development using genome editing in hPSCs (Zhu *et al.*, 2016).

18.4 Challenges and Concluding Remarks

While the application of genome editing technologies in human pluripotent stem cells has achieved significant progress, several challenges remain that hinder the broader use of genome editing. One major challenge is the low efficiency of precise genome editing in hPSCs. As hPSCs prefer to use the NHEJ rather than the HDR pathway to repair DNA DSB damage, editing events often lead to insertions and deletions rather than the intended precise modification. However, to create an isogenic cell line in disease modeling, or in most therapeutic settings, the generation of a modified allele with high frequency will be essential. Despite the efficiency of precise editing being increased by several editing technologies, it is currently not able to fulfill the need for clinic applications.

Another challenge is the lack of robust differentiation of hPSCs into functional cell types. For most cell types, no protocol has been developed for directed differentiation, or the efficiency of differentiation is low. For others, current differentiation strategies generally yield embryonic rather than adult human cell types. Therefore, the development of highly reproducible protocols to differentiate hPSCs into more mature cell types is the major task in the future.

Acknowledgement

This work is supported by the National Natural Science Foundation of China (81670093, 81372494, 81430032, U1601222), Thousand Young Talents Program of China, the Program of the Guangdong Provincial Department of Science and Technology (2014B020212018), the Program of the Guangzhou Science Technology and Innovation Commission (201508020120), the Guangdong Provincial Key Laboratory of Cancer Immunotherapy, the Guangdong Province Key Special Science and Technology Project (2015B020225004) and the South Wisdom Valley Innovative Research Team Program (2014) No. 365.

REFERENCES

Abudayyeh OO, Gootenberg JS, Konermann S, *et al.* 2016. C2c2 is a single-component programmable RNA-guided RNA-targeting CRISPR effector. *Science* **353**: aaf5573.

Avior Y, Sagi I, Benvenisty N. 2016. Pluripotent stem cells in disease modelling and drug discovery. *Nat Rev Mol Cell Biol* **17**: 170–182.

Barrangou R, Doudna JA. 2016. Applications of CRISPR technologies in research and beyond. *Nat Biotech* **34**: 933–941.

Beerli RR, Segal DJ, Dreier B, Barbas CF III. 1998. Toward controlling gene expression at will: specific regulation of the erbB-2/HER-2 promoter by using polydactyl zinc finger proteins constructed from modular building blocks. *Proc Natl Acad Sci USA* **95**: 14628–14633.

Bhakta MS, Henry IM, Ousterout DG, *et al.* 2013. Highly active zinc-finger nucleases by extended modular assembly. *Genome Res* **23**: 530–538.

Bibikova M, Beumer K, Trautman JK, Carroll D. 2003. Enhancing gene targeting with designed zinc finger nucleases. *Science* **300**: 764.

Boch J, Scholze H, Schornack S, *et al.* 2009. Breaking the code of DNA binding specificity of TAL-type III effectors. *Science* **326**: 1509–1512.

Bolukbasi MF, Gupta A, Oikemus S, *et al.* 2015. DNA-binding-domain fusions enhance the targeting range and precision of Cas9. *Nat Methods* **12**: 1150–1156.

Braam SR, Tertoolen L, Van De Stolpe A, *et al.* 2010. Prediction of drug-induced cardiotoxicity using human embryonic stem cell-derived cardiomyocytes. *Stem Cell Res* **4**: 107–116.

Capecchi MR. 1989. Altering the genome by homologous recombination. *Science* **244**: 1288–1292.

Carroll D. 2014. Genome engineering with targetable nucleases. *Annu Rev Biochem* **83**: 409–439.

Chung SK, Zhu S, Xu Y, Fu X. 2014. Functional analysis of the acetylation of human p53 in DNA damage responses. *Protein Cell* **5**: 544–551.

Davis RP, Ng ES, Costa M, *et al.* 2008. Targeting a GFP reporter gene to the MIXL1 locus of human embryonic stem cells identifies human primitive streak-like cells and enables isolation of primitive hematopoietic precursors. *Blood* **111**: 1876–1884.

De Los Angeles A, Ferrari F, Xi R, *et al.* 2015. Hallmarks of pluripotency. *Nature* **525**: 469–478.

Dominguez AA, Lim WA, Qi LS. 2016. Beyond editing: repurposing CRISPR-Cas9 for precision genome regulation and interrogation. *Nat Rev Mol Cell Biol* **17**: 5–15.

Doudna JA, Charpentier E. 2014. Genome editing: the new frontier of genome engineering with CRISPR-Cas9. *Science* **346**: 1258096.

East-Seletsky A, O'Connell MR, Knight SC, *et al.* 2016. Two distinct RNase activities of CRISPR-C2c2 enable guide-RNA processing and RNA detection. *Nature* **538**(7624): 270–273.

Fu X, Rong Z, Zhu S, *et al.* 2014a. Genetic approach to track neural cell fate decisions using human embryonic stem cells. *Protein Cell* **5**: 69–79.

Fu Y, Sander JD, Reyon D, Cascio VM, Joung JK. 2014b. Improving CRISPR-Cas nuclease specificity using truncated guide RNAs. *Nat Biotechnol* **32**: 279–284.

Gersbach CA, Gaj T, Gordley RM, Mercer AC, Barbas CF III. 2011. Targeted plasmid integration into the human genome by an engineered zinc-finger recombinase. *Nucleic Acids Res* **39**: 7868–7878.

Guilinger JP, Thompson DB, Liu DR. 2014. Fusion of catalytically inactive Cas9 to FokI nuclease improves the specificity of genome modification. *Nat Biotechnol* **32**: 577–582.

Gupta A, Christensen RG, Rayla AL, *et al.* 2012. An optimized two-finger archive for ZFN-mediated gene targeting. *Nat Methods* **9**: 588–590.

Hsu PD, Lander ES, Zhang F. 2014. Development and applications of CRISPR-Cas9 for genome engineering. *Cell* **157**: 1262–1278.

Khan IF, Hirata RK, Russell DW. 2011. AAV-mediated gene targeting methods for human cells. *Nat Protoc* **6**: 482–501.

Khan IF, Hirata RK, Wang PR, *et al.* 2010. Engineering of human pluripotent stem cells by AAV-mediated gene targeting. *Mol Ther* **18**: 1192–1199.

Kleinstiver BP, Pattanayak V, Prew MS, *et al.* 2016. High-fidelity CRISPR-Cas9 nucleases with no detectable genome-wide off-target effects. *Nature* **529**: 490–495.

Koller BH, Smithies O. 1992. Altering genes in animals by gene targeting. *Annu Rev Immunol* **10**: 705–730.

Konermann S, Brigham MD, Trevino AE, *et al.* 2015. Genome-scale transcriptional activation by an engineered CRISPR-Cas9 complex. *Nature* **517**: 583–588.

Lancaster MA, Knoblich JA. 2014. Organogenesis in a dish: modeling development and disease using organoid technologies. *Science* **345**: 1247125.

Liu GH, Qu J, Suzuki K, *et al.* 2012. Progressive degeneration of human neural stem cells caused by pathogenic LRRK2. *Nature* **491**: 603–607.

Liu GH, Suzuki K, Qu J, *et al.* 2011. Targeted gene correction of laminopathy-associated LMNA mutations in patient-specific iPSCs. *Cell Stem Cell* **8**: 688–694.

Maeder ML, Thibodeau-Beganny S, Osiak A, *et al.* 2008. Rapid "open-source" engineering of customized zinc-finger nucleases for highly efficient gene modification. *Mol Cell* **31**: 294–301.

Moscou MJ, Bogdanove AJ. 2009. A simple cipher governs DNA recognition by TAL effectors. *Science* **326**: 1501.

Porteus MH, Baltimore D. 2003. Chimeric nucleases stimulate gene targeting in human cells. *Science* **300**: 763.

Ramirez CL, Foley JE, Wright DA, *et al.* 2008. Unexpected failure rates for modular assembly of engineered zinc fingers. *Nat Methods* **5**: 374–375.

Ran FA, Hsu PD, Lin CY, *et al.* 2013. Double nicking by RNA-guided CRISPR Cas9 for enhanced genome editing specificity. *Cell* **154**: 1380–1389.

Rong Z, Fu X, Wang M, Xu Y. 2012. A scalable approach to prevent teratoma formation of human embryonic stem cells. *J Biol Chem* **287**: 32338–32345.

Rong Z, Wang M, Hu Z, *et al.* 2014. An effective approach to prevent immune rejection of human ESC-derived allografts. *Cell Stem Cell* **14**: 121–130.

Sander JD, Dahlborg EJ, Goodwin MJ, *et al.* 2011. Selection-free zinc-finger-nuclease engineering by context-dependent assembly (CoDA). *Nat Methods* **8**: 67–69.

Schwartz SD, Regillo CD, Lam BL, *et al.* 2015. Human embryonic stem cell-derived retinal pigment epithelium in patients with age-related macular degeneration and Stargardt's macular dystrophy: follow-up of two open-label phase 1/2 studies. *Lancet* **385**: 509–516.

Shalem O, Sanjana NE, Zhang F. 2015. High-throughput functional genomics using CRISPR-Cas9. *Nat Rev Genet* **16**: 299–311.

Shechner DM, Hacisuleyman E, Younger ST, Rinn JL. 2015. Multiplexable, locus-specific targeting of long RNAs with CRISPR-Display. *Nat Methods* **12**: 664–670.

Shmakov S, Abudayyeh OO, Makarova KS, *et al.* 2015. Discovery and functional characterization of diverse class 2 CRISPR-Cas systems. *Mol Cell* **60**: 385–397.

Slaymaker IM, Gao L, Zetsche B, *et al.* 2016. Rationally engineered Cas9 nucleases with improved specificity. *Science* **351**: 84–88.

Soldner F, Laganiere J, Cheng AW, *et al.* 2011. Generation of isogenic pluripotent stem cells differing exclusively at two early onset Parkinson point mutations. *Cell* **146**: 318–331.

Song H, Chung SK, Xu Y. 2010. Modeling disease in human ESCs using an efficient BAC-based homologous recombination system. *Cell Stem Cell* **6**: 80–89.

Symington LS, Gautier J. 2011. Double-strand break end resection and repair pathway choice. *Annu Rev Genet* **45**: 247–271.

Tebas P, Stein D, Tang WW, *et al.* 2014. Gene editing of CCR5 in autologous CD4 T cells of persons infected with HIV. *N Engl J Med* **370**: 901–910.

Thakore PI, Black JB, Hilton IB, Gersbach CA. 2016. Editing the epigenome: technologies for programmable transcription and epigenetic modulation. *Nat Methods* **13**: 127–137.

Trounson A, Dewitt ND. 2016. Pluripotent stem cells progressing to the clinic. *Nat Rev Mol Cell Biol* **17**: 194–200.

Tsai SQ, Wyvekens N, Khayter C, *et al.* 2014. Dimeric CRISPR RNA-guided FokI nucleases for highly specific genome editing. *Nat Biotechnol* **32**: 569–576.

Urnov FD, Rebar EJ, Holmes MC, Zhang HS, Gregory PD. 2010. Genome editing with engineered zinc finger nucleases. *Nat Rev Genet* **11**: 636–646.

Xue H, Wu S, Papadeas ST, *et al.* 2009. A targeted neuroglial reporter line generated by homologous recombination in human embryonic stem cells. *Stem Cells* **27**: 1836–1846.

Yang Y, Wang L, Bell P, *et al.* 2016. A dual AAV system enables the Cas9-mediated correction of a metabolic liver disease in newborn mice. *Nat Biotechnol* **34**: 334–338.

Yant SR, Huang Y, Akache B, Kay MA. 2007. Site-directed transposon integration in human cells. *Nucleic Acids Res* **35**: e50.

Yusa K, Rashid ST, Strick-Marchand H, *et al.* 2011. Targeted gene correction of alpha1-antitrypsin deficiency in induced pluripotent stem cells. *Nature* **478**: 391–394.

Zalatan JG, Lee ME, Almeida R, *et al.* 2015. Engineering complex synthetic transcriptional programs with CRISPR RNA scaffolds. *Cell* **160**: 339–350.

Zetsche B, Gootenberg JS, Abudayyeh OO, *et al.* 2015. Cpf1 is a single RNA-guided endonuclease of a class 2 CRISPR-Cas system. *Cell* **163**: 759–771.

Zhu Z, Li QV, Lee K, *et al.* 2016. Genome editing of lineage determinants in human pluripotent stem cells reveals mechanisms of pancreatic development and diabetes. *Cell Stem Cell* **18**: 755–768.

Zwaka TP, Thomson JA. 2003. Homologous recombination in human embryonic stem cells. *Nat Biotechnol* **21**: 319–321.

19 Principles for Targeting Adult Tissue Stem Cells to Achieve Durable Gene and Gene Editing Therapies

James L. Sherley

19.1 Introduction

The importance of targeting long-lived adult tissue stem cells in order to achieve durable curative treatments after transduction of therapeutic genes has been recognized since the earliest days of gene therapy development (Fischer *et al.*, 2013; Wirth *et al.*, 2013; Yu *et al.*, 2016; Zepeda *et al.*, 1995). Despite this knowledge, the essential role of tissue stem cells for effective gene therapy is often underemphasized or overlooked in discussions of how to accelerate progress in the development of innovative gene therapy treatments (Szybalski, 2013; Weber, 2013). This underemphasis extends both to current efforts to improve conventional genetic therapies and to the newly emerging discipline of gene editing therapeutics (Xiong *et al.*, 2016; Yu *et al.*, 2016).

Since the earliest gene therapy studies and trials, the importance of integration of therapeutic gene vectors into cells to achieve sustained gene expression effects has been well appreciated (Gouze *et al.*, 2007; Jacomino *et al.*, 1996; Szybalski, 2013; Weber, 2013; Wirth *et al.*, 2013; Zepeda *et al.*, 1995). The vast majority of mammalian tissue cells, including human, are continuously maturing, dying or being lost, and being replaced by younger cells. So, in early gene therapy studies, it became evident that after a transduced gene was introduced into most cells *in situ*, its expression time was limited by the turnover rates of the cells into which it was integrated (Gouze *et al.*, 2007; Jacomino *et al.*, 1996; Zepeda *et al.*, 1995).

This understanding of the cellular basis for the relative short lifetime of expression of therapeutic genes in early gene therapy studies led to a focus on hematopoietic stem cells (HSCs) as ideal targets for genetic modification treatments (Fischer *et al.*, 2013; Yu *et al.*, 2016). The parallel discipline of HSC transplantation medicine had already demonstrated the long-lived quality of HSCs and developed robust methods for their isolation and enrichment. Therefore, the earliest attempts at durable gene therapies focused on diseases that could be reduced by genetic correction or manipulation of genes in HSC-enriched cell populations that were returned to patients after therapeutic gene transduction (Fischer *et al.*, 2013).

Of course, looking back, we now know that the early clinical trials in gene therapy were soon to be undone by the cancers that would emerge from insertional mutagenesis or insertion of therapeutic gene vector promoters near endogenous oncogenes (Chira *et al.*, 2015; Mukherjee and Thrasher, 2013; Wirth *et al.*, 2013; Yu *et al.*, 2016). These tragic outcomes were devastating to the field of gene therapy; but the discipline persisted in sorting out the molecular missteps and proceeded to look for solutions to get around them (Chira *et al.*, 2015; Mukherjee and Thrasher, 2013; Wirth *et al.*, 2013; Yu *et al.*, 2016).

Two major advances occurred in the past decade that reduced the problem of insertional mutagenesis and promoter insertion carcinogenesis in gene therapeutics. The first advance was directly motivated by the needs of gene therapy. New gene expression constructs were developed that no longer depended on genomic integration of powerful viral transcription promoters and, more importantly, inactivated them upon genome insertion. Other strategies used episomal vectors with greatly reduced insertion rates, but gene expression from these vectors is not as stable, presumably due to their dilution by the division of their carrier cells (Chira *et al.*, 2015; Mukherjee and Thrasher, 2013; Wirth *et al.*, 2013; Yu *et al.*, 2016; Zhou *et al.*, 2010).

The second, and more recent advance, is the potential to dispense with gene expression vectors altogether by deploying gene editing technologies. Using technologies like CRISPR/Cas9 and engineered variants with idealized properties, the future promises vectors that either correct defective genes in isolated cells or by direct editing *in situ*. As is often the case in medicine, the fix for one problem may introduce new problems of its own. This situation arises for gene editing in the potential for editing errors, in particular off-target edits at other places in the genome. This property has raised concern about the introduction of initially silent events that will manifest later in treated patients as a devastating disease, like recurring cancers. Once again technological advances have greatly reduced this problem in two ways. First, editing machines have been engineered to have greatly reduced off-target rates, and second, the efficiency of DNA sequencing makes it feasible to interrogate the genetic integrity of edited cell clones for added insurance against carcinogenic rogue edits (Mukherjee and Thrasher, 2013; Yu *et al.*, 2016).

As the molecular genetics technologies for gene therapy and gene editing advance to no longer being major factors limiting the safety and efficacy of treatments, the importance of tissue stem cell targeting will take on greater importance as a barrier to the success of gene therapy as an effective medical intervention for heritable diseases. Therefore, the long-standing challenges that remain unaddressed for stem cell medicine will soon become the challenges that gene therapy and gene editing therapeutics must also address for continued progress towards their envisioned promise.

The primary challenges for stem cell medicine are technical needs for manipulation of tissue stem cells per se. They include identification of tissue stem cells for *in vivo* targeting with therapeutic gene expression and gene editing constructs, production of tissue stem cells to increase the efficiency of *ex vivo* targeting of tissue stem cells for transplantation, and quantification of tissue stem cells to monitor their *ex vivo* manipulation and to specify their treatment dose. A major secondary challenge, which involves aspects of both tissue stem cells and their *in vivo* environment, is physiological delivery of therapeutic genetically engineered cells. The natural "liquid" character of the hematopoietic system presents low macroscopic barriers to access to stem cell niches, but for solid organs with complex three-dimensional cellular architectures, effective non-disruptive routes for stable engraftment of therapeutic stem cells have not been determined. Even in the case of HSC, conditioning regimens using toxic chemotherapy agents are often still required to achieve effective levels of engraftment of transplanted cells (Miao, 2016).

The introduced considerations are presented to revive emphasis on the fundamental duality of genetic therapies and stem cell medicine. Gene therapy and gene editing therapeutics both fall under the rubric of stem cell medicine. Although previously the stem cell medicine underpinning has been understandably understated as these genetic disciplines have wrestled with dire molecular genetics problems, now, with the molecular genetics better managed, attention must turn to the stem cell biomedical challenges that remain, as firmly entrenched as ever. In Chapter 9, the biological aspects of these challenges are highlighted to provide a conceptual framework for meeting them.

19.2 Mammalian Tissue Cell Turnover Kinetics

On the time scale of the average human lifespan, the dynamic nature of human tissues is nearly invisible. In general, although tissues appear static, their cellular constituents are in a continual state of flow that integrates changes in developmental state with changes in microanatomical location. Two exceptions to this unnoticed process are hair and nails. These appendages of the human skin are well recognized for their readily apparent continuous development and growth. They are differentiated cell extensions of the continuous asymmetric self-renewal divisions of tissue stem cells in hair follicles and nail beds, respectively. The hair follicle, in particular, has been well characterized for its hair growth cycle and the presence of hair-shaft-renewing stem cells located within the follicle (Ouspenskaia *et al.*, 2016).

Similar dynamic cell production vectors exist in many other tissues of the body. They originate from long-lived tissue stem cells that undergo asymmetric self-renewal divisions, maintaining stable stem cell numbers while simultaneously producing cells that are committed precursors for differentiated cell lineages (Sherley, 2005, 2006, 2013). In the case of hair follicles and nail beds, descendant cells differentiate into hair and nails; in skin, they differentiate to form the keratinized skin surface; for the limbus of the eye, they differentiate and produce the corneal epithelium; in the case of the liver they differentiate into bile epithelium cells and hepatocytes. This cell kinetics format appears over and over again in many different tissues with termination as highly specialized acellular structures or expired highly specialized cells that are lost by wear or death by apoptosis.

Because turnover kinetics occur in the majority of human tissues, when their function is compromised by a heritable genetic mutation, effecting a genetic cure of the deficiency necessarily requires targeting of tissue stem cells. Genetic modification of tissue stem cells is essential, because theirs is the tissue cell kinetics hierarchy generation whose genetic information persists in the tissue for the longest amount of time. In contrast, committed progenitor cells and the cells of their differentiated lineages have shorter tissue lifetimes that are determined by the turnover rates of the tissue, this being the time it takes for a single stem cell to divide sufficiently to restore an entire tissue cell turnover unit. When measured or estimated, these rates have been on the order of days to weeks, and, more importantly, only a small fraction of the average human lifespan. Therefore, if gene and gene editing treatments occurred in the genomes of committed and differentiating cells, the lifetime of treatment effects would be correspondingly short-lived as well.

Although it is generally accepted that tissue stem cells might be long-lived on the time scale of the human lifespan, the actual lifetime of a tissue stem cell has not been determined. As will be considered later in this chapter, thinking about how to measure the lifetime of tissue stem cells is complicated by a long-standing biological uncertainty that ends in an unresolved lifetime semantic. At this point in this discussion, it would be instructive to consider stem cell kinetic aspects of a specific organ that is a major therapeutic target for inherited diseases, the liver (Aravalli and Steer, 2016).

19.3 The Liver: An Example of Tissue Stem Cell-based Tissue Cell Turnover Kinetics in a Solid Organ

The basic unit of cellular construction in the liver is the radially arrayed liver lobule or acinus; Figure 19.1 provides a schematic of its organization. The predominant cells of the liver are the hepatocytes, which account for 70–90% of liver cellularity. Hepatocytes are organized in sheets or plates arrayed in cylindrical acini, like cards in a Rolodex™. Each liver Rolodex™ card is made of either one plane or two planes of closely juxtaposed hepatocytes. The hepatocyte cards split, fuse and merge like an Escher painting. The tightly opposed surfaces between hepatocytes form their apical surfaces. Bile acids produced by hepatocytes are secreted into channels between these opposed apical surfaces. These channels, called bile canaliculi, form a network of passages through which bile acids flow to the periphery of the liver Rolodex™ to join larger tubules formed by biliary epithelial cells. These tubules coalesce into larger bile ducts through which bile flows to the gall bladder.

The hepatocytes' basal surfaces (two for hepatocytes in single-cell layer plates), on each side of the hepatocyte card, are covered by fenestrated endothelial cells. This endothelium allows for high fluxes of solutes, which are carried by the blood plasma that flows between hepatocyte cards from the periphery to the center of the liver Rolodex™, to and from the hepatocytes. These loosely organized blood vessels are the hepatic sinusoids. Between the endothelial cells and the

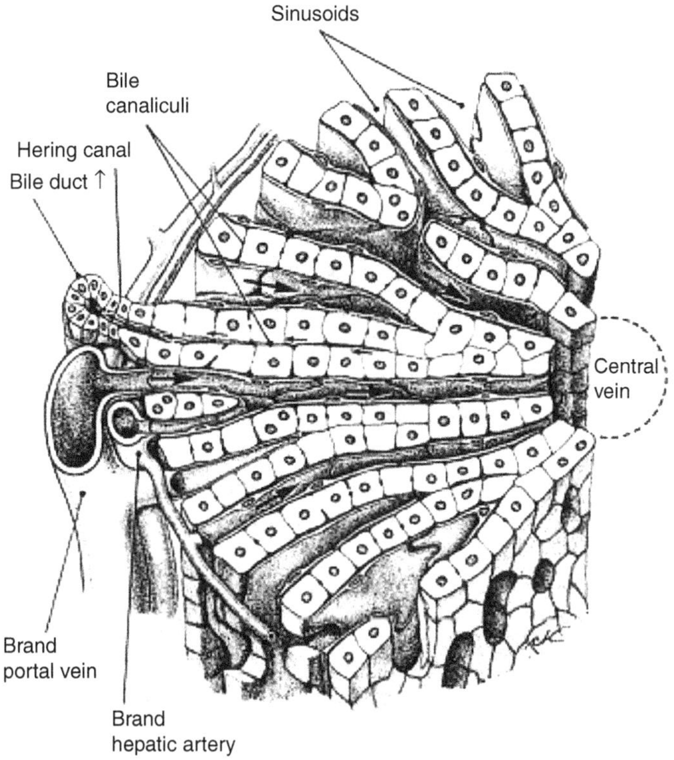

Figure 19.1 Schematic of the organization of the mammalian liver acinus. Liver hepatocyte vector turnover is based on the birth of cells from the division of progenitor cells located in the first 1–4 cell positions radially from the periportal region, followed by their subsequent division arrest and progressive migration to the ~16 cell position proximal to the central hepatic vein. Apoptosis in normal liver is limited to the two cell positions nearest the central vein. Arrows show the direction (inward) of sinusoidal blood flow and cell streaming. Bile flows in the opposite (outward) direction. (Source of image: Figure 11 from Le Lay J and Kaestner KH. 2010. The FOX genes in the liver: from organogenesis to functional integration. *Physiol Rev* 90(1): 1–22. [Modified from Bloom and Fawcett. *A Textbook of Histology, 10th edn.* Philadelphia, PA: Saunders, 1975.] Reproduced with permission.)

hepatocytes lies the space of Disse, which is filled with a specialized extracellular matrix (ECM). Along the length of hepatocyte plates, the composition of the ECM changes in a characteristic radial gradient that plays a role in hepatocyte function (McClelland *et al.*, 2008a).

The directional landmarks for liver cell turnover are the vascular inlets and outlets for sinusoidal blood flow. Blood enters the sinusoids from two sources, oxygenated blood from arteriole branches of the hepatic artery and nutrient-laden blood from venule branches of the portal vein. Together with the efferent bile ducts, these afferent blood vessels form the portal triads and constitute the periportal regions of the acinus. There are on average six portal triads per acinus, configured like equidistant posts around a hexagonal hepatocyte Rolodex™ cylinder. One additional cell structure located near the portal triads has been of particular interest to liver stem cell biologists. Called the canals of Hering, these are hepato-biliary transition zones between periportal hepatocytes and adjoining biliary epithelial cells that form portal triad bile ducts. These regions are the active zones of cell division in the liver, and they are the locations of cells with phenotypic markers indicative of hepatic stem cells (Schmelzer *et al.*, 2007). Blood flows from the periportal vessels via the hepatic sinusoids inward and centrally to the acinar central vein (i.e. the hub of the liver Rolodex™) for return to the general circulation.

Hepatocyte acinar plates are defined as having three histologically distinct, radially organized anatomical zones. Acinar zone I corresponds to cell positions 1–6, which are proximal to the portal venules, zone II corresponds to cell positions 7–11, which are intermediate between the portal triad and the central vein, and zone III corresponds to cell positions 12–16 proximal to the central vein (Rappaport, 1973). Based on cell kinetic properties, hepatic plates can be more simply delineated as a two-compartment system similar to other renewing epithelia (Sherley, 2006, 2013). In normal liver, hepatocyte division is restricted to zone I from cell positions 1 to 3 or 4, and all other hepatocytes maintain a stable cell division arrest for their normal lifetime. This form of cell kinetics compartmentalization in renewing tissues is the hallmark of the mammalian cell turnover unit format (Arber *et al.*, 1988; Sherley 2006, 2013; Zajicek *et al.*, 1985). Hepatocytes are born in the periportal region from the division of progenitor cells, they undergo a stable division arrest as they migrate into zone II and continue a radial migration to the central vein where they undergo apoptosis at a rate that matches their rate of production in zone I (Arber *et al.*, 1988; Benedetti *et al.*, 1988; Zajicek *et al.*, 1985).

Blikkendaal-Lieftnck and colleagues (1977) performed one of the earliest demonstrations of liver hepatocyte turnover by tracing the microanatomical fate of the radioactive DNA base analog ^{3}H-thymidine by autoradiography of liver tissue sections. This method is the long-standing gold standard for cell "streaming" analyses in a variety of mammalian tissues (Messier and Leblond, 1960; Potten *et al.*, 1979). Within 1 hour of injection of ^{3}H-thymidine, labeled hepatocytes are only found in zone I in cell positions 1–4. This finding indicates that cells in zones II and III are not cycling, or at best are cycling at a significantly slower rate than cells in zone I. Thereafter, at later times, labeled cells are observed progressively at positions in zone II and finally zone III, while no longer appearing in zone I. The loss of labeled cells in zone I cannot be explained simply as a result of dilution by cell division. Cells appearing in zones II and III show only modest reductions in radioactivity, indicating that they must be migrating arrested cells, which were originally produced by division of cycling cells in zone I.

Zajicek and colleagues (Arber and Zajicek, 1990; Arber *et al.*, 1988; Schwartz-Arad *et al.*, 1989; Zajicek and Schwartz-Arad, 1990; Zajicek *et al.*, 1985; 1988; 1989; 1991) conducted the most extensive hepatocyte ^{3}H-thymidine tracer studies to independently confirm hepatocyte turnover. In addition, the observations of other groups support their conclusions. Even after four days of continuous labeling, Wright and Allison (1984) showed that cells proximal to the central vein do not label. Drugs like methyltestosterone in patients (Paradinas *et al.*, 1977) and dimethyl-nitrosamine in hamsters (Ungar, 1984), which increase hepatic cell turnover to a rate that is faster than the apoptotic process, result in central vein prolapse due to accumulation of hepatocytes. Apoptosis in human and rat liver is limited to terminal hepatocytes in zone III adjoining the central vein, consistent with vectored aging of hepatocytes along a linear vector of cell turnover (Benedetti *et al.*, 1988).

Zajicek and colleagues also used data from ^{3}H-thymidine tracer experiments to quantify key cell-streaming kinetics parameters and microanatomical dimensions. These continue today as the best descriptors of mammalian liver cell vector turnover. Although defined primarily for rat liver, they are applicable to human liver as well (McClelland *et al.*, 2008a; Schmelzer *et al.*, 2007; Sigal *et al.*, 1992). The mean hepatocyte diameter is about 29 μm (Zajicek *et al.*, 1988). Hepatic progenitor cells are smaller, on the order of 8–12 μm (McClelland *et al.*, 2008b; Sigal *et al.*, 1992) and mature hepatocytes may reach sizes as great as 63 μm (Zajicek *et al.*, 1985). Thus, there is a gradient of increasing hepatocyte size (Schwartz-Arad *et al.*, 1989) and age (MacDonald, 1961), inwardly along the portocentral radius.

There are also characteristic differences in hepatocyte form and function along the portocentral vector. Hepatocyte mean DNA content and chromosomal ploidy increase (Schwartz-Arad *et al.*, 1989), culminating in bi-nucleated cells in zone III. Younger hepatocytes in zone I primarily express enzyme activities for gluconeogenesis, whereas zone III cells are restricted to glycolysis (Gumucio and Miller, 1981; Gumucio *et al.*, 1986). The expression of several well-characterized genes is highly zone-specific, indicative of the phenotypic gradient of hepatocyte function. For example, whereas cells in all zones express the albumin gene (Nahon, 1987), only a rare subset of cells in zone I weakly express the α-fetoprotein gene (Schmelzer *et al.*, 2007) and glutamine synthetase expression is limited to the small set of zone III cells that border the central vein (Fahrner *et al.*, 1993). The expression patterns of the proteins encoded by these three genes are ideal indicators of normal functioning of hepatocyte cell turnover.

Zajicek and colleagues referred to the vector turnover of hepatocytes as "streaming," an excellent description of the mammalian tissue cell kinetic architecture that integrates cell age, cell position and cell tissue lifetime. They divided streaming hepatocytes into two compartments, the P-compartment containing dividing progenitor cells and the Q-compartment containing quiescent maturing functional hepatocytes (Arber *et al.*, 1988; Zajicek *et al.*, 1988). The P-compartment extends approximately 220 μm from the portal venule and is approximately 3 or 4 hepatocyte positions deep. Thus, it is encompassed by zone I, but also contains the portal triad bile duct and canal of Hering. The Q-compartment has a mean distance from the portal venule of about 470 μm, which is the mean liver acinar radius. However, the Q-compartment can extend as far as 882 μm (Zajicek *et al.*, 1985). It contains on average about 12 hepatocytes for an average total of 16 along the portocentral radius of hepatocyte plates.

The average generation time of hepatocytes in the P-compartment was estimated to be 33–40 days in the normal liver (Arber *et al.*, 1988; Zajicek *et al.*, 1985; 1988). Hepatocyte streaming occurs in lock-step with this division rate, resulting in inward migration of cells produced from the P-compartment at a mean rate of ~2 μm per day, or ~1 cell position per 12 days (Arber *et al.*, 1988). These rates are consistent with the well-described slow turnover of hepatocytes in normal liver. They correspond to an overall mean hepatocyte transit time from the P–Q border to the central vein of ~200 days (Zajicek *et al.*, 1985; 1988). This value is in good agreement with independent estimates that hepatocyte lifespan ranges from 151–453 days (MacDonald, 1961).

In addition to hepatocytes and hepatic stem cells, there are four other significant cell populations resident in the liver. Biliary epithelial cells and sinusoidal endothelial cells were mentioned earlier. In addition, there are phagocytic Kupffer cells and fibrogenic stellate cells. Interestingly, both biliary epithelial cells and sinusoidal epithelial cells show properties indicative of cell streaming. Bile duct epithelial cells appear to stream from the smallest ducts in the liver parenchyma to larger portal triad ducts (Arber and Zajicek, 1990). Thus, they stream away from the hepatobiliary junction in the canal of Hering in the opposite direction to hepatocytes. In contrast, endothelial cells appear to stream in the same direction as hepatocytes, though their point of origin is unclear. It has been suggested that via tight junctions, hepatocytes and endothelial cells stream as a unit, along with the ECM in the space of Disse between them (Zajicek *et al.*, 1985; 1988). This organization suggests that a significant determinant of hepatocyte phenotype may not be position per se, but age since production in the P-compartment.

The most definitive evidence for tissue cell streaming was the demonstration of the existence of stem cells that have the property of asymmetric self-renewal in P-compartments (Loeffler and Potten, 1997). Implicit in the tissue cell streaming concept is the requirement that cells in P-compartments divide to produce two different types of cells at a balanced rate, P-stem cells and Q-cells. Replacement P-stem cells continue to renew the tissue unit, while Q-cells are destined to stream away from the P-compartment while undergoing differentiation, performing specialized tissue functions, aging and finally succumbing to death and exfoliation or apoptosis. In some tissues, like the small intestinal epithelium, Q-cells may undergo a limited number of "transient amplifying divisions" as they stream (Potten and Morris, 1988). However, the final Q-cell descendants of transient amplification undergo a terminal division arrest and continue streaming with aging to the point of apoptosis or loss due to death and tissue wear.

Three groups have reported experimental evidence of asymmetrically self-renewing stem cells in normal adult liver. Schmelzer and others (Schmelzer *et al.*, 2006; 2007) defined human hepatic stem cells as rare cells in adult liver that express the epithelial cell adhesion molecule, but not α-fetoprotein (i.e. $EpCAM^+$/AFP^-). In human adult liver histological sections, $EpCAM^+$ cells are detected exclusively in the canals of Hering. Many of these cells are AFP^-. Near the ends of the canals of Hering, small clusters of cells and individual cells with an $EpCAM^+$/AFP^+ phenotype are found. These cells correspond to committed hepatoblast precursors, which are produced by the division of hepatic stem cells that give rise to both mature hepatocytes and biliary epithelium cells (Figure 19.2).

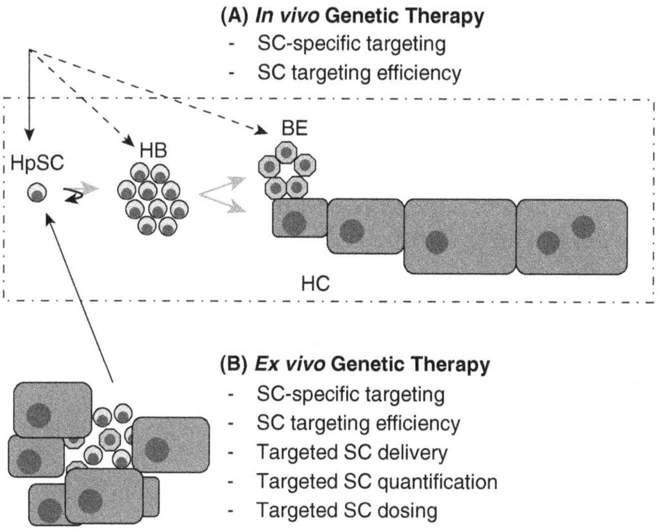

Figure 19.2 Tissue stem cell-related challenges for genetic therapies. Based on a schematic example of the liver hepatic stem cell turnover unit, several significant challenges to success in genetic therapies are indicated. *In vivo* (A) and *ex vivo* (B) therapeutic approaches may require specific targeting of tissue stem cells to avoid unsafe effects of ectopic actions of gene expression or gene editing constructs in other cell types. In either approach, achieving high-efficiency stem cell targeting, whether or not selectively, is essential to avoid only short-term treatment as a result of modifying only the short-lived cells (i.e. HB, BE and HC) in the stem cell turnover unit. The efficacy of treatments will depend on the number of tissue stem cells successfully targeted. *Ex vivo* therapies must also address the challenge of delivering targeted stem cells into tissue compartments that support effective asymmetric self-renewal. Development of *ex vivo* genetic therapies would benefit greatly from the ability to quantify tissue stem cells and determine their effective dose. HpSC, asymmetrically self-renewing hepatic stem cell; HB, bivalent hepatoblast; BE, mature biliary epithelial cells; HC, mature hepatocytes.

Like hepatic stem cells, they express lineage markers for both hepatocytes (e.g. albumin) and biliary epithelial cells (e.g. cytokeratin 19; Schmelzer *et al.*, 2007). These well-described AFP^+ hepatoblasts are rare in normal liver (<0.01%; Schmelzer *et al.*, 2007). The author's research groups have derived $EpCAM^+/AFP^-$ cells from rat and human liver that exhibit deterministic asymmetric self-renewal and retain the ability to produce progeny cells with hepatoblast, biliary epithelium and mature hepatocyte properties (Lee *et al.*, 2003; Panchalingam *et al.*, 2016; Sherley and Panchalingam, 2010; unpublished data).

19.4 Principles for Improving Gene and Gene Editing Therapeutics from Liver Cell Turnover Kinetics

The general discussion of tissue cell turnover kinetics and the specific case for human liver tissue illustrate the important role that tissue stem cells must play in any genetic therapy development strategy that has the goal of achieving single-dose, durable, curative treatments of inherited genetic diseases. As alluded to earlier, an important biological semantic operates within this requirement. The conundrum of attempting to define the significance of longevity for tissues cells that are continually renewing by cell division has been considered and resolved previously (Sherley, 2008, 2013). It is not the cell lifetime that is the crucial metric, but instead the lifetime and integrity of the genetic information that resides in the stem cell compartment. Basically, stem cell compartments, or niches, can be considered to have evolved to prevent the initiating genomes of tissue units from turning over, thereby rendering tissue genetic information more durable. The asymmetric self-renewal kinetics of the stem cell niche maintain genetic fidelity, while simultaneously allowing less durable cellular constituents to be renewed continuously. As a consequence of this universal vertebrate tissue renewal program, unless therapeutic genetic corrections are made in stem cell genomes, they will be lost from the treated tissue at the cellular turnover rate of the tissue.

It is noteworthy that all of the discussion herein is predicated on the *assumption* that the lifetimes of genomes in stem cell niches are in fact long, on the time scale of the human lifespan. The author is aware of no direct proof of this long-standing concept. In fact, a continuing debate on the mathematical form of asymmetric self-renewal in tissue stem cell niches bears heavily on this issue (Klein and Simons, 2011; Sherley, 2013). If stem cell niches were maintained by one or more stem cells undergoing individual deterministic asymmetric self-renewal divisions, then many stem cell genomes might, in fact, persist for the entire human lifespan. If, on the other hand, stem cell niches were maintained by the "neutral competition" of populations of stem cells balancing stochastic symmetric self-renewal divisions vs. stochastic differentiation (Klein and Simons, 2011), then a different treatment outcome would be predicted. Unless all the stem cells in a niche were corrected, on average, the genetic corrections would be predicted to decay exponentially with a half-life related to the differentiation rate of the stem cells in the niche. Some stem cell niches could probabilistically become stable for the genetic correction, but the number doing so might be too low to provide an effective treatment.

Of course, the continuing debate over the exact mathematical form of tissue-renewing stem cell kinetics becomes purely an academic one if all stem cells have a significant finite extinction rate (because of stochastic differentiation kinetics) or cell death rate. In fact, if stem cell turnover rates due to either differentiation, extinction or death approached the turnover rate of tissue cell units, targeting tissue stem cells would provide no advantage for genetic therapies. Presently, since the average lifetime of stem cells in humans and other mammals is not known, whether in humans it approaches the time scale of the human lifespan remains a matter of conjecture. The answer to this question is a crucial factor for expectations of the level of durability that is possible for genetic therapies.

A first clue to the stem cell lifetime question may have already surfaced. There was a recent report of an apparently stable effective genetic treatment that suddenly failed after seven years in a hematopoietic stem cell-targeted gene therapy clinical trial (Carroll, 2015). The cause of the failure is unknown. Possible causes include: (1) a molecular process that muted curative genes in targeted stem cells (e.g. gene methylation), (2) corrective targeting limited to committed progenitor cells with a surprisingly long tissue lifetime or (3) extinction or death of a small fraction of targeted hematopoietic stem cells that were responsible for the initial therapeutic effect. Clearly, sorting out the cause will be an important step for designing future gene therapy trials.

Uncertainties regarding the natural turnover rates of tissue stem cells may not have been considered previously during the development of genetic therapies. However, several other stem cell-related challenges to achieving safe and efficacious genetic therapies are well appreciated. These are long-standing challenges to the fields of tissue stem cell research and stem cell medicine. Figure 19.2 lists these barriers to effective stem cell-targeted gene therapeutics, and they are outlined and discussed briefly below.

19.4.1 Specific Targeting

Even with the new caution that the extent of tissue stem cell lifetimes is not actually known, the short-term advantages of targeting gene corrections to tissue stem cells still warrant persuing them. The lifetimes of other tissue cells are established to be short and the stem cell is the only cell in the tissue cell hierarchy whose modification is rapidly transmitted to all other cells. Initiation of corrective gene expression in stem cells may also be important for proper developmental expression and similarly targeting the stem cell for a specific type of tissue unit is likely to be important for normal genetic restoration. Conversely, ectopic correction in an unnatural tissue or cell type could be ineffective or even detrimental (Weber, 2013).

Currently selective genetic modification of tissue stem cells, whether directly *in vivo* or *in vitro* after their *ex vivo* isolation and enrichment, faces many challenges. The low fraction of the cells in either approach is a major impediment to both selective and efficient targeting (see below). Molecular tools are not available for either targeting by specific adhesion or by specific gene expression. Respectively, there are no described stem cell-specific receptors or stem cell-specific regulatory elements. Recently described candidates for tissue stem cell-specific biomarkers, identified with a different basis from previous attempts, may one day provide better molecular tools to address the selective targeting problem (Noh *et al.*, 2011). In the case of the hematopoietic tissues, methods exist to prepare highly enriched populations for stem cells and committed progenitor cells. These well-defined preparations with much clinical experience are the current standard for *ex vivo* stem cell-targeted gene therapy trials. Unfortunately, similar methods are not available for solid organs like liver.

19.4.2 Efficient Targeting

As noted above, even in the case of hematopoietic cells – for which enrichment of isolated stem cells is possible – because even in the best situations committed progenitors may outnumber stem cells by as much as 50 to 1, the lack of tools for selective targeting greatly impacts targeting efficiency. In both *in vivo* and *ex vivo* gene therapy approaches, sufficient stem cells must be modified to achieve a therapeutic effect. For *in vivo* strategies, non-selective gene constructs must be supplied in sufficient amounts to ensure that enough find their way into infrequent tissue stem cells to achieve a durable effect. Even with organ-specific targeting, which is possible for the liver, special retrograde injection strategies (e.g. portal vein or gall bladder duct) might be needed to get genetic modification vectors to stem cell compartments before being sopped up by vastly more abundant mature liver cells.

In *ex vivo* approaches, the targeting efficiency problem can be solved by increasing the number of tissue stem cells with respect to other cell types present in treatment preparations. However, for many years the selective expansion of adult tissue stem cells was a formidable challenge in stem cell biology. The author's research teams were the first to show that asymmetrically self-renewing tissue stem cells could be selectively grown from mammalian liver and propagated indefinitely in culture without neoplastic transformation (Lee *et al.*, 2003). This advance was accomplished by using a method called suppression of asymmetric cell kinetics (SACK) to reversibly shift asymmetrically self-renewing liver stem cells to symmetric self-renewal. During symmetric self-renewal, stem cells self-renew exponentially without the production of differentiating cells. The effects of SACK treatment, achieved by medium supplementation with guanine ribonucleoside precursors like xanthine and xanthosine are reversible. Thus, the SACK approach allows extensive production of hepatic stem cells that can be reversed to asymmetric self-renewal division to allow the natural production of cells that undergo hepatic differentiation. The SACK method has been adapted to derive asymmetrically self-renewing cell strains from normal human liver specimens from child and adult donors (Panchalingam *et al.*, 2016; Sherley and Panchalingam, 2010). The method has also proven to be effective for the expansion of tissue stem cells for other tissues, including the pancreas (Paré and Sherley, 2013) and hair follicles (Huh *et al.*, 2011). General implementation of SACK expansion in *ex vivo* gene therapy trials could provide significant increases in targeting efficiency.

19.4.3 Targeted Cell Delivery

After specific and efficient targeting of stem cells is achieved in the *ex vivo* therapy setting for diseases that can be treated with engineered hematopoietic stem cells (HSCs), successful therapy should follow straightforwardly. Although there can be some toxic side effects, including chronic deficits, from the conditioning that improves long-term HSC engraftment, they are generally well tolerated. Although there is still significant clinical research ongoing to improve further specific aspects of HSC transplantation medicine, it is a very mature medical discipline. The same cannot be said for most other organs and tissues. In addition to having stem cells that were not evolved to survive mobilization through the blood circulation and other unnatural tissue routes (e.g. lymphatics), solid organs do not have natural paths of access to normal tissue niches. This could lead to the development of ectopic aberrant tissue structures from stem cell treatment. In the future, this currently minimally investigated challenge may motivate the development of engineered renewing tissue devices designed with patients' *ex vivo* corrected autologous stem cells from disease-affected organs and tissues.

19.4.4 Quantification

Both *in vivo* and *ex vivo* genetic therapy clinical trials now operate literally blind to one of the most important factors for the success of their treatments. The stem cells in the trials are invisible in the analyses conducted. With the present absence of stem cell-specific biomarkers, it is not possible to either identify or quantify the number of tissue stem cells manipulated for genetic treatments. Thus, trials proceed with neither exact nor precise knowledge of the number of initial stem cells isolated; though in the case of HSCs, historical data based on hematopoiesis repopulation analyses in mice do provide a marginal substitute for the initial stem cell number. However, after genetic modifications are completed, there is no practical basis to determine the number of successfully modified stem cells of any type, including HSCs, because the much more numerous modified committed progenitor cells obscure the stem cells. Being unable to quantify the number of stem cells, there are also no effective means to monitor adverse effects of the genetic modification procedures.

19.4.5 Dosing

As there are no effective, specific means for counting tissue stem cells, it follows that in the present gene therapy clinical trials, the dose of treating stem cells is unknown. Of course, the same is true for all current stem cell medical treatments, whether in approved clinical trials or unregulated private clinics (Sherley, 2016). This situation means that it is not possible to relate the dose of genetically targeted stem cells to treatment outcomes, to compare different arms of clinical trials or even to compare the outcome of one patient to another, since in the autologous transplant setting patients do now share treatment samples. Moreover, until the dosing problem is solved, it will be very difficult to build on the results of one clinical trial to improve the next.

19.5 Final Words

In this chapter, the future success of genetic therapeutics is considered with respect to its high dependency on continued progress in tissue stem cell biotechnology. Although the discipline of genetic therapeutics is realizing marvelous advances with rapidly emerging facile molecular genetics techniques, such as efficient and precise gene editing, that address earlier catastrophic safety issues, it now must confront longer-standing challenges in tissue stem cell biology. This condition arises because of the predicted necessity that curative genetic corrections are incorporated into long-lived tissue stem cells in order to achieve single-dose durable cures for inherited diseases. This high dependency means that the long-standing and unsolved problems that have plagued stem cell medicine now demand greater attention from gene therapy and gene editing therapeutics. Perhaps the current energy and excitement for achieving major therapeutic successes with the newer gene editing technologies will also reignite efforts to solve the stem cell barriers that presently bar the door to both stem cell medicine and gene therapeutics. Tissue stem cell production, identification, quantification and dosing – when these are routine applications, the door to genetic therapies and stem cell medicine on the whole will finally swing open for waiting patients.

Acknowledgement

The author thanks the editor, Dr. K.A., for the gracious invitation to contribute this chapter.

REFERENCES

Aravalli RN, Steer, CJ. 2016. Gene editing technology as an approach to the treatment of liver diseases. *Expert Opin Biol Ther* **16**: 595–608.

Arber N, Zajicek G. 1990. Streaming liver VI: streaming intra-hepatic bile ducts. *Liver* **10**: 205–208.

Arber N, Zajicek G, Ariel I. 1988. The streaming liver II. Hepatocyte life history. *Liver* **8**: 80–87.

Benedetti A, Jézéquel AM, Orlandi F. 1988. Preferential distribution of apoptotic bodies in the acinar zone 3 of normal human and rat liver. *J Hepatol* **7**: 319–324.

Blikkendaal-Lieftinck LF, Kooij M, Kramer MF, Otter WD. 1977. Cell kinetics in the liver of rats under normal and abnormal dietary conditions. An autoradiographic study by means of [^{3}H] thymidine. *Exp Mol Path* **26**: 184–192.

Carroll J. 2015. Bluebird shares dented after gene therapy patient experiences a setback. *FierceBiotech*. www.fiercebiotech.com/r-d/bluebird-shares-dented-after-gene-therapy-patient-experiences-a-setback (accessed November 2017).

Chira S, Jackson CS, Oprea I, *et al.* 2015. Progresses towards safe and efficient gene therapy vectors. *Oncotarget* **6**: 30675–30703.

Fahrner J, Labruyere WT, Gaunitz C, *et al.* 1993. Identification and functional characterization of regulatory elements of the glutamine synthetase gene from rat liver. *Eur J Biochem* **213**: 1067–1073.

Fischer A, Hacein-Bey-Abina S, Cavazzana-Calvo M. 2013. Gene therapy of primary T cell immunodeficiencies. *Gene* **525**: 170–173.

Gouze E, Gouze JN, Palmer GD, *et al.* 2007. *Mol Ther* **15**: 1114–1120.

Gumucio JJ, May M, Dvorak C, Chianale J, Massey V. 1986. The isolation of functionally heterogeneous hepatocytes of the proximal and distal half of the liver acinus in the rat. *Hepatology* **6**: 932–944.

Gumucio JJ, Miller DL. 1981. Functional implications of liver cell heterogeneity. *Gastroenterology* **80**: 393–403.

Huh YH, King J, Cohen J, Sherley JL. 2011. SACK-expanded hair follicle stem cells display asymmetric nuclear lgr5 expression with non-random sister chromatid segregation. *Sci Rep* **1**: 175.

Jacomino M, Lau C, James SZ, Henning SJ. 1996. Gene transfer into fetal rat intestine. *Hum Gene Ther* **7**: 1757–1762.

Klein AM, Simons BD. 2011. Universal patterns of stem cell fate in cycling adult tissues. *Development* **138**: 3103–3111.

Le Lay J, Kaestner KH. 2010. The FOX genes in the liver: from organogenesis to functional integration. *Physiol Rev* **90**: 1–22.

Lee H-S, Crane GG, Merok JR, *et al.* 2003. Clonal expansion of adult rat liver epithelial stem cells by suppression of asymmetric cell kinetics (SACK). *Biotech Bioeng* **83**: 760–771.

Loeffler M, Potten CS. 1997. Stem cells and cellular pedigrees: a conceptual introduction. In *Stem Cells*, CS Potten, ed., San Diego, CA: Harcourt Brace & Co., pp. 1–28.

MacDonald RA. 1961. Lifespan of liver cells. *Arch Intern Med* **107**: 335–343.

McClelland R, Wauthier E, Uronis J, Reid L. 2008a. Gradients in the liver's extracellular matrix chemistry from periportal to pericentral zones: influence on human hepatic progenitors. *Tissue Eng Part A* **14**: 59–70.

McClelland R, Wauthier E, Zhang L, *et al.* 2008b. Ex vivo conditions for self-replication of human hepatic stem cells. *Tissue Eng Part C* **14**: 341–351.

Messier B, Leblond CP. 1960. Cell proliferation and migration as revealed by radioautography after injection of thymidine-H^3 into male rats and mice. *Am J Anat* **106**: 247–285.

Miao CH. 2016. Hemophilia A gene therapy via intraosseous delivery of factor VIII-lentiviral vectors. *Thrombosis J* **14**(Suppl. 1): 93–99.

Mukherjee S, Thrasher AJ. 2013. Gene therapy for PIDs: progress, pitfalls, and prospects. *Gene* **525**: 174–181.

Nahon J-L. 1987. The regulation of albumin and α-fetoprotein gene expression in mammals. *Biochimie* **69**: 445–459.

Noh M, Smith JL, Huh YH, Sherley JL. 2011. A resource for discovering specific and universal biomarkers for distributed stem cells. *PLoS One* **6**(7): e22077.

Ouspenskaia T, Matos I, Mertz AF, Fiore VF, Fuchs E. 2016. WNT-SHH antagonism specifies and expands stem cells prior to niche formation. *Cell* **164**: 156–169.

Panchalingam K, Noh M, Hu YH, Sherley JL. 2016. Distributed stem cell kinetotoxicity: a new concept to account for the human carcinogenicity of non-genotoxic toxicants. In *Human Stem Cell Toxicology, Issues in Toxicology No. 29*, JL Sherley, ed., London: Royal Society of Chemistry.

Paradinas FJ, Bull TB, Westaby D, Murray-Lyon IM. 1977. Hyperplasia and prolapse of hepatocytes into hepatic veins during longterm methyltestosterone therapy: possible relationships of these changes to the development of peliosis hepatis and liver tumors. *Histopathology* **1**: 225–246.

Paré J-F, Sherley JL. 2013. Ex vivo expansion of human pancreatic distributed stem cells by suppression of asymmetric cell kinetics (SACK). *J Stem Cell Res Ther* **3**: 149.

Potten CS, Morris RJ. 1988. Epithelial stem cells in vivo. *J Cell Sci Suppl* **10**: 45–62.

Potten CS, Schofield R, Lajtha LG. 1979. A comparison of cell replacement in bone marrow, testis, and three regions of surface epithelium. *Biochim et Biophys Acta* **560**: 281–299.

Rappaport AM. 1973. The microcirculatory hepatic unit. *Microvasc Res* **6**: 212–228.

Schmelzer E, Wauthier E, Reid LM. 2006. The phenotypes of pluripotent human hepatic progenitors. *Stem Cells* **24**: 1852–1858.

Schmelzer E, Zhang L, Bruce A, *et al.* 2007. Human hepatic stem cells from fetal and postnatal donors. *J Exp Med* **204**: 1973–1987.

Schwartz-Arad D, Zajicek G, Bartfeld E. 1989. The streaming liver IV. DNA content of the hepatocyte increases with age. *Liver* **9**: 93–99.

Sherley JL. 2005. Asymmetric self-renewal: the mark of the adult stem cell. In *Stem Cell Repair and Regeneration*, NA Habib, MY Gordon, N Levicar, G Jiao, L Thomas-Black, eds., London: Imperial College Press, pp. 21–28.

Sherley JL. 2006. Mechanisms of genetic fidelity in mammalian adult stem cells. In *Tissue Stem Cells*, CS Potten, RB Clarke, J Wilson, AG Renehan, eds., New York: Taylor Francis, pp. 37–54.

Sherley JL. 2008. A new mechanism for aging: chemical "age spots" in immortal DNA strands in distributed stem cells. *Breast Dis* **29**: 37–46.

Sherley JL. 2013. New cancer diagnostics and therapeutics from a 9th "hallmark of cancer": symmetric self-renewal by mutated distributed stem cells. *Expert Rev Mol Diagn* **13**: 797–810.

Sherley JL. 2016. Advancing stem cell medicine by supplying private stem cell clinics. *Clinical Trials Arena Supply Chain.* www.clinicaltrialsarena.com/news/supply-chain/advancing-stem-cell-medicine-by-supplying-private-stem-cell-clinics-5669514 (accessed November 2017).

Sherley JL, Panchalingam K. 2010. Methods for ex vivo propagation of adult hepatic stem cells. US Patent No. 7,824,912.

Sigal SH, Brill S, Fiorino AS, Reid LM. 1992. The liver as a stem cell and lineage system. *Am J Physiol* **263**: G139–G148.

Szybalski W. 2013. The 50th anniversary of gene therapy: beginnings and present realities. *Gene* **525**: 151–154.

Ungar H. 1984. Primary portal venopathy in the golden hamster treated with low doses of dimethyl nitrosamine. *Liver* **4**: 244–254.

Weber GF. 2013. Gene therapy: why can it fail? *Med Hypoth* **80**: 613–616.

Wirth T, Parker N, Ylä-Herttuala S. 2013. History of gene therapy. *Gene* **525**: 162–169.

Wright N, Alison M. 1984. *The Biology of Epithelial Cell Populations*, Vol. 2. Oxford: Clarendon Press, pp. 880–980.

Xiong X, Chen M, Lim WA, Zhao D, Qi LS. 2016. CRISPR/Cas9 for human genome engineering and disease research. *Annu Rev Genom Hum Genet* **17**: 131–154.

Yu K-R, Natanson H, Dunbar CE. 2016. Gene editing of human hematopoietic stem can progenitor cells: promise and potential hurdles. *Hum Gene Ther* **27**: 729–740.

Zajicek G, Arber N, Schwartz-Arad D. 1991. Streaming liver VIII: cell production rates following partial hepatectomy. *Liver* **11**: 347–351.

Zajicek G, Ariel I, Arber N. 1988. The streaming liver III. Littoral cells accompany the streaming hepatocyte. *Liver* **8**: 213–218.

Zajicek G, Oren R, Weinreb M Jr. 1985. The streaming liver. *Liver* **5**: 293–300.

Zajicek G, Swhwartz-Arad D. 1990. Streaming liver VII: DNA turnover in acinus zone-3. *Liver* **10**: 137–140.

Zajicek G, Swhwartz-Arad D, Bartfeld E. 1989. The streaming liver V. Time and age-dependent changes of hepatocyte DNA content, following partial hepatectomy. *Liver* **9**: 164–171.

Zepeda ML, Chinov MR, Wilson JM. 1995. Characterization of stem cells in human airway capable of reconstituting a fully differentiated bronchial epithelium. *Somat Cell Mol Genet* **21**: 61–73.

Zhou S, Mody D, DeRavin SS, *et al.* 2010. A self-inactivating lentiviral vector for SCID-XI gene therapy that does not activate LMO2 expression in human T cells. *Blood* **116**: 900–908.

20 Therapeutic Genome Editing in Human Hematopoietic Stem and Progenitor Cells

Rasmus O. Bak, Daniel P. Dever and Matthew H. Porteus

20.1 Introduction to Hematopoiesis

Human hematopoiesis is the process by which all blood and immune cells are generated from a hierarchal tree-like system with repopulating and self-renewing hematopoietic stem cells (HSCs) residing at the apex (Doulatov *et al.*, 2012). This highly dynamic and regulated process is centered on the idea that HSCs give rise to non-self-renewing multipotent progenitors (MPPs) (and also more HSCs), which subsequently generate common myeloid (CMP) and common lymphoid progenitors (CLPs) that have the capacity to produce differentiated mature blood cells and adaptive and innate immune cells (Figure 20.1A). Because of the high self-renewing capacity of the hematopoietic system and the fact that numerous genetic diseases are associated with the hematopoietic system, many hematopoietic diseases can be cured by allogeneic hematopoietic stem cell transplantation (allo-HSCT), where HSCs from healthy donors replace the patient's HSCs. Unfortunately, compatible donors can be difficult to find and there are serious complications associated with this type of therapy, which include graft-versus-host disease (GVHD), graft rejection and susceptibility to life-threating pathogens because of the broad immunodeficient state that occurs after allo-HSCT. However, when allo-HSCT is successful, the patient is cured for life, highlighting the great therapeutic benefit of using genetically "corrected" HSCs to cure genetic diseases of the blood system. An even more promising approach would be autologous transplantation using the patient's own HSCs that have been gene edited *ex vivo* (Figure 20.1B) as using autologous cells would eliminate the immunologic problems associated with allo-HSCT and would mean that every patient would have a potential donor – their own cells.

Since immunophenotypic identification of repopulating and self-renewing HSCs is still an ongoing area of investigation, the CD34$^+$ cell surface marker has been readily adopted to identify a very heterogeneous population of human cells, commonly referred to as hematopoietic stem and progenitor cells (HSPCs). CD34$^+$ HSPCs contain a small fraction of HSCs (0.1–1%) that are frequently divided into cells with short-term (ST-HSCs) and long-term (LT-HSCs) repopulation capacity,

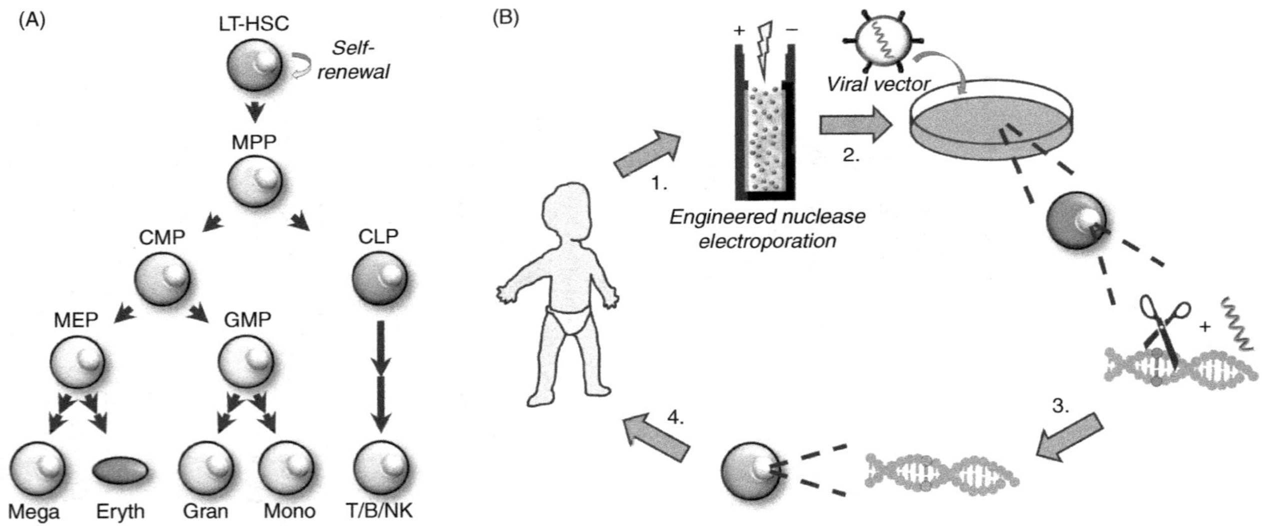

Figure 20.1 Genome editing in hematopoietic stem and progenitor cells. (A) Schematic of the hematopoietic hierarchy in which long-term repopulating hematopoietic stem cells (LT-HSCs) with self-renewal capacity give rise to multipotent progenitors (MPPs), which in turn give rise to common myeloid progenitors (CMPs) and common lymphoid progenitors (CLPs). CMPs give rise to myeloid-erythroid progenitors (MEPs) and granulocyte-monocyte progenitors (GMPs), which through several differentiation steps form fully differentiated hematopoietic cells, megakaryocytes, erythrocytes, granulocytes and monocytes. The CLPs can, through multiple differentiation steps, give rise to T cells, B cells and natural killer (NK) cells. (B) Schematic showing the putative process of autologous transplantation of genome-edited patient-derived $CD34^+$ HSPCs. Step 1: $CD34^+$ HSPCs with the genetic mutation (depicted in red) can be derived from different sources: bone marrow harvested from the iliac crest, $CD34^+$ cells mobilized into the peripheral blood, or from cord blood if the patient's cord blood was cryopreserved. Cells are cultured overnight with cytokines that promote survival and proliferation of HSCs and engineered nucleases are delivered to the cells either as mRNA or protein by electroporation. Step 2: cells recover before transplantation, or for therapies entailing homologous recombination (HR), cells are added viral vectors for donor DNA delivery or alternatively, single-stranded oligodeoxynucleotides can be included in the electroporation. Step 3: in the nuclei of electroporated cells, the nuclease facilitates a specific double-strand break at the desired site (red base pair) and either induces random sequence insertion or deletions (INDELs) leading to disruption of the locus, or in the presence of a homologous donor (green DNA) induces HR leading to gene correction or transgene integration (green base pair). Step 4: cells containing the edited locus are transplanted back into the patient to give rise to hematopoiesis of corrected cells. (A black-and-white version of this figure will appear in some formats. For the color version, please refer to the plate section.)

experimentally defined by the ability to reconstitute the human hematopoietic system in immunodeficient mouse models, such as the non-obese diabetic (NOD)-SCID-gamma (NSG) mice, which lack innate and adaptive immune surveillance allowing human cell engraftment into the host bone marrow. Several surface markers have been identified that increase the fraction of HSCs in the defined population. For example, Majeti *et al.* reported that in a population defined by the surface markers $CD34^+/CD38^-/CD90^+/CD45RA^-$, as few as ten cells could give rise to long-term multi-lineage engraftment when transplanted into immunodeficient mice (Majeti *et al.*, 2007). While a multitude of other markers have been identified that enrich for LT-HSCs, for example CD49f (Notta *et al.*, 2011), no set of markers uniquely identifies a pure LT-HSC population. This presents a great challenge for HSC research since the effect of any manipulation on the LT-HSC population is difficult to analyze unless long-term repopulation experiments in immunodeficient mice are performed. While ongoing clinical trials are looking into enriching stem-like features from subfractions of $CD34^+$ HSPCs, the CD34 cell surface marker is frequently used for purifying human HSPCs for autologous transplants in gene therapy clinical trials.

The genesis of gene therapy to cure hematopoietic monogenic diseases focus on using autologous transplants, where purified HSPCs are extracted from the patient and then copies of the functional gene are inserted at random locations in the cells' chromosomes using integrating retroviral gene vectors. Several gene therapy trials using this process have been performed with success, but serious genotoxic events have also occurred in several patients who acquired cancer due to the uncontrolled integration of the therapeutic gene causing oncogene activation (Porteus *et al.*, 2006). In contrast, gene editing has the potential to precisely modify the genome with single nucleotide precision, thereby avoiding problems associated with genotoxicity (Porteus, 2016). In the following sections we will describe recent advances in preclinical therapeutic HSPC genome editing and outline some of the challenges ahead for clinical translation of genome editing for curing diseases of the blood and immune system.

20.2 Genome Editing Reagents for Hematopoietic Stem and Progenitor Cells

A wide range of monogenic diseases of the hematopoietic system that benefit from allo-HSCT could potentially be cured by genome editing in HSPCs in an autologous transplant setting, such as myeloid/erythroid-based hemoglobinopathies and lymphoid-based primary immunodeficiencies. The two main outcomes of genome editing are: (1) gene disruption through non-homologous end joining (NHEJ)-mediated creation of insertions and deletions (indels) and (2) gene insertion/repair through homologous recombination (HR), with each having applications in therapeutic HSPC genome editing. Both genome editing events rely on engineered nucleases to create double-strand breaks in the genome that can be repaired by NHEJ when no homologous donor is present, or HR when a DNA homologous donor is present. For a number of reasons, gene editing in LT-HSCs is a challenging task. HSCs are very sensitive to DNA damage and cellular stress, and are prone to undergo apoptosis when faced with stressful conditions (Milyavsky *et al.*, 2010; Rossi *et al.*, 2007; van Galen *et al.*, 2014; Yahata *et al.*, 2011). Furthermore, LT-HSCs preferentially repair DSBs by NHEJ and their quiescence or slow cycling is also believed to be a barrier to HR (Mohrin *et al.*, 2010). Finally, HSPCs can recognize foreign DNA by innate immune DNA-sensing pathways in the cytoplasm, and subsequently launch a type I interferon (IFN) response. Triggering the innate immune system in HSPCs has been shown to induce proliferation and differentiation with subsequent loss of primitiveness (Nagai *et al.*, 2006; Yanez *et al.*, 2009). This obviously poses a challenge in terms of cell manipulations, and nuclease and donor template delivery for genome editing in human HSPCs.

20.2.1 Nuclease Delivery to Hematopoietic Stem and Progenitor Cells

Although seminal studies showed induction of DSBs in human cells using plasmid-based ZFNs, TALENs and CRISPR/Cas9, delivery of plasmid-based double-stranded DNA (dsDNA) elicits a strong innate immune response that is toxic and reduces editing efficiencies (Genovese *et al.*, 2014). Nonetheless, experiments have been described showing DSB induction in LT-HSCs using *CCR5* ZFNs delivered as plasmids, and while there was toxicity associated with this method, it proves that the editing of HSCs can be accomplished (Holt *et al.*, 2010). Since then, more sophisticated methods have been developed. For example, mRNA delivery of the nuclease has shown to be less toxic and express higher nuclease levels (DiGiusto *et al.*, 2016; Genovese *et al.*, 2014). For the CRISPR system, Cas9 can be delivered as mRNA or directly as protein precomplexed with the sgRNA before delivery. The short-lived nature of protein delivery has shown to lead to less off-target activity while retaining high on-target activity (Hendel *et al.*, 2015). Furthermore, adding chemical modifications to the sgRNA increases resistance towards cellular exonucleases, which dramatically enhances DSB induction in HSPCs (Hendel *et al.*, 2015).

20.2.1.1 Donor Delivery for Homologous Recombination

Homologous donors must be in the form of DNA to serve as a template during HR in the nucleus. Researchers have employed single-stranded oligodeoxynucleotides (ssODNs) that carry smaller homology arms than other donor types (Hoban *et al.*, 2015). Since ssODNs can be made by standard DNA synthesis, they obviate time-consuming plasmid cloning. ssODNs may bypass innate sensing pathways for dsDNA as well as avoid the immunostimulatory motifs in the bacterial backbone of plasmid DNA. However, they can only be synthesized in sizes of a few hundred base pairs, but this suffices for correction of disease-causing SNPs and recent studies have employed ssODNs for HR in $CD34^+$ HSPCs to correct the mutation in the β globin gene responsible for sickle cell disease (DeWitt *et al.*, 2016). Since various viruses have developed superior mechanisms for carrying genetic cargo into the cell's nucleus, while evading innate immune recognition, they are ideal candidates for donor DNA delivery. Indeed, to date lentiviral vectors

(LVs) and adeno-associated virus serotype 6 (AAV6) vectors have been successfully used to deliver donor templates to HSPCs without significant toxicity from innate immune sensing (De Ravin *et al.*, 2016; Dever *et al.*, 2016; Genovese *et al.*, 2014; Sather *et al.*, 2015; Wang *et al.*, 2015). Both viruses have been gutted of most of their genomic content, making them incapable of replicating. Furthermore, for donor delivery, LVs have been made integration defective (IDLV) by introducing mutations in the viral Integrase enzyme.

20.2.2 NHEJ-based Therapeutic Genome Editing

Without the presence of a homologous donor template, cells repair a DSB through NHEJ, in which the two ends of cut DNA are stitched back together in an error-prone process. Below are examples and brief descriptions of some of the most common monogenic diseases for which therapeutic NHEJ-based genome editing in HSPCs is pursued.

20.2.2.1 Human Immunodeficiency Virus (HIV)

In spite of it not being an inherited genetic disease, HIV has been a great focus of the genome editing community because a single human gene, chemokine (C-C motif) receptor 5 (*CCR5)*, can be disrupted to inhibit infection by the human immunodeficiency virus (HIV). HIV infects primarily $CD4^+$ T cells using the CD4 as the main receptor and either CCR5 or CXCR4 as co-receptors. Co-receptor dependency marks two different types of HIV (R5- and X4-tropic HIV) with R5 viruses being almost exclusively responsible for person-to-person transmission, and X4 viruses emerging at later stages of HIV infection and progression into AIDS. A 32 bp deletion in *CCR5* (Δ32) is present in the population with about a 10% heterozygous prevalence and 1% homozygous. People homozygous for Δ32 are resistant to HIV infection and in a first proof-of-principle clinical trial, a bone marrow transplant from a homozygous Δ32 donor to an HIV-infected patient functionally cured the patient with no detectable HIV levels nine years after transplant, while taking no anti-HIV drugs (Hütter, 2016). This has primed investigators to try and knock out *CCR5* in HSCs to generate a life-long functional cure for HIV, in which progeny T cells from edited HCSs would be protected against HIV infection.

Several studies have reported on *CCR5* knockout in primary human T cells using different flavors of engineered nucleases, and shown resistance to HIV infection in edited cells (DiGiusto *et al.*, 2016; Mussolino *et al.*, 2014; Perez *et al.*, 2008; Wang *et al.*, 2014; Ye *et al.*, 2014). This approach has even been tried in a first-in-man clinical trial for genome editing using zinc-finger nuclease (ZFN) mRNA in autologous T cells from 12 patients (Tebas *et al.*, 2014). Though the procedure was safe and modified T cells were shown to have a longer half-life than unmodified T cells during treatment interruption, NHEJ frequencies were not high enough to obtain a significant number of cells with biallelic disruption of *CCR5*. A T-cell-based therapy for HIV is, however, not durable, so for a life-long functional cure for HIV, *CCR5* must be disrupted in LT-HSCs. This was first pursued by Holt *et al.* who delivered *CCR5*-targeting ZFNs by plasmid-electroporation into cord-blood-derived CD34⁺ HSPCs. In the CD34⁺ population they reported indel frequencies of 17%, with an estimated 5–7% biallelic modification rate (Holt *et al.*, 2010). Twelve weeks after transplantation of edited CD34⁺ HSPCs into NSG mice, mice were challenged with R5-tropic HIV. Human *CCR5*-disrupted CD4⁺ T cells were found to be enriched rapidly to an overall *CCR5* disruption frequency exceeding the detection limit of the assay (70–80%). Sangamo BioSciences are currently initiating clinical trials using the same strategy of employing ZFNs to target *CCR5* in CD34⁺ HSPCs (DiGiusto *et al.*, 2016).

Another NHEJ-based approach for knocking out *CCR5* in HSCs has been reported using the CRISPR/Cas9 system with two sgRNAs in combination, which induces an indel at one or both sites or a deletion between the two cut sites (Hendel *et al.*, 2015; Mandal *et al.*, 2014; Nelson *et al.*, 2016). The first study using this approach was based on a single plasmid expressing Cas9 and both sgRNAs, and Cas9 was linked to GFP expression by a 2A peptide. This allowed for enrichment of transfected cells, yielding up to 48% of *CCR5* alleles disrupted. Our group has since published evidence that the CRISPR/Cas9 platform is very inefficient in CD34⁺ HSPCs when expressed from plasmids (Hendel *et al.*, 2015). In experiments measuring the functional stability of electroporated synthetic sgRNAs, it was evident that the sgRNAs are very unstable, and by stabilizing the ends of the sgRNAs using modified nucleotides with increased resistance towards exonucleases we increased indel frequencies significantly. Using the two sgRNA approach, we obtained frequencies of *CCR5* disruption in HSPCs of 43% without applying any form of selection (Hendel *et al.*, 2015), and have recently optimized this platform and achieved 75–90% indel frequencies using a single sgRNA at *CCR5* in T cells and CD34⁺ HSPCs by electroporation of Cas9 protein precomplexed with modified sgRNAs (Bak and Porteus, 2017). This represents the most state-of-the art approach to date, and for an HSC-based functional cure for R5-tropic viruses these high editing rates are a potential game changer.

20.2.2.2 Hemoglobinopathies

Hemoglobinopathies are inherited genetic diseases with mutations affecting either the function or production of hemoglobin subunits responsible for forming functional oxygen-carrying hemoglobin tetramers. The β-globin gene is most frequently mutated in hemoglobinopathies. β-thalassemia is a group of disorders associated with diminished or lack of production of β-globin leading to anemia. Sickle cell disease (SCD), on the other hand, is caused by a point mutation in at least one allele of the β-globin gene changing a single amino acid from glutamic acid (E) to valine (V) (HgbS). This change makes β-globin subunits polymerize under deoxygenated conditions, which distorts the shape of the red blood cells to the classic "sickle" shape. Sickle cell anemia is the most common form of SCD that manifests from the homozygous E6V genotype. In addition to severe anemia, SCD causes vaso-occlusive crises with painful ischemia and necrosis in the restricted organ. SCD patients have a life expectancy of around 45 years, highlighting the need for a life-long cure. SCD can be treated with hydroxyurea, which increases levels of fetal hemoglobin by induction of γ-globin expression, which is otherwise turned off during infancy. Higher levels of fetal hemoglobin persist in some adults and have been shown to reduce the severity of SCD as the γ-globin interferes with the pathologic polymerization process caused by HgbS. Gene editing approaches in HSPCs using CRISPR/Cas9 have successfully reinduced γ-globin by either disrupting expression of a γ-globin repressor protein (BCL11A) (Canver *et al.*, 2015), or disrupting a region in the γ-globin promoter that recruits transcriptional repressor proteins (Traxler *et al.*, 2016). These studies offer gene editing clinical approaches to cure hemoglobinopathies through error-prone mechanisms; however, whether fetal hemoglobin would persist remains to be determined. A more sophisticated approach would be to utilize HR to correct the genetic mutations or to express a full-length cDNA, as discussed below.

20.2.3 HR-based Therapeutic Genome Editing

HR-based genome editing relies on the delivery of both a nuclease and a DNA donor, which serves as repair template during HR, and two main approaches can be pursued, depending on the genetic nature of the disease. For genes with a single disease-causing SNP at a single site in the gene, the SNP can be corrected by HR. The donor template should be designed with homology arms on both sides of the changed nucleotide, and it's important to also introduce synonymous mutations in the nuclease target site of the donor to prevent recutting and introduction of an indel after HR. Also, if the cut site of the nuclease is not directly at the site of the desired mutation, one should include synonymous mutations between the cut site and the modified nucleotide to prevent premature crossover and incomplete HR.

If the mutations are scattered throughout the gene, a more practical approach would entail designing a one-size-fits-all HR approach, in which the full cDNA is integrated immediately after the start codon of the gene, thereby bypassing transcription of the endogenous mutated gene (Hubbard *et al.*, 2016; Voit *et al.*, 2014). The integrated cDNA would be expressed from the endogenous promoter and subject to normal regulatory control, in contrast to existing gene therapies using retroviral integration, where strong exogenous promoters drive transgene expression. For cDNA knockin after the start codon of the endogenous gene, the cDNA must be sequence-diverged with synonymous mutations throughout the cDNA. As for a SNP correcting donor, this is necessary to mutate the nuclease target site and to prevent incomplete HR. If supraphysiological cellular levels of transgene expression are desired, an exogenous strong promoter can be employed and the transgene can be integrated into so-called "safe harbors" in the genome, e.g. *CCR5* and *AAVS1*, which have tolerability of biallelic disruption as well as high transcriptional competence (Lombardo *et al.*, 2011).

20.2.3.1 Human Immunodeficiency Virus

The main limitation of the previously described approach of knocking out *CCR5* is that it is only applicable to patients with R5-tropic viruses and indeed, *CCR5*-disrupted T cells are not resistant to X4-tropic HIV strains (Holt *et al.*, 2010; Voit *et al.*, 2013). While *CXCR4* cannot be disrupted in HSCs due to its major role in

HSC homeostasis, Voit *et al.* have shown that integrating HIV restriction factor gene cassettes into *CCR5* in a T cell line can render cells resistant to both R5- and X4-tropic HIV. A multi-pronged therapy like this, such as is used in antiretroviral drug therapy, is believed to be necessary to effectively combat both R5- and X4-tropic HIV, which is a rapidly mutating virus with an extraordinary ability to adapt to negative selection pressure.

The first proof of principle of targeted integration by HR in cord-blood-derived $CD34^+$ HSPCs was published by Lombardo *et al.* in 2007 using IDLVs to deliver both ZFNs and donors (Lombardo *et al.*, 2007). The highest rates achieved were a modest 0.11%, and while HR rates in LT-HSCs were not assessed, on-target integration was confirmed in 80–90% of targeted cells. Wang *et al.* recently employed ZFN mRNA with AAV6 donor delivery and reported 15–25% GFP HR rates at *CCR5* in $CD34^+$ HSPCs, which translated to 1–5% in the bone marrow of mice 16 weeks after transplant (Wang *et al.*, 2015). The edited cells were not challenged with HIV, nor were HIV restriction factors introduced, but the study serves as a proof-of-concept that clinically relevant *CCR5* editing efficiencies can be achieved in LT-HSCs.

20.2.3.2 Hemoglobinopathies

For SCD, precise gene repair in HSCs to correct the disease-causing SNP would be the optimal way to cure patients. Furthermore, since corrected red blood cells would have a longer half-life than sickle red blood cells (120 vs. 12 days), even low correction rates around 10% are expected to be curative. This approach has been tested by several groups including our own. Hoban *et al.* used ZFN mRNA delivery with either IDLV or ssODNs carrying the homologous donor (Hoban *et al.*, 2015). Experiments showed 10–20% correction and 20–30% indels in the bulk $CD34^+$ HSPCs before transplantation, but after 16 weeks in NSG mice rates were decreased to 0.2–2.1% correction and 3–6% indels in the human population, highlighting the difficulty of targeting LT-HSCs for HR at the *HBB* gene. Our group and Dewitt *et al.* recently used Cas9 protein delivery by electroporation in conjunction with rAAV6 or ssODNs, respectively, and obtained efficiencies of up to 50% targeting in HSPCs *in vitro* (Dever *et al.*, 2016; DeWitt *et al.*, 2016). When modified cells were transplanted into NSG mice, an average of 2–4% of human cells were targeted, suggesting again that LT-HSCs are difficult to target for HR. We therefore devised an enrichment paradigm that uses a cDNA knockin approach and takes advantage of higher expression of a reporter gene following successful HR. Using this methodology, strikingly higher frequencies of modified cells in the transplanted mice were observed (up to 90%; Dever *et al.*, 2016). Thus, this enrichment methodology has the potential to overcome the problem of inefficient HSC targeting and to be curative for hemoglobinopathies.

20.2.3.3 Primary Immunodeficiencies

Primary immunodeficiencies are a large group of genetic diseases that include more than 250 distinct diseases with defects in the development or function of immune cells, which can result in recurrent and life-threatening infections.

The most common class of immunodeficiency is the severe combined immunodeficiencies (SCID), with the most prevalent being X-linked SCID (X-SCID) caused by mutations in the interleukin-2 receptor subunit gamma (*IL2RG*). Because this subunit is part of receptors for multiple interleukins, a non-functional *IL2RG* gene has a widespread effect on immune system development and activity, and untreated X-SCID patients rarely survive beyond childhood.

Genovese *et al.* edited the *IL2RG* gene using ZFN mRNA and IDLV donor in cord-blood-derived $CD34^+$ HSPCs (Genovese *et al.*, 2014). Integration of a GFP reporter into *IL2RG* yielded 5% GFP^+ cells *in vitro*, but 12–23 weeks after transplantation into NSG mice, this frequency had dropped to 1.5% (median < 0.5%). However, using an *IL2RG* cDNA (exon 5–8) donor knockin strategy at exon 5 in bone-marrow-derived $CD34^+$ cells, the authors showed 6% integration frequencies *in vitro* and 5% in human cells 17 weeks after transplant. Importantly, functional reconstitution of *IL2RG* was confirmed in the corrected lymphoid progeny, laying the foundation for a genome editing strategy for curing X-SCID by cDNA knockin.

X-linked chronic granulomatous disease (CGD) is another immunodeficiency that affects phagocytic cells, and patients have a high susceptibility towards fungal and bacterial infections. It is caused by mutations in the *gp91phox* gene, which encodes a subunit of the NADPH oxidase that produces reactive oxygen species to fight endocytosed pathogens. Consequently, patients suffer from life-threatening abscesses and granulomas in various tissues. While studies have not pursued *gp91phox* correction in $CD34^+$ HSPCs, De Ravin *et al.* used ZFNs and AAV6 donor to integrate *gp91phox* cDNA into the proposed safe harbor locus *AAVS1* (De Ravin *et al.*, 2016). AAV6 donor vectors were used with either an MND promoter driving *gp91phox* cDNA expression or a promoterless gp91phox donor relying on the endogenous *PPP1R12C* promoter residing in the AAVS1 locus. Targeted integration was achieved for both vectors in patient-derived $CD34^+$ HSPCs with 15% of cells expressing gp91phox protein. However, as expected, the exogenous MND promoter gave rise to markedly higher levels of gp91phox protein than the endogenous promoter, and only the MDN promoter construct gave rise to NADPH oxidase activity close to normal on a per cell basis. After transplant, 4% human cells expressed gp91phox (a decrease from 15% *in vitro*), and it is important to note that these levels were assessed only eight weeks following transplant. Thus, editing rates in LT-HSCs cannot be evaluated and may display lower editing frequencies than 4%. It is encouraging, however, that targeted integration of *gp91phox* cDNA in patient-derived $CD34^+$ HSPCs is feasible and that edited cells display clinically relevant levels of oxidase activity.

20.3 Concluding Remarks

Gene therapy has revolutionized the treatment of otherwise fatal hematopoietic monogenic disorders. Patients, for whom no suitable HLA-matched related stem cell donor is available, can now be treated by autologous transplantation of *ex vivo* HSPCs transduced with integrating viral vectors. Recently, such therapy for the treatment of ADA-SCID was approved for marketing in Europe. Several genetic diseases have been treated using this approach, but it has been evident that the near-random integration pattern of these viral vectors can cause oncogenesis with fatal outcomes. Though newer generations of integrating viral vectors have displayed a safer integration profile, the risks associated with uncontrolled integration remain. Genome editing using designer nucleases can not only facilitate locus-specific transgene integration, but also introduce disrupting indels, which may be beneficial for some therapeutic purposes because of the high efficiencies in LT-HSCs. While studies are still ongoing trying to fully characterize the specificity of engineered nucleases and understand the significance of off-target activity in preclinical models, history has taught us that preclinical models can be elusive, and that safety and efficacy of cell and gene therapies must stand the true test of clinical trials. Hence, it is very encouraging that the first clinical trials using genome-edited autologous T cells have proven safe (Tebas *et al.*, 2014). As described above, one bottleneck in genome editing is in the HR editing frequencies in LT-HSCs. In diseases with selective advantage of corrected cells, this will be less of a problem, but other diseases might require frequencies that are not attainable with current editing methodologies. Recently, we have identified a way of enriching HSPCs with targeted integration based on expression of a clinically relevant cell surface marker (Dever *et al.*, 2016). Such enrichment does not increase absolute numbers of edited LT-HSCs, but transplantation of an enriched edited population should reduce competition by unedited cells during repopulation. *Ex vivo* HSC expansion protocols have also been devised, some of which are in clinical trials (Butler *et al.*, 2012; Fares *et al.*, 2014), and HSC expansion may prove useful for expanding an enriched edited population of LT-HSCs for better transplant outcomes.

Genome editing has enabled a new paradigm within gene therapy, in which we can rewrite the genetic code with single nucleotide precision. The path to clinical translation is set forth, and as our genome editing toolkits develop and our understanding of HSC biology broadens, the next era in precision medicine is imminent.

REFERENCES

Bak RO, Porteus MH. 2017. CRISPR-mediated integration of large gene cassettes using AAV donor vectors. *Cell Rep* **20**(3): 750–756.

Butler JM, Gars EJ, James DJ, *et al.* 2012. Development of a vascular niche platform for expansion of repopulating human cord blood stem and progenitor cells. *Blood* **120**: 1344–1347.

Canver MC, Smith EC, Sher F, *et al.* 2015. BCL11A enhancer dissection by Cas9-mediated in situ saturating mutagenesis. *Nature* **527**: 192–197.

De Ravin SS, Reik A, Liu PQ, *et al.* 2016. Targeted gene addition in human CD34(+) hematopoietic cells for correction of X-linked chronic granulomatous disease. *Nature Biotechnol* **34**: 424–429.

Dever DP, Bak RO, Reinisch A, *et al.* 2016. CRISPR/Cas9 beta-globin gene targeting in human haematopoietic stem cells. *Nature* **539**: 384–389.

DeWitt MA, Magis W, Bray NL, *et al.* 2016. Selection-free genome editing of the sickle mutation in human adult hematopoietic stem/progenitor cells. *Sci Transl Med* **8**: 360ra134.

DiGiusto DL, Cannon PM, Holmes MC, *et al.* 2016. Preclinical development and qualification of ZFN-mediated CCR5 disruption in human hematopoietic stem/progenitor cells. *Mol Ther Methods Dev* **3**: 16067.

Doulatov S, Notta F, Laurenti E, Dick JE. 2012. Hematopoiesis: a human perspective. *Cell Stem Cell* **10**: 120–136.

Fares I, Chagraoui J, Gareau Y, *et al.* 2014. Cord blood expansion. Pyrimidoindole derivatives are agonists of human hematopoietic stem cell self-renewal. *Science* **345**: 1509–1512.

Genovese P, Schiroli G, Escobar G, *et al.* 2014. Targeted genome editing in human repopulating haematopoietic stem cells. *Nature* **510**: 235–240.

Hendel A, Bak RO, Clark JT, *et al.* 2015. Chemically modified guide RNAs enhance CRISPR-Cas genome editing in human primary cells. *Nat Biotechnol* **33**: 985–989.

Hoban MD, Cost GJ, Mendel MC, *et al.* 2015. Correction of the sickle cell disease mutation in human hematopoietic stem/progenitor cells. *Blood* **125**: 2597–2604.

Holt N, Wang J, Kim K, *et al.* 2010. Human hematopoietic stem/progenitor cells modified by zinc-finger nucleases targeted to CCR5 control HIV-1 in vivo. *Nat Biotechnol* **28**: 839–847.

Hubbard N, Hagin D, Sommer K, *et al.* 2016. Targeted gene editing restores regulated CD40 L function in X-linked hyper-IgM syndrome. *Blood* **127**: 2513–2522.

Hutter G. 2016. HIV+ patients and HIV eradication: allogeneic transplantation. *Exp Rev Hematol* **9**: 615–616.

Lombardo A, Cesana D, Genovese P, *et al.* 2011. Site-specific integration and tailoring of cassette design for sustainable gene transfer. *Nat Methods* **8**: 861–869.

Lombardo A, Genovese P, Beausejour CM, *et al.* 2007. Gene editing in human stem cells using zinc finger nucleases and integrase-defective lentiviral vector delivery. *Nat Biotechnol* **25**: 1298–1306.

Majeti R, Park CY, Weissman IL. 2007. Identification of a hierarchy of multipotent hematopoietic progenitors in human cord blood. *Cell Stem Cell* **1**: 635–645.

Mandal PK, Ferreira LM, Collins R, *et al.* 2014. Efficient ablation of genes in human hematopoietic stem and effector cells using CRISPR/Cas9. *Cell Stem Cell* **15**: 643–652.

Milyavsky M, Gan OI, Trottier M, *et al.* 2010. A distinctive DNA damage response in human hematopoietic stem cells reveals an apoptosis-independent role for p53 in self-renewal. *Cell Stem Cell* **7**: 186–197.

Mohrin M, Bourke E, Alexander D, *et al.* 2010. Hematopoietic stem cell quiescence promotes error-prone DNA repair and mutagenesis. *Cell Stem Cell* **7**: 174–185.

Mussolino C, Alzubi J, Fine EJ, *et al.* 2014. TALENs facilitate targeted genome editing in human cells with high specificity and low cytotoxicity. *Nucleic Acids Res* **42**: 6762–6773.

Nagai Y, Garrett KP, Ohta S, *et al.* 2006. Toll-like receptors on hematopoietic progenitor cells stimulate innate immune system replenishment. *Immunity* **24**: 801–812.

Nelson CE, Hakim CH, Ousterout DG, *et al.* 2016. In vivo genome editing improves muscle function in a mouse model of Duchenne muscular dystrophy. *Science* **351**: 403–407.

Notta F, Doulatov S, Laurenti E, *et al.* 2011. Isolation of single human hematopoietic stem cells capable of long-term multilineage engraftment. *Science* **333**: 218–221.

Perez EE, Wang J, Miller JC, *et al.* 2008. Establishment of HIV-1 resistance in CD4+ T cells by genome editing using zinc-finger nucleases. *Nat Biotechnol* **26**: 808–816.

Porteus M. 2016. Genome editing: a new approach to human therapeutics. *Annu Rev Pharmacol Toxicol* **56**: 163–190.

Porteus MH, Connelly JP, Pruett SM. 2006. A look to future directions in gene therapy research for monogenic diseases. *PLoS Genet* **2**: e133.

Rossi DJ, Bryder D, Seita J, *et al.* 2007. Deficiencies in DNA damage repair limit the function of haematopoietic stem cells with age. *Nature* **447**: 725–729.

Sather BD, Romano Ibarra GS, Sommer K, *et al.* 2015. Efficient modification of CCR5 in primary human hematopoietic cells using a megaTAL nuclease and AAV donor template. *Sci Transl Med* **7**: 307ra156.

Tebas P, Stein D, Tang WW, *et al.* 2014. Gene editing of CCR5 in autologous CD4 T cells of persons infected with HIV. *N Engl J Med* **370**: 901–910.

Traxler EA, Yao Y, Wang YD, *et al.* 2016. A genome-editing strategy to treat beta-hemoglobinopathies that recapitulates a mutation associated with a benign genetic condition. *Nat Med* **22**(9): 987–990.

van Galen P, Kreso A, Mbong N, *et al.* 2014. The unfolded protein response governs integrity of the haematopoietic stem-cell pool during stress. *Nature* **510**: 268–272.

Voit RA, Hendel A, Pruett-Miller SM, Porteus MH. 2014. Nuclease-mediated gene editing by homologous recombination of the human globin locus. *Nucleic Acids Res* **42**: 1365–1378.

Voit RA, McMahon MA, Sawyer SL, Porteus MH. 2013. Generation of an HIV resistant T-cell line by targeted "stacking" of restriction factors. *Mol Ther* **21**: 786–795.

Wang J, Exline CM, DeClercq JJ, *et al.* 2015. Homology-driven genome editing in hematopoietic stem and progenitor cells using ZFN mRNA and AAV6 donors. *Nat Biotechnol* **33**: 1256–1263.

Wang W, Ye C, Liu J, *et al.* 2014. CCR5 gene disruption via lentiviral vectors expressing Cas9 and single guided RNA renders cells resistant to HIV-1 infection. *PLoS One* **9**: e115987.

Yahata T, Takanashi T, Muguruma Y, *et al.* 2011. Accumulation of oxidative DNA damage restricts the self-renewal capacity of human hematopoietic stem cells. *Blood* **118**: 2941–2950.

Yanez A, Murciano C, O'Connor JE, Gozalbo D, Gil ML. 2009. *Candida albicans* triggers proliferation and differentiation of hematopoietic stem and progenitor cells by a MyD88-dependent signaling. *Microbes Infect* **11**: 531–535.

Ye L, Wang J, Beyer AI, *et al.* 2014. Seamless modification of wild-type induced pluripotent stem cells to the natural CCR5Delta32 mutation confers resistance to HIV infection. *Proc Natl Acad Sci USA* **111**: 9591–9596.

Part V

Genome Editing in Disease Biology

21 CRISPR/Cas9-based *In Vivo* Models of Cancer

Pranav V. Lalgudi, Monte M. Winslow and Ian P. Winters

21.1 Introduction to Cancer Modeling

The twenty-first century has been marked by many notable advances in molecular biology and medicine, yet cancer remains one of the most significant global health burdens of modern times; indeed, new cancer cases per year are expected to exceed 20 million by 2030 (Thun *et al.*, 2010). Cancer is characterized by a massive number of diverse genetic and epigenetic modifications. Over the past several decades, experimental models of human cancer have played a critical role in our understanding of the mechanisms underlying cancer initiation, progression and metastasis. The use of increasingly sophisticated systems that give rise to more rapid, cost-effective and accurate models of cancer will accelerate our understanding of the molecular mechanisms that drive carcinogenesis and provide novel platforms for preclinical studies.

21.1.1 History of Cancer Modeling

Although correlations between specific environmental factors and cancer incidence were noted as early as the eighteenth century, cancer modeling began in the early 1900s with the direct induction of cancer in animals. Initial approaches to induce cancer involved exposing animals to chemical carcinogens, radiation and oncogenic viruses (Fujiki, 2014; Shimkin and Stoner, 1975) In 1976, the discovery that an altered version of a normal cellular gene could induce cancer provided a potential molecular explanation for the formation of this disease (Stehelin *et al.*, 1976). Subsequent experiments lent credence to this hypothesis; other cellular genes harboring specific mutations could also transform cells in culture, and recurrent alterations in endogenous genes were observed in human tumors. However, initial experimental techniques lacked the ability to create defined mutations in normal cells within their native tissue context. A critical step towards a molecular understanding of cancer was the development of genetically engineered mouse models, which provided researchers with a powerful reverse genetics approach to determine the functional impact of specific genetic alterations in cancer.

Genetically modified mice were first generated in the early 1980s using methods for the germline modification of mice by random genomic integration of a foreign DNA "transgene" into mouse embryos (Brinster *et al.*, 1981; Costantini and Lacy, 1981; Gordon and Ruddle, 1981). The ability to stably introduce foreign DNA into mouse cells meant that cancer biologists could, for the first time, test whether overexpression of specific putative oncogenes could initiate cancer in a natural *in vivo* context (Hutchinson and Muller, 2000; Macleod and Jacks, 1999). Transgenic mice enabled the functional characterization of some of the first and most important oncogenes discovered in human cancer, but the random transgene integration and non-physiologic expression, among other factors, did not enable all types of alterations observed in human cancer to be properly recapitulated.

A major breakthrough in cancer modeling came with the development of gene-targeting techniques that allowed the introduction of defined alterations into specific loci in the mouse genome. The ability to precisely introduce gain-of-function or loss-of-function mutations into endogenous genes allowed for the development of knockin and knockout mice. Knockout mice have been extensively used to investigate the function of tumor-suppressor genes during carcinogenesis (Walrath *et al.*, 2010). Conversely, knockin models have been used to study the roles of oncogenic mutations in cancer, as well as to model germline cancer susceptibility syndromes (Olive *et al.*, 2004). Over the past several decades, studies in genetically engineered mouse models of cancer have uncovered key molecular mechanisms by which many genes and pathways promote or inhibit tumorigenesis.

Further improvements on gene-targeted mouse models have included the generation of conditional alleles that leverage site-directed recombinase systems such as Cre/Lox. While conventional knockout and knockin mice possess engineered alterations in all cells of the animal from embryonic development through adulthood, the Cre/Lox system allows for conditional activation and inactivation of genes in defined cells in adult mice. The fine-tuned spatio-temporal control afforded by these conditional alleles improved cancer modeling and enabled the development of models of many cancer types that more closely reproduced the initiating events leading to human cancer (Flemr *et al.*, 2015; Frese and Tuveson, 2007).

For the last quarter century, the use of genetically engineered mouse models has been fundamental to our understanding of the mechanisms that drive tumorigenesis. However, the usefulness and applicability of these models is severely limited by the financial and temporal burden of generating mice with novel germline modifications. Site-directed nucleases, including ZFNs (zinc-finger nucleases; Bibikova *et al.*, 2003) and TALENs (transcription activator-like effector nucleases; Li *et al.*, 2011), which induce DSBs (double-strand breaks) at specific sites in the genome, have been utilized for more efficient germline modification of mice (Carbery *et al.*, 2010; Sommer *et al.*, 2015). However, the generation of efficient TALENs and ZFNs is tedious. Ideally, next-generation mouse models of cancer could easily be applied to study novel candidate cancer genes or specific aspects of tumorigenesis at a small fraction of the time and cost of conventional systems.

21.1.2 The CRISPR/Cas9 System

The discovery of the bacterial RNA-guided nuclease system, CRISPR/Cas9 (clustered regularly interspaced palindromic repeats/CRISPR-associated protein 9), and its ease and versatility of implementation in mammalian cells have heralded a new era of genetically modified mouse models (Cong *et al.*, 2013). CRISPR/Cas9-mediated DNA cleavage at targeted sites activates cellular DNA repair machinery, which can operate through several pathways. The most common of these are non-homologous end joining (NHEJ) by error-prone ligation and homology-directed repair (HDR) by precise repair of the damaged strand in the presence of a template homologous to the targeted locus. NHEJ often generates frameshift insertions or deletions (indels) at the cleavage site, making it well suited for engineering loss-of-function mutations (Chiou *et al.*, 2015), while HDR can be employed to introduce gain-of-function mutations using a DNA template for HDR (Xue *et al.*, 2014).

CRISPR/Cas9-based cancer modeling in mice through genomic modification of somatic cells requires the delivery of two components to the desired cell type: the Cas9 nuclease and the directing single-guide RNA (sgRNA; Figure 21.1). Cas9 can be expressed in cells of interest through direct delivery of naked plasmid (Xue *et al.*,

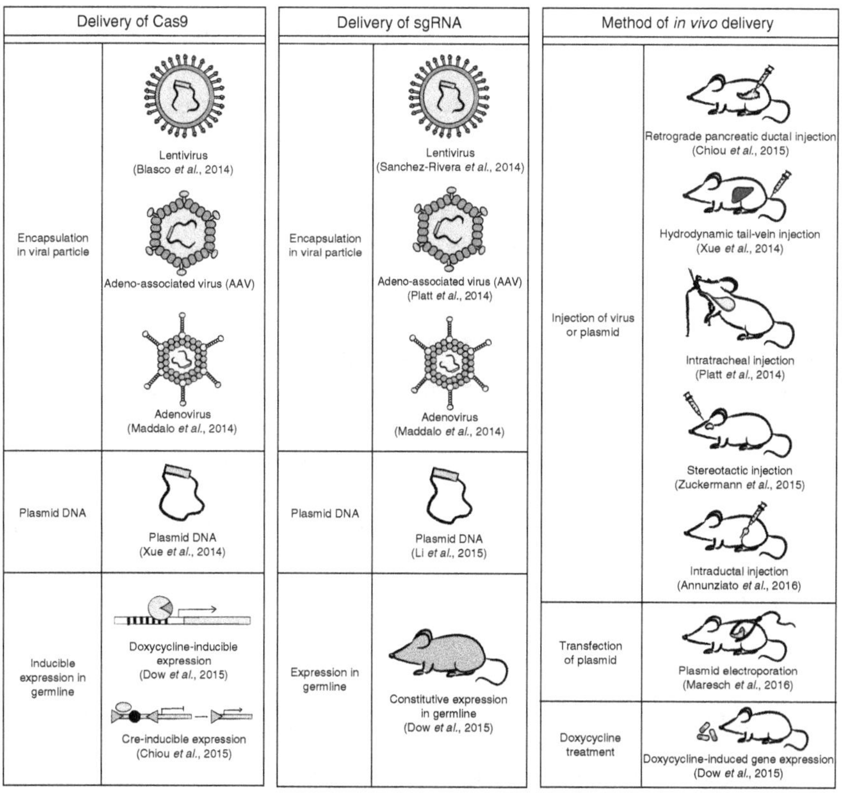

Figure 21.1 The CRISPR/Cas9 toolkit for *in vivo* genome editing. A variety of vectors and *in vivo* methods can be used to deliver nucleic acids encoding Cas9 and sgRNAs to the desired cell and tissue types. Example references are indicated.

2014) or viral vectors including lentivirus (Blasco *et al.*, 2014), adenovirus (Maddalo *et al.*, 2014) and adeno-associated virus (AAV; Li *et al.*, 2015). Cas9 can also be conditionally expressed from either Cre or doxycycline-inducible transgenes (Dow *et al.*, 2015; Platt *et al.*, 2014). The sgRNA can be delivered in the form of a plasmid (Li *et al.*, 2015) or viral vector (Chiou *et al.*, 2015), or this too can be expressed from a germline transgene (Dow *et al.*, 2015).

When these components are provided on plasmids or viral vectors, it is necessary to deliver them to the proper cell type in the target organ to initiate the development of a specific type of cancer. Different methods of *in vivo* delivery are often used, and are cell- and tissue type-dependent. These approaches include intratracheal administration of viruses to the lungs (Platt *et al.*, 2014), hydrodynamic tail vein injection to the liver (Xue *et al.*, 2014), stereotactic injection into the brain (Zuckermann *et al.*, 2015), retrograde pancreatic ductal injection to the pancreas (Chiou *et al.*, 2015) and intraductal injection of the mammary gland (Annunziato *et al.*, 2016). These techniques allow for efficient, localized introduction of precise genetic modifications *in vivo*, enabling the production of accurate models of cancer that recapitulate its phenotypic changes in a natural context.

Using CRISPR/Cas9 for genome editing to generate *in vivo* models of cancer has several advantages over previous methods, such as site-directed ZFNs and TALENs. CRISPR/Cas9 provides higher editing efficiency with regards to cleavage and indel formation, which can be attributed to factors including sensitivity of TALENs to CpG methylation (Valton *et al.*, 2012) and the more cumbersome binding of protein to DNA needed for other targeted nucleases. Moreover, low cytotoxicity and ease of design of CRISPR-based systems make them well suited for *in vivo* modeling. CRISPR's compatibility with multiplexed sgRNAs (Heckl *et al.*, 2014) also enables the rapid development of novel mouse models of cancer with multiple genetic modifications, a significant advantage over the arduous process of breeding and genotyping generations of mice when using conventional knockout or Cre-regulatable alleles. Although non-specific binding of sgRNAs in the CRISPR/Cas9 system can lead to modification of off-target loci, this issue can be alleviated through careful sgRNA design, the use of variant CRISPR systems (Dow *et al.*, 2015), and confirmation of *in vivo* phenotypes using additional independent sgRNAs targeting the same gene.

The flexibility and efficiency of CRISPR/Cas9 systems make them an exciting method with which to model cancer in living organisms. In the few years since it was first established as a tool for genome editing, the CRISPR/Cas9 system has already begun to be leveraged for this purpose. In the following sections, we examine the approaches taken to date in the creation of next-generation, CRISPR-based models of cancer.

21.2 CRISPR/Cas9 Systems and Cancer Modeling

Several approaches are used to generate genetic models of cancer using CRISPR/Cas9. First, CRISPR/Cas9-mediated genome editing enables the modification of murine germline cells to generate specific alleles. Second, inducing alterations in cell lines *in vitro* allows for observation of cellular phenotypes both in culture and after transplantation into recipient mice. Finally, and perhaps most significantly, the CRISPR/Cas9-system enables rapid, inexpensive and precise editing of target genes in somatic cells of adult organs. This advance enables the characterization of genetic alterations in tumors at a higher throughput than previously possible.

21.2.1 Germline Modifications in Models of Cancer

Germline modification of mouse embryonic stem cells or oocytes can be used to generate mice with desired heritable genetic changes. The production of such genetically engineered mouse models has been accelerated by CRISPR/Cas9 technology. The CRISPR system has had success modeling both loss-of-function mutations via NHEJ (Li *et al.*, 2014; Wang *et al.*, 2013) and gain-of-function changes by HDR-based editing (Li *et al.*, 2014) in the germline. The broad spectrum of genetic alterations that can be induced using these systems will facilitate and accelerate the generation of novel genetically engineered mouse models of cancer.

Furthermore, unlike the cumbersome genetic crosses previously required to generate mice with desired modifications in combinations of alleles, one-step approaches can be used to create homozygous alterations and simultaneously alter multiple genes (Wang *et al.*, 2013; Yang *et al.*, 2013). Additionally, Cas9-mediated germline modification has been successfully implemented in many organisms including rats (Li *et al.*, 2013), zebrafish (Xiao *et al.*, 2013) and cynomolgus monkeys (Niu *et al.*, 2014).These experiments highlight the versatility of the CRISPR/Cas9 system and suggest that other organisms may also be amenable to the analysis of tumorigenesis.

While the application of CRISPR to create germline cancer models has been relatively limited to date, the generation of specific germline-modified models of cancer will likely continue to benefit from these techniques. Genetically engineered mouse models with germline alterations in cancer genes have been particularly useful for testing therapeutics *in vivo* (Chen *et al.*, 2012; Engelman *et al.*, 2008), and thus it is likely that the generation of these models via CRISPR will contribute to such applications. These CRISPR-based germline models provide a powerful way to efficiently generate mice and advance our understanding of cancer development and treatment.

21.2.2 *Ex Vivo* Genetic Modification and Transplantation-based Models

In vitro systems often serve as basic models of cancer at the cellular level. These methods have been used in conjunction with CRISPR/Cas9 to interrogate gene function (Togashi *et al.*, 2015), induce oncogenic chromosomal rearrangements (Torres *et al.*, 2014) and conduct genetic screens in cell culture (Shalem *et al.*, 2014), among many other applications. Due to the limited nature of cell culture systems, transplantation-based systems in which genetically modified cells are introduced into a living host generally serve as more accurate models of disease. *Ex vivo* modification of cells or cell lines followed by transplantation into recipient animals can provide a better understanding of many aspects of cancer initiation, progression, drug response and metastasis (Table 21.1).

Table 21.1 *Ex vivo* cancer modeling studies using the CRISPR/Cas9 system.

Type of cancer	Cas9 expression	sgRNA expression	Genetic modifications	Summary
Leukemia	Cas9 Plasmid DNA	Plasmid DNA	Loss of function mutations in *Mll3*	Modeling loss of function mutations in *Mll3* uncovered its function as a haploinsufficient tumor suppressor. (Chen *et al.*, 2014)
Leukemia	Lentiviral delivery of Cas9	Lentiviral delivery of sgRNAs	Loss of function mutations in *Runx1* Combinatorial loss of function mutations in *Tet2*, *Runx1*, *Dnmt3a*, *Nf1*, *Ezh2*, and *Smc3*	Combinatorial mutations in 6 genes implicated in leukemias were generated, modifying up to 5 genes at once in single cells to drive tumorigenesis. (Heckl *et al.*, 2014)
Lymphoma and Leukemia	Retroviral delivery of dCas9	Lentiviral delivery of sgRNAs	Transcriptional repression of *p53* Transcriptional activation of *Mgmt* and *Chek2* 25 gene screen at the transcriptional level	CRISPRi and CRISPRa were successfully demonstrated, both on the scale of individual genes and in larger screens, to model tumor progression and treatment resistance. (Braun *et al.*, 2016)
Lymphoma	Lentiviral and retroviral delivery of Cas9	Retroviral delivery of sgRNAs	Loss of function mutations in *p53*	An *in situ* functional assay was conducted to characterize positive selection for *p53* mutants in lymphomas resistant to chemotherapy. (Malina *et al.*, 2013)
Lymphoma	Lentiviral delivery of Cas9	Lentiviral delivery of doxycycline-inducible sgRNAs	Loss of function mutations in *Mcl1* and *p53*	A novel doxycycline-inducible lentiviral CRISPR system was used to independently engineer LOF mutations in *p53* and *Mcl1* in models of lymphoma. (Aubrey *et al.*, 2015)
Colorectal cancer	Cas9 Plasmid DNA	Plasmid DNA	Gain of function mutations in *Kras* Loss of function mutations in *p53, Smad4* and *Apc*	Mutations in 5 genes were introduced in intestinal organoid culture, demonstrating the need for accumulation of mutations before metastatic progression. (Matano *et al.*, 2015)
Colorectal cancer	Cas9 Plasmid DNA	Plasmid DNA	Gain of function mutations in *Kras* Loss of function mutations in *p53, Smad4* and *Apc*	Sequential mutations in 4 genes were introduced in intestinal organoid culture, successfully modeling invasive carcinomas following transplantation. (Drost *et al.*, 2015)

Colorectal cancer	Cas9 and nCas9 Plasmid DNA	Plasmid DNA	Correcting loss of function mutations in *PKCβ* to wild-type	After classification of mutations in *PKCβ* as LOF or GOF, its role as a tumor suppressor was confirmed through correction of a LOF allele *in vitro*, resulting in reduced tumor growth after transplantation. (Antal *et al.*, 2015)
Lung cancer	Lentiviral delivery of Cas9	Lentiviral delivery of sgRNA library	Attempted genome scale knockout screen using mGeCKOa library with 67405 sgRNAs and a micropool of 524 sgRNAs LOF mutations in *Nf2*, *Pten*, *Trim72*, *Cdkn2a*, *Fga*, *Cryba4*, *miR-152*, and *miR-345*	A screen to identify genes regulating tumor growth and metastasis identified tumor suppressor genes. A smaller screen then confirmed their roles, and the functions of specific genes were individually validated. (Chen *et al.*, 2015)
Breast cancer	nCas9 Plasmid DNA	Plasmid DNA	Loss of function mutations in *Cripto-1*	*Cripto-1* was identified as a potential therapeutic target after generation of homozygous LOF mutations led to onset of tumorigenesis. (Castro *et al.*, 2015)

These methods can be used as a complementary approach to *in vivo* studies by genetically engineering populations of cells outside a host, followed by transplantation into adult organisms. While these studies do not fully recapitulate the metastatic niche of tumorigenesis, they can be used to model specific aspects of tumorigenesis across a number of cancer types.

One of the first *ex vivo* models of cancer using CRISPR targeted the *Mll3* gene, which is often mutated or lost in cancers (Chen *et al.*, 2014). Deletion of this gene and transplantation of modified cells into sublethally irradiated mice accelerated the development of leukemia. These leukemias were resistant to conventional chemotherapy, but were sensitive to a BET inhibitor. In this instance, a CRISPR-based *ex vivo* model was used to rapidly probe gene function and to uncover a potential therapeutic vulnerability. The usefulness of the CRISPR system for *ex vivo* cancer modeling was also demonstrated through the creation of xenograft models of Burkitt's lymphoma and B-cell lymphoma by independently knocking out two genes, *Bim* and *Mcl-1* (Aubrey *et al.*, 2015). Similar studies have extended these genetically complex *ex vivo* models through sequential mutation of multiple genes to model colorectal cancer in human intestinal organoids, resulting in cancerous growths as well as chromosomal instabilities and metastasis following transplantation (Drost *et al.*, 2015; Matano *et al.*, 2015). Another study described a CRISPR-based model of acute myeloid leukemia in which up to five genes were simultaneously mutated in individual cells using combinatorial sgRNAs, demonstrating the capacity of CRISPR/Cas9 to begin to model the genetic complexity of human cancers using *ex vivo* studies (Heckl *et al.*, 2014).

Ex vivo CRISPR/Cas9-based strategies have also been used to uncover novel roles of individual genes in cancer. The successful correction of a loss-of-function mutation suppressed cancer-associated growth in a mouse model of colon cancer, confirming the role of this gene as a tumor suppressor (Antal *et al.*, 2015). *Ex vivo* studies have also been used to validate the function of a putative therapeutic target for triple negative breast cancer. In that study, inactivation of *Cripto-1* inhibited the growth and metastatic ability of a cell line injected into the mammary glands of mice, evincing its potential as a target for this cancer (Castro *et al.*, 2015). These studies illustrate how CRISPR/Cas9-based *ex vivo* models of cancer can be directly used to interrogate gene function and identify novel therapeutic targets.

Variant CRISPR systems can also be used *ex vivo* to modulate transcriptional gene activity for cancer modeling (Braun *et al.*, 2016). Transcriptional activation of an oncogene and suppression of a tumor suppressor using a Cas9 variant in murine cells resulted in the development of lymphoma. These cells were resistant to DNA damage, consistent with the biological roles of the targeted genes. The success of these *ex vivo* studies using alternative CRISPR endonucleases demonstrates the power of these methods to extend beyond genetic modification systems for cancer modeling.

Ex vivo modeling of cancer has also enjoyed some success with genome-wide screens and functional assays. In a recent study, the CRISPR/Cas9 system was used to generate a spectrum of indels in *p53* in murine cells. The modified cells were then transplanted into mice and treated with doxorubicin to pinpoint specific mutants that confer resistance to the drug (Malina *et al.*, 2013). Another *ex vivo* selection-based screen attempted to use a genome-wide library of sgRNAs to identify genes that play a role in cancer and metastasis (Chen *et al.*, 2015). This study successfully identified well-known tumor-suppressor genes that regulate primary tumor growth. An additional screen was conducted in a similar manner to test genes for putative roles in treatment response and relapse (Braun *et al.*, 2016). CRISPR-based transcriptional activation of a pool of 25 genes conducted in a colon cancer cell line transplantation, followed by treatment with temozolomide, uncovered genes involved in mediating response to this drug. These results could be rapidly validated by independent *ex vivo* studies using CRISPR/Cas9 to inactivate top candidate genes.

Ex vivo studies cannot only be used to successfully interrogate gene function at the individual level, but also at much larger scales, advancing our understanding of cancer onset, progression and treatment response in an organismal context. Nevertheless, these models have significant drawbacks, since they do not accurately recapitulate many aspects of the human disease. One of the main contributors to tumorigenesis is the microenvironmental niche in which the cell of origin accrues the genetic alterations that lead to transformation (Balkwill *et al.*, 2012). Transplantation of *ex vivo*-modified cells into an adult mouse cannot capture this process, even when the cells are transplanted orthotopically. Thus, although *ex vivo* models offer numerous applications, cancer initiation from somatic cells allows the development and progression of the disease to occur in an entirely natural, *in vivo* context. These types of cancer models may offer important advantages in studying tumor initiation and the interaction of cancer cells with their microenvironment, as well as important processes related to cancer cell invasion and metastasis.

21.2.3 Direct *In Vivo* Somatic Genome Editing in Cancer Models

Many recent advances have contributed to methods for the genetic modification of somatic cells using CRISPR/Cas9. In these models, somatic cells are directly manipulated in their natural tissue context in adult mice (Figure 21.2), allowing for the induction of specific genetic alterations in a fashion analogous to spontaneous mutations that initiate and promote human cancer. Furthermore, desired alterations can be introduced into cells in specific organs through targeted delivery methods. Thus, these *in vivo* systems offer excellent means by which to study the mechanisms and molecular etiology of human cancers, as well as to discover and test therapeutics (Table 21.2).

The first proof-of-principle experiments in this space demonstrated the feasibility of using CRISPR to model liver cancer *in vivo* (Xue *et al.*, 2014). Delivery of plasmids encoding sgRNAs and Cas9 to the liver produced cells with desired mutations, and led to the development of tumors that phenocopied those generated with traditional floxed alleles. This study highlighted the efficacy of the CRISPR/Cas9 system to inactivate genes of interest in somatic cells. Furthermore, activating mutations were also introduced into mouse liver cells by HDR, demonstrating the potential for CRISPR to generate gain-of-function alleles.

Similar studies demonstrated additional capabilities of the CRISPR/Cas9 system to model several different cancer types. Lentiviral-mediated delivery of Cas9 and sgRNAs was used to introduce multiple genetic modifications into a model of lung

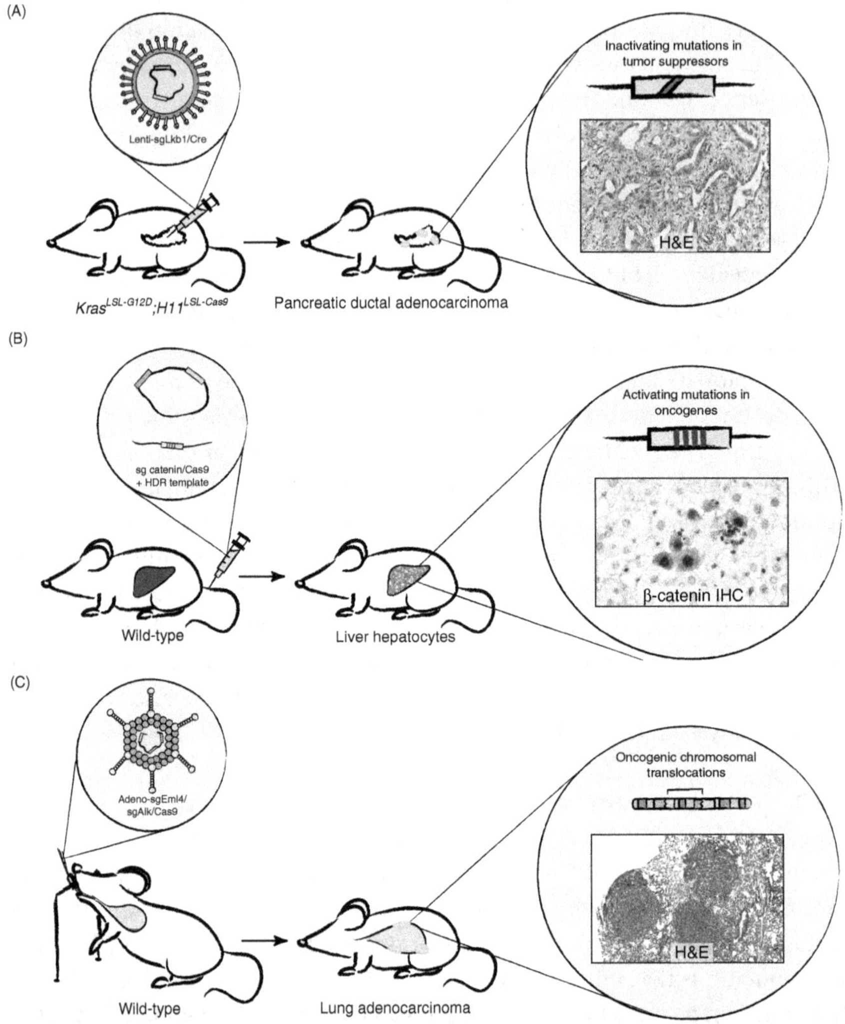

Figure 21.2 Examples of *in vivo* cancer modeling using CRISPR/Cas9 systems. (A) Tumor suppressor inactivation during carcinogenesis can be modeled using an individual sgRNA targeting the desired gene. Lentiviruses encoding both Cre and an sgRNA targeting *Lkb1* were delivered to the pancreas of a mouse with Cre-inducible *Cas9* and $Kras^{G12D}$ alleles by retrograde pancreatic ductal injection. This led to the formation of pancreatic tumors and metastases harboring *Lkb1* indels at the targeted locus (data provided by Ian Winters and Shin-Heng Chiou). (B) Modeling activating mutations in oncogenes can be achieved through the additional delivery of a template for homology-directed repair (HDR). A plasmid encoding Cas9 and an sgRNA targeting β-*catenin*, as well as a ssDNA β-*catenin* HDR template with activating mutations were delivered to the mouse liver by hydrodynamic tail vein injection. This resulted in cells with nuclear β-*catenin* expression as detected by IHC demonstrating the feasibility of this approach for modeling oncogene-induced liver cancer. (Data provided by Wen Xue and Tyler Jacks). (C) Chromosomal rearrangements in cancer can be generated in somatic cells using the CRISPR-Cas9 system. Cas9 along with two sgRNAs targeting the *Eml4* and *Alk* genes were packaged into an adenoviral vector and delivered to mouse lung by intratracheal instillation. Resulting tumors that formed in the lung harbored the desired chromosomal *Eml4-Alk* translocations (data provided by Andrea Ventura). (A black-and-white version of this figure will appear in some formats. For the color version, please refer to the plate section.)

Table 21.2 *In vivo* cancer modeling studies that have used somatic CRISPR/Cas9 genome editing in mice

Type of cancer	Cas9 expression	sgRNA expression	Method of *in viva* delivery	Genetic modifications	Summary
Lung cancer	Cre-inducible expression of Cas9 in germline	AAV delivery of sgRNAs	Intranasal and intratracheal delivery	Loss of function mutations in *p53* and *Lkb1* Gain of function mutations in *Kras*	Modeling LOF mutations in *p53* and *Lkb1* and GOF mutations in *Kras* led to tumor formation in mouse lung. (Platt *et al.*, 2014)
Lung cancer	Adenoviral delivery of Cas9	Adenoviral delivery of sgRNAs	Intratracheal delivery	Chromosomal inversion of *Eml4* and *Alk*	Chromosomal inversions of *Eml4* and *Alk* were created with multiple sgRNAs simultaneously introduced into mouse lung, modeling lung adenocarcinoma and treatment response. (Maddalo *et al.*, 2014)
Lung cancer	Lentiviral delivery of Cas9	Lentiviral delivery of sgRNAs	Intratracheal delivery	Loss of function mutations in *Pten, Apc* and *Nkx2-1*	LOF mutations in genes enabled their functional characterization in a *p53* deficient *Kras*^{G12D} model of lung adenocarcinoma. (Sanchez-Rivera *et al.*, 2014)
Lung cancer	Lentiviral delivery of Cas9	Lentiviral delivery of sgRNAs	Intratracheal and intrapulmonary delivery	Chromosomal inversion of *Eml4* and *Alk*	Chromosomal inversions of *Eml4* and *Alk* were modeled in *p53* deficient mouse lung through multiplexed sgRNA delivery, driving tumor growth. (Blasco *et al.*, 2014)
Liver cancer	Cas9 and nCas9 Plasmid DNA	Plasmid DNA	Hydrodynamic tail-vein injection	Loss of function mutations in *Pten* and *p53* Gain of function mutations in *β-catenin*	Combinatorial modeling of LOF mutations in *p53* and *Pten* generated a model of liver cancer, while GOF mutations in *β-catenin* demonstrated the potential for modeling precise point mutations in oncogenes. (Xue *et al.*, 2014)
Liver cancer	Cas9 Plasmid DNA	Plasmid DNA	Hydrodynamic tail-vein injection	Chromosomal inversions and deletions of *Pten*	Chromosomal inversions and deletions of *Pten* were induced in mouse liver. (Li *et al.*, 2015)

Continued

Table 21.2 (cont.)

Type of cancer	Cas9 expression	sgRNA expression	Method of *in viva* delivery	Genetic modifications	Summary
Liver cancer	Cas9 Plasmid DNA	Plasmid DNA	Hydrodynamic tail-vein injection	Selection based combinatorial LOF screens with 10 and 18 sgRNAs	*In vivo* selection screens in *Kras*^{G12D} mouse liver identified functional drivers of liver cancer through combinatorial modification of several genes. Chromosomal aberrations were also observed. (Weber *et al.*, 2015)
Pancreatic cancer	Cre-inducible expression of Cas9 in germline	Lentiviral delivery of sgRNAs	Retrograde pancreatic ductal injection	Loss of function mutations in *Lkb1*	Successful modeling of pancreatic ductal adenocarcinoma was achieved using retrograde pancreatic ductal injection. (Chiou *et al.*, 2015)
Pancreatic cancer	Cas9 Plasmid DNA	Plasmid DNA	Laparotomy, followed by intraparenchymal injection and electroporation	Selection based combinatorial LOF screen	A negative selection screen identified functional drivers of pancreatic cancer using combinatorial genetic modification. Chromosomal aberrations were also observed. (Maresch *et al.*, 2016)
Brain cancer	Cas9 Plasmid DNA	Plasmid DNA	PEI-mediated transfection by stereotactic injection	Loss of function mutations in *Ptch1*	Modeling of medulloblastoma in mouse brain was acheived by introduction of LOF mutations in *Ptch1* in *p53* deficient mice. (Zuckermann *et al.*, 2015)
Breast cancer	Cre-inducible expression of Cas9 in germline	Lentiviral delivery of sgRNAs	Intraductal injection	Loss of function mutations in *Pten*	Initial Cas9 expression elicited an immune response; therefore constitutive Cre-inducible Cas9 expressing mice were used to target *Pten* and model invasive lobular carcinoma. (Annunziato *et al.*, 2016)

These studies have successfully recapitulated many aspects of the genetic diversity and molecular characteristics of cancer in a variety of organs. This has enabled rapid interrogation of gene function in tumorigenesis, and allowed for the generation of new cancer models.

adenocarcinoma, confirming results obtained using floxed alleles (Sanchez-Rivera *et al.*, 2014). Successful multiplexed biallelic inactivation of genes in the intestine gave rise to marked histological changes and differences in downstream signaling (Dow *et al.*, 2015). In an experiment from our laboratory, retrograde pancreatic ductal injection of lentivirus achieved genetic modification of cells in the pancreas (Chiou *et al.*, 2015). *In vivo* somatic editing of genes in the mammary gland resulted in the onset of invasive lobular breast adenocarcinoma (Annunziato *et al.*, 2016). Similarly, CRISPR-induced genetic modification of mouse brain cells also successfully generated models of medulloblastoma and glioblastoma (Zuckermann *et al.*, 2015). The rapid emergence of these systems for modeling genetic aberrations that directly lead to cancer highlights the efficiency and broad applicability of CRISPR-based *in vivo* models to characterize gene function and produce novel models of prevalent human cancers.

In addition to inducing specific inactivating genetic alterations, the CRISPR/Cas9 system has also been used to introduce oncogenic chromosomal rearrangements *in vivo*. This has been achieved by introducing two sgRNAs in parallel to create two double-strand breaks, facilitating large-scale manipulations that recapitulate the alterations observed in human cancers. The generation of translocations was first achieved in a study that modeled an oncogenic inversion observed in lung adenocarcinoma (Blasco *et al.*, 2014). The production of a fusion protein was sufficient to induce tumors in mice, demonstrating the ability of this system to engineer oncogenic translocations. Other experiments produced oncogenic inversions and similar fusion proteins in the lung (Maddalo *et al.*, 2014), as well as deletions and inversions of a tumor-suppressor gene in the mouse liver (Li *et al.*, 2015).These results underscore the capacity of these systems to model large-scale chromosomal alterations that have been extremely difficult to generate using conventional Cre-Lox methods.

Selection-based screens using *in vivo* models of cancer have also been used to characterize gene function and uncover novel functional pathways. The ability of the CRISPR/Cas9 system to deliver multiple sgRNAs in parallel opens the door for functional screens for genes that are involved in tumorigenesis. These screens have employed expression of pools of sgRNAs to compare distributions of modified genes in tumors and therefore determine the effect of inactivating certain genes on cancer growth. This technique was utilized in a high-throughput *in vivo* screen to identify genes that play a role in hepatobiliary cancer (Weber *et al.*, 2015). Following the introduction of the system and multiple associated vectors with sgRNAs targeting genes in the mouse liver, significant selection was observed for and against specific genes in tumors. The resulting data mirrored the spectrum of genetic mutations observed in human hepatobiliary cancers, also offering functional evidence of the involvement of several pathways in this cancer type. This method was employed in a similar screen to characterize genes in pancreatic cancer (Maresch *et al.*, 2016). Combinatorial delivery of sgRNAs to pancreatic tissue uncovered many genes that seemed to be selected for and against in this cancer type. These studies illustrate the broad scope of CRISPR-based *in vivo* screens in somatically induced cancers, and contribute to our understanding of the differential mechanisms of tumorigenesis in different organs.

Optimized *in vivo* delivery systems for CRISPR/Cas9 components will allow for greater spatio-temporal control of genetic modifications. Importantly, several variants of the CRISPR/Cas9 system have been described that enable inducible gene editing *in vivo*. In particular, the Cre/LoxP system allows for inducible modification of genes through selective expression of Cre. The effective integration of CRISPR/Cas9 and Cre/LoxP systems was demonstrated through the generation of mice with conditional, Cre-dependent Cas9 alleles (Annunziato *et al.*, 2016; Chiou *et al.*, 2015; Platt *et al.*, 2014). Mice possessing these alleles do not initially express Cas9 due to an upstream transcriptional/translational STOP cassette flanked by LoxP sites. However, expression of Cre recombinase induces Cas9 expression by recombining the LoxP sites and thus removing the STOP element. This enables tissue-specific Cas9 expression and genetic modification through the co-expression of an sgRNA. These inducible systems not only offer spatial control of gene modification, but also allows for simultaneous alteration of additional conditional Cre/LoxP alleles bred into the mouse germline. This system also alleviates issues associated with the *in vivo* delivery of Cas9, a large gene that complicates the use of viral vectors, and likely eliminates immunogenicity that has been reported with somatic delivery of Cas9 *in vivo* (Annunziato *et al.*, 2016). Regulatable CRISPR systems have been used successfully in several other studies (Annunziato *et al.*, 2016; Chiou *et al.*, 2015; Sanchez-Rivera *et al.*, 2014), and will likely play a prominent role in the future of cancer modeling with CRISPR.

Another inducible approach used to model cancer *in vivo* is the doxycycline-inducible system (Dow *et al.*, 2015). In this system, Cas9 downstream of a tetracycline response element is activated following treatment with doxycycline in cells that also express the transactivator rtTA. Therefore, control over the cell type in which rtTA is expressed and the timing of doxycycline treatment also allows for temporal and cell type-specific modulation of Cas9 expression.

The efficiency and flexibility of CRISPR-based systems has facilitated the rapid creation of complex cancer models *in vivo*. This in turn has allowed – and will continue to allow – for the discovery of novel mechanisms underlying the process of tumorigenesis and development.

21.3 Challenges and the Future of Cancer Modeling with CRISPR

The CRISPR/Cas9 system has tremendous potential for modeling cancer, and several recent advances have improved its efficiency, specificity and flexibility. Current systems allow for unprecedented control over genetic circuits that govern cancer, solidifying CRISPR's place in the future of cancer modeling. Nevertheless, there are still several challenges with this system that must be adequately addressed before the full potential of the CRISPR/Cas9 system for cancer modeling is realized.

21.3.1 Challenges with CRISPR-based Cancer Modeling

Off-target effects are an issue of potential importance associated with the CRISPR/Cas9 system and other genetic engineering systems. In particular, the sgRNAs that direct Cas9 to target sites can bind to sequences of imperfect complementarity, which can result in unwanted modifications in other genomic regions (Cho *et al.*, 2014). In reality, off-target cutting events are not particularly numerous and are often not within genes; therefore the probability of these events detectably altering cancer phenotypes is likely low. Nonetheless, a variant of Cas9 (nCas9) that generates single-stranded breaks or "nicks" in DNA (Ran *et al.*, 2013) can be used with two sgRNAs to create pairs of nicks at the site, which results in a double-strand break. This allows for cleavage of the desired locus with minimal off-target effects due to the increased specificity conferred by the requirement of dual nCas9/sgRNA binding for DNA cleavage. This approach has been successfully used to model cancer, both by inducible germline expression of nCas9 (Dow *et al.*, 2015) and in somatic engineering *in vivo* (Xue *et al.*, 2014). Thus, the nCas9 system may be a useful tool in the genome engineering toolkit for applications that are especially sensitive to the off-target effects of CRISPR.

When CRISPR-based systems are used to introduce specific gain-of-function mutations using the HDR pathway with a synthetic template, the DNA repair machinery of the cell often uses the NHEJ pathway instead, introducing indels into the gene of interest. These indels create potentially unwanted null alleles rather than the desired specific HDR-mediated alterations. As such, it is necessary to develop methods to improve efficiency and selective use of certain DNA repair pathways to increase the likelihood that desired oncogenic alterations are introduced. To increase the rate of the less efficient HDR pathway, small molecule inhibitors of NHEJ or genetic alterations that reduce NHEJ can be employed. Several small molecular inhibitors have been shown to promote HDR efficiency *in vitro* and in mouse zygotes (Maruyama *et al.*, 2015). Thus, inhibition of NHEJ could prove useful in increasing the efficiency of generating gain-of-function mutations through HDR *in vivo*.

For some organs in mice, it has proven difficult to develop tissue-specific methods to effectively deliver the components of the CRISPR/Cas9 system. Despite the establishment of efficient delivery techniques to the liver by hydrodynamic tail vein injection (Dow *et al.*, 2015; Li *et al.*, 2015; Platt *et al.*, 2014; Weber *et al.*, 2015; Xue *et al.*, 2014) and to the lung by intratracheal administration (Blasco *et al.*, 2014; Maddalo *et al.*, 2014; Platt *et al.*, 2014; Sanchez-Rivera *et al.*, 2014), it remains a challenge to develop methods to effectively administer the CRISPR/Cas9 system to other organs. Specifically, optimized methods of delivery to organs such as the colon are needed to develop viral- or plasmid-based models of these cancers. In addition to the aforementioned inducible germline approaches that increase spatio-temporal control of expression, it is anticipated that novel surgical methods, such as the recently developed retrograde pancreatic ductal injection to target the pancreas (Chiou *et al.*, 2015) and intraductal injection techniques to modify the mammary gland (Annunziato *et al.*, 2016), will allow for models of many other important cancer types. Further developments that may aid the delivery process may include the modification of viral vectors expressing antibodies against cell-specific markers to enable targeted *in vivo* delivery (Buchholz *et al.*, 2015) and Cas9-sgRNA complexes delivered by cationic lipid systems to enable increased editing efficiency (Zuris *et al.*, 2015). Another technique that may increase bioavailability and improve delivery is the use of different types of nanoparticles (Gelperina *et al.*, 2005), which have successfully been used to modify pulmonary and cardiovascular epithelia *in vivo* (Platt *et al.*, 2014). Improved *in vivo* electroporation (Dean *et al.*, 2003), currently optimized for delivery to embryos and adult tissues, including the pancreas (Maresch *et al.*, 2016), may also be an effective method to deliver Cas9 and sgRNAs to other organs in a plasmid form, which may eliminate issues with the low carrying capacities of some of the viruses. Thus, several modifications and novel delivery procedures, coupled with the CRISPR/Cas9 system, will open the door to more efficient and diverse engineering of *in vivo* models of cancer.

A final current issue associated with the use of the CRISPR/Cas9 system to model cancer *in vivo* is the application of multiplexed somatic mutagenesis to generate genetically complex and diverse lesions. The simultaneous use of multiple sgRNAs necessary to achieve such modifications has occasionally produced undesirable chromosomal aberrations that could complicate analysis of the resulting phenotype (Maresch *et al.*, 2016; Weber *et al.*, 2015). Being cognizant of this issue, targeted cells should be tested for unwanted alterations resulting from multiplexed sgRNAs. Methods that allow sequential rather than simultaneous expression of sgRNAs could resolve this issue and allow for temporal control over additional genetic alterations.

21.3.2 The Future of CRISPR-based Cancer Modeling

In addition to the research that is being undertaken to effectively address the challenges associated with *in vivo* modeling of cancer using CRISPR, much investigation centers around advancements to make the system more efficient and enable a wider range of applications, including research on therapeutic targets.

The development of conditional CRISPR systems has enabled spatio-temporal control of genetic modifications. As mentioned above, examples of these systems include the use of doxycycline-inducible systems (Aubrey *et al.*, 2015; Dow *et al.*, 2015), the Cre/LoxP system (Chiou *et al.*, 2015; Platt *et al.*, 2014; Sanchez-Rivera *et al.*, 2014) and even light-based promoters (Nihongaki *et al.*, 2015) to control both localization of expression via tissue-specific promoters and the timing of genetic manipulations. Such systems that enable inducible Cas9 expression may eliminate many of the delivery challenges associated with the large size of the *Cas9* gene, as well as issues with immunogenicity observed after exogenous expression of Cas9 *in vivo* (Annunziato *et al.*, 2016).

CRISPR-based genome engineering research is also starting to make use of several variants of the basic system, including modified nucleases such as the aforementioned nCas9 and chimeric protein systems to enable various alternate modifications (Figure 21.3). While many of these systems have only been employed in cell culture, the application of these newer systems *in vivo* may

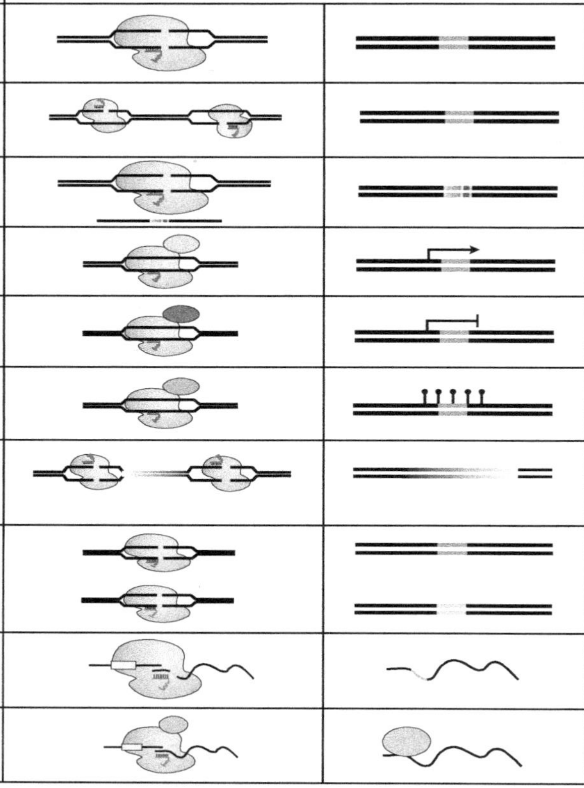

Technique	Purpose in modeling cancer
NHEJ mediated knockout	Inactivation of tumor suppressors (Chiou *et al.*, 2015)
nCas9 mediated knockout	Inactivation of tumor suppressors Decreased off-target cleavage (Xue *et al.*, 2014)
HDR mediated knockin	Activation of oncogenes (Xue *et al.*, 2014)
dCas9 mediated transcriptional activation	Transcriptional activation of oncogenes (Braun *et al.*, 2016)
dCas9 mediated transcriptional repression	Transcriptional inactivation of tumor suppressors (Braun *et al.*, 2016)
dCas9 mediated epigenetic modification	Modeling epigenetic alterations in cancer
Multiplexing guides for chromosomal rearrangements	Creating oncogenic chromosomal rearrangements (Maddalo *et al.*, 2014)
Multiplexing guides for simultaneous knockout of multiple genes	Generation of complex models of cancer with alterations in multiple genes (Matano *et al.*, 2015)
RCas9 mediated RNA modification	Post-transcriptional inactivation of genes of interest
dRCas9 mediated RNA targeting	Post-transcriptional modulation of genes of interest

Figure 21.3 Variants of the CRISPR/Cas9 system and their potential applications to cancer modeling. In addition to enabling the induction of gain-of-function and loss-of-function genetic mutations, variants of the CRISPR/Cas9 system allow for transcriptional, post-transcriptional, and epigenetic control over gene expression. These systems will enable the generation of *in vivo* models of cancer that can recapitulate a much wider spectrum of the diverse alterations observed in human cancer.

allow for greater control over gene expression, which in turn may facilitate detailed investigations of gene function and underlying molecular mechanisms of cancer.

The nuclease-dead version of Cas9 (dCas9) is a DNA-binding protein that can be repurposed to target other domains and proteins to specific sites of DNA (Qi *et al.*, 2013). In particular, the synthetic CRISPRi and CRISPRa systems are composed of dCas9 in conjunction with transcriptional repressors such as KRAB and activators including VP64, respectively, enabling regulation of gene expression at the transcriptional level (Gilbert *et al.*, 2014).

Furthermore, complexing dCas9 with other modulators such as methyltransferases and acetyltransferases allows for control of gene expression at the epigenetic level (Vojta *et al.*, 2016). These variations of the CRISPR/Cas9 system allow for detailed interrogation of gene function and modeling in relation to cancer, where epigenetic and transcriptional alterations are frequently observed. Thus, it is anticipated that these additional Cas9 variants will find key applications in *in vivo* cancer modeling.

The ability to regulate gene expression at the mRNA level could also enhance our ability to uncover gene function. The RCas9 variant was generated to target RNA with a synthetic ssDNA PAMmer that provides the PAM sequence necessary for target recognition and cleavage (O'Connell *et al.*, 2014). Thus, RCas9 can enable specific regulation of mRNAs at the post-transcriptional level. Such a system could also evolve into a targetable RNA-binding protein by combining the modifications used to generate the dCas9 and RCas9 variants of Cas9. This variant could theoretically be combined with activators, repressors, beads, protein domains or fluorescent tags to allow for novel functionalities including post-transcriptional modulation of gene expression.

These variants of the CRISPR/Cas9 system allow for widespread and precise control of gene expression at various cellular and molecular levels. They are well suited for modeling cancer and will likely allow for a deeper understanding of gene function through complementary approaches, from direct genetic changes to modulation of transcription and chromatin state, which in turn may reveal promising new therapeutic drug targets. Additional synthetic methods initially developed *in vitro* to perturb genome function and gene expression will likely be rapidly applied *in vivo* and will contribute to the ever-growing arsenal of methods with which to model and investigate cancer tumorigenesis.

21.3.3 CRISPR and Cancer Therapeutics

An important application of *in vivo* models of cancer is to uncover therapeutically actionable insights into the mechanisms that drive carcinogenesis, drug resistance and relapse. Several cancer models engineered using CRISPR have already been used to validate individual therapeutic targets and pathways (Quick *et al.*, 2016). A variety of drugs and inhibitors already in use in the clinic have also been tested in CRISPR-based models to determine their efficacy as treatments against specific genetic classes of cancer (Chen *et al.*, 2014; Lok *et al.*, 2016; Maddalo *et al.*, 2014; Togashi *et al.*, 2015). These and similar approaches may uncover new therapeutic possibilities and advance strategies for personalized cancer treatments informed by genomic analyses.

The simplicity of the CRISPR/Cas9 systems has made it possible to model some of the most prevalent cancer types at a fraction of the time and cost of conventional systems. This makes it feasible to induce multiple *in vivo* genomic alterations in a single animal, simultaneously. The ability to test gene function in parallel in the same organism not only enables much higher-throughput screens, but also facilitates direct comparisons between targeted genes. This may enable large-scale screens in the future to identify promising novel therapeutic targets, specific mutations that modulate response to treatment (Malina *et al.*, 2013; Shalem *et al.*, 2014), and conduct preclinical testing of candidate drugs, all at clinically relevant scales (Hong *et al.*, 2016; Shi *et al.*, 2015).

21.4 Concluding Remarks

The ability to rapidly, cheaply and accurately recapitulate the alterations observed in human cancers in somatic cells of living organisms has the potential to exponentially increase our knowledge of the fundamental mechanisms that drive cancer and strategies to treat this disease. CRISPR/Cas9 technology heralds great improvements for cancer modeling, enabling the study of increasingly complex and accurate genetic modifications at an unprecedented scale. We anticipate that improved *in vivo* models of cancer will greatly contribute to our growing understanding of cancer and help usher in a new era of cancer therapies and personalized medicine.

REFERENCES

Annunziato S, Kas SM, Nethe M, *et al.* 2016. Modeling invasive lobular breast carcinoma by CRISPR/Cas9-mediated somatic genome editing of the mammary gland. *Genes Dev* **30**: 1470–1480.

Antal CE, Hudson AM, Kang E, *et al.* 2015. Cancer-associated protein kinase C mutations reveal kinase's role as tumor suppressor. *Cell* **160**: 489–502.

Aubrey BJ, Kelly GL, Kueh AJ, *et al.* 2015. An inducible lentiviral guide RNA platform enables the identification of tumor-essential genes and tumor-promoting mutations in vivo. *Cell Rep* **10**: 1422–1432.

Balkwill FR, Capasso M, Hagemann T. 2012. The tumor microenvironment at a glance. *J Cell Sci* **125**: 5591–5596.

Bibikova M, Beumer K, Trautman JK, Carroll D. 2003. Enhancing gene targeting with designed zinc finger nucleases. *Science* **300**: 764.

Blasco RB, Karaca E, Ambrogio C, *et al.* 2014. Simple and rapid in vivo generation of chromosomal rearrangements using CRISPR/Cas9 technology. *Cell Rep* **9**: 1219–1227.

Braun CJ, Bruno PM, Horlbeck MA, *et al.* 2016. Versatile in vivo regulation of tumor phenotypes by dCas9-mediated transcriptional perturbation. *Proc Natl Acad Sci USA* **113**: E3892–E3900.

Brinster RL, Chen HY, Trumbauer M, *et al.* 1981. Somatic expression of herpes thymidine kinase in mice following injection of a fusion gene into eggs. *Cell* **27**: 223–231.

Buchholz CJ, Friedel T, Buning H. 2015. Surface-engineered viral vectors for selective and cell type-specific gene delivery. *Trends Biotechnol* **33**: 777–790.

Carbery ID, Ji D, Harrington A, *et al.* 2010. Targeted genome modification in mice using zinc-finger nucleases. *Genetics* **186**: 451–459.

Castro NP, Fedorova-Abrams ND, Merchant AS, *et al.* 2015. Cripto-1 as a novel therapeutic target for triple negative breast cancer. *Oncotarget* **6**: 11910–11929.

Chen C, Liu Y, Rappaport AR, *et al.* 2014. MLL3 is a haploinsufficient 7q tumor suppressor in acute myeloid leukemia. *Cancer Cell* **25**: 652–665.

Chen S, Sanjana NE, Zheng K, *et al.* 2015. Genome-wide CRISPR screen in a mouse model of tumor growth and metastasis. *Cell* **160**: 1246–1260.

Chen Z, Cheng K, Walton Z, *et al.* 2012. A murine lung cancer co-clinical trial identifies genetic modifiers of therapeutic response. *Nature* **483**: 613–617.

Chiou SH, Winters IP, Wang J, *et al.* 2015. Pancreatic cancer modeling using retrograde viral vector delivery and in vivo CRISPR/Cas9-mediated somatic genome editing. *Genes Dev* **29**: 1576–1585.

Cho SW, Kim S, Kim Y, *et al.* 2014. Analysis of off-target effects of CRISPR/Cas-derived RNA-guided endonucleases and nickases. *Genome Res* **24**: 132–141.

Cong L, Ran FA, Cox D, *et al.* 2013. Multiplex genome engineering using CRISPR/Cas systems. *Science* **339**: 819–823.

Costantini F, Lacy E. 1981. Introduction of a rabbit beta-globin gene into the mouse germ line. *Nature* **294**: 92–94.

Dean DA, Machado-Aranda D, Blair-Parks K, Yeldandi AV, Young, JL. 2003. Electroporation as a method for high-level nonviral gene transfer to the lung. *Gene Ther* **10**: 1608–1615.

Dow LE, Fisher J, O'Rourke KP, *et al.* 2015. Inducible in vivo genome editing with CRISPR-Cas9. *Nat Biotechnol* **33**: 390–394.

Drost J, Van Jaarsveld RH, Ponsioen B, *et al.* 2015. Sequential cancer mutations in cultured human intestinal stem cells. *Nature* **521**: 43–47.

Engelman JA, Chen L, Tan X, *et al.* 2008. Effective use of PI3 K and MEK inhibitors to treat mutant Kras G12D and PIK3CA H1047 R murine lung cancers. *Nat Med* **14**: 1351–1356.

Flemr M, Buhler M. 2015. Single-step generation of conditional knockout mouse embryonic stem cells. *Cell Rep* **12**: 709–716.

Frese KK, Tuveson DA. 2007. Maximizing mouse cancer models. *Nat Rev Cancer* **7**: 645–658.

Fujiki H. 2014. Gist of Dr. Katsusaburo Yamagiwa's papers entitled "Experimental study on the pathogenesis of epithelial tumors" (I to VI reports). *Cancer Sci* **105**: 143–149.

Gelperina S, Kisich K, Iseman MD, Heifets L. 2005. The potential advantages of nanoparticle drug delivery systems in chemotherapy of tuberculosis. *Am J Respir Crit Care Med* **172**: 1487–1490.

Gilbert LA, Horlbeck MA, Adamson B, *et al.* 2014. Genome-scale CRISPR-mediated control of gene repression and activation. *Cell* **159**: 647–661.

Gordon JW, Ruddle FH. 1981. Integration and stable germ line transmission of genes injected into mouse pronuclei. *Science* **214**: 1244–1246.

Heckl D, Kowalczyk MS, Yudovich D, *et al.* 2014. Generation of mouse models of myeloid malignancy with combinatorial genetic lesions using CRISPR-Cas9 genome editing. *Nat Biotechnol* **32**: 941–946.

Hong AL, Tseng YY, Cowley GS, *et al.* 2016. Integrated genetic and pharmacologic interrogation of rare cancers. *Nat Commun* **7**: 11987.

Hutchinson JN, Muller WJ. 2000. Transgenic mouse models of human breast cancer. *Oncogene* **19**: 6130–6137.

Li F, Cowley DO, Banner D, *et al.* 2014. Efficient genetic manipulation of the NOD-Rag1-/-IL 2RgammaC-null mouse by combining in vitro fertilization and CRISPR/Cas9 technology. *Sci Rep* **4**: 5290.

Li T, Huang S, Jiang WZ, *et al.* 2011. TAL nucleases (TALNs): hybrid proteins composed of TAL effectors and FokI DNA-cleavage domain. *Nucleic Acids Res* **39**: 359–372.

Li W, Teng F, Li T, Zhou Q. 2013. Simultaneous generation and germline transmission of multiple gene mutations in rat using CRISPR-Cas systems. *Nat Biotechnol* **31**: 684–686.

Li Y, Park AI, Mou H, *et al.* 2015. A versatile reporter system for CRISPR-mediated chromosomal rearrangements. *Genome Biol* **16**: 111.

Lok BH, Gardner EE, Schneeberger VE, *et al.* 2016. PARP inhibitor activity correlates with SLFN11 expression and demonstrates synergy with temozolomide in small cell lung cancer. *Clin Cancer Res* **23**(2): 523–535.

Macleod KF, Jacks T. 1999. Insights into cancer from transgenic mouse models. *J Pathol* **187**: 43–60.

Maddalo D, Machado E, Concepcion CP, *et al.* 2014. In vivo engineering of oncogenic chromosomal rearrangements with the CRISPR/Cas9 system. *Nature* **516**: 423–427.

Malina A, Mills JR, Cencic R, *et al.* 2013. Repurposing CRISPR/Cas9 for in situ functional assays. *Genes Dev* **27**: 2602–2614.

Maresch R, Mueller S, Veltkamp C, *et al.* 2016. Multiplexed pancreatic genome engineering and cancer induction by transfection-based CRISPR/Cas9 delivery in mice. *Nat Commun* **7**: 10770.

Maruyama T, Dougan SK, Truttmann MC, *et al.* 2015. Increasing the efficiency of precise genome editing with CRISPR-Cas9 by inhibition of nonhomologous end joining. *Nat Biotechnol* **33**: 538–542.

Matano M, Date S, Shimokawa M, *et al.* 2015. Modeling colorectal cancer using CRISPR-Cas9-mediated engineering of human intestinal organoids. *Nat Med* **21**: 256–262.

Nihongaki Y, Yamamoto S, Kawano F, Suzuki H, Sato M. 2015. CRISPR-Cas9-based photo-activatable transcription system. *Chem Biol* **22**: 169–174.

Niu Y, Shen B, Cui Y, *et al.* 2014. Generation of gene-modified cynomolgus monkey via Cas9/RNA-mediated gene targeting in one-cell embryos. *Cell* **156**: 836–843.

O'Connell MR, Oakes BL, Sternberg SH, *et al.* 2014. Programmable RNA recognition and cleavage by CRISPR/Cas9. *Nature* **516**: 263–266.

Olive KP, Tuveson DA, Ruhe ZC, *et al.* 2004. Mutant p53 gain of function in two mouse models of Li-Fraumeni syndrome. *Cell* **119**: 847–860.

Platt RJ, Chen S, Zhou Y, *et al.* 2014. CRISPR-Cas9 knockin mice for genome editing and cancer modeling. *Cell* **159**: 440–455.

Qi LS, Larson MH, Gilbert LA, *et al.* 2013. Repurposing CRISPR as an RNA-guided platform for sequence-specific control of gene expression. *Cell* **152**: 1173–1183.

Quick L, Young R, Henrich IC, *et al.* 2016. Jak1-STAT3 signals are essential effectors of the USP6/TRE17 oncogene in tumorigenesis. *Cancer Res* **76**: 5337–5347.

Ran FA, Hsu PD, Lin CY, *et al.* 2013. Double nicking by RNA-guided CRISPR Cas9 for enhanced genome editing specificity. *Cell* **154**: 1380–1389.

Sanchez-Rivera FJ, Papagiannakopoulos T, Romero R, *et al.* 2014. Rapid modelling of cooperating genetic events in cancer through somatic genome editing. *Nature* **516**: 428–431.

Shalem O, Sanjana NE, Hartenian E, *et al.* 2014. Genome-scale CRISPR-Cas9 knockout screening in human cells. *Science* **343**: 84–87.

Shi J, Wang E, Milazzo JP, *et al.* 2015. Discovery of cancer drug targets by CRISPR-Cas9 screening of protein domains. *Nat Biotechnol* **33**: 661–667.

Shimkin MB, Stoner GD. 1975. Lung tumors in mice: application to carcinogenesis bioassay. *Adv Cancer Res* **21**: 1–58.

Sommer D, Peters AE, Baumgart AK, Beyer M. 2015. TALEN-mediated genome engineering to generate targeted mice. *Chromosome Res* **23**: 43–55.

Stehelin D, Varmus HE, Bishop JM, Vogt PK. 1976. DNA related to the transforming gene(s) of avian sarcoma viruses is present in normal avian DNA. *Nature* **260**: 170–173.

Thun MJ, Delancey JO, Center MM, Jemal A, Ward EM. 2010. The global burden of cancer: priorities for prevention. *Carcinogenesis* **31**: 100–110.

Togashi Y, Mizuuchi H, Tomida S, *et al.* 2015. MET gene exon 14 deletion created using the CRISPR/Cas9 system enhances cellular growth and sensitivity to a MET inhibitor. *Lung Cancer* **90**: 590–597.

Torres R, Martin MC, Garcia A, *et al.* 2014. Engineering human tumour-associated chromosomal translocations with the RNA-guided CRISPR-Cas9 system. *Nat Commun* **5**: 3964.

Valton J, Dupuy A, Daboussi F, *et al.* 2012. Overcoming transcription activator-like effector (TALE) DNA binding domain sensitivity to cytosine methylation. *J Biol Chem* **287**: 38427–38432.

Vojta A, Dobrinic P, Tadic V, *et al.* 2016. Repurposing the CRISPR-Cas9 system for targeted DNA methylation. *Nucleic Acids Res* **44**: 5615–5628.

Walrath JC, Hawes JJ, Van Dyke T, Reilly KM. 2010. Genetically engineered mouse models in cancer research. *Adv Cancer Res* **106**: 113–164.

Wang H, Yang H, Shivalila CS, *et al.* 2013. One-step generation of mice carrying mutations in multiple genes by CRISPR/Cas-mediated genome engineering. *Cell* **153**: 910–918.

Weber J, Ollinger R, Friedrich M, *et al.* 2015. CRISPR/Cas9 somatic multiplex-mutagenesis for high-throughput functional cancer genomics in mice. *Proc Natl Acad Sci USA* **112**: 13982–13987.

Xiao A, Wang Z, Hu Y, *et al.* 2013. Chromosomal deletions and inversions mediated by TALENs and CRISPR/Cas in zebrafish. *Nucleic Acids Res* **41**: e141.

Xue W, Chen S, Yin H, *et al.* 2014. CRISPR-mediated direct mutation of cancer genes in the mouse liver. *Nature* **514**: 380–384.

Yang H, Wang H, Shivalila CS, *et al.* 2013. One-step generation of mice carrying reporter and conditional alleles by CRISPR/Cas-mediated genome engineering. *Cell* **154**: 1370–1379.

Zuckermann M, Hovestadt V, Knobbe-Thomsen CB, *et al.* 2015. Somatic CRISPR/Cas9-mediated tumour suppressor disruption enables versatile brain tumour modelling. *Nat Commun* **6**: 7391.

Zuris JA, Thompson DB, Shu Y, *et al.* 2015. Cationic lipid-mediated delivery of proteins enables efficient protein-based genome editing in vitro and in vivo. *Nat Biotechnol* **33**: 73–80.

22 Inducible CRISPR-based Genome Editing for the Characterization of Cancer Genes

Nalani Sachan, Madison Miller, Nitin H. Shirole, Serif Senturk, Vincenzo Corbo, Justin B. Kinney and Raffaella Sordella

22.1 Introduction

Recent advances in sequencing technology have enabled us to study the molecular alterations in tumors to an unprecedented degree of granularity. At the same time, CRISPR/Cas9 systems have proven to be extraordinary tools that may be used to generate genetic models in animal and cell culture systems that improve our knowledge of tumor formation, development and progression. Importantly, these systems have also facilitated the identification of genes required for the survival of cancer cells. Compared to other gene editing tools, CRISPR/Cas9 methods are easy to implement and allow for highly efficient and specific gene editing. These superior features of CRISPR/Cas9 methods have spurred the exceptional development of this technology and its widespread adoption. Ultimately, CRISPR/Cas9 methods have played a critical role in narrowing the gap between the genomic and functional characterization of tumors.

In general, CRISPR/Cas9 methods rely on a *Streptococcus pyogenes* Type II CRISPR system in which the central components are a guide RNA and a Cas9 enzyme (Doudna and Charpentier, 2014; Doudna and Sontheimer, 2014; Jinek *et al.*, 2012; Wright *et al.*, 2016). In order for the Cas9 to be activated, the single-guide RNA (sgRNA) has to be fused to two key RNA units: the CRISPR RNA and the transactivating CrRNA (Doudna and Charpentier, 2014; Doudna and Sontheimer, 2014; Jinek *et al.*, 2012; Wright *et al.*, 2016). When introduced to target cells, the sgRNA promptly recognizes complementary sequences of DNA, unwinds the double-stranded DNA and enables its cleavage by Cas9 (Doudna and Charpentier, 2014; Doudna and Sontheimer, 2014; Jinek *et al.*, 2012; Wright *et al.*, 2016). Consequently, the generation of double-stranded DNA breaks triggers error-prone non-homologous end joining (NHEJ) repair or, in some cases, homology-directed repair (HDR), leading to the formation of indels in the target DNA (Doudna and Charpentier, 2014; Doudna and Sontheimer, 2014; Jinek *et al.*, 2012; Wright *et al.*, 2016).

Cancer mutations can result from single nucleotide variations, small insertions or deletions, large deletions and structural alterations (e.g. chromosomal copy number alterations, translocations and inversions). To model these diverse genetic aberrations, novel CRISPR/Cas9 applications have been developed and adapted with different degrees of success (Wright *et al.*, 2016). In addition, although CRISPR/Cas9 methods have been traditionally used to generate stable genetic alterations, the integration of a mutated, enzymatically dead Cas9 has enabled the repression or activation of the expression of genes of interest (Chavez *et al.*, 2015; Konermann *et al.*, 2015). This is a particularly interesting implementation of the gene editing tool, as it allows the reversible expression of genes of interest.

Since 2013 when CRISPR/Cas9 was first reported as a tool to induce gene editing, a plethora of studies have verified the efficient and robust employment of this technology in cancer biology (Doudna and Charpentier, 2014; Sternberg *et al.*, 2014; Wang *et al.*, 2015). The CRISPR/Cas9 system has mostly been utilized for loss-of-function studies that facilitate the identification of novel tumor-suppressor genes and genes that are essential for tumor cells' survival (Doudna and Charpentier, 2014; Sternberg *et al.*, 2014; Wang *et al.*, 2015).

In addition to *in vitro* studies, CRISPR/Cas9 methods have also been used to generate genetic mouse models of both inherited germline mutations and somatic mutations (Singh *et al.*, 2015; Wright *et al.*, 2016; Yang *et al.*, 2014). The creation of genetically engineered mouse models (GEMMs) via the manipulation of embryonic stem (ES) cells has become the standard for generating cancer models over the last three decades and has been easily adapted to the use of CRISPR/Cas9 methods (Singh *et al.*, 2015; Wright *et al.*, 2016; Yang *et al.*, 2014). Although very powerful, one of the major drawbacks of traditional gene targeting has been the low efficiency of double-strand breaks (DSB) at targeting sites. CRISPR/Cas9 circumvents this limitation by increasing the frequency of DSB at specific loci. In 2013, Yang *et al.* (2013) highlighted the advantages of this GEMM generation technique by showing the feasibility of concurrently disrupting five genes in ES cells using a CRISPR/Cas9-based method. Furthermore, the use of tissue-specific and inducible CRISPR/Cas9 systems has also enabled the development of highly sophisticated models based on tissue-specific and temporal genetic manipulations of genes of interest, as well as models based on the constitutive all-body expression/deletion of genes of interest (Dow *et al.*, 2015; Utomo *et al.*, 1999).

More recently, it has been demonstrated that CRIPSR/Cas9 could be used to introduce gene editing directly into zygotes at pronuclear states with an efficiency of over 80% (Yang *et al.*, 2013; 2014). This has been a major breakthrough, as this approach doesn't require the cumbersome manipulation of ES cells and facilitates the more rapid generation of animal models. Moreover, while the generation of GEMMs using traditional gene-targeting methods in ES cells requires several rounds of breeding over the span of months, the CRISPR/Cas9-based approach allows the generation of mice with the desired genotype in just a few weeks. Another advantage of this approach is the possibility of creating GEMMs with animals that formerly could not be used for genetically modified modeling, including pigs, rats and non-human primates.

Although faster and more affordable, the traditional CRISPR/Cas9 system mostly relies on NHEJ repair mechanisms. This depedence impacts the nature of the mutations that can be generated, as NHEJ-based gene disruption can carry different indels. Consequently, different mutations can be introduced to the two alleles of a target gene, and mice in the same litter can harbor very different mutations. To overcome this problem, HDR-based CRISPR/Cas9 methods have been developed in which single-stranded DNA templates with sgRNA and Cas9 mRNA have been co-injected into zygotes to introduce a defined point mutation (Yang *et al.*, 2014). Although this process is less efficient compared to NHEJ-based CRISPR/Cas9 methods, it allows targeted mutation insertions to be achieved with 10–50% efficiency depending on the size of the fragments that are being inserted (Yang *et al.*, 2014).

In addition to the aforementioned applications of CRISPR/Cas9, multiple groups have developed efficient methods for using CRISPR/Cas9 for somatic gene editing *in vivo*. In one such study, Bell *et al.* leveraged the unique capability of liver cells to uptake naked DNA and demonstrated the possibility of transducing Cas9 and sgRNAs in liver cells to induce indel formation using hydrodynamic tail injection methods (Bell *et al.*, 2007). As this particular method is practically liver-specific, alternative techniques have been developed to enable gene editing in other organs based on viral delivery. These techniques involve the construction of adeno-associated-based vectors (AAV), as well as lentiviral vectors (Swiech *et al.*, 2015). As the size of Cas9 can limit the packaging of the viral particles, several groups have now generated mice expressing Cas9, either constitutively or conditionally, using lox-stop-lox strategies or using doxycycline-inducible promoters (Dow *et al.*, 2015, Utomo *et al.*, 1999). This approach has been shown to be more efficient as it enables the viral-mediated transduction of only the sgRNAs.

Recent technical advances have expanded the use of CRISPR/Cas9 for the generation of mutations other than indels (Platt *et al.*, 2014). The Jacks Lab, for example, have shown that KrasG12D mutations could be generated in the lungs of mice upon adenoviral delivery of sgRNA, although only with an efficiency of 0.1% (DuPage *et al.*, 2009). Similar results were obtained in the livers of mice upon hydrodynamic delivery (Xue *et al.*, 2014). For the introduction of larger deletions, the Ventura Lab was able to generate deletions of up to 1.5 MB (Maddalo *et al.*, 2014). Specifically, they successfully introduced concomitant double-strand DNA breaks at intron 14 of *Eml4* (corresponds to intron 13 of *EML4*) and at intron 19 of *Alk* using a CRISPR/Cas9 system in order to model the most common *EML4–ALK* variant in human non-small-cell lung cancers (NSCLC; Xue *et al.*, 2014).

As an alternative strategy to model genetic mutations *in vivo*, mosaic models in which isolated cells can be genetically modified *in vitro* and then transplanted into recipient mice have been developed (Dow *et al.*, 2012). These methods are notable because, much like siRNA-based systems, they are amenable for unbiased multiplex large-scale genomic screening.

Tumors harbor diverse genetic aberrations that accumulate during the process of tumorigenesis (Hanahan and Weinberg, 2011). While some of these mutations accumulate during clonal expansions, others contribute to the evolution and establishment of particular clonal populations (Hanahan and Weinberg, 2011). Therefore, a major challenge in tumor biology is elucidating the contributions of each of these cancer mutations during cancer initiation and progression. Moreover, several of these mutations have become essential for cancer cell survival, rendering the characterization of these mutations difficult – especially *in vivo*. Nevertheless, the study of essential genes in cancer is particularly salient as it could lead to the identification of novel therapeutic targets.

To overcome these hurdles, CRISPR/Cas9-inducible systems have been developed (Davis *et al.*, 2015; Dow *et al.*, 2015; Nihongaki *et al.*, 2015; Wright *et al.*, 2015). These systems are based on the transcriptional regulation of Cas9, for example, through the use of a DOX-induced sgRNA or on post-transcriptional methods. In the latter case multiple strategies have been employed including: (1) the generation of a split Cas9 protein that can be restored and activated in the presence of rapamycin, (2) the insertion of a 4-hydroxytamoxifen (4-HT)-dependent intein that can be spliced-out upon treatment with tamoxifen or (3) by fusing the light-inducible heterodimerizing proteins CRY2 and CIB1 to Cas9 or a mutated version of Cas9 (Davis *et al.*, 2015; Dow *et al.*, 2015; Nihongaki *et al.*, 2015; Wright *et al.*, 2015).

In addition, methods based on the temporal control of CRISPR/Cas9 activity by conditional destabilization of Cas9 have proved promising (Senturk *et al.*, 2017). The first conditional destabilization method of proteins based on the destabilization domain (DD)-based technology stemmed from the identification of specific destabilizing mutations in the human FKBP12 protein, and the concurrent development of a family of highly specific synthetic FKBP12 ligands that are cell permeable and non-toxic in cultured cells and animals (Banaszynski *et al.*, 2006; 2008). The small-molecule ligand Shield-1 was shown to bind mutant FKBP12 more tightly than the wild-type by three orders of magnitude, and to prevent proteasome-induced degradation of DD-GFP and other fusion proteins (Bonger *et al.*, 2011; Clackson *et al.*, 1998). This DD system has been demonstrated to work well for a variety of proteins, including kinases, cell cycle regulatory proteins and small GTPases expressed in cultured cells and mice (Banaszynski *et al.*, 2006; 2008).

By fusing the FKBP12-derived destabilizing domain to Cas9 (DD-Cas9), Senturk *et al.* demonstrated that this means of conditionally regulating protein stability could be exploited for rapid and reversible Cas9 expression *in vitro* (Senturk *et al.*, 2017).

Specifically, they showed that fusing an FKBP12-derived destabilizing domain to Cas9 (DD-Cas9) enables conditional Cas9 expression and temporal control of gene editing in the presence of an FKBP12 synthetic ligand (Figure 22.1A; Senturk *et al.*, 2017). This technique could be employed to conditionally target a variety of genes controlling diverse biological processes such as oncogenic transformation (EGFR), DNA damage responses (TP53), mitochondria metabolism (CypD) and DNA replication/repair (RPA3). By targeting the RPA3 gene and EGFR in "EGFR-addicted" cells, they demonstrated the ability of the DD-Cas9 system to identify genes that are essential for sustained cell growth and survival (Senturk *et al.*, 2017).

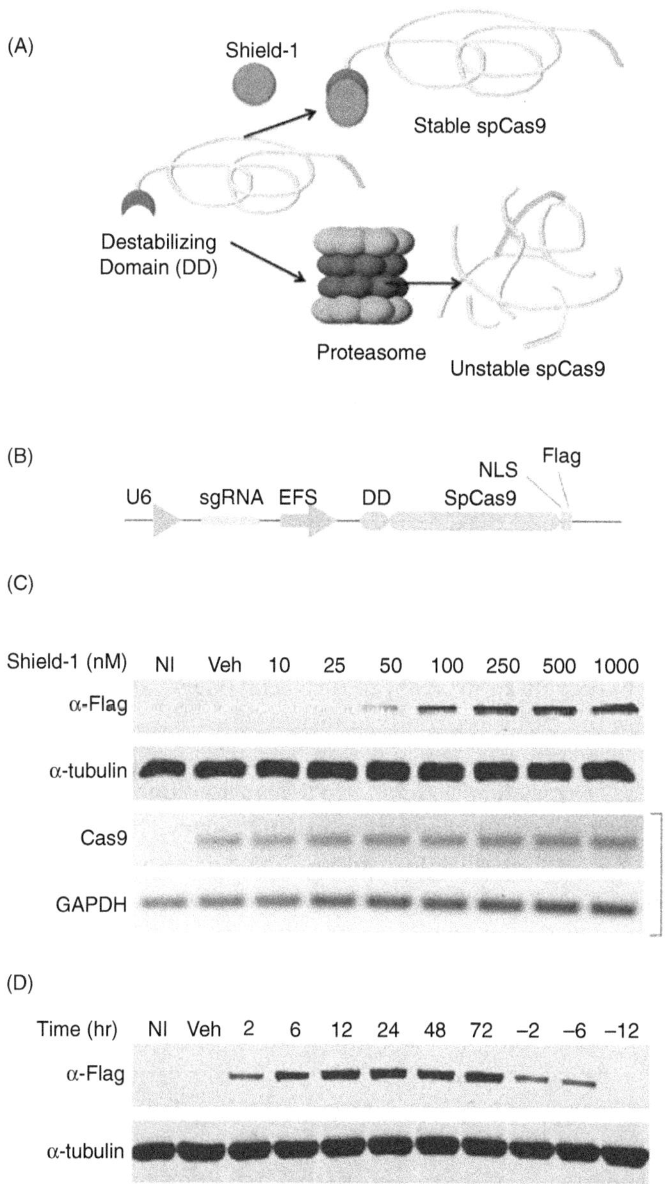

Figure 22.1 Establishment of an inducible lentiviral guide RNA expression system based on conditionally destabilized Cas9. (A) Diagram of conditionally destabilized Cas9 (DD-Cas9). Cas9 is fused to the destabilizing domain of mutant FKBP12 that triggers the rapid degradation of the fused-protein. The synthetic ligand (Shield-1) prevents this destabilizing effect and thus degradation of Cas9. Note, a Flag tag is present at the C-terminal of Cas9. (B) Schematic of the lentiviral construct utilized in our studies. (C) Ligand-dependent stabilization of DD-Cas9. Cells expressing destabilized Cas9 (DD-Cas9) and uninfected cells (NI) were either mock-treated (vehicle) or treated with increasing concentrations of Shield-1, as indicated. To measure the level of expression of Cas9, lysates were analyzed by western blot with an anti-Flag antibody. Various degrees of Shield-1-induced stabilization could be observed in a dose-dependent manner. DD-Cas9 was not detected in lysate from mock-treated cells. To verify that the destabilization of DD-Cas9 occurs post-transcriptionally, DD-Cas9 expression was also verified by RT-PCR analysis using Cas9-specific primers and GAPDH as the loading control. (D) Rapid and reversible regulation of destabilized DD-Cas9. A549 cells transduced with the DD-Cas9 lentiviral vector and uninfected cells were treated with Shield-1 (200 nM) for different time points as indicated. Samples were assayed by western blot analysis at the indicated time points (Senturk *et al.*, 2017). (A black-and-white version of this figure will appear in some formats. For the color version, please refer to the plate section.)

One unique feature of this method is that the Cas9 protein's expression is conditionally regulated independently of its mRNA expression. Hence, this vector system can be easily adapted to co-express – from the same promoter – DD-Cas9 with any other gene of interest, without the latter being co-modulated. Senturk *et al.* demonstrated that this approach could be used for generating traceable systems and for interrogating genes' functional interactions *in vivo* (Senturk *et al.*, 2017). Moreover, when they coupled DD-Cas9 with a conditional Cre allele (Cre-ER^{T2}), they showed this system could be utilized to facilitate the analysis of genes that modulate disease onset and progression in a variety of pre-existing mouse models of human disease based on the Cre-Lox system (Senturk *et al.*, 2017).

22.2 Methodology

Fusing Cas9 to a destabilizing domain provides a highly efficient and easily scalable new modality for the temporal control of gene editing. In principle, this technique is suited for use in a broad spectrum of *in vitro* and *in vivo* models. In this section, we will provide an overview of this technology and describe its possible application for the characterization of cancer genes.

22.2.1 Controlled Cas9 Expression by Fusion to a Destabilized Domain

Previous studies demonstrated that the presence of an FKBP12-derived destabilizing peptide (DD-) induces the rapid degradation of the fused-protein by the proteasome system (Banaszynski *et al.*, 2006; 2008). Yet, in the presence of a synthetic ligand, Shield-1, the peptide and the fusion partner are effectively "shielded" from degradation (Banaszynski *et al.*, 2006; 2008). To this end, we developed a dual lentiviral vector system consisting of: (1) a first module, in which the sgRNA cassette is driven by the constitutive U6 promoter and (2) a second module, in which Cas9 is fused at its N-terminus with a ligand-responsive destabilizing domain derived from an engineered FKBP12 mutant (DD-Cas9) under the control of the elongation factor-1α short (EFS) promoter (Figure 22.1A and B).

This vector was engineered based on the LentiCRISPR V2 (Addgene) backbone. The high-affinity ligand-dependent DD of a mutant FKBP12 protein (F36V, L106P) was inserted at the N-terminus of SpCas9. The DD-sequence was amplified from a degron-donor backbone (pMK backbone). The BsmBI digestion site in the DD-sequence was replaced with silent mutations via site-directed mutagenesis. All PCR-based amplifications of cloning templates were performed with Phusion High Fidelity DNA Polymerase (M0530, New England Biolabs). Primers used in cloning procedures and in sequence analysis are listed in Table 22.1. To package the Virus Hek-293T cells we transiently co-transfected on a 10 cm culture dish with 3 μg of the construct encoding the genes of interest and sgRNA, along with 6 μg of the packaging plasmid psPAX2 and 3 μg of the envelope plasmid pMD2.G (Addgene) using 30 μL of the Lipofectamine 2000 reagent (Life Technologies). 10 mL viral particles were collected after 48 hours of transfection by clarifying the supernatant through 0.45 μm filter membrane (GE Healthcare). Virus

Table 22.1 Complementary oligonucleotides used for cloning sgRNAs and PCR primers for surveyor assay

Oligo	Sequence
CypD g.131 F	5′- CACCgCAGGTACACGAGCGGGTTCC -3′
CypD g.131 R	5′- AAACGGAACCCGCTCGTGTACCTGc -3′
CypD g.150 F	5′- CACCgTCCCGTTGGCGTCCACGTCC -3′
CypD g.150 R	5′- AAACGGACGTGGACGCCAACGGGAc -3′
EGFR g.10 F	5′- CACCgGCTGCCCCGGCCGTCCCGGA -3′
EGFR g.10 R	5′- AAACTCCGGGACGGCCGGGGCAGCc -3′
EGFR g.60 F	5′- CACCgTCCTCCAGAGCCCGACTCGC -3′
EGFR g.60 R	5′- AAACGCGAGTCGGGCTCTGGAGGAc -3′
RPA3 g.25 F	5′- CACCgCCGGCGTTGATGCGCGACCT -3′
RPA3 g.25 R	5′- AAACAGGTCGCGCATCAACGCCGGc -3′
RPA3 g.44 F	5′- CACCgGATGAATTGAGCTAGCATGC -3′
RPA3 g.44 R	5′- AAACGCATGCTAGCTCAATTCATCc -3′
TP53 g.140 F	5′- CACCgCCATTGTTCAATATCGTCCG -3′
TP53 g.140 R	5′- AAACCGGACGATATTGAACAATGGc -3′
tdTOM g.230 F	5′- CACCgTTGTAATCGGGGATGTCGGC -3′
tdTOM g.230 R	5′- AAACGCCGACATCCCCGATTACAAc -3′
tdTOM g.282 F	5′- CACCgGGAGCGCGTGATGAACTTCG -3′
tdTOM g.282 R	5′- AAACCGAAGTTCATCACGCGCTCCc -3′
tdTOM g.292 F	5′- CACCgGAACTTCGAGGACGGCGGTC -3′
tdTOM g.292 R	5′- AAACGACCGCCGTCCTCGAAGTTCc -3′
EGFR-SVR-F	5′- TTGGCTCGACCTGGACATAG -3′
EGFR-SVR-R	5′- GGGGAAAGTGAGGGAAGAAA -3′
RPA3-SVR-F	5′- ATCCTGTGATCGCAGAAAGG -3′
RPA3-SVR-R	5′- CAAGACTTTGGGCCTGTTTC -3′
CYPD-SVR-F	5′- AGGGGTAGTCCACGGACAG -3′
CYPD-SVR-R	5′- CCGCACTACTGCTGGAAAC -3′
TP53-SVR-F	5′- GGGTTGGAAGTGTCTCATGC -3′
TP53-SVR-R	5′- TCAAAAGCCAAGGAATACACG -3′

transduction was optimized in order to achieve low MOI transduction. Typically, 500 μL fresh virus particles from 10 mL stock were used to infect 1x106 cells on a 10 cm culture dish in 10 mL total volume of culture medium. Virus aliquots were stored at –80°C.

To test for the feasibility of this approach, we transduced our engineered lentiviral construct into A549 cells and we measured the levels of Cas9 in the presence and absence of Shield-1 by western blot and RT-PCR analyses (Figure 22.1C and D). Shield-1, obtained from Cheminpharma, was solubilized in pure ethanol and added to culture media.

An antibody against Flag was used to gauge Cas9 levels indirectly. Although Cas9 was not detected in uninfected cells or in lysates of mock-treated cells, cells treated with Shield-1 showed strong expression of the expected fusion protein in a dose-dependent manner. Importantly, all experimental conditions presented with similar levels of expression of Cas9 mRNA in the presence or absence of synthetic ligand (Figure 22.1C).

We next assayed the destabilizing domain for the kinetics of protein degradation. Cells were treated with Shield-1 and harvested 2, 6, 12, 24, 48 and 72 hours after ligand treatment. Upon just 2 hours of treatment with Shield-1, we observed a rapid stabilization of Cas9 protein compared with uninfected and mock-control cells (Figure 22.1D). This effect was reversible upon 2 hours of ligand withdrawal from the media, with protein levels becoming negligible within 12 hours (Figure 22.1D). The fact that Cas9 was not detectable in mock-control cells or upon Shield-1 withdrawal indicates that the fusion of Cas9 with the engineered FKBP12 destabilized mutant holds Cas9 expression under tight control (Figure 22.1C and D).

Taken together, these data demonstrated that – when fused to Cas9 – FKBP12-derived destabilizing domains enable tight, rapid and reversible control of Cas9 expression, making DD-Cas9 a suitable tool for the generation of inducible genome editing systems.

22.2.2 Temporal Control of Gene Editing by Destabilized Cas9

To assess whether DD-Cas9 was competent for gene editing, we targeted multiple genes controlling diverse biological processes such as oncogenic transformation (EGFR), DNA damage responses (TP53), mitochondria metabolism (CypD) and DNA replication/repair (RPA3). A549 cells were transduced with lentiviral constructs expressing DD-Cas9 and locus-specific guide RNAs (Figure 22.2A). Cells were treated with the ligand 24 hours after lentiviral infection. Cell extracts were collected after six days and immunoblotted to verify gene editing (Figure 22.2A). The efficient formation of on-target indel mutations in cells treated with Shield-1 was confirmed via the SURVEYOR assay (Figure 22.2B) and in the case of TP53 by deep sequencing (Figure 22.2C–E).

Cells were infected with lentivirus and incubated at 37 °C for indicated times as described. Genomic DNA was extracted using mini-spin columns following the manufacturer's instructions (Macherey-Nagel). In brief, cell pellets were resuspended in lysis buffer and incubated at 56 °C for 3 hours, followed by 70 °C for 10 minutes. SpCas9-induced mutations were detected using the SURVEYOR Mutation Detection Kit (Transgenomic/IDT). An approximately 900 bp region surrounding the CRISPR mutation target site was PCR-amplified using Phusion High-Fidelity DNA Polymerase (NEB). Single band PCR products were purified using USB PrepEase Gel Extraction kit (Affymetrix). 450 ng purified PCR fragments mixed with 2 μL 10× Taq DNA Polymerase PCR buffer in a total volume of 20 μL were subjected to a series of melt and reanneal temperature cycles with gradual increments lowered in each thermal cycle. SURVEYOR nuclease and SURVEYOR enhancer were then added to the reaction mixture in order to selectively digest heteroduplex DNA substrates. Digested products were run and visualized on a 1.2% agarose gel. Band intensities were quantified using ImageJ software. Indel percentage was calculated using the formula $100 \times (1 - (1 - (b + c)/(a + b + c))^{1/2})$, where a stands for the integrated intensity of the undigested product and b and

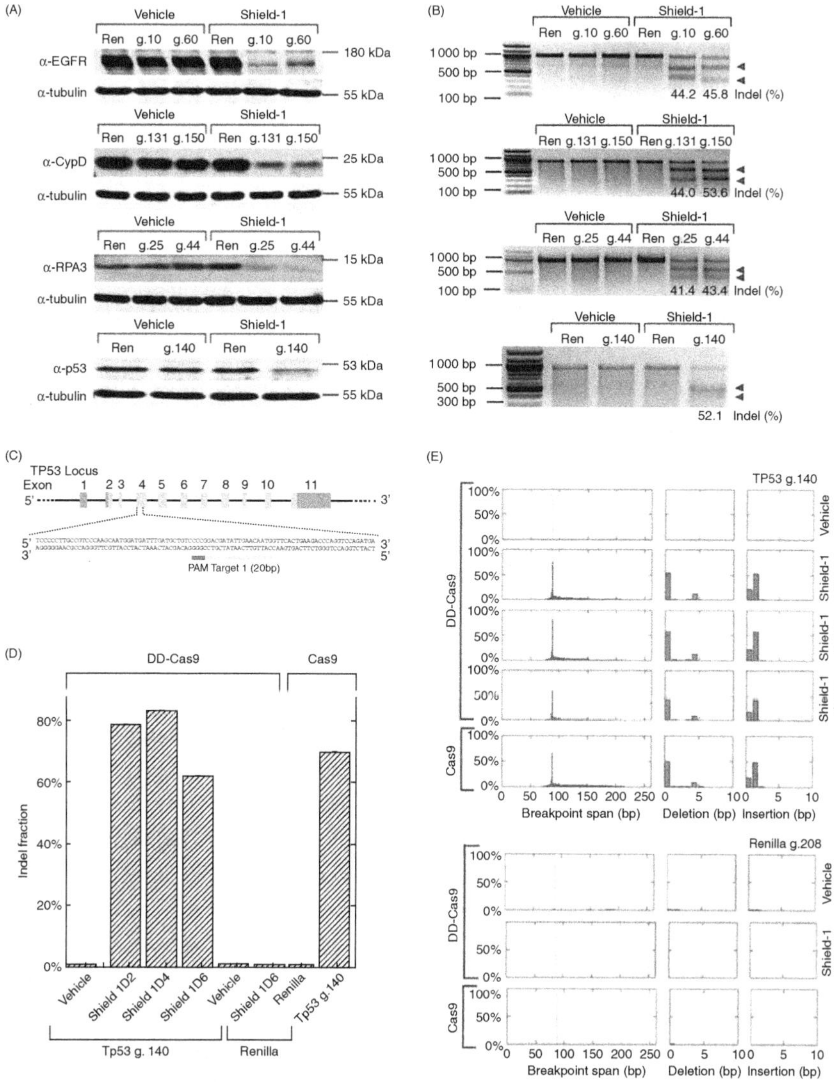

Figure 22.2 Validation of DD-Cas9 system with multiple methods for gene editing efficiency and also comparing its activity with constitutive Cas9. (A) Targeting of *EGFR, CypD, RPA3* and *TP53* genes in human A549 cells by DD-Cas9. A549 cells were transduced with DD-Cas9 lentiviral vector harboring sgRNAs targeting the indicating genes. Western blot analysis was used to assess the expression of the indicated genes before and after treatment with Shield-1 (200 nM). Alpha-tubulin was used as a loading control. Protein expression analysis was conducted following three days of Shield-1 treatment. (B) Representative SURVEYOR assays for DD-Cas9-mediated on-target indels generation at the human *EGFR, CypD, RPA3* and *TP53* loci in the presence and absence of Shield-1. DNA extracts were prepared similar to those described for the above protein expression analysis. Arrowheads indicate expected SURVEYOR fragments. (C) Schematic of location of sgRNA on the TP53 locus utilized in our study. (D) Insertion/deletion efficiencies. The percentage of reconstructed sequences in each sample that differ in length from the wild-type sequence is shown. Note that this count excludes Cas9-edited sequences that have an insertion and deletion of the same length. (E) Insertion/deletion summary statistics. Summary statistics for each sample are shown. Left-most plots show the fraction of sequences in which each nucleotide position was edited, either through a deletion spanning that base or through an adjacent insertion. Middle plots report the fraction of sequences having a deletion of the specified length. Right plots report the fraction of sequences having an insertion of the specified length. Note that individual sequences can possess both an insertion and a deletion (Senturk *et al.*, 2017).

c are the integrated intensities of each cleaved PCR product. Primers used for SURVEYOR assay are listed in Table 22.1.

Target region primers (F-CCTGGTCCTCTGACTGCTCT, R-CTGCCCTGGTAG GTTTTCTG) surrounding the TP53 guide RNA (gRNA.140) recognition site were designed using the batch version of Primer3. The genomic locus was amplified with Phusion High-Fidelity DNA Polymerase (NEB) using 100 ng genomic DNA as input and 20 cycles of PCR. Amplicons were purified using Nucleospin Gel and PCR Clean-up (Macherey-Nagel). Sequencing libraries were prepared using the NEBNext Ultra DNA Library Preparation kit for Illumina (cat# E7370), strictly following the manufacturer's instructions. Briefly, 50 ng of input template was end-repaired, tagged with Illumina adaptors and treated with USER enzyme, followed by limited-cycle PCR with indexing primers. After this amplification, DNA was purified with Agencourt AMPure XP beads (Beckman Coulter). Then, barcoded libraries were pooled in equimolar ratios, quantified by a qPCR library quantification kit (KAPA Biosystems) and subjected to massively parallel 150 bp pair-end sequencing using Miseq Instrument with Miseq Reagent 300 Cycles Kit v2 (Illumina).

To understand the sequencing output, llumina reads were processed as previously described, with minor modifications (Shi *et al.*, 2015). To summarize, paired forward and reverse Illumina reads were stitched together if they overlapped by at least 15 bp at their 3′ ends and contained at least 40 bp of wild-type sequence in their 5′ ends. These "observed sequences" were then aligned with the wild-type sequence starting from both ends and moving inward. Forward and reverse breakpoints were defined for each observed sequence as positions where the aligned wild-type sequence and observed sequence first began to differ by at least two consecutive mismatches. The region of wild-type sequence between these breakpoints was defined as the "deletion," and the region of observed sequence between these breakpoints was defined as the "insertion." Figure 22.2D and E illustrates a pile-up of the observed deletion sequences and provides a histogram showing the lengths of the insertion and deletion regions.

Figure 22.2E also illustrates an alignment of the 20 most frequently observed patterns near the Cas9 target sequence, as well as the counts of these patterns for each of the samples. Interestingly, the sequence alignment shows that some observed sequences had very long insertions. These long insertions usually came paired with long deletions, a phenomenon that likely represents an artifact of the breakpoint calling algorithm arising from a region of low sequence quality in the reverse reads. This artifact, however, does not affect the conclusions drawn from this data set.

Overall the data analyses indicated that induction of Cas9 with Shield-1 led to efficient gene editing, and resulted in a significant reduction in the expression of EGFR, CypD, RPA3 and p53 proteins. Notably a comparison between inducible DD-Cas9 and constitutive Cas9 also indicated similar levels of efficiency of gene editing (Figure 22.2E).

22.2.3 Functional Characterization of Cancer Genes

Next, to test whether DD-Cas9 could be used to functionally characterize genes *in vitro* and *in vivo*, we targeted the essential RPA complex. RPA is a heterotrimeric single-stranded DNA-binding complex composed of three proteins – RPA1, 2 and 3 – which stabilize replication forks by coating unwound DNA (Zou and Elledge, 2003; Zou *et al.*, 2006). The RPA complex also binds ssDNA at sites of DNA damage and recruits cell cycle checkpoint kinases (Dutta and Stillman, 1992; Zou *et al.*, 2006). Previous work showed that targeting the smallest RPA subunit, RPA3, induces degradation of all three members of the RPA complex, and is associated with decreased cell proliferation and increased cell lethality (McJunkin *et al.*, 2011; Zuber *et al.*, 2011).

A549 cells were transduced with lentiviral constructs expressing DD-Cas9 and two independent RPA3 sgRNAs (RPA3g.25 and g.44) and with Renilla sgRNA (g.208) as a control (Figure 22.3A). Cells infected with lentiviruses were plated in equal numbers in 12-well plates (BD falcon). The expression of Cas9 was stabilized using 200 nM of Shield-1 and incubated at 37 °C for indicated times, as described. Cells were rinsed twice with PBS to eliminate any floating cells, and were fixed in 4% formaldehyde in PBS (v/v) for 10–15 minutes. Fixed cells were stained by staining solution (0.1% Crystal Violet in 10% ethanol) for 20 minutes. Staining solution was aspirated from the wells, and the cells were washed with water three times in order to remove extra stain. Stained cells were air dried and imaged using the Odyssey Imaging System (LICOR). To quantify the relative cell numbers, cells were de-stained with 10% acetic acid and the absorbance of de-stained solution was measured at 590 nm at appropriate dilutions.

In the absence of Shield-1, we did not observe a difference in the number of cells when comparing uninfected cells and cells infected with lentiviruses targeting RPA3 or Renilla (Figure 22.3A). Yet, upon treatment with Shield-1, a drastic decrease in cell number was apparent in cells infected with the lentivirus harboring the RPA3, but not the Renilla guide (Figure 22.3A). Of note, analysis of the onset of this effect indicated that just 48 hours of exposure to Shield-1 compound was sufficient to induce robust decreases in cell viability.

We previously showed that the non-small-cell lung cancer (NSCLC)-derived PC9 cells harbor an oncogenic-EGFR mutation and are highly dependent on EGFR for their survival, unlike EGFR wild-type cells (i.e. A549 cells; Yao *et al.*, 2010). We took advantage of this observation to test whether DD-Cas9 could be used for the identification of genotype-driven vulnerabilities. PC9 cells were transduced with lentiviral constructs carrying the guide RNAs targeting the EGFR locus (EGFR g.10 and g.60) (Figure 22.3B). As demonstrated with Crystal Violet staining and cell counting, treatment of PC9 cells with Shield-1 resulted in a prominent and rapid decrease of viable cells (Figure 22.3C and D). This was not observed in A549 cells transduced with constructs targeting Renilla or the EGFR locus; however, western blot analysis of cell extracts confirmed decreased expression of EGFR in these cells.

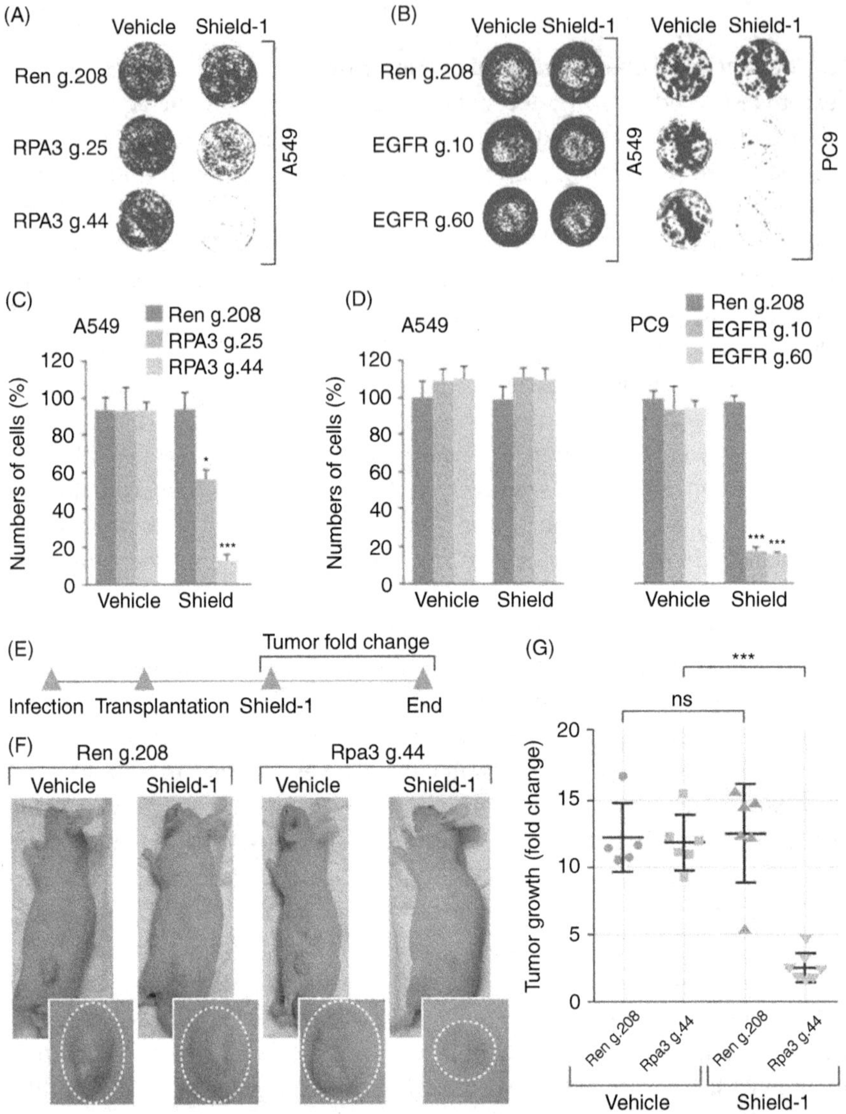

Figure 22.3 DD-Cas9 can drive robust synthetic ligand-dependent gene editing in mammalian cells and enables the identification of essential genes *in vitro* and *in vivo*. Targeting RPA3 in A549 cells and EGFR in PC9 cells resulted in a dramatic decrease in the number of viable cells following induction with Shield-1. (A) A549 cells were transduced with the DD-Cas9 vector targeting the RPA3 locus and, as a control, the Renilla gene. (B) PC9 cells and A549 cells were transduced with DD-Cas9 vector targeting the EGFR locus. In all cases cells were treated 24 hours after infection with Shield-1 (200 nM) for six days. The number of cells was assessed after six days by Crystal Violet staining. (C, D) The data in the charts represent quantification of Panels A and B, respectively, and are expressed as mean ± standard deviation upon treatment with Shield-1 and vehicle control relative to uninfected cells ($n = 4$, * $p < 0.05$, ** $p < 0.01$, *** $p < 0.005$) Student's t-test. (E) Schematic of the transplantable model system used in this study. (F) A549 cells were transduced with DD-Cas9 targeting RPA3 and as control Renilla and transplanted sub-cutaneously in immune-deficient mice. When the tumors reached an approximate size of 4–5 mm in diameter, mice were treated with Shield-1 (1 μg). Fold change in tumor growth was determined ten days after treatment. Pictures illustrate representative tumor sizes. The chart in panel (G) illustrates combined quantification of tumor volume in the indicated cohorts upon Shield-1 treatment. Data represent mean ± standard deviation and are shown relative to the negative control (Cas9 negative) *** $p < 0.0001$ (RPA3 g.44 + Shield-1 vs. – Shield-1) Student's t-test (Senturk *et al.*, 2017). (A black-and-white version of this figure will appear in some formats. For the color version, please refer to the plate section.)

The Shield-1 compound has been shown to be non-toxic and to display favorable pharmacodynamics and kinetic properties in mice models (Banaszynski *et al.*, 2006). Having shown ligand-dependent gene knockout in *in vitro* cell culture, we reasoned that this platform could be exploited for assessing gene function *in vivo* as well. As a proof of principle, we transplanted A549 cells transduced with the DD-Cas9 vector targeting either RPA3 or Renilla into immune-compromised mice via subcutaneous injection (Figure 22.3E–G). When tumors reached an approximate size of 4–5 mm in diameter, mice were treated with Shield-1 or vehicle for four days and tumor size was assessed for the next ten days. We observed a drastic decrease in tumor growth in tumors expressing RPA3 sgRNA upon Shield-1 treatment (Figure 22.3F and G). To confirm the decreased expression of RPA3, we used immunostaining (not shown).

22.2.4 Design of a Traceable System Based on Destabilized Cas9

An advantage of the described system is the conditional regulation of Cas9 protein stability independent of its mRNA expression. This implies that any gene of interest can be co-expressed with destabilized Cas9 from the same promoter. Among various strategies employed to construct bi-cistronic or multicistronic vectors, an internal ribosomal entry site (IRES) has been widely used. More recently, a self-cleaving 2A peptide has been shown to be an efficacious alternative. To verify the possibility of expressing a gene of interest together with conditionally destabilized Cas9, we cloned the fluorescent protein Venus – located downstream of DD-Cas9 – and separated them with the self-cleaving 2A peptide (P2A) (Figure 22.4A). To achieve this, we replaced P2A-linked puromycin with P2A-linked Venus – a modified version of green fluorescent protein (GFP) – for easy visualization of viral transduction efficiency as well as competition assays. To do so, P2A-Puro-WPRE-3′LTR cassette on the LentiCRISPR V2 was excised with BamHI/PmeI restriction enzymes, then a PCR product containing only WPRE-3′ LTR piece was religated, which was again surrounded by BamHI/PmeI. Following this step, a PCR product containing the P2A-Venus cassette surrounded by BamHI digestion sites on both ends was amplified and cloned into the BamHI site on the plasmid in the correct orientation.

As shown in Figure 22.4B, expression of the modified-GFP fluorescent protein (Venus) was constitutively expressed, despite fusion to DD-Cas9, and was not altered by Shield-1 treatment. Similarly, the mRNA expression was found to be independent of ligand treatment (Figure 22.4B).

An important consideration when developing therapeutic strategies based on decreased expression of a gene by viral delivery is the potential impact on the physiology and viability of the transduced cells. This effect can be particularly relevant in longer time-course experiments, where subtle differences in cell growth rates can have a major impact.

In this regard, "competition assays" have been shown to be powerful systems to assess subtle cell growth changes (Zuber *et al.*, 2011). These assays are based on the competitive cell growth between transduced and non-transduced cells. The mixed culture obtained after gene transduction can be observed in the ratio of

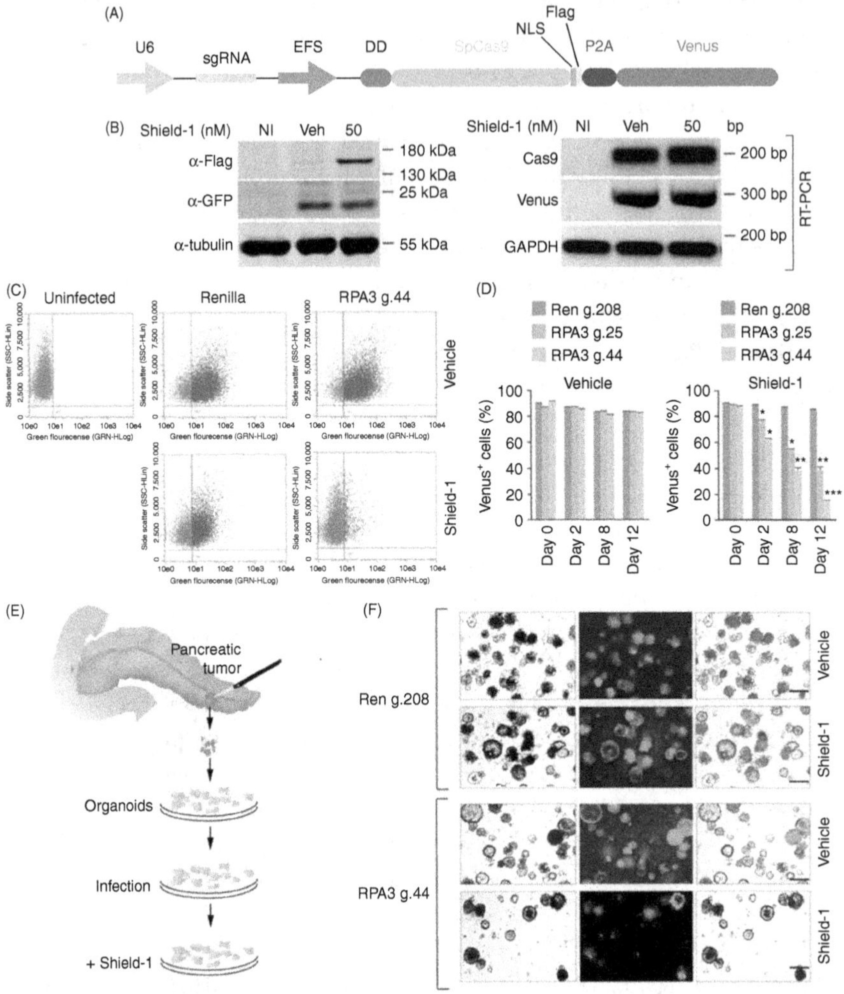

Figure 22.4 DD-Cas9 can be coupled to a modified fluorescent protein to generate a traceable system. (A) Schematic of the lentiviral DD-Cas9 plasmid containing U6 promoter-driven single-guide RNA (sgRNA) and EFS promoter-driven DD-Cas9 pieces. Note, a FLAG tag is present at the Cas9 C-terminal moiety, and 2A self-cleaving peptide (P2A) separates both DD-Cas9 and the gene of interest, in this case the modified fluorescent protein Venus. (B) Shield-1-independent expression of Venus. A549 cells were transduced with the DD-Cas9/P2A/Venus lentiviral vector and treated with Shield-1 for three days. Expression of Cas9 and Venus were assayed by western blot and RT-PCR. (C) Competition assay. A549 cells were transduced at low MOI with the DD-Cas9/P2A/Venus lentiviral vector. Cells were then treated with Shield-1 or vehicle control for 12 days and quantified by FACS. The panels represent typical FACS dot plots from A549 sorted cells. (D) A chart illustratrating the percentage of Venus$^+$ cells during time in A549 cells transduced with the DD-Cas9/P2A/Venus lentiviral vector targeting the indicated genes upon treatment with Shield-1 and vehicle control, error bars are standard deviation ($n = 3$, $^* p < 0.05$, $^{**} p < 0.01$, $^{***} p < 0.005$) Student's t-test. (E) Schematic representing our test for cell vulnerabilities in pancreatic tumor organoids. (F) DD-Cas9/P2A/Venus could be used to infect primary human tumor-derived cultures and to assess their cellular vulnerabilities. Human pancreatic cancer-derived organoids were infected with the DD-Cas9/P2A/Venus lentiviral vector targeting RPA3 and as control Renilla. Organoids, cultured in 24-well plates, were treated with Shield-1 or vehicle control 72 hours post-infection. The number of Venus$^+$ organoids was determined 72 hours after infection and six days after Shield-1 treatment (Senturk *et al.*, 2017). (A black-and-white version of this figure will appear in some formats. For the color version, please refer to the plate section.)

Venus$^+$ and Venus$^-$ cells over time to identify minor cell growth defects. When compared to established methods of counting cells and viability assays (e.g. MTT and ATP bioluminescence), competition assays were shown to have better sensitivity.

To verify that our platform could also be used in a competition assay set-up, we infected A549 cells with the DD-Cas9/P2A/Venus lentiviral vectors targeting either RPA3 or, as a negative control, the Renilla gene at low MOI. This ensured the generation of Venus$^+$ and Venus$^-$ mixed culture (Figure 22.4C). By tracking the ratio of Venus$^+$ and Venus$^-$ cells infected with the RPA3-targeting vector on a time-course of 12 days, we observed a Shield-1-dependent depletion of Venus$^+$ cells (Figure 22.4C and D).

The recently developed 3D "organoid" culture system allows for cultivation of both normal and patient-derived primary tumors *in vitro* (Boj *et al.*, 2015). We tested whether the DD-Cas9/P2A/Venus platform could also be applicable to this culture system (Figure 22.4E and F). We followed methods to establish and propagate human organoid cultures that were previously described by the Tuveson and Clevers laboratories (Boj *et al.*, 2015).

In brief, organoids were grown in 24-well culture plates for two days before infection. One well per infection was used. On the day of infection, organoids were dissociated into small fragments by first triturating them in media through a fire-polished glass pipette, and then by a 5-minute enzymatic digestion with TrypLE (Life Technologies) at 37 °C. The resulting small cell clusters were then resuspended with low MOI lentivirus and spinoculated at 600 RCF for 1 h at room temperature.

Upon infection, we observed approximately 80% Venus$^+$ organoids. Strikingly, in organoids derived from two different tumors, after treatment with Shield-1 this fraction was decreased to less than 20% for the RPA3-targeting sgRNA (Figure 22.4F). This result clearly demonstrates the utility and efficiency of the DD-Cas9 system as a screening tool in primary tumor cells. Despite these promising results, further experiments (i.e. independent sgRNAs) will be required to validate RPA3 as a gene that is essential for primary pancreatic cells growth and/or survival.

22.2.5 DD-Cas9 Linked to Cre Allows Genetic Interaction Studies

Recent studies indicated that a CRISPR/Cas9-based approach could be used for the functional investigation of candidate genes in well-established mouse models of cancer (Maddalo *et al.*, 2014; Xue *et al.*, 2014). In particular, these studies showed that Cre-dependent somatic activation of oncogenic Kras (G12D) could be combined with CRISPR/Cas9-mediated genome editing of tumor-suppressor genes by using a lentiviral-based system that delivers both the CRISPR/Cas9 system and the Cre recombinase.

We reasoned that by generating a bi-cistronic vector for the expression of destabilized Cas9 and tamoxifen-inducible Cre, we could uncouple gene editing and Cre-mediated recombination. By providing temporal control of gene editing and activation/inactivation of LOX alleles, this approach, in principle, could bolster precision in the study of functional interactions among genes in a variety of pre-existing mouse models derived from the Cre-Lox system. Especially in the context of cancer models, it could be used to dissect more precisely the role of mutations identified in onco-genomic studies during tumor onset and progression. Given that cancer is a multi-step process that involves the progressive accumulation of mutations, this approach could be invaluable for the identification of tumor cell vulnerabilities during the evolution of tumors from preneoplastic to fully malignant diseases.

To achieve this, we engineered the DD-Cas9/P2A/Venus backbone to express Cre-ER^{T2} downstream of Cas9 (Figure 22.5A). More specifically, we inserted an internal ribosome entry site (IRES) downstream of the DD-Cas9 open reading frame (ORF) which was then coupled to tamoxifen-inducible Cre (Cre-ER^{T2}). In this context, gene editing could be induced independently of Cre-mediated recombination. The Cre-ER^{T2} backbone was generated by introducing a stop codon at the end of the Cas9 ORF, and cloning IRES and Cre-ER^{T2} ORFs downstream of the Cas9 cassette with NLS and a Flag tag (Figure 22.5A). The Cre-ER^{T2} protein codon optimized for human expression with three naturally existing BsmBI restriction enzyme sites removed by silent mutations was synthesized, cloned into a pUC57 cloning vector and sequence-verified by Genscript. This vector served as a template for subsequent cloning experiments. IRES sequence was PCR-amplified from a TTI-GFP.

To test this system, we used genetically engineered mouse lines that express the fluorescent reporter protein tdTomato in a Cre-dependent manner from the constitutive CAG promoter (tdTom^{LSL}; Ai14). In this setting, following a Cre-mediated deletion of the floxed stop cassette, tdTomato can be readily expressed. We derived mouse embryonic fibroblasts (MEFs) and transduced them with our lentiviral vector. Three days after infection, MEFs were acutely treated with 4-hydroxytamoxifen to induce activation of Cre (Figure 22.5B). We assessed expression of tdTomato by fluorescent microscopy and FACS. TdTomato positive cells were readily observed only in MEFs infected with DD-Cas9/IRES/Cre-ER^{T2} upon 4-hydroxytamoxifen treatment (Figure 22.5C).

Next, to assess the gene editing efficiency of our system, we targeted the Cre-activated tdTomato knockin reporter allele with lentiviruses expressing sgRNAs against tdTomato (sgTom) or Renilla as a control. When we quantified sgRNA-transduced tdTomato positive MEF cells over time, we observed a substantial reduction in tdTomato expression only in those cells treated with Shield-1 compound (Figure 22.5D). Thus, this result suggests that the unique system that we have developed is a functional tool for the gene editing of an endogenous allele, and that Cre and Cas9 activity may be successfully uncoupled.

Consummately, these results demonstrate the potential of using destabilizing domain technology to generate scalable and efficient platforms for the temporal control of CRISPR/Cas9-based gene editing. Moreover, the described method enables the testing of co-operativity between genetic events and the identification of tumor cell vulnerabilities in both *in vitro* and *in vivo* models.

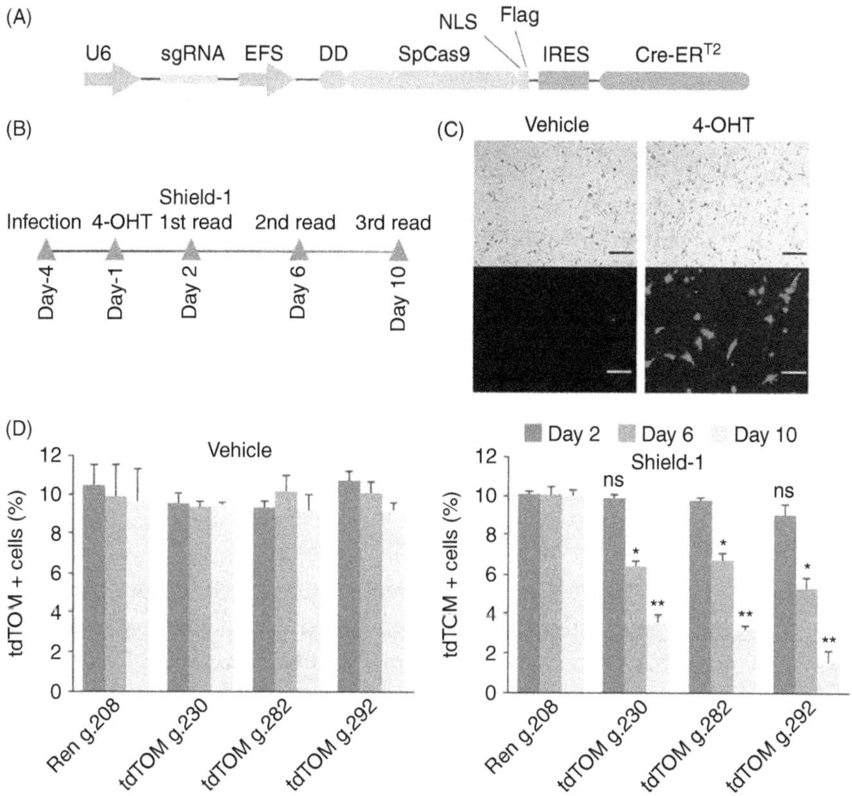

Figure 22.5 DD-Cas9 can be combined with tamoxifen-inducible Cre (Cre-ER^{T2}) to facilitate the study of genetic interactions *in vivo* in pre-existing mouse models based on the Cre-lox system. (A) Diagram of the DD-Cas9/IRES/Cre-ER^{T2} vector utilized in this study. (B) Schematic for experimental design. MEF cells obtained from CAG-LSL-tdTomato mice were transduced with the indicated vectors. Induction of Cre-mediated recombination was induced upon 4-hydroxytamoxifen treatment (5 μM) for three days. To test for efficiency of gene inactivation, cells were treated with Shield-1 (200 nM). (C) Bright-field and fluorescence images of LSL-tdTomato cells treated with 4-hydroxytamoxifen. (D) The chart represents the percentage of tdTomato-positive MEFs transduced with DD-Cas9/IRES/Cre-ER^{T2} lentiviruses carrying the indicated sgRNAs over time after exposure to Shield-1 and vehicle. Numbers of tdTomato-positive and -negative cells were determined by FACS. Each bar represents the average of three individual replicates, error bars are standard deviation ($n=3$, $^{*}p<0.05$, $^{**}p<0.01$) Student's t-test (Senturk *et al.*, 2017). (A black-and-white version of this figure will appear in some formats. For the color version, please refer to the plate section.)

22.3 Discussion

The possibility of acutely inactivating genes in cells and in organisms is essential for our understanding of gene function. Lethality is the most conclusive phenotype defining a gene as "essential." In the context of cancer research, studying essential genes is particularly important as they could serve as potential therapeutic targets. Yet, the fact that their inactivation leads to cell death renders them the most difficult class of genes to study.

CRISPR/Cas9 technology has been widely utilized to create heritable changes in the genome (Doudna and Charpentier, 2014). However, the inactivation of genes often results in cell lethality, functional deficits and developmental defects, limiting the utility of such approaches for studying tumor gene function. Additionally, constitutive expression of Cas9 may result in toxicity and in the generation of off-target effects. To overcome these limitations, we generated a temporally controlled Cas9 expression system by fusing Cas9 to an engineered mutant of the human FKBP12 protein (DD-Cas9).

22.3.1 The Choice of an Inducible Vector

Many different approaches have been developed and used in the past for the identification and characterization of essential genes. These methods include: (1) the generation of temperature-sensitive mutants, (2) the generation of conditional deletion alleles in which a gene of interest is flanked by loxP recombination sites to enable acute gene inactivation upon expression of Cre recombinase and (3) inducible gene knockdown based on shRNA-mediated silencing (Premsrirut *et al.*, 2011; Tyner *et al.*, 2002; Ventura *et al.*, 2007). More recently, methods to temporally control CRISPR/Cas9-based genome editing technologies have also been developed. These systems are based either on the transcriptional control of Cas9 and/or sgRNA or on Cas9 post-transcriptional/translational activation (Davis *et al.*, 2015; Dow *et al.*, 2015; Wright *et al.*, 2015). Regarding the former, multiple methods have been developed based on the use of DOX-induced sgRNA and Cas9 expression. In the latter case, Cas9 activity is regulated either by generating a split Cas9 enzyme that could be restored upon treatment with rapamycin (a light-inducible Cas9) or by inserting a 4-HT-intein in the CAs9 transcript that could be spliced by treatment with tamoxifen. As an alternative to these existing CRISPR/Cas9 techniques, we utilized a conditionally destabilized Cas9. By exploiting a strategy developed by the Wandless lab, we showed that by conjugating a genetically encoded destabilizing domain to Cas9 we could achieve a highly specific conditional regulation of Cas9 expression and gene editing (Senturk *et al.*, 2017). Compared to other methods, this approach enables the generation of a system that would: (1) provide potent, robust and temporally controlled gene editing, (2) be applicable to a broad spectrum of cell types and tissues, (3) facilitate high-throughput manipulation and (4) be traceable.

As the small-molecule DD ligand Shield-1 is highly permeable and non-toxic in cultured cells and animals, we also demonstrated our methods could be utilized in transplantable model systems and in mosaic mouse models (Senturk *et al.*, 2017). In particular, by coupling DD-Cas9 with tamoxifen-inducible Cre, we showed that our lentiviral vector could, in principle, facilitate the study of genetic interactions *in vivo* in a variety of pre-existing mouse models of human diseases based on the Cre-Lox system.

Recent studies have shown that a similar lentiviral vector based on tandem expression of Cas9 and Cre recombinase could be used *in vivo* to rapidly evaluate human cancer genome candidates (Senturk *et al.*, 2017). Our approach enables the temporal and independent control of gene editing and Cre-mediated recombination. Therefore, in principle, this method uniquely informs the investigation of the roles of putative cancer genes in tumorigenesis and in driving tumor vulnerabilities.

Although we have described here the use of CRISPR/Cas9 technology for the characterization of cancer genes, this same approach could be easily implemented for the identification of genes that are required for the survival or progression of tumors.

22.3.2 Future Development

CRISPR/Cas9-based gene editing is an ever-evolving field, and novel Cas9 mutants with different properties are constantly being generating. The efficiency and simplicity of the Shield-1 degron Cas9 approach facilitates the easy adaptation of this system to explore novel research objectives. The amenability of this system has been demonstrated, for example, by the design and use of catalytically inactive Cas9 to activate or repress transcription (Chavez *et al.*, 2015).

In addition to the Shield-1-based system, many other destabilizing domains have been developed. In principle, these other approaches can be adapted for the generation of novel inducible Cas9 systems. In preliminary experiments, we have shown that the *E. coli* dihydrofolate reductase (ecDHFR) could be fused to Cas9 to destabilize it (Iwamoto *et al.*, 2010). Similar to our observations when we treated cells with Shield-1, our data suggest that cells treated with the small-molecule ecDHFR ligand trimethoprim (TMP) stabilized the ecDHFR-Cas9 protein in a rapid, reversible and dose-dependent manner. Compared to Shield-1, TMP is inexpensive and capable of more efficiently traversing the blood–brain barrier; therefore, its use will enable the regulation of gene editing in the mammalian central nervous system (Iwamoto *et al.*, 2010).

22.4 Concluding Remarks

Cancer is a multi-step process characterized by the accumulation of multiple genetic aberrations over time. The development of CRISPR/Cas9 technologies has both simplified and enhanced the efficiency of gene editing approaches. By rapidly introducing mutations in cell lines, specific tissues and organisms, the CRISPR/Cas9 system has enabled whole-genome and targeted loss-of-function and gain-of-function cancer screening. The implementation of CRISPR/Cas9 systems has already facilitated the identification and characterization of multiple novel oncogenes, tumor-suppressor genes and cancer non-coding regulatory regions. CRISPR/Cas9 approaches have also allowed the development of animal models of cancer at an unprecedented pace. The generation of conditionally destabilized Cas9 could prove to be an important tool enabling the temporal control of gene editing. This novel platform is robust, easily scalable and, therefore, suitable for the identification and/or characterization of essential genes and the interrogation of the functional interaction of genes *in vitro* and *in vivo*.

REFERENCES

Banaszynski LA, Chen LC, Maynard-Smith LA, Ooi AG, Wandless TJ. 2006. A rapid, reversible, and tunable method to regulate protein function in living cells using synthetic small molecules. *Cell* **126**(5): 995–1004.

Banaszynski LA, Sellmyer MA, Contag CH, Wandless TJ, Thorne SH. 2008. Chemical control of protein stability and function in living mice. *Nat Med* **14**(10): 1123–1127.

Bell JB, Podetz-Pedersen KM, Aronovich EL, *et al.* 2007. Preferential delivery of the Sleeping Beauty transposon system to livers of mice by hydrodynamic injection. *Nat Protoc* **2**(12): 3153–3165.

Boj SF, Hwang CI, Baker LA, *et al.* 2015. Organoid models of human and mouse ductal pancreatic cancer. *Cell* **160**(1–2): 324–338.

Bonger KM, Chen LC, Liu CW, Wandless TJ. 2011. Small-molecule displacement of a cryptic degron causes conditional protein degradation. *Nat Chem Biol* **7**(8): 531–537.

Chavez A, Scheiman J, Vora S, *et al.* 2015. Highly efficient Cas9-mediated transcriptional programming. *Nat Methods* **12**(4): 326–328.

Clackson T, Yang W, Rozamus LW, *et al.* 1998. Redesigning an FKBP-ligand interface to generate chemical dimerizers with novel specificity. *Proc Natl Acad Sci USA* **95**(18): 10437–10442.

Davis KM, Pattanayak V, Thompson DB, Zuris JA, Liu DR. 2015. Small molecule-triggered Cas9 protein with improved genome-editing specificity. *Nat Chem Biol* **11**(5): 316–318.

Doudna JA, Charpentier E. 2014. Genome editing: the new frontier of genome engineering with CRISPR-Cas9. *Science* **346**(6213): 1258096.

Doudna JA, Sontheimer EJ. 2014. Methods in enzymology. The use of CRISPR/Cas9, ZFNs, and TALENs in generating site-specific genome alterations. Preface. *Methods Enzymol* **546**: xix–xx.

Dow LE, Fisher J, O'Rourke KP, *et al.* 2015. Inducible in vivo genome editing with CRISPR-Cas9. *Nat Biotechnol* **33**(4): 390–394.

Dow LE, Lowe SW. 2012.Life in the fast lane: mammalian disease models in the genomics era. *Cell* **148**(6): 1099–1109.

DuPage M, Dooley AL, Jacks T. 2009. Conditional mouse lung cancer models using adenoviral or lentiviral delivery of Cre recombinase. *Nat Protoc* **4**(7): 1064–1072.

Dutta A, Stillman B. 1992. cdc2 family kinases phosphorylate a human cell DNA replication factor, RPA, and activate DNA replication. *EMBO J* **11**(6): 2189–2199.

Hanahan D, Weinberg RA. 2011. Hallmarks of cancer: the next generation. *Cell* **144**(5): 646–674.

Iwamoto M, Bjorklund T, Lundberg C, Kirik D, Wandless TJ. 2010. A general chemical method to regulate protein stability in the mammalian central nervous system. *Chem Biol* **17**(9): 981–988.

Jinek M, Chylinski K, Fonfara I, *et al.* 2012. A programmable dual-RNA-guided DNA endonuclease in adaptive bacterial immunity. *Science* **337**(6096): 816–821.

Konermann S, Brigham MD, Trevino AE, *et al.* 2015. Genome-scale transcriptional activation by an engineered CRISPR-Cas9 complex. *Nature* **517**(7536): 583–588.

Maddalo D, Manchado E, Concepcion CP, *et al.* 2014. In vivo engineering of oncogenic chromosomal rearrangements with the CRISPR/Cas9 system. *Nature* **516**(7531): 423–427.

McJunkin K, Mazurek A, Premsrirut PK, *et al.* 2011. Reversible suppression of an essential gene in adult mice using transgenic RNA interference. *Proc Natl Acad Sci USA* **108**(17): 7113–7118.

Nihongaki Y, Kawano F, Nakajima T, Sato M. 2015. Photoactivatable CRISPR-Cas9 for optogenetic genome editing. *Nat Biotechnol* **33**(7): 755–760.

Platt RJ, Chen S, Zhou Y, *et al.* 2014. CRISPR-Cas9 knockin mice for genome editing and cancer modeling. *Cell* **159**(2): 440–455.

Premsrirut PK, Dow LE, Kim SY, *et al.* 2011. A rapid and scalable system for studying gene function in mice using conditional RNA interference. *Cell* **145**(1): 145–158.

Senturk S, Shirole NH, Nowak DG, *et al.* 2017. Rapid and tunable method to temporally control gene editing based on conditional Cas9 stabilization. *Nat Commun* **8**: 14370.

Shi J, Wang E, Milazzo JP, *et al.* 2015. Discovery of cancer drug targets by CRISPR-Cas9 screening of protein domains. *Nat Biotechnol* **33**(6): 661–667.

Singh P, Schimenti JC, Bolcun-Filas E. 2015. A mouse geneticist's practical guide to CRISPR applications. *Genetics* **199**(1): 1–15.

Sternberg SH, Redding S, Jinek M, Greene EC, Doudna JA. 2014. DNA interrogation by the CRISPR RNA-guided endonuclease Cas9. *Nature* **507**(7490): 62–67.

Swiech L, Heidenreich M, Banerjee A, *et al.* 2015. *In vivo* interrogation of gene function in the mammalian brain using CRISPR-Cas9. *Nat Biotechnol* **33**(1): 102–106.

Tyner SD, Venkatachalam S, Choi J, *et al.* 2002. p53 mutant mice that display early ageing-associated phenotypes. *Nature* **415**(6867): 45–53.

Utomo AR, Nikitin AY, Lee WH. 1999. Temporal, spatial, and cell type-specific control of Cre-mediated DNA recombination in transgenic mice. *Nat Biotechnol* **17**(11): 1091–1096.

Ventura A, Kirsch DG, McLaughlin ME, *et al.* 2007. Restoration of p53 function leads to tumour regression in vivo. *Nature* **445**(7128): 661–665.

Wang T, Birsoy K, Hughes NW, *et al.* 2015. Identification and characterization of essential genes in the human genome. *Science* **350**(6264): 1096–1101.

Wright AV, Nunez JK, Doudna JA. 2016. Biology and applications of CRISPR systems: harnessing nature's toolbox for genome engineering. *Cell* **164**(1–2). 29–44.

Wright AV, Sternberg SH, Taylor DW, *et al.* 2015. Rational design of a split-Cas9 enzyme complex. *Proc Natl Acad Sci USA* **112**(10): 2984–2989.

Xue W, Chen S, Yin H, *et al.* 2014. CRISPR-mediated direct mutation of cancer genes in the mouse liver. *Nature* **514**(7522): 380–384.

Yang H, Wang H, Jaenisch R. 2014. Generating genetically modified mice using CRISPR/Cas-mediated genome engineering. *Nat Protoc* **9**(8): 1956–1968.

Yang H, Wang H, Shivalila CS, *et al.* 2013. One-step generation of mice carrying reporter and conditional alleles by CRISPR/Cas-mediated genome engineering. *Cell* **154**(6): 1370–1379.

Yao Z, Fenoglio S, Gao DC, *et al.* 2010. TGF-beta IL-6 axis mediates selective and adaptive mechanisms of resistance to molecular targeted therapy in lung cancer. *Proc Natl Acad Sci USA* **107**(35): 15535–15540.

Zou L, Elledge SJ. 2003. Sensing DNA damage through ATRIP recognition of RPA-ssDNA complexes. *Science* **300**(5625): 1542–1548.

Zou Y, Liu Y, Wu X, Shell SM. 2006. Functions of human replication protein A (RPA): from DNA replication to DNA damage and stress responses. *J Cell Physiol* **208**(2): 267–273.

Zuber J, Shi J, Wang E, *et al.* 2011. RNAi screen identifies Brd4 as a therapeutic target in acute myeloid leukaemia. *Nature* **478**(7370): 524–528.

23 Genome Editing for Retinal Diseases

Vinod Jaskula-Ranga

The revolutionary new genome editing technology known as CRISPR is transforming biological research and ushering in a new era for genetic and personalized medicine. CRISPR allows researchers to target disease-associated mutations with unprecedented precision, treating diseases at the true root cause. For the first time, advances in genome editing have made the development of therapeutics a near-term reality. Many of the initial targets for this revolutionary technology will be for retinal disorders in the eye.

23.1 Introduction

The eye is a highly developed organ with distinctive properties that has led to seminal advances in the development of both gene-based and cell-based therapies. These properties are key to understanding why this unique milieu is poised for the development of genome editing therapeutics. On a basic level, the eye is easily accessible and it is contained. Being compartmentalized from the rest of the body affords a significant advantage over other organs, as delivery of a small volume of a therapeutic near the cells of interest can yield a high concentration without systemic circulation (Dalkara *et al.*, 2016). Importantly, while minimizing systemic leakage, a blood–retinal barrier also ensures ocular immune-privilege. Although blood–ocular barriers are not absolute, the barrier provides a significant immunologic tolerance for foreign antigens (Perez *et al.*, 2013; Streilein, 2003). This immune-privilege provides a safer environment for the expression of foreign proteins, such as genome editing nucleases.

The ability to easily and quantitatively evaluate clinical outcomes is another key feature of the eye (Scholl *et al.*, 2016). Visual function and retinal structure can be easily monitored in real-time using numerous non-invasive and highly sensitive techniques; enhanced imaging techniques can even permit resolution of the retina down to the cellular level (Williams, 2011). In cases where there might be potential complications, this can be quickly detected and addressed, and conveniently, given that many ocular diseases are bilaterally symmetrical, the contralateral eye can be used for internal controls (Dalkara *et al.*, 2016). These properties, along with technological advancements that allow for close monitoring, quantitative measures and standardization, have brought the eye to the forefront of emerging therapeutics, making eye diseases a model target for the next revolution of CRISPR-based therapeutics.

Within the eye, the retina has been a particularly attractive target for gene therapies and it is likely be one of the initial areas for *in vivo* genome editing therapeutics (Figure 23.1). The retina is spatially arranged in a laminar configuration consisting of six major types of neurons: photoreceptors (rods and cones), horizontal cells, bipolar cells, ganglion cells, amacrine cells and Müller glia cells; for over 100 years, the patterning and organization of the retina has been well appreciated (Cajal, 1892). Vision begins with the capture of light quanta by visual pigments in photoreceptor cells. This process results in a signal transduction that is integrated and processed by bipolar, horizontal, retinal ganglion and amacrine cells, and ultimately transmitted to the brain via the optic nerve. Photoreceptor cells belong to one of two classes: rods, which function under dim light conditions, and cones, which function in bright light and mediate color and high acuity vision. At the back of the neural retina is a monolayer of supportive epithelial cells that form part of the blood–retinal barrier known as the retinal pigment epithelium (RPE) (Perez *et al.*, 2013). The apical membrane of the RPE interfaces with the outer segments of photoreceptors, while the basolateral membrane faces the choriocapillaris blood vessels. The RPE performs critical functions that support photoreceptor viability and mutations affecting these functions often lead to photoreceptor degeneration and subsequent vision loss (da Cruz *et al.*, 2007).

The eye is unique in that technological advancements have also made possible combined genome editing and *ex vivo* approaches (Gregory-Evans *et al.*, 2012). Recent advances in cell-based/stem cell therapies are being developed for the treatment of advanced retinal degenerations with cell loss. Landmark clinical trials have demonstrated the ability to successfully transplant RPE cells derived from human embryonic stem cells for the treatment of Stargardt disease and age-

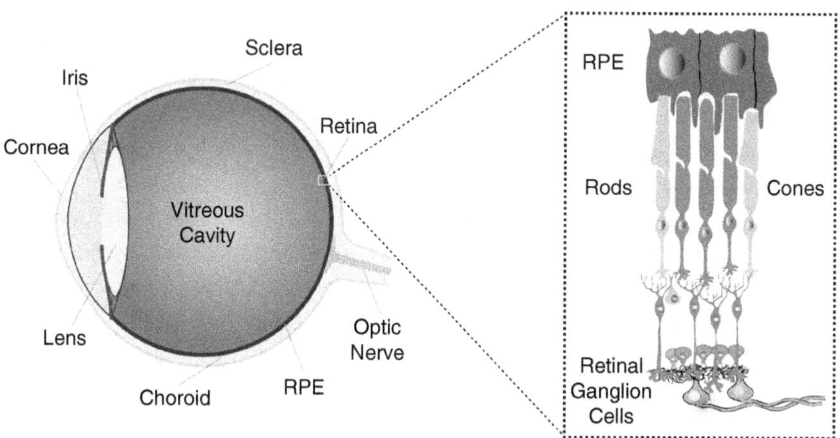

Figure 23.1 Anatomy of the eye and retina.

related macular degeneration (Schwartz *et al.*, 2012; 2015). Here, genome editing approaches can complement the use of human-induced pluripotent stem cells (iPSCs) for autologous cell therapies. First, somatic cells or fibroblasts are isolated from patients and then converted into iPSCs. Genome editing is then used to correct the mutation. These engineered iPSCs can then be differentiated into RPE cells in culture, which can be delivered back to the patient. The additional advantage of correcting patient-derived cells is the reduced potential for tissue rejection and immune responses (Kamao *et al.*, 2014).

While numerous studies across the world continue to demonstrate that disease mutations can be efficiently targeted using CRISPR-Cas9 technology *in vitro*, the development of CRISPR-based therapeutics has been challenged in large part to delivery constraints. The CRISPR-Cas9 system is composed of a guide RNA (gRNA) that targets the Cas9 nuclease in a sequence-specific fashion. Cleavage by the CRISPR system requires complementary base pairing of the gRNA to a DNA sequence and the requisite protospacer-adjacent motif (PAM), a short nucleotide motif found 3′ to the target site (Doudna and Charpentier, 2014; Hsu *et al.*, 2014). Currently, the least restrictive and most commonly used protein is from *S. pyogenes* (SpCas9), which recognizes the sequence NGG.

CRISPR-Cas9, and more generally RNA-guided nucleases, ultimately require the formation of a ribonucleoprotein (RNP) complex within the target cell. This can be accomplished by DNA delivery (plasmid), RNA (mRNA/gRNA) or direct delivery of the assembled Cas9:gRNA ribonucleotide protein complex (RNP). For *ex vivo* indications Cas9 mRNA and gRNA electroporation is generally considered the preferred approach. This avoids issues of low transfection efficiency seen in primary cells, and safety concerns regarding insertional mutagenesis and delivery of plasmid backbone sequences. RNP delivery has also been shown to be highly effective (Staahl *et al.*, 2017). Nanoparticles are another emerging delivery vehicle that is being developed for genome editing tools (Wang *et al.*, 2016). These typically suffer from very low targeting efficiencies and are limited to use in liver diseases; however, one benefit to note for direct or transient delivery methods is reduced exposure to nuclease activity and a reduced propensity for generating off-target hits (Komor *et al.*, 2017).

For *in vivo* applications, delivery is the most significant hurdle for genome editing therapeutics (Figure 23.2). For wide-ranging gene therapy applications, the preferred vehicle has been adeno-associated virus (AAV). AAV vectors are by far the most frequently and successfully used viral vectors in ocular gene therapy injection (Dalkara and Sahel, 2014; Day *et al.*, 2014; Dinculescu *et al.*, 2005; Willett and Bennett, 2013). Several features make AAV the most attractive choice: the virus is non-pathogenic, it infects both dividing and non-dividing cells, and AAV has a particularly noteworthy history of safety, efficacy and a general lack of toxicity in clinical trials (Samulski and Muzyczka, 2014). Additionally, the use of AAV serotypes is highly effective at discriminating different cell types within the retina (Watanabe *et al.*, 2013). For example, the AAV5 serotype has been shown to mediate highly specific rod photoreceptor transduction after subretinal injection in non-human primates, and is capable of mediating retinal therapy (Dyka *et al.*, 2014).

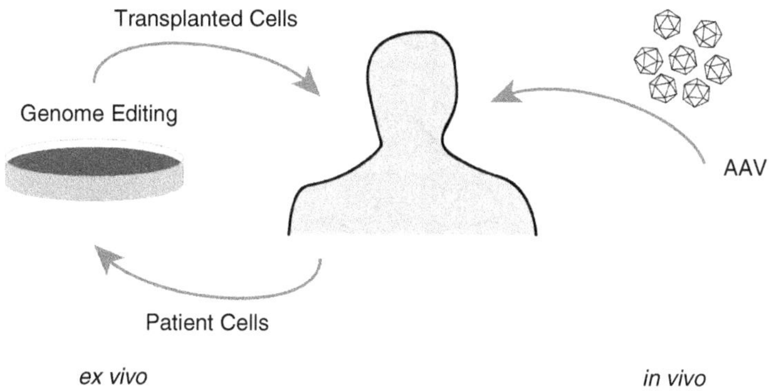

Figure 23.2 *Ex vivo* vs. *in vivo* therapeutic strategies. For *ex vivo* therapeutic strategies, stem or progenitor cells are extracted from the patient, and genetically modified in the laboratory. Gene editing components can be delivered by physical or chemical methods, or by viruses, and genetically modified cells are then reintroduced into the patient. For *in vivo* therapeutic strategies, cells are directly targeted in the patient. Delivery methods can be lipid nanoparticles, but are most commonly viruses like AAV.

While AAV vectors provide a safe means of delivery, there is one major technical obstacle that limits its utility in delivering CRISPR-based therapeutics: the AAV packaging capacity (Gaj *et al.*, 2016; Moser and Hirsch, 2016). Wild-type AAV genomes are ~4.7 kb in length, and recombinant viruses have been shown to package up to a maximum of ~5.2 kb (Dong *et al.*, 2010; Wu *et al.*, 2010; Figure 23.3). This packaging capacity defines the upper limit of DNA that can be used for a single viral vector, and includes the requisite AAV inverted terminal repeats (ITRs), which together are ~300 bp in length. The commonly used Cas9 from *S. pyogenes* is ~4.2 kb, meaning that it alone occupies over 85% of the available payload. Thus, considering the additional DNA sequences required to express SpCas9 and a gRNA by conventional methods, the 5.2 kb packaging capacity of AAV is exceeded: Pol II promoter (~0.5 kb), SpCas9 (~4.2 kb), Pol II terminator (~0.2 kb), Pol III promoter (~0.3 kb), gRNA (~0.1 kb) and AAV ITRs (~0.3 kb).

The size limitation of AAV is a well-known factor in gene therapy as diseases caused by mutations in large genes cannot be delivered; many recessive mutations causing retinal dystrophies, such as ABCA4, CEP290, MYO7A or USH2A, surpass the AAV capacity (Thompson *et al.*, 2015). Some alternative strategies that have been developed to overcome these limitations in gene therapy have been applied to genome editing purposes, as well. One approach to deliver CRISPR components is through a dual-vector delivery, where one AAV vector is used to deliver SpCas9, and another AAV vector for the gRNA (Swiech *et al.*, 2015). This approach relied on the use of the mouse Mecp2 promoter, a uniquely small pol II promoter, to express SpCas9. While Mecp2 expression has been found in many cell types, including some retinal cells, expression is not observed in rod photoreceptors (Jain *et al.*, 2010; Song *et al.*, 2014). This suggests that use of this promoter could fail to target the clear majority of retinal degeneration mutations *a priori*. Furthermore, reconstitution of the active CRISPR complex necessitates co-transduction, which is

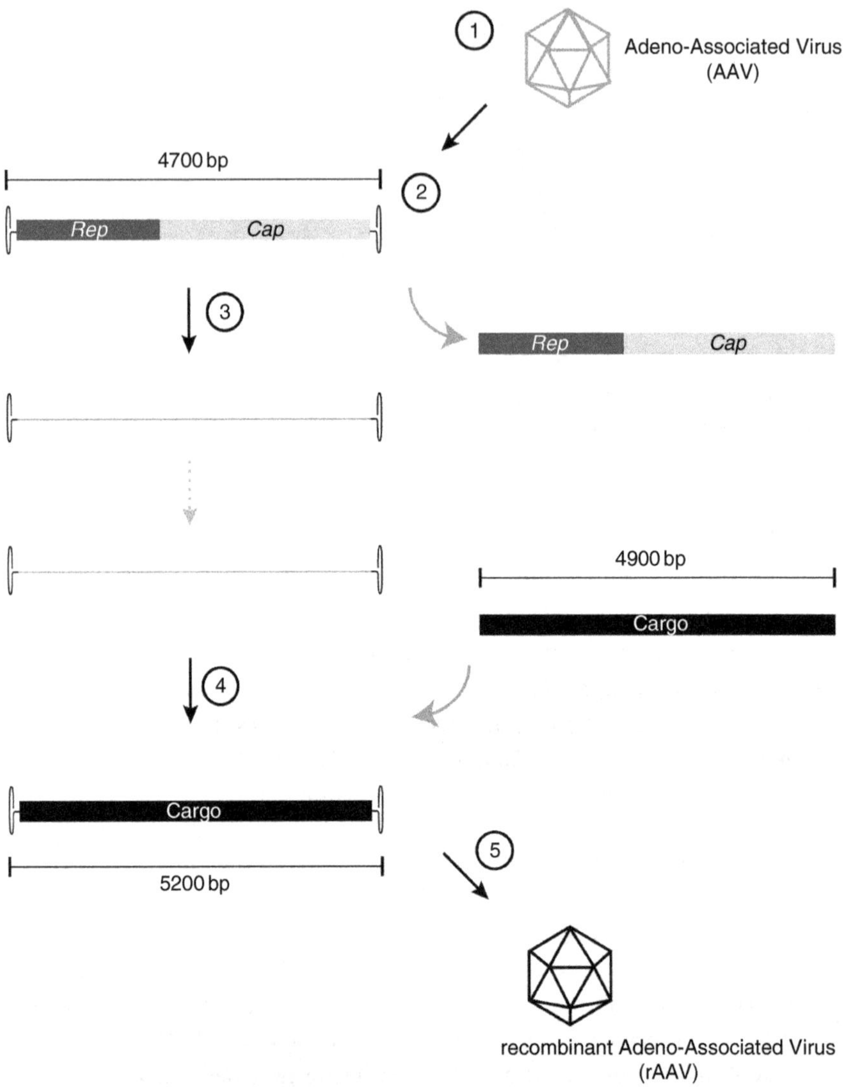

Figure 23.3 AAV is the gold standard delivery vehicle. 1. Adeno-associated virus (AAV) is the safest and most commonly used vector in gene therapy. The virus has one drawback: it is very small. 2. Naturally, it consists of a single-stranded DNA genome of 4700 nucleotides. The ends of the genome are inverted repeat sequences known as ITRs. 3. These sequences (~150 bases each) are the only required elements for packaging DNA into AAV, so the viral sequences (*Rep and Cap*) can be gutted. 4. Any cargo can be inserted in between the ITRs and the virus can be stuffed with up to 5200 nucleotides total, meaning that the virus can accommodate cargo up to 4900 nucleotides. 5. Recombinant adeno-associated virus (rAAV) is used as a high-efficiency vehicle for delivering cargo into a cell.

less efficient than a single transduction. This requires a larger viral dose, which could potentially induce a larger immune response and associated toxicity. It is likely that required delivery of a second viral vector in human trials could lead to additional challenges with safety and regulatory approval. For these reasons, this is unlikely to be pursued for many clinical applications. Other dual-vector approaches include use of trans-splicing and split inteins to generate functional genome editing complexes in the cell (Fine *et al.*, 2015; Truong *et al.*, 2015).

Another more promising approach that recently emerged is the use of smaller Cas9 proteins. Cas9 orthologs like *N. meningitidis* (NmCas9) or *S. aureus* (SaCas9), which are ~3.2 kb and significantly smaller than SpCas9 (Friedland *et al.*, 2015; Hou *et al.*, 2013; Ran *et al.*, 2015). The use of SaCas9 has been shown to be highly effective in several disease models, and is being pursued for the treatment of Leber's congenital amaurosis 10 (LCA10) (Maeder and Gersbach, 2016). One limitation in using the SaCas9 or NmCas9 is due to their PAM requirements (NNGRRT and NNNNGATT, respectively). The number of unique genomic targeting sites is significantly less with these orthologs due to the longer PAM sequence (~4-fold and ~13-fold, respectively; Jaskula-Ranga and Zack, 2016). An interesting observation to note regarding the use of smaller orthologs is that smaller Cas9 proteins tend to have longer PAM restrictions (Fonfara *et al.*, 2014). This suggests that Cas9 proteins that can be delivered by AAV would likely have fewer genomic targets, which in turn means a statistically lower likelihood of targeting any given mutation.

Aside from AAV, viral vectors for ocular delivery generally include adenovirus and lentivirus. Adenoviral vectors have a very large cargo capacity, and are maintained as episomes, but the vector has key disadvantages. Although the RPE can be targeted, photoreceptor cells, where the majority of inherited retinopathy genes are expressed, are not effectively transduced and attempts to use various serotypes and pseudotypes have not been effective (Puppo *et al.*, 2014). Compounded with cytotoxic effects and transient gene expression following adenoviral transduction, these vectors have not been used extensively for targeting retinal cells (Reichel *et al.*, 1998; Stilwell and Samulski, 2004). Lentiviruses are single-stranded retroviruses that are competent for transducing both mitotic and post-mitotic cells. While efficient RPE transduction has been reported, photoreceptors are generally recalcitrant to transduction (Bainbridge *et al.*, 2001; Lipinski *et al.*, 2014; Yanez-Munoz *et al.*, 2006). The most significant concern regarding lentiviral vectors is insertional mutagenesis, which can cause gene disruption, altered gene expression and potentially lead to oncogenesis (Rothe *et al.*, 2014).

While some of the delivery methods offer advantages, the vast majority of retinal gene therapy studies have used AAV, making it the gold standard delivery method. Furthermore, promising new advancements have been made with engineered AAV variants that allow for retinal transduction following intravitreal injection, making viral delivery even easier (Dalkara *et al.*, 2013; Petrs-Silva *et al.*, 2011). Several recent studies using structure-guided protein engineering have generated SpCas9 variants that exhibit low-to-undetectable off-target mutagenesis (Kleinstiver *et al.*, 2016; Slaymaker *et al.*, 2016). These variants represent key advances for the use of CRISPR in the clinic, as SpCas9 specificity has reached such a high threshold that off-target mutagenesis with these variants is undetectable by current sequencing technology. The development of effective methods to deliver SpCas9 and variants is key for widespread use of CRISPR-based therapeutics.

23.2 Application to Retinal Diseases

Retinal degenerations represent a leading cause of irreversible blindness. Inherited retinopathies represent a group of blinding diseases that affect approximately 1 in 2000 people worldwide. Most of these diseases are monogenic and are characterized by dysfunction of photoreceptor cells, or the RPE (Berger *et al.*, 2010). To date, over 250 genes causing retinal diseases have been identified, mostly in photoreceptor cells (RetNet: www.sph.uth.tmc.edu/RetNet/). The targeting of dominant disease mutations is the most simple and effective approach for genome editing technologies. Ideally, the mutation itself can be targeted to provide allele specificity, although linked SNPs can also be used to disrupt gene function as well. While there are many retinal diseases that are potential targets for genome editing therapeutics, the following focuses on some key areas that are likely to be at the forefront of emerging genome editing therapeutics.

23.3 Genome Editing for Retinitis Pigmentosa

Retinitis pigmentosa (RP) is an inherited retinal degenerative disease in which dysfunction and death of retinal photoreceptor cells (rods and cones) leads to vision loss or blindness. There are both autosomal recessive and autosomal dominant genetic forms of RP (arRP and adRP, respectively). For adRP only one copy is mutated, resulting in a toxic or harmful gain-of-function protein. Here, gene replacement strategies are generally ineffective, as introduction of a functional gene is not sufficient to overcome the dominant allele. Therefore, most gene therapy approaches use two steps to accomplish suppression and replacement therapy: (1) eliminate the function of both copies, typically through siRNA or knockdown approaches and (2) introduce a knockdown-resistant version of the wild-type gene (Gorbatyuk *et al.*, 2005; Kiang *et al.*, 2005; Millington-Ward *et al.*, 2011; O'Reilly *et al.*, 2007).

In contrast, the targeting of dominant disease mutations is the most simple and effective approach for genome editing technologies. Ideally, the mutation itself would be targeted to provide allele specificity, although linked SNPs can also be used to disrupt gene function. The autosomal dominant form of retinitis pigmentosa (adRP) constitutes approximately 30–40% of all cases of RP, and among adRP patients the most commonly mutated RP-associated gene is the one that encodes the rod visual pigment rhodopsin (Dryja *et al.*, 1990a; 1990b). While this approach may result in expression of only 50% of the wild-type level of rhodopsin, animal data suggest that this is sufficient to provide clinically useful rod function (Liang *et al.*, 2004). The use of non-cutting genome editing approaches should also be noted as a possible strategy, as they can be effective at reducing expression of the dominant allele through transcriptional downregulation (Botta *et al.*, 2016). Previous work using zinc-finger-based transcriptional repression of rhodopsin demonstrated effective treatment by selective downregulation of the dominant allele in a model of adRP (Mussolino *et al.*, 2011).

Some recent preclinical proof-of-concept studies have demonstrated the ability to rescue retinal structure and function in various animal models. One study demonstrated the ability to target outside of the mutation, achieving allele-specific targeting of rat rhodopsin by using a single base change that resulted in a novel PAM site linked in *cis* (Bakondi *et al.*, 2016). Another study demonstrated *in vivo* editing of the human rhodopsin gene in transgenic mice (Latella *et al.*, 2016). Both studies delivered plasmids by electroporation following subretinal injection, demonstrating the effectiveness of CRISPR to target adRP mutations, while simultaneously highlighting the need for clinically relevant delivery methods.

23.4 Genome Editing for LCA10

Leber's congenital amaurosis (LCA) is comprised of a group of early onset childhood retinal dystrophies that are characterized by severe retinal dysfunction and severe visual impairment, or blindness during the first months of life. LCA, an orphan disease, is the most common cause of inherited blindness, constituting as much as 5% of all known hereditary retinal degenerative diseases. The most common cause for LCA10 is a deep intronic mutation in the CEP290 gene (Figure 23.4; den Hollander *et al.*, 2006). The IVS26c.2991+1655 A>G mutation results in a *de novo* strong splice donor site and the subsequent inclusion of a cryptic exon (exon X) into the CEP290 mRNA. CEP290, centrosomal protein located in the connecting cilium of photoreceptors, plays a role in ciliogenesis and ciliary trafficking (Craige *et al.*, 2010; Drivas *et al.*, 2013; Tsang *et al.*, 2008). Inclusion of Exon X, which contains a premature stop codon, results in a truncated form of CEP290 and eventual photoreceptor degeneration. The CEP290 mutation is unattractive as a therapeutic target for CRISPR-Cas9 technology (Maeder and Gersbach, 2016). The gene is too large for AAV delivery, and therefore remains untreatable by gene therapy approaches. Moreover, photoreceptors are highly sensitive to protein levels (Olsson *et al.*, 1992; Seo *et al.*, 2013; Tan *et al.*, 2001), and transgenic experiments have demonstrated photoreceptor toxicity from CEP290 overexpression (Burnight *et al.*, 2014). Because this is an intronic mutation, current strategies are now looking to remove a portion of the sequence containing the cryptic exon.

23.5 Recessive Mutations and Enhancing Gene Therapy

Recessive mutations causing retinal degenerations are also potential targets for genome editing. The gene correction approach generally requires the delivery of an exogenous template adding another level of complexity that will likely push these therapeutics further into the future. However, building on the success of gene therapy, genome editing might have the potential to produce more effective cures. In arRP, mutations occur in both copies of the gene, and in such cases, gene therapy or gene supplementation has been the preferred strategy. Elegant examples of this approach are the ongoing human treatment studies for Leber

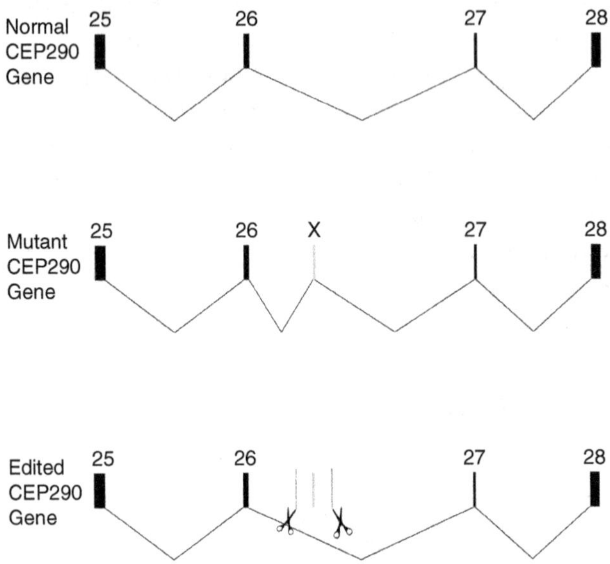

Figure 23.4 LCA10 is caused by a point mutation in CEP290. A mutation in CEP290, which results in the incorporation of a cryptic exon (exon X) between exons 26 and 27, results in LCA10. A therapeutic strategy for LCA10 involves using gene editing technology to remove the cryptic exon by two flanking cuts in the intronic region.

Congenital Amaurosis 2 (LCA2) in which gene therapy with an AAV is being used to replace the function of the defective RPE65 gene that causes the disease (Maguire *et al.*, 2008). Interestingly, although responses were noted, the effects were not durable potentially due to loss of transgene expression (Azvolinsky, 2015; Schimmer and Breazzano, 2015). In such cases, genome editing therapeutics may surpass gene therapy by providing permanent cures; correction of the endogenous locus would appear to provide the greatest chance of success.

23.6 Conclusion

For more than a century, the prevailing model of medical intervention has been by means of using drugs to alleviate symptoms. For some genetic disorders, medications can help manage symptoms, but none offer an actual cure, and for the roughly 7000 known genetic disorders, far too many diseases simply lack any effective treatment at all. The development of CRISPR technology has revolutionized the field of gene editing, unleashing new and exciting possibilities for disease treatment. Disease-associated mutations can be targeted with unprecedented precision, meaning that cures for a wide range of diseases is on the horizon. While challenges remain, the rapid developments in genome editing therapeutics is showing great promise and the eye is uniquely positioned to be at the forefront of this revolution.

REFERENCES

Azvolinsky A. 2015. Gene therapy "cure" for blindness wanes. *Nat Biotechnol* **33**: 678.

Bainbridge JW, Stephens C, Parsley K, *et al.* 2001. In vivo gene transfer to the mouse eye using an HIV-based lentiviral vector; efficient long-term transduction of corneal endothelium and retinal pigment epithelium. *Gene Ther* **8**: 1665–1668.

Bakondi B, Lv W, Lu B, *et al.* 2016. In vivo CRISPR/Cas9 gene editing corrects retinal dystrophy in the S334ter-3 rat model of autosomal dominant retinitis pigmentosa. *Mol Ther* **24**: 556–563.

Berger W, Kloeckener-Gruissem B, Neidhardt J. 2010. The molecular basis of human retinal and vitreoretinal diseases. *Prog Retin Eye Res* **29**: 335–375.

Botta S, Marrocco E, De Prisco N, *et al.* 2016. Rhodopsin targeted transcriptional silencing by DNA-binding. *Elife* **5**: e12242.

Burnight ER, Wiley LA, Drack AV, *et al.* 2014. CEP290 gene transfer rescues Leber congenital amaurosis cellular phenotype. *Gene Ther* **21**: 662–672.

Cajal SRY. 1892. *La Rétine des Vértebrés (La Cellule, English Trans.; S. Thorpe and M. Glickstein, trans., 1972).* Springfield, IL: Charles C. Thomas.

Craige B, Tsao CC, Diener DR, *et al.* 2010. CEP290 tethers flagellar transition zone microtubules to the membrane and regulates flagellar protein content. *J Cell Biol* **190**: 927–940.

Da Cruz L, Chen FK, Ahmado A, Greenwood J, Coffey P. 2007. RPE transplantation and its role in retinal disease. *Prog Retin Eye Res* **26**: 598–635.

Dalkara D, Byrne LC, Klimczak RR, *et al.* 2013. In vivo-directed evolution of a new adeno-associated virus for therapeutic outer retinal gene delivery from the vitreous. *Sci Transl Med* **5**: 189ra76.

Dalkara D, Goureau O, Marazova K, Sahel JA. 2016. Let there be light: gene and cell therapy for blindness. *Hum Gene Ther* **27**: 134–147.

Dalkara D, Sahel JA. 2014. Gene therapy for inherited retinal degenerations. *C R Biol* **337**: 185–192.

Day TP, Byrne LC, Schaffer DV, Flannery JG. 2014. Advances in AAV vector development for gene therapy in the retina. *Adv Exp Med Biol* **801**: 687–693.

Den Hollander AI, Koenekoop RK, Yzer S, *et al.* 2006. Mutations in the CEP290 (NPHP6) gene are a frequent cause of Leber congenital amaurosis. *Am J Hum Genet* **79**: 556–561.

Dinculescu A, Glushakova L, Min SH, Hauswirth WW. 2005. Adeno-associated virus-vectored gene therapy for retinal disease. *Hum Gene Ther* **16**: 649–663.

Dong B, Nakai H, Xiao W. 2010. Characterization of genome integrity for oversized recombinant AAV vector. *Mol Ther* **18**: 87–92.

Doudna JA, Charpentier E. 2014. Genome editing: the new frontier of genome engineering with CRISPR-Cas9. *Science* **346**: 1258096.

Drivas TG, Holzbaur EL, Bennett J. 2013. Disruption of CEP290 microtubule/membrane-binding domains causes retinal degeneration. *J Clin Invest* **123**: 4525–4539.

Dryja TP, McGee TL, Hahn LB, *et al.* 1990a. Mutations within the rhodopsin gene in patients with autosomal dominant retinitis pigmentosa. *N Engl J Med* **323**: 1302–1307.

Dryja TP, McGee TL, Reichel E, *et al.* 1990b. A point mutation of the rhodopsin gene in one form of retinitis pigmentosa. *Nature* **343**: 364–366.

Dyka FM, Boye SL, Chiodo VA, Hauswirth WW, Boye SE. 2014. Dual adeno-associated virus vectors result in efficient in vitro and in vivo expression of an oversized gene, MYO7A. *Hum Gene Ther Methods* **25**: 166–177.

Fine EJ, Appleton CM, White DE, *et al.* 2015. Trans-spliced Cas9 allows cleavage of HBB and CCR5 genes in human cells using compact expression cassettes. *Sci Rep* **5**: 10777.

Fonfara I, Le Rhun A, Chylinski K, *et al.* 2014. Phylogeny of Cas9 determines functional exchangeability of dual-RNA and Cas9 among orthologous type II CRISPR-Cas systems. *Nucleic Acids Res* **42**: 2577–2590.

Friedland AE, Baral R, Singhal P, *et al.* 2015. Characterization of *Staphylococcus aureus* Cas9: a smaller Cas9 for all-in-one adeno-associated virus delivery and paired nickase applications. *Genome Biol* **16**: 257.

Gaj T, Epstein BE, Schaffer DV. 2016. Genome engineering using adeno-associated virus: basic and clinical research applications. *Mol Ther* **24**: 458–464.

Gorbatyuk MS, Pang JJ, Thomas J Jr., Hauswirth WW, Lewin AS. 2005. Knockdown of wild-type mouse rhodopsin using an AAV vectored ribozyme as part of an RNA replacement approach. *Mol Vis* **11**: 648–656.

Gregory-Evans K, Bashar AM, Tan M. 2012. Ex vivo gene therapy and vision. *Curr Gene Ther* **12**: 103–115.

Hou Z, Zhang Y, Propson NE, *et al.* 2013. Efficient genome engineering in human pluripotent stem cells using Cas9 from *Neisseria meningitidis*. *Proc Natl Acad Sci USA* **110**(39): 15644–15649.

Hsu PD, Lander ES, Zhang F. 2014. Development and applications of CRISPR-Cas9 for genome engineering. *Cell* **157**: 1262–1278.

Jain D, Singh K, Chirumamilla S, *et al.* 2010. Ocular MECP2 protein expression in patients with and without Rett syndrome. *Pediatr Neurol* **43**: 35–40.

Jaskula-Ranga V, Zack DJ. 2016. grID: a CRISPR-Cas9 guide RNA database and resource for genome-editing. *bioRxiv*, 097352.

Kamao H, Mandai M, Okamoto S, *et al.* 2014. Characterization of human induced pluripotent stem cell-derived retinal pigment epithelium cell sheets aiming for clinical application. *Stem Cell Rep* **2**: 205–218.

Kiang AS, Palfi A, Ader M, *et al.* 2005. Toward a gene therapy for dominant disease: validation of an RNA interference-based mutation-independent approach. *Mol Ther* **12**: 555–561.

Kleinstiver BP, Pattanayak V, Prew MS, *et al.* 2016. High-fidelity CRISPR-Cas9 nucleases with no detectable genome-wide off-target effects. *Nature* **529**(7587): 490–495.

Komor AC, Badran AH, Liu DR. 2017. CRISPR-based technologies for the manipulation of eukaryotic genomes. *Cell* **168**: 20–36.

Latella MC, Di Salvo MT, Cocchiarella F, *et al.* 2016. In vivo editing of the human mutant rhodopsin gene by electroporation of plasmid-based CRISPR/Cas9 in the mouse retina. *Mol Ther Nucleic Acids* **5**: e389.

Liang Y, Fotiadis D, Maeda T, *et al.* 2004. Rhodopsin signaling and organization in heterozygote rhodopsin knockout mice. *J Biol Chem* **279**: 48189–48196.

Lipinski DM, Barnard AR, Charbel Issa P, *et al.* 2014. Vesicular stomatitis virus glycoprotein- and Venezuelan equine encephalitis virus-derived glycoprotein-pseudotyped lentivirus vectors differentially transduce corneal endothelium, trabecular meshwork, and human photoreceptors. *Hum Gene Ther* **25**: 50–62.

Maeder ML, Gersbach CA. 2016. Genome-editing technologies for gene and cell therapy. *Mol Ther* **24**: 430–446.

Maguire AM, Simonelli F, Pierce EA, *et al.* 2008. Safety and efficacy of gene transfer for Leber's congenital amaurosis. *N Engl J Med* **358**: 2240–2248.

Millington-Ward S, Chadderton N, O'Reilly M, *et al.* 2011. Suppression and replacement gene therapy for autosomal dominant disease in a murine model of dominant retinitis pigmentosa. *Mol Ther* **19**: 642–649.

Moser RJ, Hirsch ML. 2016. AAV vectorization of DSB-mediated gene editing technologies. *Curr Gene Ther* **16**: 207–219.

Mussolino C, Sanges D, Marrocco E, *et al.* 2011. Zinc-finger-based transcriptional repression of rhodopsin in a model of dominant retinitis pigmentosa. *EMBO Mol Med* **3**: 118–128.

O'Reilly M, Palfi A, Chadderton N, *et al.* 2007. RNA interference-mediated suppression and replacement of human rhodopsin in vivo. *Am J Hum Genet* **81**: 127–135.

Olsson JE, Gordon JW, Pawlyk BS, *et al.* 1992. Transgenic mice with a rhodopsin mutation (Pro23His): a mouse model of autosomal dominant retinitis pigmentosa. *Neuron* **9**: 815–830.

Perez VL, Saeed AM, Tan Y, Urbieta M, Cruz-Guilloty F. 2013. The eye: a window to the soul of the immune system. *J Autoimmun* **45**: 7–14.

Petrs-Silva H, Dinculescu A, Li Q, *et al.* 2011. Novel properties of tyrosine-mutant AAV2 vectors in the mouse retina. *Mol Ther* **19**: 293–301.

Puppo A, Cesi G, Marrocco E, *et al.* 2014. Retinal transduction profiles by high-capacity viral vectors. *Gene Ther* **21**: 855–865.

Ran FA, Cong L, Yan WX, *et al.* 2015. In vivo genome editing using *Staphylococcus aureus* Cas9. *Nature* **520**(7546): 186–191.

Reichel MB, Ali RR, Thrasher AJ, *et al.* 1998. Immune responses limit adenovirally mediated gene expression in the adult mouse eye. *Gene Ther* **5**: 1038–1046.

Rothe M, Schambach A, Biasco L. 2014. Safety of gene therapy: new insights to a puzzling case. *Curr Gene Ther* **14**: 429–436.

Samulski RJ, Muzyczka N. 2014. AAV-mediated gene therapy for research and therapeutic purposes. *Annu Rev Virol* **1**: 427–451.

Schimmer J, Breazzano S. 2015. Investor outlook: significance of the positive LCA2 gene therapy phase III results. *Hum Gene Ther Clin Dev* **26**: 208–210.

Scholl HP, Strauss RW, Singh MS, *et al.* 2016. Emerging therapies for inherited retinal degeneration. *Sci Transl Med* **8**: 368rv6.

Schwartz SD, Hubschman JP, Heilwell G, *et al.* 2012. Embryonic stem cell trials for macular degeneration: a preliminary report. *Lancet* **379**: 713–720.

Schwartz SD, Regillo CD, Lam BL, *et al.* 2015. Human embryonic stem cell-derived retinal pigment epithelium in patients with age-related macular degeneration and Stargardt's macular dystrophy: follow-up of two open-label phase 1/2 studies. *Lancet* **385**: 509–516.

Seo S, Mullins RF, Dumitrescu AV, *et al.* 2013. Subretinal gene therapy of mice with Bardet-Biedl syndrome type 1. *Invest Ophthalmol Vis Sci* **54**: 6118–6132.

Slaymaker IM, Gao L, Zetsche B, *et al.* 2016. Rationally engineered Cas9 nucleases with improved specificity. *Science* **351**: 84–88.

Song C, Feodorova Y, Guy J, *et al.* 2014. DNA methylation reader MECP2: cell type- and differentiation stage-specific protein distribution. *Epigenetics Chromatin* **7**: 17.

Staahl BT, Benekareddy M, Coulon-Bainier C., *et al.* 2017. Efficient genome editing in the mouse brain by local delivery of engineered Cas9 ribonucleoprotein complexes. *Nat Biotechnol* **35**(5): 431–434.

Stilwell JL, Samulski RJ. 2004. Role of viral vectors and virion shells in cellular gene expression. *Mol Ther* **9**: 337–346.

Streilein JW. 2003. Ocular immune privilege: therapeutic opportunities from an experiment of nature. *Nat Rev Immunol* **3**: 879–889.

Swiech L, Heidenreich M, Banerjee A, *et al.* 2015. In vivo interrogation of gene function in the mammalian brain using CRISPR-Cas9. *Nat Biotechnol* **33**: 102–106.

Tan E, Wang Q, Quiambao AB, *et al.* 2001. The relationship between opsin overexpression and photoreceptor degeneration. *Invest Ophthalmol Vis Sci* **42**: 589–600.

Thompson DA, Ali RR, Banin E, *et al.* 2015. Advancing therapeutic strategies for inherited retinal degeneration: recommendations from the Monaciano Symposium. *Invest Ophthalmol Vis Sci* **56**: 918–931.

Truong DJ, Kuhner K, Kuhn R, *et al.* 2015. Development of an intein-mediated split-Cas9 system for gene therapy. *Nucleic Acids Res* **43**: 6450–6458.

Tsang WY, Bossard C, Khanna H, *et al.* 2008. CP110 suppresses primary cilia formation through its interaction with CEP290, a protein deficient in human ciliary disease. *Dev Cell* **15**: 187–197.

Wang L, Li F, Dang L, *et al.* 2016. In vivo delivery systems for therapeutic genome editing. *Int J Mol Sci* **17**(5): E626.

Watanabe S, Sanuki R, Ueno S, *et al.* 2013. Tropisms of AAV for subretinal delivery to the neonatal mouse retina and its application for in vivo rescue of developmental photoreceptor disorders. *PLoS One* **8**: e54146.

Willett K, Bennett J. 2013. Immunology of AAV-mediated gene transfer in the eye. *Front Immunol* **4**: 261.

Williams DR. 2011. Imaging single cells in the living retina. *Vision Res* **51**: 1379–1396.

Wu Z, Yang H, Colosi P. 2010. Effect of genome size on AAV vector packaging. *Mol Ther* **18**: 80–86.

Yanez-Munoz RJ, Balaggan KS, MacNeil A, *et al.* 2006. Effective gene therapy with nonintegrating lentiviral vectors. *Nat Med* **12**: 348–353.

24 Manipulation of Long Non-coding RNAs in Cardiovascular Disease Using Genome Editing Technology

Stefan Haemmig and Mark W. Feinberg

24.1 Introduction

The traditional view on non-coding RNA (ncRNA) as "junk of the genome" has been revolutionized to how we view eukaryotic genome complexity. Transcriptomic analysis revealed that around three-quarters of the human genome is pervasively transcribed, of which protein-coding genes represent only 1.5% (Djebali *et al.*, 2012; Kapranov *et al.*, 2007; Okazaki *et al.*, 2002). So what about the remaining 98.5% of transcribed sequences? The debate regarding futility vs. function of this pervasive transcription has clearly shifted towards the recognition that ncRNAs exert critical functions and do not simply act as templates for protein translation. Three types of regulatory ncRNAs exist, classified by their size: (1) short ncRNA with transcripts smaller than 30 nucleotides (nt); (2) medium-sized ncRNAs and (3) long ncRNAs (lncRNAs) with transcripts usually longer than 200 nt (Farazi *et al.*, 2013). The most widely studied class of ncRNAs are microRNAs (miRNAs), representing part of the smallest class of ncRNAs with 19–25 nt short, functionally active mature transcripts (Bartel, 2004). Within the last two decades, miRNAs have clearly proven their regulatory role on gene regulation in nearly all biological processes and in an abundant number of diseases. Emerging studies demonstrate that lncRNAs regulate protein-coding genes in a tight spatial and temporal manner. While the number of miRNAs for most species has plateaued, lncRNAs are being identified in growing numbers among most species (Hwang and Mendell, 2006). The NONCODE consortium is an interactive database aiming to represent a complete collection of annotated ncRNAs with a special focus on lncRNAs (http://noncode.org). From their first release in 2004 encompassing 1557 lncRNAs out of 5339 ncRNAs, the number of lncRNAs has skyrocketed within the last decade to 527 336 lncRNA genes across 16 species out of 723 849 ncRNAs. While the number of ~25 000 human protein-coding genes has plateaued, the number of human lncRNAs has grown from 37 049 in 2012 to more than 101 700 (Figure 24.1; Bu *et al.*, 2012; He *et al.*, 2008; Liu *et al.*, 2005; Xie *et al.*, 2014; Zhao *et al.*, 2016). Most lncRNAs are predicted to lack protein-coding potential; however, they may encode for

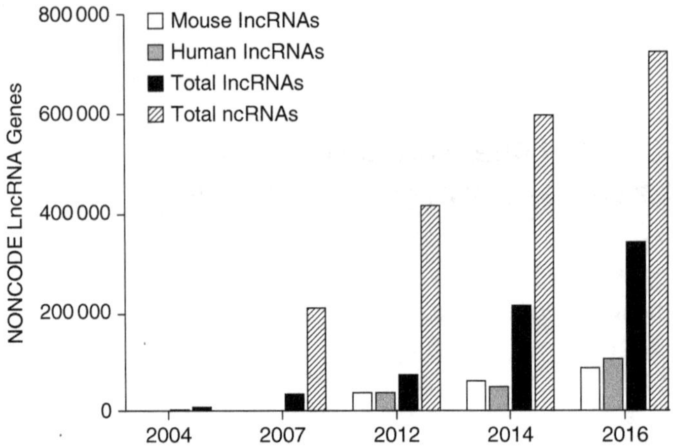

Figure 24.1 Discovery of lncRNAs over the last decade. Representation of human and mouse lncRNAs of the NONCODE data collection. Total counts across 16 species.

micropeptides or act as precursors for miRNAs (Anderson *et al.*, 2015; Andrews and Rothnagel, 2014). lncRNAs are defined as mRNA-like transcripts that are longer than 200 nt and often display features such as splicing and polyadenylation; their genomic loci may exist in close proximity to protein-coding genes. Although emerging studies have begun to elucidate our understanding of their function, expression and link to diseases, the precise function of most lncRNAs remains poorly understood.

This chapter provides an overview on lncRNAs in cardiovascular diseases (CVDs) and specifically highlights tools used for loss-of-function studies, with a special interest in genome editing technologies such as CRISPR/Cas9.

24.2 lncRNAs in Cardiovascular Diseases

24.2.1 lncRNAs Regulating Cardiac Hypertrophy and Heart Failure

Pathological cardiac hypertrophy is characterized by thickening of the ventricle wall, subsequently reducing the contractile capacity of the heart to pump blood. Prolonged hypertrophy of cardiomyocytes leads to heart failure (Hill and Olson, 2008). A cluster of antisense non-coding RNAs to the *Myh7* gene (i.e. *Mhrt*) was recently discovered to interfere with cardiac hypertrophy. As a consequence of pressure-induced myocardial injury, *Mhrt* expression was reduced. Gain-of-function experiments demonstrated a protective role of *Mhrt* to pathophysiological heart failure in mice. Although *Mhrt* loci lie antisense to *Myh7*, no direct interaction of *Mhrt* and *Myh7* could be observed. Interestingly, *Mhrt* binds to Brg1, a important member of the chromatin repressor complex and hinders Brg1 from chromatin binding, thereby inhibiting chromatin targeting and gene regulation by Brg1 (Han *et al.*, 2014b). More recently, a whole-genome lncRNA profile identified the lncRNA named cardiac hypertrophy-associated transcript (*Chast*) as being highly expressed in hypertrophic heart tissue from mice. Expression of

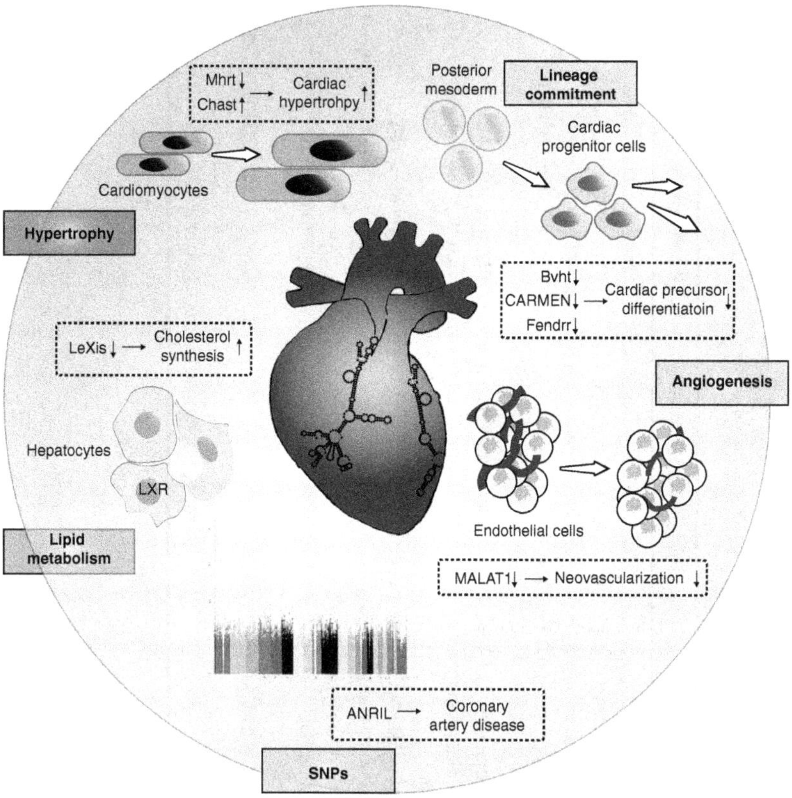

Figure 24.2 LncRNAs in cardiovascular disease. Selection of lncRNAs involved in hypertrophy, lipid metabolism, cardiac lineage development, vascular endothelium and SNPs associated with coronary artery disease. (A black-and-white version of this figure will appear in some formats. For the color version, please refer to the plate section.)

the human *Chast* homolog was induced in hypertrophic heart tissue from aortic stenosis patients. Silencing of *Chast* by GapmeRs prevented pathological cardiac remodeling in mice and thus could prevent potential heart failure by cardiac hypertrophy (Figure 24.2; Viereck *et al.*, 2016). *Chaer* and *Chrf*, two additional lncRNAs described in cardiac hypertrophy, are not discussed further (Wang *et al.*, 2014; 2016; Table 24.1). While therapeutic manipulation of *Mhrt* and *Chast* in species beyond mice is yet to be shown, a range of RNA-based chemically modified antisense oligonucleotide (ASO) molecules may provide new therapeutic options (see section on transient techniques for lncRNA silencing).

24.2.2 LncRNAs Regulating Lipid Metabolism

The formation of the lipid core by foam cells represents a hallmark during the progression of atherosclerosis (Libby *et al.*, 2011). The liver X receptors (LXRs) on macrophages function as nuclear cholesterol sensors, activated in response to elevated intracellular cholesterol. Active LXRs induce a variety of genes involved in cholesterol absorption, efflux and transport. LXR agonists are effective for treatment of murine atherosclerosis, diabetes and Alzheimer's disease, since they lower the

Table 24.1 LncRNAs in the cardiovascular system and used techniques for loss- and gain-of-function studies

LincRNA	Species	CVDs	Targets	Loss of function	Gain of function	Reference
ANRIL	*Mus musculus*	Smooth muscle cell function	CDKN2A/B	siRNA	–	Congrains *et al.*, 2012
Bvht	*Mus musculus*	Cardiovascular lineage development	SUZ12/CNBP	Lentiviral shANAs/ CAISPR for functional unit	Aosa26 locus *Bhvt*	Klattenhoff *et al.*, 2013; Xue *et al.*, 2016
CARMEN	*Mus musculus*	Cardiovascular lineage development	SUZ12/EZH2	Murine GapmeR/ pKL0.1 shRNAi	–	Ounzain *et al.*, 2015
Chaer	*Mus musculus*	Cardiac hypertrophy	PRC2	siRNA knockout CR1SPR/Cas9	Adenoviral *Chaer in vivo*	Wang *et al.*, 2016
CHAST/ Chast	*Homo sapiens Mus musculus*	Cardiac hypertrophy	Plekm1	Murine GapmeR intraperitoneal/ *in vitro*	Adenoviral *Chast in viva/* Lentiviral *CHAST in vitro*	Viereck *et al.*, 2016
Chrf	*Mus musculus*	Cardiac hypertrophy	miR-489	Adenoviral shRNA to jugular vein	–	Wang *et al.*, 2014
Fendrr	*Mus musculus*	Cardiovascular lineage development	PRC2/TrxG/MLL	Knockout premature poly(A) knockout Lacz replacement	–	Grote *et al.*, 2013; Sauvageau *et al.*, 2013
LeXis	*Mus musculus*	Lipid metabolism	RALY	Knockout, ASO	Adenoviral tail vein-injection	Sallam *et al.*, 2016
MALAT1	*Homo sapiens Mus musculus*	Endotheilal cell function	Cdk1, CCNA2/ p21	GapmeR, siRNA/ knockout, ASO	–	Michalik *et al.*, 2014; Gutschner *et al.*, 2013
Mhrt	*Mus musculus*	Cardiac hypertrophy	Brg1	–	Transgenic mouse model	Han *et al.*, 2014b

cholesterol level in serum and liver and hence inhibit the progression of those diseases (Zelcer and Tontonoz, 2006). LXR agonist GW3965 promotes not only cholesterol efflux, but also inhibits cholesterol biosynthesis. In order to elucidate this mechanism, Sallam *et al.* (2016) performed RNA-Seq in primary mouse hepatocytes treated with GW3965 or control. The most induced gene was surprisingly a lncRNA named liver-expressed LXR-induced sequence (*LeXis*). Gain- and loss-of-function studies were able to induce or reduce cholesterol, respectively. Mechanistically, they identified the ribonucleoprotein termed RALY as a mediator of *LeXis*. The binding of *LeXis* to RALY prevented its cholesterol synthesis-promoting activity via transcriptional control of a subset of genes. Total serum cholesterol was reduced in mice upon adenoviral administration of *LeXis*. It remains unknown if a human homolog exists and has a similar prominent role in lipid metabolism (Figure 24.2; Table 24.1).

24.2.3 lncRNAs Involved in Cardiac Lineage Commitment

The role of lncRNAs in CVD has just begun to be uncovered in recent years. Since lncRNAs are differentially expressed among different cell types, three lncRNAs named Braveheart (*Bvht*), fetal-lethal non-coding development regulatory RNA (*Fendrr*) and cardiac mesoderm enhancer-associated lncRNA (*CARMEN*) have been shown to regulate cardiac-specific lineage commitment (Grote *et al.*, 2013; Klattenhoff *et al.*, 2013; Ounzain *et al.*, 2015; Sauvageau *et al.*, 2013). Because the heart arises from the posterior mesoderm, *Bvht* and *Fendrr* regulate core cardiovascular networks upstream of mesoderm posterior 1 (MesP1), which is a key player for cardiovascular lineage development (Bondue *et al.*, 2008). *Bvht* binds to SUZ12, a component of the polycomb repressive complex 2 (PRC2), suggesting that *Bvht* acts via epigenetic regulation (Klattenhoff *et al.*, 2013). Interestingly, the locus for *CARMEN* neighbors *Bvht* in the mouse genome. Loss-of-function experiments of *CARMEN* inhibited differentiation into cardiac precursor cells by regulation of PRC2 (Figure 24.2; Table 24.1; Ounzain *et al.*, 2015). These findings regarding *Bvht*, *Fendrr* and *CARMEN* provide new avenues for future myocardial regenerative and cell-based therapies.

24.2.4 LncRNAs Associated with a Locus Implicated in Cardiovascular Disease

In order to stratify personalized treatment options, genetic markers have gained tremendous interest in the last decade. Genome-wide association studies (GWAS) are being performed, in which large cohorts of unrelated individuals are genotyped for millions of small nucleotide polymorphisms (SNPs) and compared to phenotypic differences (Ward and Kellis, 2012). For example, one of the first susceptibility loci associated with coronary artery disease on chromosome 9p21 was identified in the last exon of the lncRNA named antisense non-coding RNA INK4 (*ANRIL*) (Figure 24.2; Samani *et al.*, 2007). Loss of function of *ANRIL* resulted in reduced cell viability of SMCs (Congrains *et al.*, 2012). Despite its expression being associated with risk for coronary and carotid atherosclerosis, its molecular mechanism remains poorly understood (Holdt *et al.*, 2010). In this manner, lncRNAs provide a new powerful class of regulatory factors that may uncover how SNPs in the non-coding genome may influence CVD susceptibility.

24.2.5 LncRNAs in the Vascular Endothelium

RNA-Seq has identified a number of hypoxia-induced, conserved lncRNAs, such as *MALAT1, TUG1, MEG3, linc00493* and *linc000657* (known as *NORAD*; Lee *et al.*, 2016) in HUVECs. Michalik *et al.* (2014) focused on *MALAT1*, demonstrating that genetic ablation or GapmeR-mediated knockdown of *MALAT1* in mice led to reduced vessel density of neonatal retina and postnatal neovascularization in response to hindlimb ischemia. *In vitro* analysis verified these findings using siRNA and GapmeRs against *MALAT1*, resulting in an antiproliferative effect *in vitro*, as well as enhanced sprouting (Figure 24.2; Table 24.1). The latter findings were unexpected, since previous studies of *MALAT1* knockout showed the opposite results in lung cancer cells (Gutschner *et al.*, 2013). This discrepancy may either be explained by different cell-specific effects or may be the result of using different approaches to lncRNA silencing.

24.3 Dissecting the Role of lncRNAs Using Genome Editing Tools

Targeted genome editing using engineered nucleases has revolutionized gene editing because of its simplicity and high efficiency. Conventional techniques using zinc-finger nucleases (ZFNs) and transcription activator-like effector nucleases (TALENs) rely on protein–DNA-based interactions, which is in contrast to clustered regularly interspaced short palindromic repeat (CRISPR) technology, which uses RNA-guided nucleases. CRISPR provides a precise applicable tool based on simple base pairing for the study of small indels, transient gene silencing, gene activation, creation of transgenic organisms, high-throughput screening, chromosomal rearrangement, gene tagging, monitoring DNA and mutation analysis, among many others (Sander and Joung, 2014). A loss-of-function study for protein-coding genes may be accomplished by a single frameshift, which is sufficient to abrogate its translation into protein. However, mutation studies are somewhat more complex for non-coding genes since they are missing an open reading frame (ORF) (Gutschner *et al.*, 2011). Apart from inherent technical considerations using gene editing techniques, we will highlight recent methods used in the field of lncRNAs and discuss methodological advantages and challenges. In particular, lncRNAs are not as well conserved across species, compared to miRNAs, which complicates the extrapolation of findings from animals to humans, and vice versa.

24.3.1 Genome Editing Approaches

24.3.1.1 Entire Loci Deletion

While excision of the entire genomic lncRNA loci becomes an attractive choice in mice, identification of whether the phenotype or functional role is ascribed to the lncRNA itself or its regulatory elements mediate interactions with DNA, RNA or protein molecules remains a challenge. Successful examples of whole-genomic loci deletion are *UCA1, lncRNA-21A* (Ho *et al.*, 2015), *NORAD* (Lee *et al.*, 2016) and *Morrbid* (Kotzin *et al.*, 2016), using two short guide RNAs (sgRNAs) flanking the first and last exon in combination with paired Cas9, which induces each a double-strand break followed by either non-homologous end joining (NHEJ) or homologous recombination (HR; Figure 24.3A; Ran *et al.*, 2013). One advantage of the dual gRNA approach is its scalability, as demonstrated for a short LncRNA of 475 bp (LncRNA-21A) or a relatively long fragment (> 5.6 Kb for UCA1) (Ho *et al.*, 2015). Since many lncRNAs are known host genes for miRNAs, snRNAs and snoRNAs, and lncRNAs may lie in an antisense orientation to regulatory DNA elements, deletion of the entire genomic loci harbors potential risks of misaligning a phenotype to its driver gene (Bassett *et al.*, 2014; da Rocha *et al.*, 2008; Katayama *et al.*, 2005; Royo and Cavaillé, 2008). In order to avoid excision of cryptically encoded miRNA genes, Han and colleagues successfully deleted up to 23 kb of the 57.8 kb long *Rian* lncRNA using two distinct sgRNAs flanking the target region. Validation of top predicted off-targets (http:/crispr .mit.edu/) did not reveal any expression changes. However, upstream transcriptional activity of six neighboring genes was increased in a tissue-specific manner (Han *et al.*, 2014a). The observed effect of *Rian* on these neighboring genes was not further investigated in this study. The fact that lncRNAs may regulate gene expression in *trans* and in this case more importantly also in *cis*, suggests that too large deletions may be critical for disrupting their expression as well (Bassett *et al.*, 2014). Moreover, transcription of the lncRNA *Airn* has been described rather than the RNA product itself, which represses its antisense counterpart Igfr2 (Latos *et al.*, 2012). This model of transcriptional interference by competing for RNA Pol II (Palmer *et al.*, 2011) highlights the importance of understanding the function of a candidate lncRNA and its silencing mechanism(s).

24.3.1.2 Deletion of a Functional Unit

In an elegant study, Xue and colleagues revealed critical elements of the secondary structure of *Bvht*, which controls cardiac lineage commitment (Klattenhoff *et al.*, 2013; Xue *et al.*, 2016). Notably, they found that a small deletion of 11 nt in the 5′ asymmetric G-rich internal loop (AGIL) motif in the modular secondary structure of *Bvht* using CRISPR/Cas9-mediated HR in mouse embryonic stem cells (ESCs) prevented its interaction with its binding partner CNBP. In order to increase efficiency of the mutation at designated loci at both alleles, two templates, including different selection markers (i.e. puromycin and hygromycin) were used for subsequent dual selection of positive clones following HR. Selection cassettes were removed later by Cre-recombinase of LoxP sites (Figure 24.3B; Xue *et al.*, 2016). This method enables the precise excision of a small domain, which is essential for mediating its function. However, prediction tools for identifying the functional important elements are not available and complementary mutational approaches are still required.

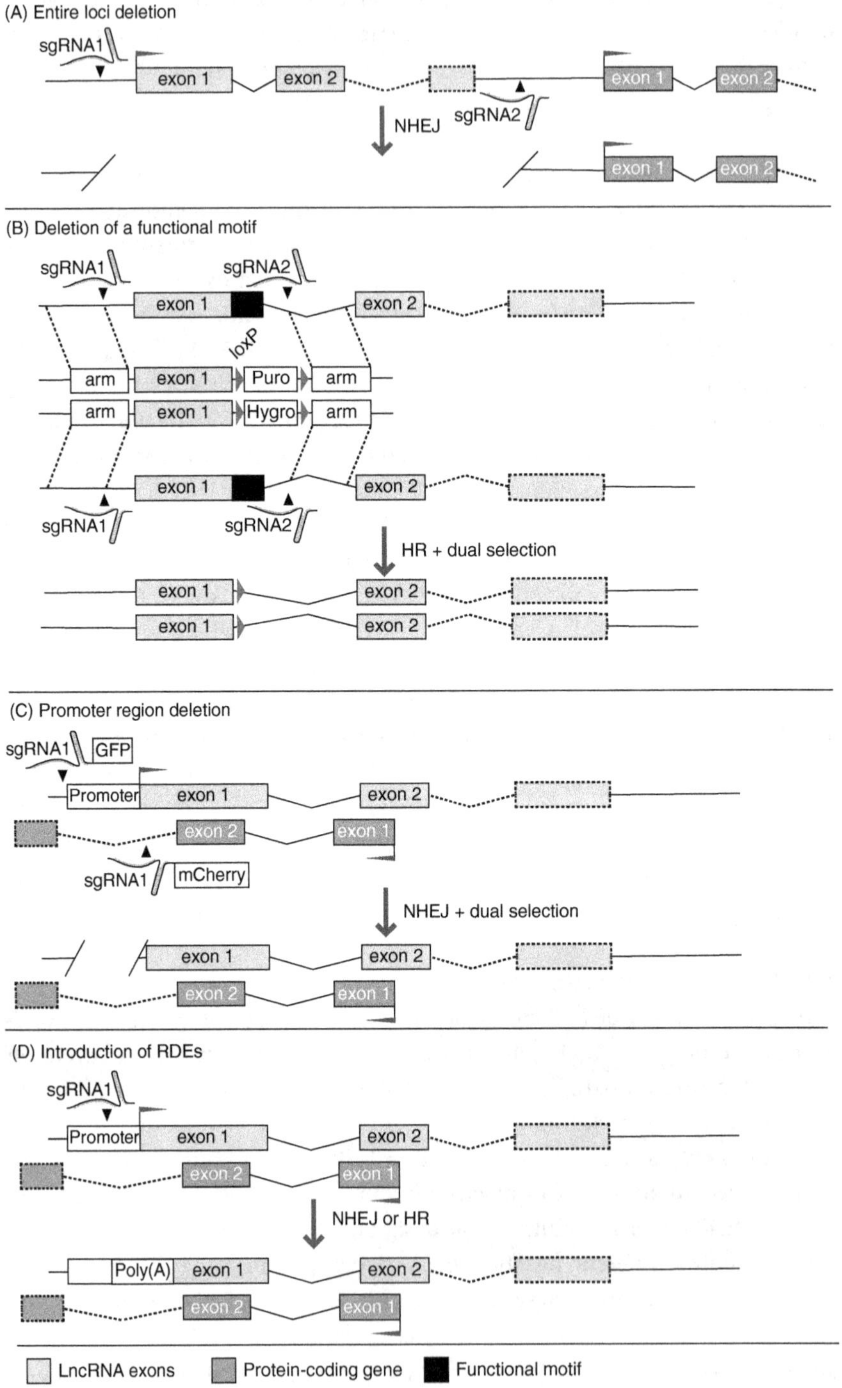

Figure 24.3 Gene editing approaches using CRISPR/Cas9. Schematic display for lncRNA silencing by (A) entire loci deletion, (B) deletion of a functional motif, (C) promoter region deletion and (D) introduction of RNA-destabilizing elements. (A black-and-white version of this figure will appear in some formats. For the color version, please refer to the plate section.)

24.3.1.3 Promoter Region Deletion

Alternatively, only targeting the promoter region provides another method for targeting lncRNA. Identification of promoter regions is facilitated by RNA-Seq data on ENCODE (Djebali *et al.*, 2012); however, lncRNAs are known for their multiple isoforms often derived from several promoters, which complicates silencing of a single lncRNA gene. *Morrbid* promoter deletion was achieved using two sgRNAs targeting the 5′ and 3′ flanking region of its promoter. sgRNAs were cloned into a Cas9 vector containing either a GFP or an mCherry selection marker, which were subsequently used for sorting of double-positive cells after transient transfection (Figure 24.3C; Kotzin *et al.*, 2016). In an attempt to avoid two Cas9 vectors, Aparicio and colleagues successfully cloned two gRNAs in tandem into a lentiviral Cas9 backbone using a single 165 bp long oligodimer for cloning. Using this technique, they excised the promoter of *MALAT1* and *UCA1* leading to a reduction of ~60% of the steady-state RNA level in the majority of cells. Single-cell sorting for homozygous deletion increased knockdown efficiency up to 95% (Aparicio-Prat *et al.*, 2015). The primary challenge that remains is the rather low efficiency of homozygote knockout clones, although deletion frequency is inversely related to deletion size (Canver *et al.*, 2014).

24.3.1.4 Introduction of RNA-destabilizing Elements (RDE)

An attractive avenue for silencing lncRNA expression is by introducing destabilizing elements such as premature stop codons (i.e. poly(A) signals). To this end, multiple polyadenlyation signals may be introduced towards the 5′ end of an ncRNA gene, leading to the stabilization of upstream transcripts by the addition of a poly(A) tail, as well as destabilizing the downstream transcript due to absent 5′-cap structure (Bond *et al.*, 2009; Sleutels *et al.*, 2002). For example, synthetic ZFNs were used in the biallelic integration of poly(A) signals into the locus of *MALAT1*. Similar to CRISPR/Cas9, ZFNs can be used to generate NHEJ or HR to knock out or integrate exogenous elements of interest (Figure 24.3D). The ZFNs integrate RDEs between the TATA box and the transcriptional start. For example, introduction of two different poly(A) sequences, i.e. (1) bovine growth hormone (bGH) and/or (2) simian virus 40 (SV40), were either in tandem or single compared to each other in the *MALAT1* locus in A459 cells. A dramatic reduction of *MALAT1* RNA by nearly 1000-fold was detected with only minor potency differences of bGH to SV40 observed (Gutschner *et al.*, 2011; 2015). The integration efficiency of homozygous clones was 15.4%, which is comparable to earlier-described whole-genomic loci deletion of *Rian* (Gutschner *et al.*, 2011; Han *et al.*, 2014b). Dedicated experiments for off-target effects as well as specificity by ZFN-mediated integration did not raise any concerns (Gutschner *et al.*, 2011). Complementary experiments are necessary to verify whether the RNA product itself mediates the functional effect or whether transcriptional interference is the dominant mechanism of action (Goff and Rinn, 2015; Palmer *et al.*, 2011). To address this question, the transgene for *Fendrr* rescued *Fendrr* loss-of-function studies by insertion of premature poly(A), suggesting that the RNA product itself is indispensable (Grote *et al.*, 2013). Other than ZFNs, Lee *et al.* (2016) used

TALENs to facilitate an HR-mediated recombination of a stop element with a puromycin-resistance cassette in HCT116 cells. Following puromycin selection, ~10% of clones were homozygous, reducing RNA levels by nearly ~98%, leading to significantly elevated levels of aneuploidy. Using Cre recombinase, the introduced cassette, initially flanked by LoxP sites, not only restored the RNA expression, but also significantly lowered levels of aneuploidy.

In summary, RDEs such as premature stop elements, miRNA binding sites, RNase P recognition sites, self-cleaving ribozymes or AU-rich elements provide an excellent repertoire of tools for manipulating lncRNA gene expression in an effective manner; however, considerations about transcriptional interference represent a caveat (Goff and Rinn, 2015; Gutschner *et al.*, 2011).

24.3.1.5 Homologous Recombination for Knockout/transgene Mice

MALAT1 (Zhang *et al.*, 2012), *LeXis* (Sallam *et al.*, 2016) and *Mhrt* (Han *et al.*, 2014b) are three examples for conventional generation of HR without any nucleases. Han and colleagues (2014b) developed a transgenic model for tetracycline-dependent restoration of *Mhrt* expression driven by a cardiac-specific promoter (i.e. Tnnt2/Tre). Transgenic Mhrt expression was not subject to Brg1 repression of the endogenous *Mhrt* locus and was thus able to reduce hypertrophy by keeping its prestress expression levels without being repressed. The entire loci of *Lexis* was removed to create a *Lexis –/–* strain with a classical vector containing FRT, PKG Neo and bGH elements (Sallam *et al.*, 2016). On the other hand, a ~3 kb region containing the *MALAT1* promoter and its 5′ end of the gene were removed using HR similar to that used for *Mhrt*. In most tissues *MALAT1* transcripts were completely deleted, except a minimal amount of *MALAT1*-associated small cytoplasmic RNA in brain. Depletion of *MALAT1* gene led to a ~twofold induction of proximal genes close to the *MALAT1* locus. This finding suggest that *MALAT1* is an example whereby its transcription per se may be driving the regulation for flanking genes in *cis* (Zhang *et al.*, 2012).

24.3.1.6 Transient Techniques for lncRNA Silencing

Transient methods have emerged as powerful scalable tools to identify potentially functional lncRNAs. RNAi-based techniques such as small interfering RNA oligos (siRNA) or vector-associated short hairpin RNA (shRNA) both utilize the RNA-induced silencing complex (RISC), whereas shRNAs are often integrated into viral backbones for stable knockdown. Depending on the lncRNA, RNAi may not be the method of choice, because the RISC is localized in the cytosol and hence nuclear lncRNA or chromatin-associated lncRNAs may not be subject to degradation (Moore *et al.*, 2010). Recently, nuclear active RNAi factors have been characterized in a subset of human cell lines, which would clearly provide new possibilities for nuclear-associated lncRNAs (Gagnon *et al.*, 2014). In order to target nuclear or chromatin-associated lncRNAs, GapmeRs might be a powerful tool leading to RNAse-H-dependent degradation, which is an endogenous enzyme. If localization of the lncRNA is unknown, GapmeRs may still present a reasonable method of choice, since the site of transcription is always in the nucleus. GapmeR are ~20 nt long LNA-modified antisense oligonucleotide sequences (ASOs) that bind by simple Watson–Crick base pairing to target RNA transcript. To assess the degree of knockdown, exon-spanning RT-qPCR primers should be designed that do not overlap with the targeted sequence of the GapmeR. Moreover, multiple RT-qPCR primers are recommended in order to verify whole-transcript degradation. Furthermore, GapmeRs are applicable for *in vivo* studies. For example, administration of GapmeRs against *MALAT1* leads to inhibition of ischemia-induced angiogenesis in mice and GapmeRs targeting *Chast* prevented pathological cardiac remodeling (Michalik *et al.*, 2014; Viereck *et al.*, 2016). In cancer research, intratumor injection of GapmeRs against *SAMMSON* drastically decreased viability of melanoma cells *in vitro* and in patient-derived xenograft models (Leucci *et al.*, 2016). Alternative methods of targeting include transcriptional inhibition or activation using CRISPR technology (i.e. CRISPRi or CRISPRa, respectively) using a catalytically inactive Cas9 (i.e. dCas9), which can be targeted to any site of interest. Known transcriptional activators such as VP64 or repressors such as KRAB have been previously described that could be introduced into target sites (Gilbert *et al.*, 2013; Maeder *et al.*, 2013; Perez-Pinera *et al.*, 2013). To date, this approach has not been applied to lncRNA-based targets, but holds strong promise.

24.3.2 The Function and Genomic Loci of a lncRNA May Define the Targeting Strategy

lncRNAs may be defined based on their genomic location to nearby protein-coding genes. Taking this into account is important for considering whether large genomic deletions might affect cryptically encoded small RNA classes and/or antisense regulatory elements of protein-coding genes. If this is the case, then small genetic modifications, for example by introduction of premature stop elements or other RDEs, towards the 5′ end of the transcription start site or promoter excision are favorable. On the other hand, if no protein-coding genes are overlapping or in close proximity to the lncRNA of choice, whole-genomic deletion bears the advantage of removing all possible regulatory elements and is scalable up to 10^6 bp (Han *et al.*, 2014b; Ho *et al.*, 2015). Another important consideration is whether the lncRNA acts in *cis* or *trans* and whether the RNA product itself is essential for fulfilling its function or if its transcription per se that underlies its function. In the case of *cis* function, targeting may be best using small genetic editing approaches similar to lncRNA that overlap with protein-coding genes. In contrast, *trans*-functioning lncRNAs are less sensitive to larger genetic deletions. However, efficiency using CRISPR/Cas9 tends to be inversely correlated with the fragment size of deletions. Genetic rescue experiments by either removing the RDE, as described for *NORAD*, or by ectopic expression using a viral system are recommended, although the latter is only applicable for *trans* acting lncRNA since integration will be random (Lee *et al.*, 2016). Finally, one should keep in mind that transcriptional interference of a lncRNA promoter to its flanking protein-coding promoters may occur, as described for *Airn* (Latos *et al.*, 2012).

While most lncRNA-targeting studies result in exploration and focus on the phenotype and function, experimental studies that focus on mechanistic insights often are acquired later. To understand lncRNA mechanism, identification of direct lncRNA-binding partners might drastically increase our phenotypic understanding of the targeted lncRNA. Thus, we next discuss the most widely used techniques to identify direct binding partners of lncRNAs that enable considerable mechanistic insights.

24.4 Identification of lncRNA-interacting Binding Partners

While analysis of neighboring flanking genes may elucidate whether a lncRNA acts in *cis*, the identification of distantly acting *trans*-lncRNAs is somehow more challenging. The functional diversity of lncRNAs arises either through its architecture, its binding affinity in a combinatorial manner by RNA–RNA, RNA–DNA or RNA–protein interactions, or both (Rinn and Chang, 2012). A few selected RNA-centric methods are briefly discussed to identify cellular components with which lncRNAs may interact.

24.4.1 dChIRP

One method of identifying chromatin-associated lncRNA interactors is termed domain-specific chromatin isolation by RNA purification (dChIRP; Quinn *et al.*, 2014), which is an improved technique from the initial ChIRP assay (Figure 24.4A; Chu *et al.*, 2011). The aim of this method is to dissect functional domains of target RNA within its cellular context. The advantage over most other techniques is the simultaneous analysis of RNA–RNA, RNA–DNA and RNA–protein interactions. The method is based on biotinylated antisense 20-mer oligos, which are divided into domain-specific pools, instead of "odd" vs. "even" pools, as described earlier in ChIRP (Chu *et al.*, 2011). Additional steps are optimized throughout the protocol, as described by Quinn *et al.* (2014). The lower resolution limit of dChIRP is proposed to be ~200–500 bp, to which the identified domains are appropriately dissected. For example, Sallam *et al.* (2016) performed an *in vivo* pull down (ChIRP) of *LeXis* lncRNA from mouse liver, and identified by mass spectrometry (MS) the ribonucleoprotein RALY as a binding partner. They focused only on the nuclear pellet, since *LeXis* is exclusively nuclear-localized. As an important control for subsequent MS, they designed probes against LacZ to reduce background noise.

24.4.2 Rap

Early studies on lncRNA reported a regulation of lncRNA with PRC2, which itself regulates chromatin states in a general fashion (Khalil *et al.*, 2009). In order to elucidate the mechanism of *Xist* to DNA, Engreitz *et al.* (2013) developed a method termed RNA antisense purification (RAP) using biotinylated antisense probes to target endogenous RNA similar to ChIRP (Chu *et al.*, 2011). Probes of 120 nt were designed, spanning the entire *Xist* transcript, separated every 15 nt and were produced through T7-specific *in vitro* transcription. Cross-linked cell lysate was then purified for biotin-labeled RNA capture probes and eluted DNA was analyzed by subsequent next-generation sequencing (NGS)

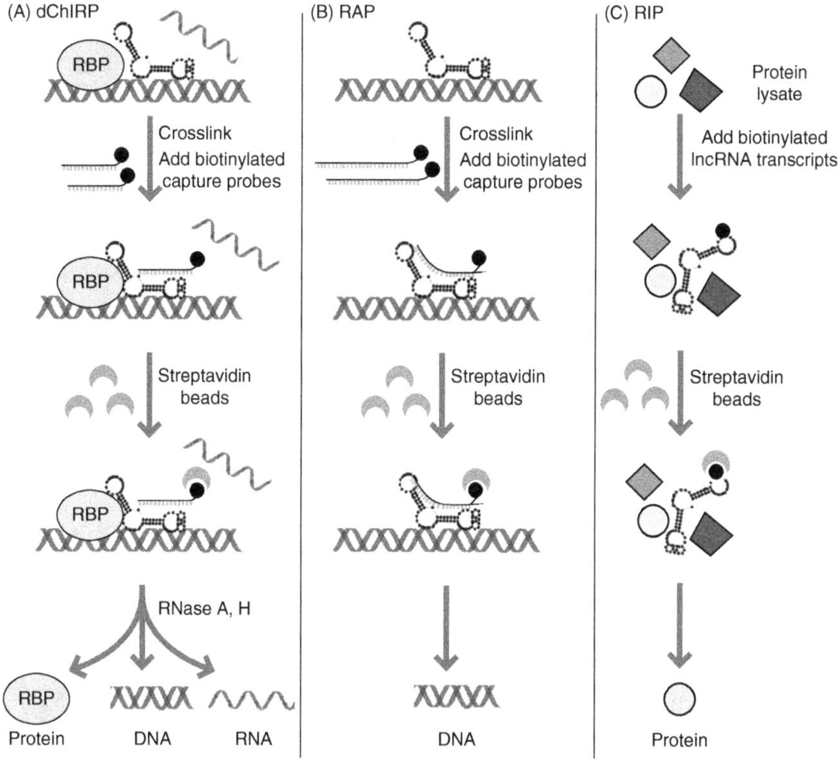

Figure 24.4 Identification of lncRNA interactors. Schematic illustration of lncRNA pulldown methods including (A) dChIRP, (B) RAP and (C) RIP.

analysis (Figure 24.4B). Enrichment of *Xist* could be confirmed by RT-qPCR over 100-fold compared to control probes, which were designed in the same way, but not complementary to *Xist*. These control probes should bind to the same non-specific double-stranded DNA that the lncRNA probe could potentially bind to, with the same efficiency, but without hybridizing to target RNA. Hence the control DNA result can be considered to be background noise. To exclude non-specific binding of capture probes, obtained DNA sequences were analyzed for complementarity to probes; however, none was found. To ensure that captured DNA reflects pre-existing interactions, non-crosslinked cellular extracts did not obtain any DNA, but *Xist* RNA was enriched as expected (Engreitz *et al.*, 2013).

To address the question of direct and indirect RNA–RNA interactions, Engreitz *et al.* (2014) further develop conventional RAP (i.e. RNA-RAP) in three related protocols: RAP-RNA$^{[AMT]}$, RAP-RNA$^{[FA]}$ and RAP-RNA$^{[FA\text{-}DSG]}$. Those three protocols mainly differ in the chemistry used for crosslinking, as 4′-aminomethyltroxalen (AMP) crosslinks uridine bases in RNA, but not proteins; formaldehyde (FA) crosslinks RNA–protein and protein–protein interactions, and a combination of FA and disuccinimidyl glutarate (DSG) is used for crosslinking protein–protein interactions to identify indirect binding.

24.4.3 Rip

In contrast to the classical RNA immunoprecipitation (RIP) assay (Gilbert and Svejstrup, 2006), the RIP assay for lncRNAs employs target RNAs that are transcribed *in vitro* using T7- and/or SP6-specific polymerase for biotin labeling and does not use FA for fixation. The *in vitro* transcribed biotin-labeled target RNA is folded into its structure by heat and is subsequently incubated with whole or fraction-specific cell lysate. MS analyzes the eluted proteins (Figure 24.4C; Feng *et al.*, 2014). For example, using this technique, PUMILIO was identified as the interacting partner of *NORAD* (Lee *et al.*, 2016).

The advantage of using dChIRP compared to RAP and RIP is the ability to simultaneously identify RNA, DNA and protein interactors. Depending on the reagent used for crosslinking, RAP may also identify RNA, DNA and protein interactors. However, RAP uses ~120 nt long capture probes vs. 20 nt long probes for dChIRP. On the one hand, a longer capture probe increases specificity; however, it may also hinder binding due to lncRNA secondary structure. RIP is an *in vitro* RNA pulldown independent of biotin-oligonucleotide probes, without crosslinking, for the identification of protein interactors only. A caveat for *in vitro* RNA pulldowns is that heating followed by slow cool down may not necessarily result in folding into the right *in vivo* structure. Another caution is the high input of biotin-labeled RNA to the lysate compared to low *in vivo* abundance of lncRNAs transcripts, which may yield artifacts (Uchida and Dimmeler, 2015).

24.5 Conclusions and Future Directions

LncRNAs hold promise as important regulators of protein-coding genes, RNA-binding proteins and chromatin-modifying factors. Indeed, accumulating studies demonstrate that lncRNAs function to alter key biological processes important for cardiovascular disease susceptibility, progression and regeneration. Several important considerations for interpretating lncRNA structure and function have been highlighted by recent studies. For example, using CRISPR/Cas9-mediated genetic editing, Xue and colleagues demonstrated that a small AGIL motif of the modular secondary structure of *Bvht* is essential for binding to the zinc-finger transcription factor CNBP and cardiac lineage development (Xue *et al.*, 2016). Combining careful genomic loci analysis with precise genetic editing techniques and consideration of RNA secondary structure will improve our understanding of lncRNAs and their precise mechanism of action with RNAs, DNAs and interacting proteins.

Nucleic-acid-based therapeutics constitute the next frontier in CVD management. Several challenges targeting other classes of RNA, such as miRNAs, also hold true for targeting lncRNAs. In contrast to targeting one gene, as highlighted using siRNAs, therapeutic targeting of lncRNA provides a unique strategy for regulating a network of genes and signaling pathways by virtue of lncRNA interacting with more than one target. While this may open potential unwanted effects if multiple targets are regulated, it may also prove to be more effective at treating complex diseases in which multiple pathways would benefit from a multi-targeted approach, especially if such targets are disease-associated genes. Similar to miRNAs, lncRNA targets may be differentially regulated in a cell-specific manner or in response to specific pathophysiological stimuli, providing unique properties for therapeutic intervention.

While lncRNAs are not well conserved across species on the level of sequence identity, tissue-specific lncRNA expression shows strong conservation across species (Washietl *et al.*, 2014). Furthermore, the presence of evolutionarily ultra-conserved regions at the level of the lncRNA locus highlights that these regulatory regions may be important for conferring expression in a cell- or stimulus-specific manner. Moreover, SNPs located in these regions may be relevant for CVD susceptibility. Future studies applying genetic editing approaches should pay close attention to lncRNA genomic location, secondary structure, proximity to surrounding elements and protein-coding genes, and mechanisms for identified interactors that may underlie phenotypic or functional lncRNA biology. Taken together, targeting lncRNAs may have tremendous effects on biological pathways, cell function and cardiovascular tissue homeostasis in health and disease.

Acknowledgements

This work was supported by the National Institutes of Health (HL115141, HL117994 and GM115605 to M.W.F.), the Arthur K. Watson Charitable Trust (to M.W.F.), the Dr. Ralph and Marian Falk Medical Research Trust (to M.W.F.) and a Swiss National Foundation Early Postdoc Mobility Fellowship (SNF# P2BEP3_162063 to S.H).

REFERENCES

Anderson DM, Anderson KM, Chang CL, *et al.* 2015. A micropeptide encoded by a putative long noncoding RNA regulates muscle performance. *Cell* **160**(4): 595–606.

Andrews SJ, Rothnagel JA. 2014. Emerging evidence for functional peptides encoded by short open reading frames. *Nat Rev Genet* **15**(3): 193–204.

Aparicio-Prat E, Arnan C, Sala I, *et al.* 2015. DECKO: single-oligo, dual-CRISPR deletion of genomic elements including long non-coding RNAs. *BMC Genomics* **16**(1): 846.

Bartel DP. 2004. MicroRNAs: genomics, biogenesis, mechanism, and function. *Cell* **116**(2): 281–297.

Bassett AR, Akhtar A, Barlow DP, *et al.* 2014. Considerations when investigating lncRNA function in vivo. *eLife* **3**: e03058.

Bond AM, Vangompel MJ, Sametsky EA, *et al.* 2009. Balanced gene regulation by an embryonic brain ncRNA is critical for adult hippocampal GABA circuitry. *Nat Neurosci* **12**(8): 1020–1027.

Bondue A, Lapouge G, Paulissen C, *et al.* 2008. Mesp1 acts as a master regulator of multipotent cardiovascular progenitor specification. *Cell Stem Cell* **3**(1): 69–84.

Bu D, Yu K, Sun S, *et al.* 2012. NONCODE v3.0: integrative annotation of long noncoding RNAs. *Nucleic Acids Res* **40**(Database issue): D210–D215.

Canver MC, Bauer DE, Dass A, *et al.* 2014. Characterization of genomic deletion efficiency mediated by clustered regularly interspaced palindromic repeats (CRISPR)/Cas9 nuclease system in mammalian cells. *J Biol Chem* **289**(31): 21312–21324.

Chu C, Qu K, Zhong FL, Artandi SE, Chang HY. 2011. Genomic maps of long noncoding RNA occupancy reveal principles of RNA-chromatin interactions. *Mol Cell* **44**(4): 667–678.

Congrains A, Kamide K, Oguro R, *et al.* 2012. Genetic variants at the 9p21 locus contribute to atherosclerosis through modulation of ANRIL and CDKN2A/B. *Atherosclerosis* **220**(2): 449–455.

da Rocha ST, Edwards CA, Ito M, Ogata T, Ferguson-Smith AC. 2008. Genomic imprinting at the mammalian Dlk1-Dio3 domain. *Trends Genet* **24**(6): 306–316.

Djebali S, Davis CA, Merkel A, *et al.* 2012. Landscape of transcription in human cells. *Nature* **489**(7414): 101–108.

Engreitz JM, Pandya-Jones A, McDonel P, *et al.* 2013. The Xist lncRNA exploits three-dimensional genome architecture to spread across the X chromosome. *Science* **341**(6147): 1237973–1237973.

Engreitz JM, Sirokman K, McDonel P, *et al.* 2014. RNA-RNA interactions enable specific targeting of noncoding RNAs to nascent pre-mRNAs and chromatin sites. *Cell* **159** (1): 188–199.

Farazi TA, Hoell JI, Morozov P, Tischl T. 2013. MicroRNAs in human cancer. In *MicroRNA Cancer Regulation*, U Schmitz, O Wolkenhauer, J Vera, eds., Dordrecht, the Netherlands: Springer, pp. 1–20.

Feng Y, Hu X, Zhang Y, *et al.* 2014. Methods for the study of long noncoding RNA in cancer cell signaling. *Methods Mol Biol* **1165**: 115–143.

Gagnon KT, Li L, Chu Y, *et al.* 2014. RNAi factors are present and active in human cell nuclei. *Cell Rep* **6**(1): 211–221.

Gilbert C, Svejstrup JQ. 2006. RNA immunoprecipitation for determining RNA-protein associations in vivo. *Curr Protoc Mol Biology* Chapter 27: Unit 27.4.

Gilbert LA, Larson MH, Morsut L, *et al.* 2013. CRISPR-mediated modular RNA-guided regulation of transcription in eukaryotes. *Cell* **154**(2): 442–451.

Goff LA, Rinn JL. 2015. Linking RNA biology to lncRNAs. *Genome Res* **25**(10): 1456–1465.

Grote P, Wittler L, Hendrix D, *et al.* 2013. The tissue-specific lncRNA Fendrr is an essential regulator of heart and body wall development in the mouse. *Dev Cell* **24**(2): 206–214.

Gutschner T. 2015. Silencing long noncoding RNAs with genome-editing tools. *Methods Mol Biol* **1239**: 241–250.

Gutschner T, Baas M, Diederichs S. 2011. Noncoding RNA gene silencing through genomic integration of RNA destabilizing elements using zinc finger nucleases. *Genome Res* **21**(11): 1944–1954.

Gutschner T, Hämmerle M, Eissmann M, *et al.* 2013. The noncoding RNA MALAT1 is a critical regulator of the metastasis phenotype of lung cancer cells. *Cancer Res* **73**(3): 1180–1189.

Han J, Zhang J, Chen L, *et al.* 2014a. Efficient in vivo deletion of a large imprinted lncRNA by CRISPR/Cas9. *RNA Biol* **11**(7): 829–835.

Han P, Li W, Lin CH, *et al.* 2014b. A long noncoding RNA protects the heart from pathological hypertrophy. *Nature* **514**(7520): 102–106.

He S, Liu C, Skogerbø G, *et al.* 2008. NONCODE v2.0: decoding the non-coding. *Nucleic Acids Res* **36**(Database issue): D170–D172.

Hill JA, Olson EN. 2008. Cardiac plasticity. *N Engl J Med* **358**(13): 1370–1380.

Ho T-T, Zhou N, Huang J, *et al.* 2015. Targeting non-coding RNAs with the CRISPR/Cas9 system in human cell lines. *Nucleic Acids Res* **43**(3): e17.

Holdt LM, Beutner F, Scholz M, *et al.* 2010. ANRIL expression is associated with atherosclerosis risk at chromosome 9p21. *Arterioscl Thromb Vasc Biol* **30**(3): 620–627.

Hwang H-WH, Mendell JTJ. 2006. MicroRNAs in cell proliferation, cell death, and tumorigenesis. *Br J Cancer* **94**(6): 776–780.

Kapranov P, Cheng J, Dike S, *et al.* 2007. RNA maps reveal new RNA classes and a possible function for pervasive transcription. *Science* **316**(5830): 1484–1488.

Katayama S, Tomaru Y, Kasukawa T, *et al.* 2005. Antisense transcription in the mammalian transcriptome. *Science* **309**(5740): 1564–1566.

Khalil AM, Guttman M, Huarte M, *et al.* 2009. Many human large intergenic noncoding RNAs associate with chromatin-modifying complexes and affect gene expression. *Proc Natl Acad Sci USA* **106**(28): 11667–11672.

Klattenhoff CA, Scheuermann JC, Surface LE, *et al.* 2013. Braveheart, a long noncoding RNA required for cardiovascular lineage commitment. *Cell* **152**(3): 570–583.

Kotzin JJ, Spencer SP, McCright SJ, *et al.* 2016. The long non-coding RNA Morrbid regulates Bim and short-lived myeloid cell lifespan. *Nature* **537**(7619): 239–243.

Latos PA, Pauler FM, Koerner MV, *et al.* 2012. Airn transcriptional overlap, but not its lncRNA products, induces imprinted Igf2 r silencing. *Science* **338**(6113): 1469–1472.

Lee S, Kopp F, Chang TC, *et al.* 2016. Noncoding RNA NORAD regulates genomic stability by sequestering PUMILIO proteins. *Cell* **164**(1–2): 69–80.

Leucci E, Vendramin R, Spinazzi M, *et al.* 2016. Melanoma addiction to the long non-coding RNA SAMMSON. *Nature* **531**(7595): 518–522.

Libby P, Ridker PM, Hansson GK. 2011. Progress and challenges in translating the biology of atherosclerosis. *Nature* **473**(7347): 317–325.

Liu C, Bai B, Skogerbø G, *et al.* 2005. NONCODE: an integrated knowledge database of non-coding RNAs. *Nucleic Acids Res* **33**(Database issue): D112–D115.

Maeder ML, Linder SJ, Cascio VM, *et al.* 2013. CRISPR RNA-guided activation of endogenous human genes. *Nat Methods* **10**(10): 977–979.

Michalik KM, You X, Manavski Y, *et al.* 2014. Long noncoding RNA MALAT1 regulates endothelial cell function and vessel growth. *Circ Res* **114**(9): 1389–1397.

Moore CB, Guthrie EH, Huang MT, Taxman DJ. 2010. Short hairpin RNA (shRNA): design, delivery, and assessment of gene knockdown. *Methods Mol Biol* **629**: 141–158.

Okazaki Y, Furuno M, Kasukawa T, *et al.* 2002. Analysis of the mouse transcriptome based on functional annotation of 60,770 full-length cDNAs. *Nature* **420**(6915): 563–573.

Ounzain S, Micheletti R, Arnan C, *et al.* 2015. CARMEN, a human super enhancer-associated long noncoding RNA controlling cardiac specification, differentiation and homeostasis. *J Mol Cell Cardiol* **89**(Pt A): 98–112.

Palmer AC, Egan JB, Shearwin KE. 2011. Transcriptional interference by RNA polymerase pausing and dislodgement of transcription factors. *Transcription* **2**(1): 9–14.

Perez-Pinera P, Kocak DD, Vockley CM, *et al.* 2013. RNA-guided gene activation by CRISPR-Cas9-based transcription factors. *Nat Methods* **10**(10): 973–976.

Quinn JJ, Ilik IA, Qu K, *et al.* 2014. Revealing long noncoding RNA architecture and functions using domain-specific chromatin isolation by RNA purification. *Nat Biotechnol* **32**(9): 933–940.

Ran FA, Hsu PD, Lin CY, *et al.* 2013. Double nicking by RNA-guided CRISPR Cas9 for enhanced genome editing specificity. *Cell* **154**(6): 1380–1389.

Rinn JL, Chang HY. 2012. Genome regulation by long noncoding RNAs. *Annu Rev Biochem* **81**(1): 145–166.

Royo H, Cavaillé J. 2008. Non-coding RNAs in imprinted gene clusters. *Biol Cell* **100**(3): 149–166.

Sallam T, Jones MC, Gilliland T, *et al.* 2016. Feedback modulation of cholesterol metabolism by the lipid-responsive non-coding RNA LeXis. *Nature* **534**(7605): 124–128.

Samani NJ, Erdmann J, Hall AS, *et al.* 2007. Genomewide association analysis of coronary artery disease. *N Engl J Med* **357**(5): 443–453.

Sander JD, Joung JK. 2014. CRISPR-Cas systems for editing, regulating and targeting genomes. *Nat Biotechnol* **32**(4): 347–355.

Sauvageau M, Goff LA, Lodato S, *et al.* 2013. Multiple knockout mouse models reveal lincRNAs are required for life and brain development. *eLife* **2**: e01749.

Sleutels F, Zwart R, Barlow DP. 2002. The non-coding Air RNA is required for silencing autosomal imprinted genes. *Nature* **415**(6873): 810–813.

Uchida S, Dimmeler S. 2015. Long noncoding RNAs in cardiovascular diseases. *Circ Res* **116**(4): 737–750.

Viereck J, Kumarswamy R, Foinquinos A, *et al.* 2016. Long noncoding RNA Chast promotes cardiac remodeling. *Sci Transl Medicine* **8**(326): 326ra22.

Wang K, Liu F, Zhou LY, *et al.* 2014. The long noncoding RNA CHRF regulates cardiac hypertrophy by targeting miR-489. *Circ Res* **114**(9): 1377–1388.

Wang Z, Zhang XJ, Ji YX, *et al.* 2016. The long noncoding RNA Chaer defines an epigenetic checkpoint in cardiac hypertrophy. *Nat Med* **22**(10): 1131–1139.

Ward LD, Kellis M. 2012. HaploReg: a resource for exploring chromatin states, conservation, and regulatory motif alterations within sets of genetically linked variants. *Nucleic Acids Res* **40**(Database issue): D930–D934.

Washietl S, Kellis M, Garber M. 2014. Evolutionary dynamics and tissue specificity of human long noncoding RNAs in six mammals. *Genome Res* **24**(4): 616–628.

Xie C, Yuan J, Li H, *et al.* 2014. NONCODEv4: exploring the world of long non-coding RNA genes. *Nucleic Acids Res* **42**(Database issue): D98–D103.

Xue Z, Hennelly S, Doyle B, *et al.* 2016. A G-rich motif in the lncRNA braveheart interacts with a zinc-finger transcription factor to specify the cardiovascular lineage. *Mol Cell* **64**(1): 37–50.

Zelcer N, Tontonoz P. 2006. Liver X receptors as integrators of metabolic and inflammatory signaling. *J Clin Invest* **116**(3): 607–614.

Zhang B, Arun G, Mao YS, *et al.* 2012. The lncRNA Malat1 is dispensable for mouse development but its transcription plays a cis-regulatory role in the adult. *Cell Rep* **2**(1): 111–123.

Zhao Y, Li H, Fang S, *et al.* 2016. NONCODE 2016: an informative and valuable data source of long non-coding RNAs. *Nucleic Acids Res* **44**(D1): D203–D208.

25 Gene Silencing, Disruption and Latency Reactivation with RNA-based and Gene Editing CRISPR/Cas, ZFN and TALEN Technologies for HIV-1/AIDS Therapies

Paige Charlins and Ramesh Akkina

25.1 Introduction

Stable genome integration into the host cell chromosome and establishment of latency pose challenges for the development of more effective treatments and a complete cure for HIV/AIDS. While the HIV-1 replication cycle is complex, there are many unique steps in the viral infection that lend themselves for novel drug targeting. These include both viral and cellular molecules. While siRNAs are shown to degrade viral and cellular transcripts, effectively mediating viral suppression *in vitro*, their specific delivery *in vivo*, in a cost-efficient manner, has been challenging. Since the nature of HIV-1 disease is chronic, spanning many years, unlike severe acute infectious diseases that progress rapidly, there is a long window of opportunity for interfering with viral-mediated pathology by novel gene therapeutic strategies. A number of gene silencing/editing and gene disruption technologies that include siRNAs, aptamers, ZFNs, TALENs and CRISPR/Cas systems are being harnessed to prevent, treat and reverse viral latency. As depicted in Figure 25.1 and Table 25.1, a number of viral genome sites identified for targeting are being investigated utilizing these technologies, as further discussed below. These innovative technologies are currently pursued to minimalize toxicity, increase efficacy and develop a functional cure for HIV/AIDS. For these experimental strategies to be successfully translated into the clinic, further research is needed to overcome a number of obstacles that include efficient delivery systems, as discussed below.

25.2 RNA-based Strategies for HIV Treatment

The main staples of current HIV-1 treatment are small-molecule drugs that interfere with different stages of viral replication, which include viral entry, reverse transcription, integration and proteolytic processing. Various combinations of these drugs led to HAART therapy, with effective treatment that suppresses viral loads to undetectable levels. However, there is no complete cure. With a view to developing alternative treatments, a number of approaches, including gene therapy strategies have been studied with some reaching clinical trials. The anti-HIV molecules

Table 25.1 Viral genome sites identified for targeting, utilizing different gene editing technologies

	Target	Function	Gene editing tool	Model	Reference
Viral	LTR	HIV-1 promoter	siRNA, RNA Aptamer, ZFN, TALEN, CRISPR/Cas9	*In vitro, in vivo*	Kaminski *et al.*, 2016a; Malecova and Morris, 2010; Sayman *et al.*, 2015; Wang *et al.*, 2016a, 2016b
	Tat/Rev	Regulates transcription export	siRNA, Aptamer, ZFN, TALEN, CRISPR/Cas9	*In vitro, in vivo*	Anderson *et al.*, 2007; Neff *et al.*, 2011; Wang *et al.*, 2016b
	Gag/Pol	Multiple functions (RT, integrase, protease)	siRNA, Aptamer, ZFN, TALEN, CRISPR/Cas9	*In vitro, in vivo*	Wang *et al.*, 2016b; Whatley *et al.*, 2013
	Env	Viral envelope	siRNA, RNA Aptamer, ZFN, TALEN, CRISPR/Cas9	*In vitro, in vivo*	Zhou *et al.*, 2009; Neff *et al.*, 2011
Host	CD4	HIV-1 primary receptor	siRNA, Aptamer	*In vitro*	Wheeler *et al.*, 2013; Kumar *et al.*, 2008
	CXCR4	Co-receptor X4-tropic	siRNA, Aptamer, ZFN, TALEN, CRISPR/Cas9	*In vitro, in vivo*	Hou *et al.*, 2015; Yuan *et al.*, 2012
	CCR5	Co-receptor R5-tropic	siRNA, Aptamer, ZFN, TALEN, CRISPR/Cas9	*In vitro, in vivo,* clinical trials	Choi *et al.*, 2016; Kim *et al.*, 2010; Li *et al.*, 2013; Perez *et al.*, 2018; Maier *et al.*, 2013; Mussolino *et al.*, 2011; Tebas *et al.*, 2014; Wheeler *et al.*, 2013; Ye *et al.*, 2014; Zhou *et al.*, 2015

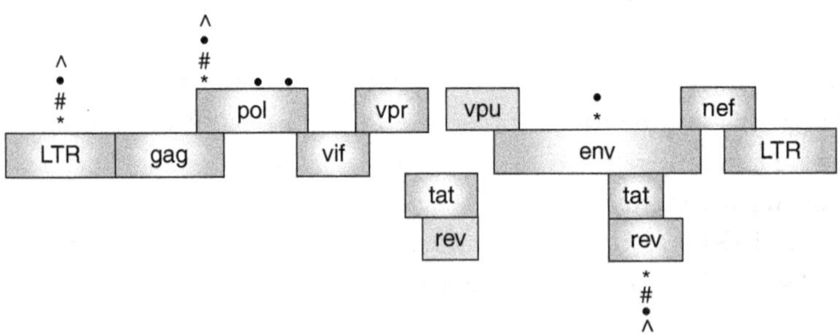

Figure 25.1 Target sites in HIV genome. *ZFN, #TALEN, • Aptamer/aptamer-siRNA, ^ CRISPR/Cas.

include protein-based constructs, such as transdominant proteins (revM10), peptide inhibitors, broadly neutralizing antibodies and RNA-based molecules such as ribozymes, siRNAs and aptamers (Berkhout, 2004; Bobbin *et al.*, 2015). A major problem with protein-based approaches is the potential for generating host immune responses to these foreign or altered proteins, ultimately leading to treatment failure. In this regard, RNA-based approaches provide some advantages, which include no immunogenicity, thus permitting prolonged treatment, and the ability to target both internal and external components of the host cell and the virus.

25.3 siRNAs

RNA interference (RNAi)-mediated gene silencing involves two pathways. The first pathway involves post-transcriptional gene silencing (PTGS), where translational suppression or degradation of mRNAs is mediated by the guide strand (antisense) displaying near-perfect or perfect complementarity to the target RNA, resulting in translation failure. In mimicking the host miRNA pathway, siRNAs and shRNAs engage endogenous host factors to mediate gene silencing. The second pathway, transcriptional gene silencing (TGS), involves siRNAs acting on the cell's transcriptional apparatus (Malecova and Morris, 2010; Saayman *et al.*, 2015). In a TGS approach, siRNA is targeted to HIV-1 LTR-mediated viral silencing. Numerous studies have previously shown the effectiveness of siRNAs and shRNAs against HIV-1 *in vitro* targeting both viral and cellular RNA targets. Viral-specific targets include *tat/rev*, LTR and Env, among others. Cellular targets include host receptor proteins CCR5, CXCR4 and CD4, with CCR5 being the most studied due to its importance in predominant viral transmission in nature by the R5-tropic HIV-1. For delivery, most *in vitro* studies used transfection of chemically synthesized siRNAs. Encapsulation of siRNAs into nanoparticles formed by the use of PAMAM dendrimers also enabled efficient and lasting delivery. For *in vivo* testing of various siRNA delivery formulations, humanized mice that permit HIV-1 infection and display helper $CD4^+$ T-cell depletion have played an important role (Akkina, 2013). For example, dendrimer formulated siRNAs against host cell molecules CD4, TNPO3 and viral transcripts *tat/rev* are administered to viremic humanized $Rag2^{-/-}\gamma c^{-/-}$ (RAG-hu) mice. Reduction of HIV-1 viremia by several orders of magnitude was noted in addition to protection from $CD4^+$ T-cell loss, demonstrating that siRNAs can be effectively delivered *in vivo* with a lasting therapeutic effect.

siRNAs can also be delivered for a much longer-lasting effect by gene therapy approaches that are produced endogenously in a constitutive manner and thus do not require repeated administrations for *in vivo* applications. Lentiviral vectors have been commonly used for siRNA gene transduction. HIV-1 co-receptor molecules such as CCR5 and viral transcripts *tat* and *rev* were among the siRNA targets. Since the generation of viral escape mutants would invariably occur with the use of single-target siRNAs, several siRNAs directed to multiple targets would be necessary to overcome any potential HIV-1 viral escape. Taking this into consideration, anti-HIV-1 siRNAs were combined with other RNA-based constructs and in some cases with protein-based inhibitors. A combinatorial construct incorporated into a lentiviral vector used *tat/rev* siRNA, a tar decoy and a CCR5 ribozyme. Based on its *in vivo* efficacy, as shown in a humanized mouse system, a Phase I human clinical trial was conducted (Anderson *et al.*, 2007). Results have shown no toxic effects with transgenic cells being detected at very low levels for up to three years. However, no clinical efficacy was seen due to low initial engraftment levels of modified hematopoietic stem cells (HSC). Newer strategies are aimed at achieving higher HSC engraftment with improved patient conditioning and the use of drug selection to enrich for engrafted cells. Other new combinatorial constructs include host restriction factors such as TRIM5α variants in addition to siRNAs that showed efficacy in hu-mouse models. More clinical trials are underway with some of the new constructs.

25.4 Aptamers Against HIV-1 and Host Factors

Aptamers, synthetic single-stranded RNA or DNA molecules with unique secondary and tertiary structures, can bind target molecules with high specificity and affinity akin to antibodies (Keefe *et al.*, 2010; Thiel and Giangrande, 2010). A process called systematic evolution of ligands by exponential enrichment (SELEX) is used to identify and enrich aptamers to desired targets. Bulk quantities of aptamers can be chemically synthesized, incorporating chemical modifications to improve nuclease resistance and increase their half-life *in vivo*. Apparent lack of immunogenicity and systemic toxicity are among desirable characteristics in addition to exquisite specificity.

A number of aptamers targeted to host cell receptors or viral proteins have been identified that interfere with HIV-1 infection. Anti-gp120 aptamers bind to the HIV-1 envelope resulting in viral neutralization, which can also potentially interfere with viral assembly and budding when delivered intracellularly. The gp120 aptamers have also been exploited for cell-specific delivery of anti-HIV-1 siRNAs into HIV-1-infected cells to inhibit viral replication. Aptamers that bind to RT were shown to hinder viral replication by competing with primer/template access, inhibiting DNA polymerization or RNase activity (Whatley *et al.*, 2013). Aptamers against other HIV-1 proteins, such as viral protease and integrase are currently being investigated. Cellular CD4 receptor and co-receptor CCR5-specific aptamers were shown to inhibit HIV-1 entry by blocking action.

25.5 SiRNA–Protein and siRNA–Aptamer Chimeric Conjugates for Cell-specific Delivery

A major problem with exogenously administered non-specific delivery methods of siRNAs for systemic therapeutic effect is that massive quantities are required to saturate the system to reach the HIV-1-susceptible cells *in vivo*. Exorbitant cost and potential off-target effects *in vivo* are among several drawbacks with this approach. Therefore, specific targeting of siRNAs to HIV-1-susceptible or already infected cells is highly desirable with a number of studies that have addressed this issue. In one approach, a single-chain antibody (scFV) to pan T-cell marker CD7 was used to deliver siRNAs into T cells (Kumar *et al.*, 2008). A modified scFVCD7Cys was conjugated to a nona-d-arginine (9R) peptide to deliver siRNAs targeted to CCR5 or HIV-1 Vif. While the results were impressive, with lower viremia in HIV-1-infected humanized mice, unfortunately all T cells were indiscriminately targeted using this approach. Also, long-term repeated dosing of these constructs is not possible due to the immunogenicity of the modified antibody–peptide conjugates. In a different study, LFA-1 integrin-targeted antibody was used to deliver CCR5 via nanoparticle formulation and showed efficacy in HIV-1-infected hu-mice (Kim *et al.*, 2010). Similarly, this approach also targets all T cells, including $CD8^+$ T cells and therefore is not selective for HIV-1-susceptible $CD4^+$ T cells. In a more recent *in vitro* study, CCR5 aptamers were used to direct anti-HIV-1 siRNAs to HIV-1-susceptible cells, which showed efficacy (Zhou *et al.*, 2015).

However, all cells expressing CCR5, including non-HIV-1 targets, took up the siRNA, thus showing a lack of selectivity.

Therefore, strategies that selectively target HIV-1-susceptible cells and virus-infected cells in particular, are necessary for a more effective therapeutic approach in HIV$^+$ patients. A subunit of the viral envelope, gp120, is expressed on the HIV-1-infected cell surface, providing a marker for siRNA targeting. A single chain gp120 antibody complexed to anti-HIV-1 siRNAs was successfully used *in vitro* in this context to target infected cells (Zhou *et al.*, 2009). However, potential host immune responses to antibody conjugates remains an issue. A novel approach to overcome this potential drawback with protein conjugates employed a HIV-1 gp120-binding aptamer to deliver an anti-HIV-1 *tat/rev* siRNA (aptamer–siRNA chimera) specifically into HIV-1-infected cells. These chimeric RNA constructs were 2′-fluoro-modified to reduce nuclease degradation *in vivo* and evaluated in HIV-1-infected humanized mice by weekly injections (Neff *et al.*, 2011). Reduction in viral loads to below undetectable levels was seen within a week and this was maintained up to three weeks after treatment cessation in the majority of mice, showing sustained efficacy. There was also protection from CD4T cell loss, the primary goal of targeted anti-HIV-1 therapy. Three different mechanisms operating simultaneously are envisaged for the antiviral effect (as depicted in Figure 25.2): (1) direct aptamer neutralization of free-floating virus in the plasma, thus preventing viral spread to new cells, (2) intracellular interference by aptamer-delivered *tat/rev* siRNA in infected cells and (3) interference by gp120-binding aptamer with viral budding and/or release. It needs to be pointed out,

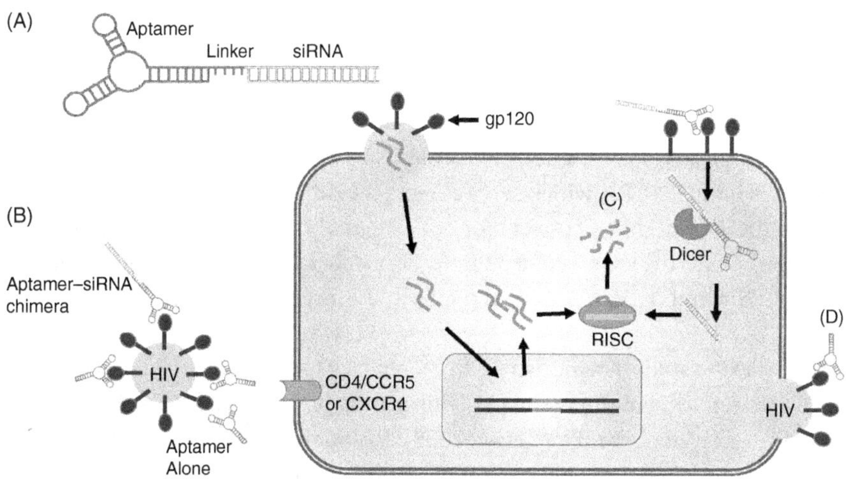

Figure 25.2 Overview of aptamer and aptamer–siRNA chimera mode of action. (A) Schematic of aptamer conjugated to siRNA via linker bridge. (B) Interaction of aptamer and aptamer–siRNA chimera for viral neutralization and entry inhibition. (C) After internalization of the aptamer–siRNA chimera, siRNA is processed by the DICER/RISC pathway, which leads to viral mRNA degradation. (D) Aptamer interaction during viral budding inhibits budding and maturation. (A black-and-white version of this figure will appear in some formats. For the color version, please refer to the plate section.)

however, that in some treated mice there was viral rebound, indicating viral escape from the gp120 aptamer. To prevent viral escape during long-term therapy it will be necessary to employ multiple aptamer–siRNA constructs with varying specificities in a combination akin to that being used in the current effective HAART therapy.

Another area of investigation employing aptamer–siRNA chimeras focused on prevention of HIV-1 transmission. Another study employed CD4 aptamers conjugated to siRNAs targeted to host cell CCR5 and viral transcripts for *vif* and *gag* (Wheeler *et al.*, 2013). When tested in BLT hu-mice against HIV-1 vaginal transmission, these chimeras prevented infection. Collectively the above findings demonstrated the feasibility of targeted siRNA delivery *in vivo* by various methods and provided promising preclinical data.

25.6 Zinc-finger Nucleases and TALENS

HIV-1 depends on both viral and cellular molecules for infection and replication. To interfere with the function of these essential molecules, gene editing technologies can be harnessed. In this regard, zinc-finger nucleases (ZFN) and transcription activator-like effector nucleases (TALENs) have been put to use primarily targeting cellular receptors for permanent disruption leading to viral resistance. ZFN proteins are a class of transcription factors that selectively bind DNA via specific motifs that are linked to the active domain of an endonuclease Fok1, creating ZFNs. The Fok1 domain activates after dimerization for cleavage function. Three to six ZF modules are typically linked together with each module recognizing three sequence-specific nucleotides.

ZFNs can be custom designed to target specific genomic regions. Cleavage of DNA results in double-strand breaks which are repaired by either non-homologous end joining (NHEJ) or homologous recombination (HR), with NHEJ being the predominant outcome resulting in gene disruption. For HIV-1 gene therapy, ZFNs were fashioned precisely for a variety of targets with a focus on CD4, CCR5 and CXCR4, which serve as receptors for viral entry. Of these, disruption of the CCR5 HIV-1 co-receptor is particularly attractive since individuals homozygous for the CCR5 delta 32 mutation were found to be resistant to infection by R5-tropic HIV-1, which represents the predominant transmitted viral strain.

Many *in vitro* studies demonstrated ZFN efficacy for CCR5 knockdown in $CD4^+$ T cells as well as hematopoietic progenitor cells ($CD34^+$ HSC) resulting in viral inhibition (Li *et al.*, 2013; Perez *et al.*, 2008). *In vivo* studies using humanized mice susceptible to HIV-1 infection provided further proof of its utility (Yuan *et al.*, 2012). Translating this, a clinical trial was conducted using autologous ZFN-treated CCR5-disrupted $CD4^+$ T cells (Maier *et al.*, 2013; Tebas *et al.*, 2014). The resulting modification showed selective advantage for gene-modified cells, with improved $CD4^+$ T-cell counts and delayed viral rebound during ART cessation. This outcome showed the clinical feasibility of this approach and set the stage for future attempts for a functional cure, especially in patients heterozygous for the CCR5 delta 32 mutation. However, for ZFNs to be fully successful on a wider scale, higher biallelic CCR5 knockout levels and elimination of potential off-target effects are necessary.

Similar to ZFNs, TALENs have also been investigated as genome editing tools for HIV-1 gene therapy. TALENs are comprised of custom-engineered TALE-derived DNA-binding domains in conjunction with a FokI endonuclease domain that dimerizes on the target double-stranded DNA region. Dimerization allows for the cleavage of the spacer region ending in a double-strand break with subsequent indel generation disrupting the targeted gene. Unlike ZFNs, each TALE repeat recognizes only one nucleotide and sets of TALE repeats are assembled together to target 14–20 nucleotides in a gene. Due to the need for intensive protein engineering, it has generally proven to be cumbersome to assemble these. Nevertheless, creative cloning techniques are being employed to overcome this difficulty. One study evaluated CCR5 disruption in a T-cell line and showed its efficacy comparable to that of ZFNs, albeit with lower cytotoxicity and also reduced off-target effects on a related CCR2 locus (Mussolino *et al.*, 2011). Due to the large size of TALENs, delivery by lentiviral vectors poses challenges, whereas adenoviral vectors can accommodate the large cargo.

25.7 CRISPR/Cas9 Systems for HIV-1 Therapies

While the above protein-based ZFN and TALEN systems helped set the stage for custom genome editing and increased our knowledge of the intricacies involved, the expense and cumbersome protein engineering required limited their wide application. In this regard, the more recent CRISPR/Cas9 system has facilitated a new revolution via a much more simplified gene editing method universally applicable to virtually any organism, permitting the expansion of its utility and application. The guide component of CRISPR/Cas9 is the single-guide RNA (sgRNA), which is comprised of the CRISPR RNA (crRNA), including the target-specific complementary sequence, and the partially complementary trans-acting RNA (tracRNA). Programmable single-guide RNAs can be easily designed for any target DNA sequence. Upon the direction of the sgRNA, Cas9 locates the target DNA sequence and mediates DS breaks, which are later repaired by either NHEJ or HR, with NHEJ being the major outcome.

Three aspects of the CRISPR/Cas9 system provide unique opportunities for HIV-1 gene therapy and activation of latent virus. First, indel generation permits disruption of either viral-specific genes or cellular genes necessary for viral infection or replication. Second, with an available option for introduction of a new external donor DNA sequence, host gene loci can be uploaded with anti-HIV-1 gene cargo. Third, a nuclease-deficient Cas9 (dCas9) selectively fused to either transcription activators or repressors of choice can be targeted to bind integrated viral DNA and act at critical regulatory regions such as the HIV-1 promoter, LTR (Figure 25.3F).

As mentioned above, CCR5 disruption for HIV-1 prevention and therapies is a method of choice. A number of recent studies successfully targeted this gene

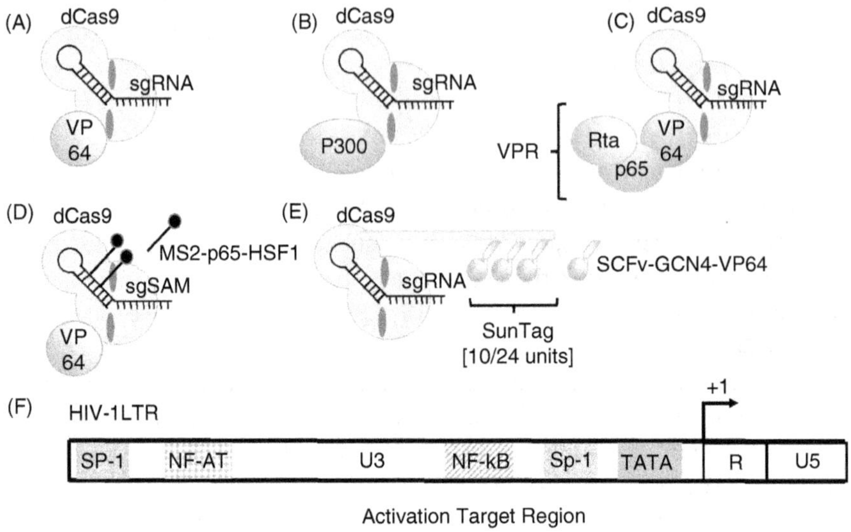

Figure 25.3 Strategies to reactivate latent HIV-1 by targeting the viral LTR. (A) Nuclease-deficient dCas9 is fused to VP64 (four copies of the herpes simplex virus VP16 activator) resulting in recruitment of TATA-binding protein, TFIIB and CBP/p300. (B) dCas9 is fused to p300, resulting in a chromatin remodeling complex. This interaction of dCas9–p300 results in induction of a chromatin state preferential for transcription. (C) dCas9–VPR includes VP64, p65 and Rta (Epstein–Barr transactivator), which acts to recruit NFkB. (D) sgRNA is modified to sgSAM; co-delivery of sgSAM and MS2–P65–HSF1 permits recruitment of dCas9 and subsequent target activation. (E) dCas9 is fused to ScFv–GCN4–VP64, which recruits up to 24 copies of green fluorescent protein to allow visualization and activation of the target site. (F) Viral activation target region.

with CRISPR/Cas9 in cell lines and primary cells. A variety of delivery systems were used, including lentiviral vectors. Ye *et al.* (2014) were able to recreate a naturally occurring CCR5 delta 32 mutation in induced pluripotent stem cells (iPSCs) employing *piggy-Bac* transposon-delivered sequences. These modified cells, when differentiated into monocytes/macrophages *in vitro*, displayed HIV-1 resistance. In another study, CCR5 was effectively disrupted in primary CD4$^+$ T cells as well as CD34$^+$ HSPCs that differentiated normally into hematopoietic lineages (Choi *et al.*, 2016).

CRISPR/Cas9 has also been shown to disrupt a second HIV-1 co-receptor, CXCR4, in primary human CD4$^+$ T cells with minimal off-target effects, leading to HIV-1 resistance (Hou *et al.*, 2015). As long-term constitutive expression of Cas in a cell is undesirable, a transient lentiviral vector expression system targeting CCR5 also showed efficacy in gene disruption (Choi *et al.*, 2016). Moreover, this system also demonstrated efficacy for disabling the HIV-1 LTR in a cell line model. Since HIV-1 proviral DNA integration into a host cell results in life-long virus persistence, an attractive strategy is to target the provirus for gene excision and removal. In a number of studies, HIV-1 viral targets used for CRISPR/Cas9-mediated silencing included the LTR, Gag, Tat and Rev regions (Figure 26.1). Viral gene disruptions were validated in both immortalized HIV-1-infected cell lines and primary cells (Kaminski *et al.*, 2016b; Zhu *et al.*, 2015). In one study, use of an adeno-associated virus 9 vector expressing Cas9 and sgRNA targeted to a region between the LTR and Gag resulted in the excision of a 978 bp proviral segment, leading to viral silencing in a transgenic mouse model (Kaminski *et al.*, 2016a). These encouraging proof-of-concept studies laid the groundwork for more in-depth *in vivo* studies and future clinical trials.

It should be noted, however, that newly emerging data are pointing to the fact that, as with many current therapies, HIV-1 can escape from CRISPR/Cas9-mediated viral suppression (Liang *et al.*, 2016). Employing stably transduced cells harboring Cas9/sgRNA constructs targeted to single sites in the viral *LTR*, *gag/pol* and *tat/rev* regions, several groups evaluated both viral inhibition and possible breakthrough. Although significant viral inhibition was observed as expected initially, viral escape occurred upon long-term culture (Ueda *et al.*, 2016). It was confirmed that initial viral inhibition was indeed due to introduction of indels in up to 30% of the proviral population (Wang *et al.*, 2016b). Further analysis revealed that the NHEJ repair pathway acts as a double-edged sword, generating indels which are lethal and, surprisingly, permit development of viral resistance. This phenomenon was also confirmed by another study showing viral evasion from sgRNA targets in LTR, *gag/pol* and *tat/rev* regions (Wang *et al.*, 2016a). Sequence analysis revealed favorable indels that permitted viral escape. The above findings portend the potential problems CRISPR/Cas9-based strategies will face in the clinic with single-target constructs, thus necessitating multi-target approaches for long-term durable efficacy.

As mentioned above, a major roadblock for a permanent HIV-1/AIDS cure is the existence of a small portion of latently infected cells in the body that do not express any viral proteins. These are invisible to the immune system and thus refractory to immune clearance by cytotoxic T lymphocytes (CTLs). Current approaches are focused on "shock and kill" strategies that aim to first reactivate the virus such that killing can occur by endogenous CTLs or alternatively by administering novel cytocidal agents directed to virus awakened cells. Latency reversing agents (LRAs) comprising small-molecule drugs, such as vorinistat, are being tested in the clinic, but thus far have not shown efficacy in significantly reducing the latent viral reservoir, with non-specific and incomplete viral activation representing the most significant challenges. In this context, the CRISPR/Cas9 system shows considerable promise to activate the latent viral reservoir. Due to its capacity to locate and bind to sgRNA-directed sites, Cas9 can be exploited to reach proviral target sites of interest, such as the transcriptional regulatory regions of the HIV-1 LTR.

A nuclease-deficient dCas9 can be put to use for transcriptional activation when fused to a transcription-activating domain. A number of recent studies explored this concept with promising results (Lin and Klase, 2016; Saayman *et al.*, 2016). As depicted in Figure 26.3, these complexes include: (1) dCas fused to VP64, four tandem copies of VP16 activator from herpes simplex virus, for initiation of transcription via recruitment of TFIIB, TATA binding protein and the CBP/p300 chromatin remodeling complex, (2) dCas conjugated to p300, a histone remodeling protein that can transform chromatinized

DNA to a more permissive state for transcription, (3) dCas–VPR, which consists of VP64, p65, an activator of NFkB, and Rta from EB virus that recruits sp1 and Oct1, (4) a modified guide RNA (sgSAM) with a co-delivery of MS-2-P65-HSF1, allowing the sSAM to recruit dCas and MS-2-P65-HSF1 protein; this complex activates NFkB via p65 and adds the capacity of HSF1 to recruit Sp1, (5) Sun-Tag system with dCas melded to a repeated epitope tag to help recruit 24 copies of an activating complex that consists of GFP and VP64. The above innovative and flexible approaches point to increased specificity and potency compared to agents such as vorinistst. Moreover, combining CRISPR/dCas activator constructs with previously validated LRAs may provide synergy, further increasing the efficiency of viral reactivation from latency (Limsirichai *et al.*, 2016).

25.8 Summary and Future Directions

Based on many recent developments, siRNAs and aptamers show considerable promise as novel drugs for HIV-1 therapies. Efficient *in vivo* delivery of these to HIV-1-susceptible or infected cells and favorable PK profiles with low systemic toxicity for a therapeutic benefit remain challenging. Gene editing technologies have opened up new avenues for HIV-1 treatment with the recent CRISPR/Cas9 system providing a more versatile and less expensive tool. Integrated viral silencing, genome excision and latency activation are important areas currently being pursued, with exciting data emerging. Many questions remain, however, with regard to whether these can be effectively translated into the clinic. Major hurdles that need to be overcome include reaching HIV-1-susceptible and -infected cells, achieving long-lasting therapeutic benefits and eliminating HIV-1 latency by effectively targeting a few latent cells that are sequestered in difficult-to-reach viral sanctuaries sites, including the brain. In this regard, hu-mice permissive to HIV-1 infection will likely play a significant role for testing promising RNA-based and gene editing constructs, thus providing important data for future clinical trials.

Acknowledgement

Work reported here in the Akkina laboratory is supported by NIH grants.

REFERENCES

Akkina R. 2013. New generation humanized mice for virus research: comparative aspects and future prospects. *Virology* **435**: 14–28.

Anderson J, Li MJ, Palmer B, *et al.* 2007. Safety and efficacy of a lentiviral vector containing three anti-HIV genes – CCR5 ribozyme, tat-rev siRNA, and TAR decoy – in SCID-hu mouse-derived T cells. *Mol Ther* **15**: 1182–1188.

Berkhout B. 2004. RNA interference as an antiviral approach: targeting HIV-1. *Curr Opin Mol Ther* **6**: 141–145.

Bobbin ML, Burnett JC, Rossi JJ. 2015. RNA interference approaches for treatment of HIV-1 infection. *Genome Med* **7**: 50.

Choi JG, Dang Y, Abraham S, *et al.* 2016. Lentivirus pre-packed with Cas9 protein for safer gene editing. *Gene Ther* **23**: 627–633.

Hou P, Chen S, Wang S, *et al.* 2015. Genome editing of CXCR4 by CRISPR/cas9 confers cells resistant to HIV-1 infection. *Sci Rep* **5**: 15577.

Kaminski R, Bella R, Yin C, *et al.* 2016a. Excision of HIV-1 DNA by gene editing: a proof-of-concept in vivo study. *Gene Ther* **23**: 696.

Kaminski R, Chen Y, Salkind J, *et al.* 2016b. Negative feedback regulation of HIV-1 by gene editing strategy. *Sci Rep* **6**: 31527.

Keefe AD, Pai S, Ellington A. 2010. Aptamers as therapeutics. *Nat Rev Drug Disc* **9**: 537–550.

Kim SS, Peer D, Kumar P, *et al.* 2010. RNAi-mediated CCR5 silencing by LFA-1-targeted nanoparticles prevents HIV infection in BLT mice. *Mol Ther* **18**: 370–376.

Kumar P, Ban HS, Kim SS, *et al.* 2008. T cell-specific siRNA delivery suppresses HIV-1 infection in humanized mice. *Cell* **134**: 577–586.

Li L, Krymskaya L, Wang J, *et al.* 2013. Genomic editing of the HIV-1 coreceptor CCR5 in adult hematopoietic stem and progenitor cells using zinc finger nucleases. *Mol Ther* **21**: 1259–1269.

Liang C, Wainberg MA, Das AT, Berkhout B. 2016. CRISPR/Cas9: a double-edged sword when used to combat HIV infection. *Retrovirology* **13**: 37.

Limsirichai P, Gaj T, Schaffer DV. 2016. CRISPR-mediated activation of latent HIV-1 expression. *Mol Ther* **24**: 499–507.

Lin A, Klase Z. 2016. A CRISPR approach for reactivating latent HIV-1. *Mol Ther* **24**: 416–418.

Maier DA, Brennan AL, Jiang S, *et al.* 2013. Efficient clinical scale gene modification via zinc finger nuclease-targeted disruption of the HIV co-receptor CCR5. *Hum Gene Ther* **24**: 245–258.

Malecova B, Morris KV. 2010. Transcriptional gene silencing through epigenetic changes mediated by non-coding RNAs. *Curr Opin Mol Ther* **12**: 214–222.

Mussolino C, Morbitzer R, Lutge F, *et al.* 2011. A novel TALE nuclease scaffold enables high genome editing activity in combination with low toxicity. *Nucleic Acids Res* **39**: 9283–9293.

Neff CP, Zhou J, Remling L, *et al.* 2011. An aptamer-siRNA chimera suppresses HIV-1 viral loads and protects from helper CD4(+) T cell decline in humanized mice. *Sci Transl Med* **3**: 66ra6.

Perez EE, Wang J, Miller JC, *et al.* 2008. Establishment of HIV-1 resistance in CD4+ T cells by genome editing using zinc-finger nucleases. *Nat Biotechnol* **26**: 808–816.

Saayman SM, Lazar DC, Scott TA, *et al.* 2016. Potent and targeted activation of latent HIV-1 using the CRISPR/dCas9 activator complex. *Mol Ther* **24**: 488–498.

Saayman S, Roberts TC, Morris KV, Weinberg MS. 2015. HIV latency and the noncoding RNA therapeutic landscape. *Adv Exp Med Biol* **848**: 169–189.

Tebas P, Stein D, Tang WW, *et al.* 2014. Gene editing of CCR5 in autologous CD4 T cells of persons infected with HIV. *N Engl J Med* **370**: 901–910.

Thiel KW, Giangrande PH. 2010. Intracellular delivery of RNA-based therapeutics using aptamers. *Ther Del* **1**: 849–861.

Ueda S, Ebina H, Kanemura Y, Misawa N, Koyanagi Y. 2016. Anti-HIV-1 potency of the CRISPR/Cas9 system insufficient to fully inhibit viral replication. *Microbiol Immunol* **60**: 483–496.

Wang G, Zhao N, Berkhout B, Das AT. 2016a. CRISPR-Cas9 can inhibit HIV-1 replication but NHEJ repair facilitates virus escape. *Mol Ther* **24**: 522–526.

Wang Z, Pan Q, Gendron P, *et al.* 2016b. CRISPR/Cas9-derived mutations both inhibit HIV-1 replication and accelerate viral escape. *Cell Rep* **15**: 481–489.

Whatley AS, Ditzler MA, Lange MJ, *et al.* 2013. Potent inhibition of HIV-1 reverse transcriptase and replication by nonpseudoknot, "UCAA-motif" RNA aptamers. *Mol Ther Nucleic Acids* **2**: e71.

Wheeler LA, Vrbanac V, Trifonova R, *et al.* 2013. Durable knockdown and protection from HIV transmission in humanized mice treated with gel-formulated CD4 aptamer-siRNA chimeras. *Mol Ther* **21**: 1378–1389.

Ye L, Wang J, Beyer AI, *et al.* 2014. Seamless modification of wild-type induced pluripotent stem cells to the natural CCR5Delta32 mutation confers resistance to HIV infection. *Proc Natl Acad Sci USA* **111**: 9591–9596.

Yuan J, Wang J, Crain K, *et al.* 2012. Zinc-finger nuclease editing of human cxcr4 promotes HIV-1 CD4(+) T cell resistance and enrichment. *Mol Ther* **20**: 849–859.

Zhou J, Satheesan S, Li H, *et al.* 2015. Cell-specific RNA aptamer against human CCR5 specifically targets HIV-1 susceptible cells and inhibits HIV-1 infectivity. *Chem Biol* **22**: 379–390.

Zhou J, Swiderski P, Li H, *et al.* 2009. Selection, characterization and application of new RNA HIV gp 120 aptamers for facile delivery of Dicer substrate siRNAs into HIV infected cells. *Nucleic Acids Res* **37**: 3094–3109.

Zhu W, Lei R, Le Duff Y, *et al.* 2015. The CRISPR/Cas9 system inactivates latent HIV-1 proviral DNA. *Retrovirology* **12**: 22.

26 Use of the CRISPR/Cas9 System for Genome Editing of Immune System Cells, Defense Against HIV-1 and Cancer Therapies

Amanda Freire Assis, Cesar Speck Hernandez, Larissa Cotrim-Sousa, Rafaela Martins Felicio, Ernna Hérida Oliveira, Lorena Alves Teixeira, Valeria Valente, Wilson Araújo Silva Jr. and Geraldo A. Passos

26.1 Introduction

It is estimated that there are approximately 20 000–25 000 functional genes in the human genome, of which only 1.5% transcribe messenger RNAs (mRNAs) (www.encodeproject.org/) that are translated into proteins. Another portion of the genome transcribes the non-coding RNAs (ncRNAs) that include microRNAs (miRNAs) and long non-coding RNAs (lncRNAs), among other RNA species (Assis *et al.*, 2014).

These ncRNAs and a large number of still non-annotated mRNAs have increasingly been the focus of research in molecular genetics, especially the role of ncRNAs in the post-transcriptional control of gene expression. Therefore, the functions of many regions of human and other mammals' genomes need to be better characterized (Jia *et al.*, 2015).

To demonstrate the effect of a gene in a given biological process or the effect of mutations in a particular gene and its association with a genetic disease in particular, researchers usually provoke perturbations or even excision of the candidate gene, followed by observation of the resulting consequences. To this end, animal models can be used, usually the mouse, *Mus musculus*, but also other models such as the zebrafish, *Danio rerio*, or even culture of human cells.

To turn off the expression of a gene, the choice method is the generation of knockout animals or cell lines. In simple terms, the "target" gene is disrupted by transfection of embryonic cells with a vector containing a non-functional homologous sequence that physically interrupts the wild-type functional gene by homologous recombination with the non-functional sequence (insertion of a non-functional stretch of DNA; Hall *et al.*, 2009).

We can also add a new feature, such as knockin models, or even generate so-called conditional mutants where only a particular tissue or organ harbors the gene alteration (Hall *et al.*, 2009). The technology for the generation of knockout animals by homologous recombination is widely used, but requires precise micromanipulation of embryonic cells using a sophisticated laboratory platform executed by specialized technicians.

Another strategy that is widely used to perturb gene expression is RNA interference (RNAi), in which cultured cells or even living animals (*in vivo* organ transfection in the mouse or in zebrafish, as examples) are transfected with an RNAi sequence that is complementary to a particular mRNA. Once within the cytoplasm, the RNAi molecules anneal with the target mRNA by base complementarity. The RNAi–mRNA hybrid is then cleaved by the cell's RNA-induced silencing complex (RISC), preventing protein translation.

The RNAi method is easier to perform since it does not require a sophisticated laboratory, but unlike gene knockout in which the gene disruption occurs in the nucleus and its phenotypic effect is complete and permanent, the mode of action of RNAi occurs in the cytoplasm, and its effect is partial and transient, i.e. it is reversible. The RNAi method is currently being evaluated for its use for clinical purposes (Alagia and Eritjia, 2016).

Short hairpin RNA (shRNA) represents another method for gene silencing by an RNAi mechanism. As the expression of shRNA is by means of plasmid vector transfection or lentiviral delivery into the recipient cells, this procedure features a lower rate of degradation and turnover in comparison with the classical naked RNAi transfection method (Burnett *et al.*, 2011).

Recently, the "clustered regularly interspaced short palindromic repeats (CRISPR)/CRISPR-associated" (CRISPR/Cas9) system has emerged as a versatile and efficient new alternative to edit genomes and identify gene function. Compared with other techniques such as RNAi, the CRISPR/Cas9 method allows generating and analyzing knockout or knockin animals or cell lines to determine whether a given region of the genome may regulate the expression of different genes (Xue *et al.*, 2015). Among various uses, the CRISPR/Cas9 system allows the introduction of different types of mutations, such as point mutations, deletions, insertions, inversions, translocations and deletions/duplications, into a specific genomic region (Kato and Takada, 2016; Figure 26.1).

Using the CRISPR/Cas9 system as a new tool for gene "editing" in its broad sense is something that has revolutionized the biological and biomedical sciences in recent years. Since this genetic technology was successfully employed to manipulate the genome of mammalian cells, including human cells, several significant examples of its potential practical use in medicine have emerged (Ahmad and Amiji, 2018; Harrison and Hart, 2017; Hochheiser *et al.*, 2018; Thurtle-Schmidt and Lo, 2018).

In this chapter, we focus on the use of the CRISPR/Cas9 system in the manipulation of immune cells, mainly T and B cells and macrophages, as well as in allograft rejection, elimination of a virus (using HIV-1 as an example) and cancer therapy.

26.2 Using the CRISPR/Cas9 System in Genome Editing of Immune System Cells

26.2.1 Genome Editing of T cells

The CRISPR/Cas9 system has been used to efficiently manipulate the genomes of several mammalian cell lines (Doudna and Charpentier, 2014; Hsu *et al.*, 2014).

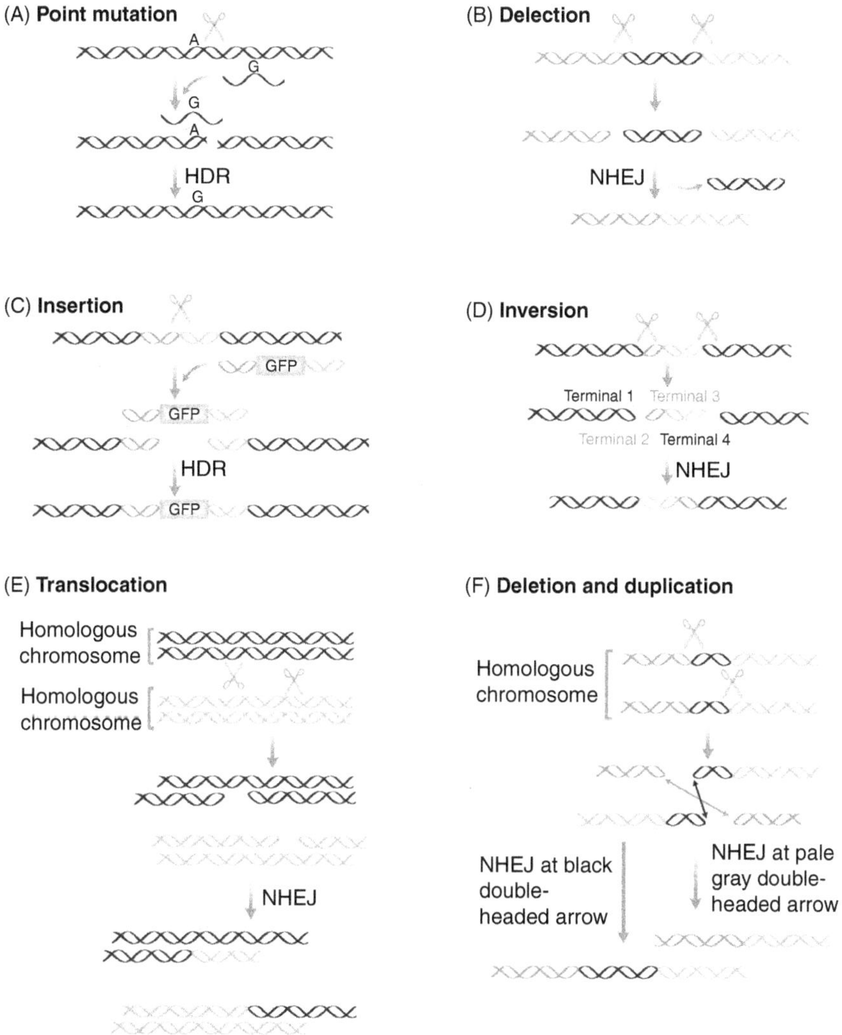

Figure 26.1 Schematic representation of the introduction of mutations by the CRISPR/Cas9 system. (A) Introduction of a point mutation. A double-strand break (DSB) is induced and the lesion is repaired by homology-directed repair (HDR) using as template a designed single-strand oligodeoxynucleotide (ssODN) containing a single altered nucleotide. Apart from substitutions, this method can also be applied for introduction of frameshift mutations through the design of ssODNs consisting of more or fewer nucleotides compared to the sequences to be replaced. (B) Introduction of a deletion. DSBs are induced; a non-homologous end joining (NHEJ) of the deep gray and pale gray fragment termini results in a deletion. (C) Introduction of an insertion. A DSB is induced and the lesion is repaired by HDR using as template a donor containing the fragment of interest flanked by homology arms, resulting in the fragment being inserted at the target genomic locus. (D) Introduction of an inversion. DSBs are induced; NHEJ of the termini of 3 and 2 fragments with those of the 1 and 4 fractions, respectively, causes an inversion. (E) Introduction of a translocation. DSBs are induced at different chromosomes. NHEJ of the termini of the black and pale gray fragments causes a translocation to take place. (F) Introduction of deletion and duplication. DSBs are induced at different positions of each homologous chromosome; NHEJ of the termini of deep gray with the termini of the pale gray fragments leads to deletions (pale gray double-headed arrow), while joining of the two black fragments leads to duplication (black double-headed arrow). Genomic DNA is represented by black or gray double helixes. The positions of gRNA recognition sequences are shown with scissors. ssODN is represented with a single helix. Dashed lines denote homologous recombination. (This figure and caption were taken from Kato T and Takada S. *In vivo* and *in vitro* disease modeling with CRISPR/Cas9. *Briefings in Functional Genomics* 2016, 1–12, with permission from the Oxford University Press to GAP).

However, T cells, specifically T $CD4^+$ cells, are proving difficult to modify by genome editing (Mandal *et al.*, 2014). In an attempt to overcome this difficulty, Schumann *et al.* (2015) transfected human T $CD4^+$ cells with Cas9:single-guide RNA ribonucleoprotein (Cas9RNPs), allowing ablation of CXCR4, a co-receptor used for HIV-1 entry. This procedure caused up to 40% of cells to lose cell surface expression of CXCR4. The results from this work established a new Cas9RNP technology that can be used in experimental and/or therapeutic genome editing of primary human T cells.

Tumor escape from the immune response is a major problem for immunotherapy against cancer. Immunosuppressive factors, i.e. molecules that block or hinder the antitumor immune response, represent an important obstacle for cancer immunotherapy. The cell-associated factor programmed death-ligand 1 (PD-L1) is one of these factors, which is expressed on dendritic cells (DCs) and on some tumor cells (Kim and Chen, 2016). PD-L1 interacts with one of its receptors, programmed death-1 (PD-1) that is expressed on T cells, thus impeding the antitumor T-cell cytotoxic activity (Chen *et al.*, 2012).

Su *et al.* (2016) used the CRISPR/Cas9 system to disrupt via gene knockout the PD-1 receptor in human T cells. In this work, the authors showed that disruption of inhibitory checkpoint gene PD-1 did not affect the viability of the genome-edited cells, which exhibited upregulation of IFN-γ production and enhanced cytotoxicity. This new approach has the potential to improve T-cell-based cancer immunotherapies.

26.2.2 Genome Editing of B Cells for Specific Antibody Production

The technology for antibody production by genome editing of B cells can also benefit from the use of the CRISPR/Cas9 system.

In a recent study (Cheong *et al.*, 2016), the authors adapted this system to induce class-switching recombination in primary mouse B cells and human B-cell lines. They used the system to engineer hybridoma cells to produce monoclonal antibodies with a chosen heavy chain or to secrete just a portion of the immunoglobulin (Ig) molecule, for example, the fragment antigen binding (Fab).

In other very promising work (Chen *et al.*, 2015), the authors showed the possibility of using CRISPR/Cas9 system to manipulate the genome of pigs. The IgM joining heavy (J_H) region that is crucial for B-cell development and differentiation was targeted in porcine fetal fibroblast cells, thereby inducing knockout within the IgM locus. They used nuclear transfer technology to produce piglets with a biallelic IgM heavy chain gene mutation; these animals showed no antibody-producing B cells due to inactivated Ig genes. These animals are preferred for the production of human polyclonal antibodies because then they are suitably engineered for this purpose.

26.3 Controlling Excessive Inflammation Through Genome Editing of Macrophages

Although macrophages play important roles in the early innate immune response by recognizing and phagocytozing invading pathogens, these cells also constitute key elements of undesirable inflammation.

The innate immune response mediated by macrophages is dependent on pattern recognition receptors (PRRs) and toll-like receptors (TLRs), which stimulate cells to detect and eliminate invading pathogens (Kaway and Akira, 2010; Kondo *et al.*, 2012). During this process, downstream signaling pathways are activated for the production of proinflammatory cytokines and type I interferons, which may present a potential problem. Depending on the degree of pathogen invasion, the host immune cells respond strongly and cause damaging inflammation through a mechanism that is not fully understood.

Researchers used the CRISPR/Cas9 system to turn off key genes associated with the exacerbation of inflammatory process by macrophages. For example, Tang *et al.* (2014), showed that Ras guanine nucleotide releasing protein 3 (RasGRP3) limits the inflammatory response by activating Rap1 during low-intensity pathogen infection, setting a threshold for preventing an excessive inflammatory response. Using CRISPR/Cas9, these authors showed that RasGRP3 limits the inflammatory response by activating Rap1 during low-intensity pathogen infection, setting a threshold for preventing an excessive inflammatory response.

The antiviral response also needs to be strictly controlled to avoid excessive inflammation. Xia *et al.* (2015) used CRISPR/Cas9 to turn off poly(rC) binding protein 2 (PCBP2), which is sumoylated by insulin receptor tyrosine kinase substrate (IRTKS) during RNA virus infection. They demonstrated that IRTKS functions as a negative modulator of excessive inflammation during viral infection.

26.4 Controlling Allograft Rejection

Transplant rejection represents another major obstacle in medicine, even with selection for HLA-matching donors and treatment with immunosuppressive drugs. Allograft rejection is mediated by host cytotoxic T $CD8^+$ cells that target graft MHC class I molecules. When activated by class II MHC molecules, the host T $CD4^+$ effector memory cells promote crucial help for the occurrence of this process (Abrahimi *et al.*, 2016). These authors showed that siRNA-mediated knockdown or CRISPR/Cas9-induced annulment of class II MHC molecules on endothelial cells impedes T $CD4^+$ effector memory cells from helping T $CD8^+$ effector memory cells develop into alloreactive cytotoxic T cells *in vitro*. In addition, they observed that when microvessels originating from CRISPR/Cas9-edited endothelial cells lacking class II MHC molecules were implanted, these structures were significantly protected from T $CD8^+$ cell-mediated cytotoxicity *in vivo*. This finding may provide new immunotherapeutic insights for controlling allograft rejection.

26.5 Elimination of the HIV-1 Genome from Reservoir T Cells

One of the applications of the CRISPR/Cas9 system that may produce a greater impact is the possibility of its use against HIV-1 infection.

When acquired immunodeficiency syndrome (AIDS) was recognized to originate from an infection with a retrovirus, human immunodeficiency virus (HIV-1) (Barré-Sinoussi *et al.*, 1983; Gallo *et al.*, 1983), most researchers and physicians suspected that there were no restrictions for infection to occur once a person is exposed to this virus.

In fact, once infected with HIV-1, the patient should be strictly followed and controlled. Although effective in controlling the viral load in infected individuals, antiretroviral therapy (ART) does not eliminate the HIV-1 proviral genome, which is inserted into the genome of latently infected CD4$^+$ T cells in patients. The integrated proviral genome persists and replicates during cell mitoses. It can be reactivated to produce new HIV-1 particles upon interruption of ART, and host T cells are activated (Chun *et al.*, 1997; 2003; Finzi *et al.*, 1997; Hermankova *et al.*, 2003; Siliciano *et al.*, 2003; Wong *et al.*, 1997).

The entry of HIV-1 into T cells is mediated by its surface GP 120 glycoprotein that binds to the cell CD4 receptor in parallel to the chemokine receptor CCR5 (R5-tropic) or CXCR4 (X4-tropic). The CCR5 receptor, which is expressed in lymphocytes, myeloid cells or CD4$^+$ T-cell subsets, is implicated in new infections during the chronic phase of HIV-1 infection. In individuals in which HIV-1 infection is established, the virus can use CXCR4 as an alternative receptor. Because of this, both the CCR5 and CXCR4 receptors represent important targets for therapies based on genome editing in the context of HIV-1 infection (Hou *et al.*, 2015).

It was in 1996 that researchers confirmed that a 32 bp deletion in the CCR5 co-receptor gene (Δ32ccr5) conferred on the carriers the phenotype of HIV-1 resistance (Liu *et al.*, 1996; Samson *et al.*, 1996). People with the Δ32ccr5/Δ32ccr5 homozygous genotype do not contract infection, even if exposed to the virus. This finding altered the concept that the human condition per se was sufficient for HIV-1 susceptibility; now, a subtle genetic variation imparted resistance to the virus.

The greater insight that the Δ32ccr5 allele polymorphism discovery brought in the context of AIDS research is that susceptibility to HIV-1 is influenced by the genetic constitution of the individual. The proof-of-concept for the protective effect of the Δ32ccr5 allele against HIV-1 infection comes from the case known as "The Berlin Patient," the first case in which a patient was cured from HIV-1 infection. A man who suffered from leukemia was treated with a bone marrow transplantation in 2007. However, this patient had concurrent HIV-1 infection, and the physicians decided to use bone marrow cells from an HLA-identical donor with the Δ32ccr5/Δ32ccr5 homozygous genotype (Hütther *et al.*, 2009). Two months after surgery, the HIV-1 viremia fell to undetectable levels, which persisted for a period of at least two years with no ART.

Nevertheless, as the frequency of the Δ32ccr5 allele, a characteristic of individuals with Caucasian European background, is low among different human populations, the susceptibility to HIV-1 infection remains highly prevalent.

As mentioned above, the Δ32ccr5/Δ32ccr5 homozygous genotype confers resistance to HIV-1 infection. In fact, the disruption of the CCR5 or CXCR4 genes by the use of a zinc-finger nuclease (ZFN) system resulted in inhibition of HIV-1 infection of T CD4$^+$ cells (Didigu *et al.*, 2014; Tebas *et al.*, 2014).

Transcription activator-like effector nucleases (TALENs) and CRISPR/Cas9 were successfully used to disrupt CCR5 in pluripotent stem cells and hematopoietic stem cells (Kang *et al.*, 2015; Mandal *et al.*, 2014; Ye *et al.*, 2014). Recently, CRISPR/Cas9 was used to introduce CXCR4 loss-of-function mutations into human T cells (Hou *et al.*, 2015), which rendered the genome-edited cells resistant to HIV-1 infection.

HIV-1-infected memory CD4$^+$ T cells are considered the main latently infected cell pool and therefore should be the focus of research directed at the elimination of the integrated HIV-1 genome (Kaminski *et al.*, 2016). Researchers have attempted to reactivate the proviral DNA in these cells to destroy HIV-1 through the use of ART drugs or immunotherapy.

However, due to the inherent limitations of these approaches, the strategy should be geared towards the elimination of the proviral genome from most cells infected with HIV-1 (Kaminski *et al.*, 2016). These researchers modified the CRISPR/Cas9 system such that it could recognize specific HIV-1 proviral DNA sequences positioned within the HIV-1 promoter spanning the 5′ long terminal sequence (5′ LTR; Hu *et al.*, 2014; Khalili *et al.*, 2015).

In a recent work (Kaminski *et al.*, 2016), the authors demonstrated the ability of the modified CRISPR/Cas9 to excise integrated copies of the proviral DNA from a latently HIV-1-infected human T-lymphoid cell line. These cells were unable to produce new viral particles, even after treatment with trichostatin A and vorinostat, which are drugs that inhibit the enzyme histone deacetylase and lead to the reactivation of latent viruses in cell lines. Moreover, they observed that CRISPR/Cas9 reduced viral replication when CD4$^+$ T cells were infected with HIV-1 in culture.

These results indicate that the strategy of genome editing by CRISPR/Cas9 has a promising potential for the eradication of HIV-1 from the reservoir cells and the prevention of the re-emergence of AIDS.

However, depending on the target sequence of the proviral DNA, which must be precisely chosen against essential viral sequences, the non-homologous end joining (NHEJ) mechanism can influence this process and can create HIV-1 escape variants that are refractory to CRISPR/Cas9. For this reason, researchers currently consider that CRISPR/Cas9 might be more efficient as an antiviral than a therapeutic in the case of HIV-1 infection (Wang *et al.*, 2016).

26.6 CRISPR/Cas9 in Cancer

Numerous large-scale sequencing efforts have generated a huge catalog of genetic alterations present in different types of human cancers, which illustrates the complexity of this multi-step disease. In this scenario, the landscape of characterized mutations makes the distinction between passenger and driver mutations, the latter being directly or indirectly implicated in the transformation of normal cells into cancer cells. Therefore, experimental advances to manipulate genomes are fundamental for the systematic analysis of the many genes involved in this process. Until recently, a majority of studies that dissected the function of oncogenes and tumor-suppressor genes was based on cDNA overexpression and/or RNAi knockdown. Although these approaches have permitted important discoveries, they present various limitations, such as the production of artifactual phenotypes due to non-physiological expression levels, uncertainty of silencing efficiency and considerable off-target effects. The CRISPR/Cas9 system emerged as a revolutionary tool for functional studies in the multi-faceted context of cancer cells. Considering its simple and accurate ability for targeted alteration of endogenous loci, this technology allows precise multi-step genome editing that fits perfectly for the study of cancer biology and certainly will allow not only the identification of driver mutations, but also the refinement of protocols for the development of novel therapies.

Several studies have demonstrated the power of CRISPR-mediated engineering to generate cells or mice harboring multiple combinations of gain-of-function and loss-of-function mutations that encompass the genetic diversity of the cancer cell genome (reviewed in Sánchez-Rivera and Jacks, 2015). This technology can be efficiently applied for genome editing in somatic cells, and germline and embryonic stem (ES) cells, allowing the establishment of new models for *ex vivo* and *in vivo* studies, with cell-based allograft/xenografts or genetically engineered mouse models (GEMMs), and non-germline GEMMs (nGEMMs).

Ex vivo CRISPR-based genome editing was successfully used in somatic cells of hematopoietic malignancies to evaluate the impact of specific genetic alterations in the phenotype of cancer cells (Chen *et al.*, 2014; Heckel *et al.*, 2014; Malina *et al.*, 2013). For example, Trp53 gene editing in $Arf^{-/-}$Eμ-*Myc* lymphoma cells, which were subsequently transplanted into syngeneic mice, revealed a considerable increase in doxorubicin resistance associated with p53 deficiency (Malina *et al.*, 2013). The flexibility of the CRISPR/Cas9 system was also exploited to perform high-throughput screens in engineered somatic cells for the identification of genes involved in specific biological phenotypes, as well as vulnerabilities and resistance mechanisms of cancer cells. Using libraries with tens of thousands of single-guide RNAs to generate complex populations of edited cells has already permitted the identification of essential genes for both cancer cell lines and pluripotent stem cells, and also novel candidates mediating resistance to vemurafenib in melanoma cells (Shalem *et al.*, 2014). Additionally, the adapted death Cas9-based activators and repressors can be used to perform complementary genome-wide screens of gene activation and repression (Gilbert *et al.*, 2014). Both approaches would allow great advances in the characterization of novel drug targets, synthetic lethality and chemotherapeutic modulators.

In vivo experiments using this technology have allowed an even more elegant analysis using plasmids (Xue *et al.*, 2014) or lentiviral (Sanchez-Rivera *et al.*, 2014) vectors as delivery systems to generate somatic mouse models of cancer. Release of plasmids encoding Cas9 and sgRNAs targeting *Pten* and *Trp53* tumor-suppressor genes in the hepatocytes of wild-type mice was able to recapitulate liver tumors of *Pten*$^{fl/fl}$*Trp53*$^{fl/fl}$ GEMMs, supporting the equivalence of these models. It was also demonstrated that gain-of-function mutations exploiting the same strategy are feasible (Xue *et al.*, 2014). More sophisticated manipulation using a lentivirus simultaneously encoding CRISPR components and Cre recombinase was used to edit various tumor-suppressor genes into *Kras*$^{LSL-G12D/+}$ or *Kras*$^{LSL-G12D/+}$*Trp53*$^{fl/fl}$ mice. Intratracheal delivery of these lentiviruses resulted in lung adenocarcinomas with histopathological and molecular features varying according to the tumor-suppressor gene targeted (Sanchez-Rivera *et al.*, 2014). Another study has also shown the availability of using CRISPR/Cas9 for modeling large chromosomal rearrangements. Animals manipulated with adenovirus designed to induce an *Eml4–Alk* (echinoderm microtubule-associated protein-like 4–anaplastic lymphoma kinase) inversion developed tumors with complete penetrance and showed crizotinib sensitivity, similar to human lung tumors harboring this aberration (Cong *et al.*, 2013).

Regarding therapeutic genetic correction, the development of the CRISPR/Cas9 technology in eukaryotic cells (Mali *et al.*, 2013a; 2013b; Reardon, 2016) brought new light to a more efficient therapy against cancer cells. Two clinical trial proposals using CRISPR/Cas9 for the treatment of cancer were announced simultaneously in June 2016 by two groups of researchers, one at the University of Pennsylvania in Philadelphia, USA (Reardon, 2016) and the other at Sichuan University in Chengdu, China (Cyranoski, 2016). The central idea of the two strategies was based on the reprogramming of T cells of the immune system to destroy the tumor cells efficiently (Figure 26.2). The protocol involves obtaining T cells from patients with different types of cancer and performing three genetic modifications: (1) inserting the receptor protein NY-ESO-1, a specific antigen of tumor cells, directing the T cells to attack the tumor, (2) removing the natural T-cell receptor, so that no interference occurs, thus ensuring the efficiency of the treatment and (3) removing the PD-1 protein gene, which negatively regulates T-cell response, enhancing the killing of tumor cells. The success of this therapeutic approach is a big step towards winning the fight against cancer.

26.7 Concluding Remarks

In this chapter, we focused on results that actually serve as proof-of-concept for the use of the CRISPR/Cas9 system in potential applications in biomedical research and medicine. Using this genetic engineering approach for genome editing of immune system cells such as T cells, B cells and macrophages, it was shown that it is possible to enhance the cytotoxicity of T cells, favoring cancer immunotherapies, to engineer B cells for chimeric antibody production, to treat cancer and, finally, to target proviral DNA and/or HIV-1 receptors, which has important implications in AIDS treatment. The CRISPR/Cas9 system is currently causing excitement in biomedicine that is comparable to the beginning of recombinant DNA, gene therapy or stem cell research.

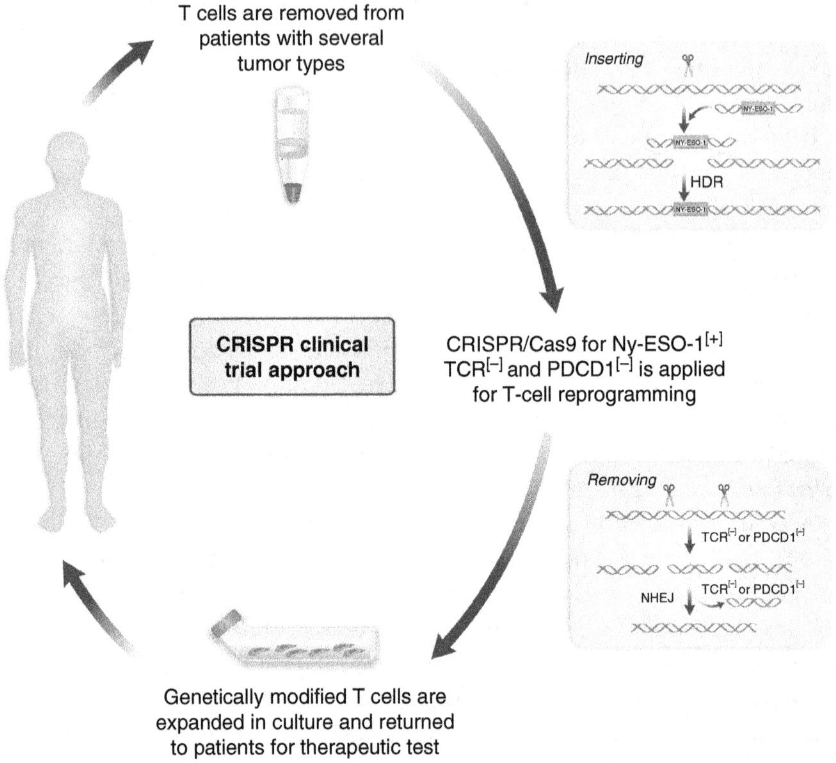

Figure 26.2 Schematic representation of the use of CRISPR/Cas9 system for cancer treatment. The protocol is based on reprograming of T cells of the immune system to destroy the tumor cells. The T cells from patients with different types of cancer are subjected to three genetic alterations: (1) inserting the receptor protein NY–ESO–1, a specific antigen of tumor cells, directing the T cells to attack the tumor, (2) removing the natural T-cell receptor, so that no interference occurs, thus ensuring the efficiency of the treatment and (3) removing the PD-1 protein gene, which negatively regulates T-cell response, enhancing the killing of tumor cells. (A black-and-white version of this figure will appear in some formats. For the color version, please refer to the plate section.)

Acknowledgements

Our laboratories are funded by Fundação de Amparo à Pesquisa do Estado de São Paulo (FAPESP, São Paulo, Brazil) (Grant # 13/17481–1 to GAP and Grant # 13/08135–2 to WASJ), Conselho Nacional de Desenvolvimento Científico e Tecnológico (CNPq, Brasília, Brazil) (Grant # 306315/2013–0 to GAP) and Coordenação de Aperfeiçoamento de Pessoal de Nível Superior (CAPES, Brasília, Brazil) (Grant # 88881.068105/2014–01 to GAP).

REFERENCES

Abrahimi P, Qin L, Chang WG, *et al.* 2016. Blocking MHC class II on human endothelium mitigates acute rejection. *JCI Insight* **1**: e85293.

Ahmad G, Amiji M. 2018. Use of CRISPR-Cas9 gene-editing tools for developing models in drug discovery. *Drug Discov Today*. Doi: 10.1016/j.drudis. 2018.01.04.

Alagia A, Eritja R. 2016. siRNA and RNAi optimization. *Wiley Interdiscipl Rev RNA* **7**: 316–329.

Assis AF, Oliveira EH, Donate PB, *et al.* 2014. What is the transcriptome and how is it evaluated? In Passos GA, ed., *Transcriptomics in Health and Disease*, Basel, Switzerland: Springer International Publishing, 344 pp.

Barré-Sinoussi F, Chermann JC, Rey F, *et al.* 1983. Isolation of a T-lymphotropic retrovirus from a patient at risk for acquired immune deficiency syndrome (AIDS). *Science* **220**: 868–871.

Burnett JC, Rossi JJ, Tiemann K. 2011. Current progress of siRNA/shRNA therapeutics in clinical trials. *Biotechnol J* **6**: 1130–1146.

Chen C, Liu Y, Rappaport AR, *et al.* 2014. MLL3 is a haploinsufficient 7q tumor suppressor in acute myeloid leukemia. *Cancer Cell* **25**: 652–665.

Chen DS, Irving BA, Hodi FS. 2012. Molecular pathways: next-generation immunotherapy-inhibiting programmed death-ligand 1 and programmed death 1. *Clin Cancer Res* **18**: 6580–6587.

Chen F, Wang Y, Yuan Y, *et al.* 2015. Generation of B cell-deficient pigs by highly efficient CRISPR/Cas 9-mediated gene targeting. *J Genet Genom* **42**: 437–444.

Cheong T-C, Compagno M, Chiarle R. 2016. Editing of mouse and human immunoglobulin genes by CRISPR-Cas 9 system. *Nat Comm* **7**: 10934.

Chun TW, Carrut L, Finzi D, *et al.* 1997. Quantification of latent tissue reservoirs and total body viral load in HIV-1 infection. *Nature* **387**: 183–188.

Chun TW, Justement JS, Lempicki RA, *et al.* 2003. Gene expression and viral production in latently infected, resting CD4+ T cells in viremic versus anemic HIV-infected individuals. *Proc Natl Acad Sci USA* **100**: 1908–1913.

Cong L, Ran FA, Cox D, *et al.* 2013. Multiplex genome engineering using CRISPR/Cas systems. *Science* **339**: 819–823.

Cyranoski D. 2016. First trial of CRISPR in people: Chinese team approved to test gene-edited cells in people with lung cancer. *Nature* **535**: 476–477.

Didigu CA, Wilen CB, Wang J. 2014. Simultaneous zinc-finger nuclease editing of the HIV-1 coreceptors CCR5 and CXCR4 protects CD4+ T cells from HIV-1 infection. *Blood* **123**: 61–69.

Doudna JA, Charpentier E. 2014. Genome editing: the new frontier of genome engineering with CRISPR-Cas 9. *Science* **346**: 1258096.

Finzi D, Hermankova M, Pierson T, *et al.* 1997. Identification of a reservoir for HIV-1 in patients in highly active antiretroviral therapy. *Science* **278**: 1295–1300.

Gallo RC, Sarin PS, Gelmann EP, *et al.* 1983. Isolation of human T-cell leukemia virus in acquired immune deficiency syndrome. *Science* **220**: 865–867.

Gilbert LA, Horlbeck MA, Adamson B, *et al.* 2014. Genome-scale CRISPR-mediated control of gene repression and activation. *Cell* **159**: 647–661.

Hall B, Limaye A, Kulkarni AB. 2009. Overview: generation of gene knockout mice. *Curr Protoc Cell Biol* Chapter 19: Unit 19.12.19.12.1–17.

Harrison PT, Hart S. 2017. A beginner's guide to gene editing. *Exp Physiol.* Doi: 10.1113/EP086047.

Heckl D, Kowalczyk MS, Yudovich D, *et al.* 2014. Generation of mouse models of myeloid malignancy with combinatorial genetic lesions using CRISPR-Cas9 genome editing. *Nat Biotechnol* **32**: 941–946.

Hermankova M, Siliciano JD, Zhou Y, *et al.* 2003. Analysis of human immunodeficiency virus type 1 gene expression in latently infected resting CD4+ T lymphocytes *in vivo*. *J Virol* **77**: 7388–7392.

Hochheiser K, Kueh AJ, Gebhardt T, *et al.* 2018. CRISPR-Cas9: a tool for immunological research. *Eur J Immunol.* Doi: 10.1002/eji.201747131.

Hou W, Fang C, Liu J, *et al.* 2015. Molecular insights into the inhibition of HIV-1 infection using a CD4 domain-1-specific monoclonal antibody. *Antivir Res* **122**: 101–111.

Hsu PD, Lander ES, Zhang F. 2014. Development and application of CRISPR-Cas9 for genome engineering. *Cell* **157**: 1262–1278.

Hu W, Kaminski R, Yang F. *et al.* 2014. RNA-directed gene editing specifically eradicates latent and prevents new HIV-1 infection. *Proc Natl Acad Sci USA* **111**: 11461–11466.

Hütther G, Nowak D, Mossner M, *et al.* 2009. Long-term control of HIV by CCR5 delta32/delta32 stem-cell transplantation. *N Engl J Med* **360**: 692–698.

Jia Y, Chen L, Ma Y, *et al.* 2015. To know how a gene works, we need to redefine it first but then, more importantly, to let the cell itself decide how to transcribe and process its RNAs. *Int J Biol Sci* **11**: 1413–1423.

Kaminski R, Chen Y, Fischer T, *et al.* 2016. Elimination of HIV-1 genomes from human T-lymphoid cells by CRISPR/Cas 9 gene editing. *Sci Rep* **6**: 22555.

Kang H, Minder P, Park MA, *et al.* 2015. CCR-5 disruption in induced pluripotent stem cells using CRISPR/Cas 9 provides selective resistance of immune cells to CCR5-tropic HIV-1 virus. *Mol Ther Nucleic Acids* **4**, e268.

Kato T, Takada S. 2016. *In vivo* and *in vitro* disease modeling with CRISPR/Cas 9. *Brief Funct Genom* **2016**: 1–12.

Kawai T, Akira S. 2010. The role of pattern-recognition receptors in innate immunity: update on toll-like receptors. *Nat Immunol* **11**: 373–384.

Khalili K, Kaminski R, Gordon J, *et al.* 2015. Genome editing strategies: potential tools for eradicating HIV-1/AIDS. *J Neurovirol* **21**: 310–321.

Kim JM, Chen DS. 2016. Immune escape to PD-L1/PD-1 blockade: seven steps to success (or failure). *Ann Oncol* **27**: 1492–1504.

Kondo T, Kawai T, Akira S. 2012. Dissecting negative regulation of toll-like receptor signaling. *Trends Immunol* **33**: 449–458.

Li C, Guan X, Jin W, *et al.* 2015. Inhibition of HIV-1 infection of primary CD4+ T-cells by gene editing of CCR5 using adenovirus-delivered CRISPR/Cas 9. *J Gen Virol* **96**: 2381–2393.

Liu R, Paxton WA, Choe S, *et al.* 1996. Homozygous defect in HIV-1 coreceptor accounts for resistance of some multiply-exposed individuals to HIV-1 infection. *Cell* **86**: 367–377.

Mandal PK, Ferreira LM, Collins R, *et al.* 2014. Efficient ablation of genes in human hematopoietic stem and effector cells using CRISPR/Cas 9. *Cell Stem Cell* **15**: 643–652.

Mali P, Aach J, Stranges PB, *et al.* 2013a. CAS9 transcriptional activators for target specificity screening and paired nickases for cooperative genome engineering. *Nat Biotechnol* **31**: 833–838.

Mali P, Yang L, Esvelt KM, *et al.* 2013b. RNA-guided human genome engineering via Cas9. *Science* **339**: 823–826.

Malina A, Mills JR, Cencic R, *et al.* 2013. Repurposing CRISPR/Cas9 for in situ functional assays. *Genes Devel* **27**: 2602–2614.

Reardon S. 2016. First CRISPR clinical trial gets green light from US panel: the technique's first test in people could begin as early as the end of the year. *Nature News*, 22 June. Doi:10.1038/nature.2016.20137.

Samson M, Libert F, Doranz BJ, *et al.* 1996. Resistance to HIV-1 infection in Caucasian individuals bearing mutant alleles of the CCR-5 chemokine receptor gene. *Nature* **382**: 722–725.

Sánchez-Rivera FJ, Jacks T. 2015. Applications of the CRISPR-Cas9 system in cancer biology. *Nat Rev Cancer* **15**: 387–395.

Sánchez-Rivera FJ, Papagiannakopoulos T, Romero R, *et al.* 2014. Rapid modelling of cooperating genetic events in cancer through somatic genome editing. *Nature* **516**: 428–431.

Schumann K, Lin S, Boyer E, *et al.* 2015. Generation of knock-in primary human T cells using Cas 9 ribonucleoproteins. *Proc Natl Acad Sci USA* **112**: 10437–10442.

Shalem O, Sanjana NE, Hartenian E, *et al.* 2014. Genome-scale CRISPR-Cas9 knockout screening in human cells. *Science* **343**: 84–87.

Siliciano JD, Kajdas J, Finzi D, *et al.* 2003. Long-term follow up studies confirm the stability of the latent reservoir for HIV-1 in resting CD4+ T cells. *Nat Med* **9**: 727–728.

Su S, Hu B, Shao J, *et al.* 2016. Crispr-Cas 9 medaited efficient PD-1 disruption on human primary T cells from cancer patients. *Sci Rep* **6**: 20070.

Tang S, Chen T, Yu Z, *et al.* 2014. RasGRP3 limits toll-like receptor-triggered inflammatory response in macrophages by activating Rap1 small GTPase. *Nat Comm* **5**: 4657.

Tebas P, Stein D, Tang W, *et al.* 2014. Gene editing of CCR5 in autologous CD4 T cells of persons infected with HIV-1. *N Engl J Med* **370**: 901–910.

Thurtle-Schmidt DM, Lo TW. 2018. Molecular biology at the cutting edge: a review on CRISPR-Cas9 gene editing for undergraduates. *Biochem Mol Biol Educ.* Doi: 10.1002/bmb.21108.

Wang G, Zhao N, Berkhout B, *et al.* 2016. CRISPR-Cas9 can inhibit HIV-1 replication but NHEJ repair facilitates virus escape. *Mol Ther* **24**: 522–526.

Wong JK, Hezaret M, Günthard HF, *et al.* 1997. Recovery of replication-competent HIV despite prolonged suppression of plasma viremia. *Science* **278**: 1291–1295.

Xia P, Wang S, Xiang Z, *et al.* 2015. IRTKS negatively regulates antiviral immunity through PCBP2 sumoylation-mediated MAVS degradation. *Nat Comm* **6**: 8132.

Xue HY, Ji LJ, Gao AM, *et al.* 2015. CRISPR-Cas9 for medical genetic screens: applications and future perspectives. *J Med Genet* **53**: 91–97.

Xue W, Chen S, Yin H, *et al.* 2014. CRISPR-mediated direct mutation of cancer genes in the mouse liver. *Nature* **514**: 380–384.

Ye L, Wang J, Beyer AI, *et al.* 2014. Seamless modification of wild-type induced pluripotent stem cells to the natural CCR5Δ32 mutation confers resistance to HIV infection. *Proc Natl Acad Sci USA* **111**: 9591–9596.

27 Harnessing the Therapeutic Potential of Long Non-coding RNAs in Immunity

Wendy J.M. Huang

27.1 Introduction

Long non-coding RNAs (lncRNAs) are involved in fundamental biological processes, such as cell growth, lineage determination, chromatin remodeling and gene expression (Maass *et al.*, 2014). They regulate gene expression at the levels of transcription, RNA processing and translation. Most lncRNAs operate by forming complexes with protein partners and some use base pairing to selectively bind and act on other nucleic acids (Cech and Steitz, 2014). Despite growing appreciation for their involvement in diverse homeostatic processes, our knowledge of lncRNAs' expressions and functions in host immunity, as well as their contributions to immunological diseases in humans has only started to be evaluated recently (Atianand and Fitzgerald, 2014). In this chapter, we will explore the recent findings on lncRNAs in both innate and adaptive immunity. Following these advances, we will discuss ways of targeting lncRNAs with the latest genome editing technology to harness their therapeutic potential towards the treatment of human immunological diseases.

27.2 Long Non-coding RNAs in Innate Immunity

The innate immune system is the first line of defense against foreign infectious agents. Cell surface receptors, such as chemokine receptors that sense environmental cues and toll-like receptors that sense pathogen-associated molecular patterns and products of tissue damage, play key roles in initiating local immune responses. These receptors act through complex signal transduction cascades, changing expressions of both coding and non-coding RNAs in a host cell. In response to interferon receptor activation upon SARS-CoV infection, specific interferon-receptor-dependent lncRNA clusters exhibit over 1.4-fold change in expression in lung tissues of infected mice (Peng *et al.*, 2010). In macrophages, stimulation with TLR2 agonist induces over 62 lncRNAs in a MyD88- and NFkB-dependent manner (Carpenter *et al.*, 2013). Researchers have just begun to identify numerous intriguing molecular mechanisms of how these dynamically regulated lncRNAs control host innate immunity.

Intracellular NFkB pathway is a major hub for integrating extracellular signals to drive cellular responses in the immune system. This pathway is well appreciated as highly regulated at the level of protein–protein interactions and post-translational modifications. LncRNAs are recently found to provide an additional layer of feedback mechanism in tightly controlling NFkB signal strength. Two negative regulators of NFkB, LncRNAs Lethe and Malat1 (Rapicavoli *et al.*, 2013; Zhao *et al.*, 2016), interact with the RelA subunit of NFkB to prevent its DNA-binding activity, consequently decreasing target gene expressions in the nucleus. In macrophages, LncRNA PACER, p50-associated Cox2 extragenic RNA, is a regulator of Cox2 (PTGS2) expression. There, PACER directly interacts with p50, the inhibitory component of NFkB complex and prevents it from binding to the promoter of the Ptgs2 gene in *cis* (Krawczyk and Emerson, 2014). In macrophages, LncRNA-Cox2 is induced by TLR4 agonist stimulation. There, it modulates the binding of NFkB subunits to the SWI/SNF complex, promoting SWI/SNF-associated chromatin remodeling and transactivation of late inflammatory gene programs (Hu *et al.*, 2016).

Other lncRNAs act as positive regulators during immune responses. Virus infections, as well as Poly I:C stimulation of TLR3, induce lncRNA NEAT1v2 expression. NEAT1 induction relocates SFPQ, a transcription repressor, from gene promoter to paraspeckles. Removal of transcription repressor facilitates the transcriptional activation of antiviral genes, including cytokine IL8 (Imamura *et al.*, 2014). In addition, lncRNAs also recruit transcription co-activators to turn on target gene expression. LncRNA NeST controls susceptibility to Theiler's virus and *Salmonella* infection in mice by recruiting WDR5 to deposit H3K4me3 marks, driving enhanced antiviral IFN-γ expression at the transcription level (Gomez *et al.*, 2013). Another lncRNA, the Natural Antisense Transcript (AS-IL1α), localizes in the nucleus and acts on its neighboring gene IL-1α promoter by recruiting RNA polymerase II (Chan *et al.*, 2015).

In addition to controlling effector functions of innate immune cells during infectious challenges, lncRNAs are required for differentiation and lineage commitment in hematopoiesis. Lnc-DC is required for the differentiation of dendritic cells. Biochemistry pull-down assays using biotinylated lnc-DC followed by mass spectrometry (MS) identified STAT3 (signal transducer and activator of transcription 3), as a lnc-DC-interacting protein. During DC differentiation, binding of lnc-DC to STAT3 protects STAT3 dephosphorylation by SHP1. When DCs are depleted of lnc-DC, reduced STAT3 phosphorylation significantly impacts the ability of DCs to activate T cells from the adaptive immune system (Wang *et al.*, 2014).

27.3 Long Non-coding RNAs in Adaptive Immunity

Together with the innate immune system, adaptive immunity provides critical immunological memory for protecting the host against recurrent infections. T cells develop in the thymus, where precursors undergo positive and negative selection, and commitment towards specific CD4$^+$ or CD8$^+$ lineage. There, hundreds of lncRNAs exhibit dynamic expression patterns at the various developmental stages (Xia *et al.*, 2014). After egressing from the thymus, mature naïve T cells circulate in the periphery, surveying for antigens presented by cells such as DCs. The recognition of antigen presented on the surface MHC molecules of DCs triggers T-cell receptor activation and intracellular signaling. Dephosphorylation and nuclear import of NFAT is a key step for relaying TCR activation signals to regulate cellular responses in the nucleus. In the cytoplasm of CD8T cells, lncRNA *NRON* containing IQGAP protein complex interacts with NFAT and serves as its negative regulator. Depletion of *NRON* increases expression of NFAT-dependent cytokines, including IFNg (Pang *et al.*, 2009; Sharma *et al.*, 2011). In CD8T cells, another lncRNA, lncRNA-CD244, is a nuclear chromatin regulator that recruits polycomb protein enhancer of zeste homolog 2 (EZH2) that mediates H3K27 trimethylation on *infg* and *tnfa* locus to maintain repressive chromatin states and inhibits target gene expression (Wang *et al.*, 2015).

For naïve CD4 T cells, TCR activation in combination with additional cytokine stimulation promotes further differentiation into various T helper subsets, including the classical Th1, Th2, Th17 and T regulatory cells, destined to carry distinct immune modulatory functions in the local tissues. A Th2-specific lncRNA, *lncR-Ccr2-5′AS*, controls expression of several chemokine receptor genes (*Ccr1, Ccr3, Ccr2* and *Ccr5*) on the same genomic locus required for promoting cell migration to the lungs *in vivo* (Hu *et al.*, 2013). In Th1 cells, lnc-MAF-4 is abundantly expressed and associated with the chromatin modifiers LSD1 and EZH2. Together this RNA–protein complex orchestrates long-distance interaction between the *linc-MAF-4* and *MAF* genomic locus to mediate *MAF* silencing (Ranzani *et al.*, 2015). Consistent with this, downregulation of lnc-MAF-4 results in T-cell differentiation skewing away from Th1 towards the Th2 phenotype.

27.4 LncRNAs in Immune Diseases

Recent studies have furthered our appreciation of the contribution of lncRNAs to autoimmune and inflammatory disease conditions in human as well as mouse models of human diseases. In one early study, researchers observed that more than 85 lncRNAs are significantly modulated upon anti-IL6 and TNFa therapy in *ex vivo* CD14$^+$ monocytes from human patients with rheumatoid arthritis (Muller *et al.*, 2014). In particular, lncRNA Gas5, glucocorticoid receptor non-coding RNA repressor, expression levels were altered in the patients with autoimmune pathologies including rheumatoid arthritis, systemic lupus erythematosus, multiple sclerosis or sarcoidosis (Mayama *et al.*, 2016). Gas5 contains a decoy RNA "glucocorticoid response element" and is a potent repressor of glucocorticoid receptors in cell culture studies. Another lncRNA genomic loci, C5T1, residing in the rheumatoid arthritis TRAF1-C5 risk allele, controls the expression of C5, a gene previously linked to rheumatoid arthritis (Messemaker *et al.*, 2016). In Crohn's disease patients, lncRNA DQ786243 is upregulated and correlates with disease severity. While in colorectal cancer cells, knockdown of lncRNA DQ786243 results in the induction of cell cycle arrest and apoptosis, the same lncRNA regulates the expression of CREB and Foxp3, and contributes to regulatory T lymphocyte functions (Qiao *et al.*, 2013; Sun *et al.*, 2016). Together, these studies illustrate that the ubiquitously expressed lncRNAs could display cell type-specific functions in the immune system.

Detailed mechanistic understanding of lncRNA action in immune cells will likely provide valuable therapeutic venues. LncRNA Rmrp is the causative allele mutated in cartilage-hair hypoplasia (CHH) patients, a rare genetic disorder characterized by early childhood onset of skeletal dysplasia, hypoplastic hair, hematological abnormalities and autoimmune pathologies in the joints and kidneys. Mutations in evolutionarily conserved lnRNA Rmrp nucleotides also result in defects in T-cell immunity. We have recently reported that lnRNA Rmrp associates with the DDX5/RORγt protein complex in T helper 17 cells. The nuclear receptor RORγt is a transcription factor expressed in several lymphoid cell types. In the thymus, RORγt promotes the survival of developing T cells; in the periphery, it is required for the development of lymphoid tissue inducer cells and class 3 innate lymphoid cells (ILC3), and directs the differentiation of Th17 cells and other types of T cells that produce IL-17 cytokines, including $CD8^+$ T cells, γδ T cells and NKT cells. T helper 17 cells are involved in multiple inflammatory and autoimmune diseases. In Th17 cells, the DEAD-box RNA helicase, DDX5, mediates the binding of lncRNA Rmrp to RORγt to provide transcriptional co-activator activity at key target sites throughout the genome (Huang *et al.*, 2015). The mutation corresponding to that in CHH patients abrogates lncRNA Rmrp's IL-17 inducing function and controls the roles of Th17 cells *in vivo* at steady state and in mouse models of autoimmunity.

In a different study, researchers identified lncRNA MORRBID to be dysregulated in hypereosinophilic syndrome patients with persistent eosinophilia that results in multiple tissue and organ damage. The authors found that MORRBID tightly controls the survival of neutrophils, eosinophils and "classical" monocytes by promoting the enrichment of the PRC2 complex on the Bcl2l1 (Bim) promoter and mediates allele-specific transcription repression locally (Kotzin *et al.*, 2016). In macrophages, lncRNA-*EPS* is an important negative regulator of inflammatory responses. *LncRNA-EPS*-deficient mice manifest enhanced inflammation *in vivo* (Atianand *et al.*, 2016). LncRNA-*EPS* associates with chromatin and interacts with hnRNPL through its *CANACA* motif (2386–2391 nt), to repress the expression of immune response gene clusters.

These recent studies suggest that lncRNAs can be targeted to boost host immunity and ameliorate inflammatory diseases in humans. For example, mutations in lncRNA Rmrp in CHH patients can be corrected by CIRSPR/Cas9-directed homologous recombination (HDR). Guide RNAs targetting the mutated loci can be delivered to patient-derived bone marrow stem cells, together with a donor template containing the corrected gene insert. Selection and expansion of HDR-corrected stem cells followed by bone marrow reconstitution will help to ameliorate hematological defects in CHH patients. For patients with hypereosinophilic syndrome, MORRBID expression in hematopoietic stem cells can be similarly corrected. For patients suffering from autoimmune pathology due to overproduction of IL17 cytokines, Th17 cell-specific delivery of guide RNAs that ablate key Rmrp residues implicated in promoting transcription activation of IL17 can be used to down-modulate excess Th17-mediated inflammation. In addition, incorporation of the CANACA motif from lncRNA-EPS into RNA-based small molecules would compete for endogenous lncRNA-EPS binding partners and boost innate immunity in susceptible hosts. Furthermore, knowing that the 3′-region (2000–2531 nt) of lncRNA-EPS is both necessary and sufficient to suppress *Cxcl10* expression, we could generate short RNA oligos containing this region of lncRNA-EPS and deliver them as anti-inflammatory agents. Better understanding of lncRNA expressions in the immune system, their functions in host immunity, and how they contribute to disorders will help identify novel therapeutic-targeting venues that can take advantage of the latest genome editing technology to combat a wide array of human immunological conditions.

REFERENCES

Atianand MK, Fitzgerald KA. 2014. Long non-coding RNAs and control of gene expression in the immune system. *Trends Mol Med* **20**: 623–631.

Atianand MK, Hu W, Satpathy AT, *et al.* 2016. A long noncoding RNA lincRNA-EPS acts as a transcriptional brake to restrain inflammation. *Cell* **165**: 1672–1685.

Carpenter S, Aiello D, Atianand MK, *et al.* 2013. A long noncoding RNA mediates both activation and repression of immune response genes. *Science* **341**: 789–792.

Cech TR, Steitz JA. 2014. The noncoding RNA revolution – trashing old rules to forge new ones. *Cell* **157**: 77–94.

Chan J, Atianand M, Jiang Z, *et al.* 2015. Cutting edge: a natural antisense transcript, AS-IL1alpha, controls inducible transcription of the proinflammatory cytokine IL-1alpha. *J Immunol* **195**: 1359–1363.

Gomez JA, Wapinski OL, Yang YW, *et al.* 2013. The NeST long ncRNA controls microbial susceptibility and epigenetic activation of the interferon-gamma locus. *Cell* **152**: 743–754.

Hu G, Gong AY, Wang Y, *et al.* 2016. LncRNA-Cox2 promotes late inflammatory gene transcription in macrophages through modulating SWI/SNF-mediated chromatin remodeling. *J Immunol* **196**: 2799–2808.

Hu G, Tang Q, Sharma S, *et al.* 2013. Expression and regulation of intergenic long noncoding RNAs during T cell development and differentiation. *Nat Immunol* **14**: 1190–1198.

Huang W, Thomas B, Flynn RA, *et al.* 2015. DDX5 and its associated lncRNA Rmrp modulate TH17 cell effector functions. *Nature* **528**: 517–522.

Imamura K, Imamachi N, Akizuki G, *et al.* 2014. Long noncoding RNA NEAT1-dependent SFPQ relocation from promoter region to paraspeckle mediates IL8 expression upon immune stimuli. *Mol Cell* **53**: 393–406.

Kotzin JJ, Spencer SP, McCright SJ, *et al.* 2016. The long non-coding RNA Morrbid regulates Bim and short-lived myeloid cell lifespan. *Nature* **537**: 239–243.

Krawczyk M, Emerson BM. 2014. p50-associated COX-2 extragenic RNA (PACER) activates COX-2 gene expression by occluding repressive NF-kappaB complexes. *eLife* **3**: e01776.

Maass PG, Luft FC, Bahring S. 2014. Long non-coding RNA in health and disease. *J Mol Med* **92**: 337–346.

Mayama T, Marr AK, Kino T. 2016. Differential expression of glucocorticoid receptor noncoding RNA repressor Gas5 in autoimmune and inflammatory diseases. *Horm Metab Res* **48**: 550–557.

Messemaker TC, Frank-Bertoncelj M, Marques RB, *et al.* 2016. A novel long non-coding RNA in the rheumatoid arthritis risk locus TRAF1-C5 influences C5 mRNA levels. *Genes Immun* **17**: 85–92.

Muller N, Doring F, Klapper M, *et al.* 2014. Interleukin-6 and tumour necrosis factor-alpha differentially regulate lincRNA transcripts in cells of the innate immune system in vivo in human subjects with rheumatoid arthritis. *Cytokine* **68**: 65–68.

Pang KC, Dinger ME, Mercer TR, *et al.* 2009. Genome-wide identification of long noncoding RNAs in CD8+ T cells. *J Immunol* **182**: 7738–7748.

Peng X, Gralinski L, Armour CD, *et al.* 2010. Unique signatures of long noncoding RNA expression in response to virus infection and altered innate immune signaling. *MBio* **1**(5): e00206-10.

Qiao YQ, Huang ML, Xu AT, *et al.* 2013. LncRNA DQ786243 affects Treg related CREB and Foxp3 expression in Crohn's disease. *J Biomed Sci* **20**: 87.

Ranzani V, Rossetti G, Panzeri I, *et al.* 2015. The long intergenic noncoding RNA landscape of human lymphocytes highlights the regulation of T cell differentiation by linc-MAF-4. *Nat Immunol* **16**: 318–325.

Rapicavoli NA, Qu K, Zhang J, *et al.* 2013. A mammalian pseudogene lncRNA at the interface of inflammation and anti-inflammatory therapeutics. *eLife* **2**: e00762.

Sharma S, Findlay GM, Bandukwala HS, *et al.* 2011. Dephosphorylation of the nuclear factor of activated T cells (NFAT) transcription factor is regulated by an RNA-protein scaffold complex. *Proc Natl Acad Sci USA* **108**: 11381–11386.

Sun L, Xue H, Jiang C, *et al.* 2016. LncRNA DQ786243 contributes to proliferation and metastasis of colorectal cancer both in vitro and in vivo. *Biosci Rep* **36**(3): e00328.

Wang P, Xue Y, Han Y, *et al.* 2014. The STAT3-binding long noncoding RNA lnc-DC controls human dendritic cell differentiation. *Science* **344**: 310–313.

Wang Y, Zhong H, Xie X, *et al.* 2015. Long noncoding RNA derived from CD244 signaling epigenetically controls CD8+ T-cell immune responses in tuberculosis infection. *Proc Natl Acad Sci USA* **112**: E3883–E3892.

Xia F, Dong F, Yang Y, *et al.* 2014. Dynamic transcription of long non-coding RNA genes during CD4+ T cell development and activation. *PLoS One* **9**: e101588.

Zhao G, Su Z, Song D, Mao Y, Mao X. 2016. The long noncoding RNA MALAT1 regulates the lipopolysaccharide-induced inflammatory response through its interaction with NF-kappaB. *FEBS Lett* **590**: 2884–2895.

Part VI

Legal (Intellectual Property) and Bioethical Issues of Genome Editing

28 Patenting of CRISPR-based Inventions

Philip Webber

28.1 Introduction

In 1953, after Watson and Crick had discovered the structure of DNA, they famously announced their discovery in The Eagle pub in Cambridge, UK. Today's inventors, however, have a greater understanding of the commercial value of biotechnological innovations. They are more likely, therefore, to contact their patent attorneys first, with a view to seeking patent protection for their inventions as a means to help fund their future research or to set up a spin-out company.

Hence, in May 2012, Jennifer Doudna's group filed their first patent application for methods of altering gene expression using CRISPR/Cas9 and this was followed in December 2012 by Feng Zhang's group's first patent application. Both of these patent applications had claims to compositions comprising CRISPR guide RNAs. Since that time, hundreds of patent applications have been filed to various different uses of CRISPR and modified Cas9 enzymes.

In order to fully understand the significance of these patent applications, and others which have been filed since then, one first has to understand the basics of the patent system and how biotech inventions such as these can be patented.

28.2 The Patent System

28.2.1 What is a Patent Application?

A patent application is, basically, a 20–100 page book that describes an invention. The patent application generally starts with a description of the problem that the invention addresses and how others have previously tried to deal with that problem. This is followed by details of the invention and examples of its preferred features, and then some technical examples of how to put the invention into practice. The patent application may also include graphs or figures to illustrate data that have been produced by the examples.

The most important section of the patent application, however, is the claims section. Patent claims are short sentences or paragraphs which explicitly define the features of the invention in words (not drawings or figures). Once the patent is granted, the claims will define the scope of the monopoly that the patent owner will have over that invention.

28.2.2 General Criteria for Patentability

The key criterion for the grant of a patent is that the invention – as defined in the patent claims – is novel and inventive over everything that is in the public domain before the date that the patent application is filed. Novelty refers to the fact that the claimed invention must be different in at least one technical detail to what was previously in the public domain. Inventiveness is defined as something that is not obvious to a person of average skill in that area of technology.

A further criterion is that the patent application must describe the invention in a sufficient level of detail in order for it to be put into practice across the whole scope of what is being claimed.

28.2.3 What Can You Patent in the Biotech Field?

Over the past 30 years, patent offices around the world have had to address issues regarding the patenting of genes,[1] plants,[2] animals[3] and, more recently, human embryonic stem cells.[4] In general, the patent offices have viewed each of these inventions as complex chemical entities and assessed their patentability accordingly. Thus isolated genes and proteins have been considered to be patentable in the same way as other isolated natural products (e.g. dyes and antibiotics). In Europe, the patentability of gene and protein sequences was explicitly confirmed in 1998 in the EU Biotech Directive.[5] Similarly, in 2001, the US Patent and Trademark Office (USPTO) revised its examination guidelines[6] to confirm the patentability of gene sequences. As a consequence of these new laws and guidelines, it is necessary for European patent applications to include details of the function of the sequence; in the US, a specific, substantial and credible utility for the sequence is required. Without this function or utility, the sequence is nothing more than a discovery and hence is not patentable.

Similarly, transgenic plants and animals have been considered by the patent offices as complex chemical compositions. For example, the introduction of a fish gene into the genome of plant produces a novel "composition" (i.e. a novel plant) and if it was not obvious to introduce such a fish gene into the plant, then the inventive step hurdle is also overcome. There are limitations, however. In Europe, the early inventions relating to human embryonic stem cells were unpatentable because they were deemed to be immoral, on grounds that human embryos had to be destroyed in order to make or use those inventions.

More recently, the US Supreme Court has ruled that genomic DNA is not patentable on grounds that it is a product of nature[7] and, at the time of writing, the USPTO and US courts are extending this decision to apply to all products of nature. Hence the validity of hundreds of previous US patents which claim DNA sequences or other natural products has recently been put into question. This decision only applies, however, in the US (although some other countries, e.g. Australia, do have similar rules).

Consequently, by the time that Jennifer Doudna and Feng Zhang came to file their first CRISPR patent applications in 2012, patent offices around the world were familiar with biotech patents and there were standardized criteria for the examination of such patent applications.

The Doudna patent applications have now been assigned to The Regents of the University of California *et al.*; Zhang's patents are now assigned to the Broad Institute and MIT. For convenience, however, they shall be referred to herein as the Doudna and Zhang patents/applications. Furthermore, whilst Doudna and Zhang have filed patent applications in many countries, this chapter will concentrate on those in Europe and the US.

28.3 The CRISPR Patent Wars

Although Doudna's group filed their CRISPR patent applications first, Zhang's group were the first to obtain granted CRIPSR patents; hence the Zhang patents will be discussed here first.

28.3.1 Zhang's Patents

Zhang's group filed a set of 12 "provisional" US patent applications between December 2012 and June 2013. The first of these contained basic details of the invention and the subsequent provisional applications added more and more details. ("Provisional" applications are called "priority" applications in non-US countries.) A number of International (PCT) patent applications (including PCT/US2013/074819, WO2014/093712) were filed in December 2013 claiming priority from these earlier provisional applications. Sets of US and European (EP) patent applications were later filed, all of which were derived from Zhang's earlier international patent applications.

Zhang's group had the foresight to request accelerated examination of some of their US and EP patent applications, and (as of September 2016) they have 13 granted US patents[8] and 7 granted EP patents.[9] They also have a number of US and EP patent applications that are still pending,[10] i.e. under examination. Claim 1 of one of Zhang's granted EP patents is given in Box 28.1.

Whilst the language of the Box 28.1 claim is rather convoluted (even for a patent claim), it can be seen that the claim relates essentially to a composition comprising a CRISPR RNA, which comprises a guide sequence, tracr mate and tracr RNA sequence. The claim refers to the fact that when the tracr mate sequence is transcribed (which is a strange choice of language given that an RNA is being discussed), the tracr mate is capable of hybridizing to the tracr RNA sequence and the guide sequence directs the RNA to a target sequence. This patent also claims uses of the CRISPR RNAs for "genome engineering." Similar claims have been granted in the US (see endnote 8).

28.3.2 Doudna's Patent Applications

Doudna's group filed four US provisional applications between May 2012 and February 2013. They then filed an international (PCT) patent application[11]

BOX 28.1 Claim 1 from Zhang's Patent EP 2771468 B1

1. A non-naturally occurring or engineered composition comprising: a Clustered Regularly Interspersed Short Palindromic Repeats (CRISPR)-CRISPR-associated (Cas) (CRISPR-Cas) system chimeric RNA (chiRNA) polynucleotide sequence, wherein the polynucleotide sequence comprises:

   (a) a guide sequence of between 10–30 nucleotides in length, capable of hybridizing to a target sequence in a eukaryotic cell,
   (b) a tracr mate sequence, and
   (c) a tracr RNA sequence

wherein (a), (b) and (c) are arranged in a 5′ to 3′ orientation,
wherein when transcribed, the tracr mate sequence hybridizes to the tracr RNA sequence and the guide sequence directs sequence-specific binding of a CRISPR complex to the target sequence,
wherein the CRISPR complex comprises a Type II Cas9 protein complexed with:

(1) the guide sequence that is hybridized to the target sequence, and
(2) the tracr mate sequence that is hybridized to the tracr RNA sequence,

wherein the tracr RNA sequence is 50 or more nucleotides in length.

in May 2013; US and EP patent applications based on that international patent application have now been filed. However, in contrast to Zhang's group, accelerated examination was not initially requested and hence Doudna's patent applications have languished in the (generally rather slow) patent system.

Claim 1 of one of Doudna's international patent applications is given in Box 28.2.

It is interesting to note that whilst both Box 28.1 (Zhang's claim) and Box 28.2 (Doudna's claim) both relate to the same CRISPR invention, the language used is remarkably different. Doudna's claim 1 is directed to a "DNA-targeting RNA" which has a first targeting component (which could be essentially any RNA sequence) and a second component which interacts with any "site-directed modifying polypeptide" (with Cas9 in mind). This claim is clearly far too broad to be patentable, but there are later claims that limit the second components to defined Cas9-binding RNA sequences.

As of the time of writing (September 2016), none of Doudna's EP or US CRISPR patent applications had been granted. In fact, third parties have filed hundreds of pages of anonymous "observations" and documents on the EP and US patent applications in order to "encourage" the patent examiners not to grant these patent applications. (Any third party has the right to file comment/observations on EP patent applications, or relevant documents on US patent applications, in

BOX 28.2 Claim 1 from Doudna's International Patent Application WO 2013/176772

1. A DNA-targeting RNA comprising:
   (i) a first segment comprising a nucleotide sequence that is complementary to a sequence in a target DNA; and
   (ii) a second segment that interacts with a site-directed modifying polypeptide.

BOX 28.3 Timeline of Zhang's and Doudna's Intertwined Provisional Applications

Zhang's US provisional applications	Doudna's US provisional applications
May 25, 2012	US 61/652,086
Oct 19, 2012	US 61/716,256
US 61/736,527	Dec 12, 2012
US 61/748,427	Jan 2, 2013
Jan 28, 2013	US 61/757,640
US 61/758,468	Jan 30, 2013
Feb 15, 2013	US 61/765,576
US 61/769,046	Feb 25, 2013

(and 8 others).

order to try to influence the examination process and to improve the quality of any granted patents.)

28.3.3 The Intertwined Provisional Applications

As mentioned above, between 2012 and 2013, Zhang's group filed twelve US provisional applications, and Doudna's group filed four. Both group's later international, EP and US patent applications are based on their earlier US provisional applications. The way in which the timelines of these two sets of US provisional applications interlock is important; this is illustrated in Box 28.3.

In most countries, the first person to file a patent application that fully describes the claimed invention is generally considered to have the rights to that invention (provided that the patent is later granted). As can be seen from Box 28.3, Doudna's first two US provisional applications were filed before any of Zhang's.

One of the key issues that is currently under dispute in both Europe and the US is which provisional application – Doudna's or Zhang's – was the first to describe full details of the CRISPR gene editing invention. In Europe, this point is relevant to whether the disclosures of Doudna's EP patent application are citable against the novelty of Zhang's, or vice versa. In the US, the key issue is ownership of the patent rights, based on who made the invention first.

28.3.4 Oppositions at the European Patent Office (EPO)

In Europe, an opposition can be filed against any granted European (EP) patent, provided that the opposition is filed within nine months of the date of grant of the EP patent. The first of Zhang's EP CRISPR patents (EP 2771468 B1) was granted on February 11, 2015 and, within the following nine months, oppositions were filed by nine different parties. The opponents include Novozymes A/S and CRISPR Therapeutics AG; the other opponents have not made their identities known.

The grounds of opposition include assertions that Zhang's granted EP patent lacks novelty and/or inventive step over earlier CRISPR publications and Doudna's patent applications; that the EP patent application does not adequately describe the claimed invention and that the claims were illegally amended during the examination process.

In accordance with the EPO's standard opposition procedures, the EPO has sent copies of all of the oppositions to Zhang's group; in reply, they filed a 142 page written response. The EPO's Opposition Division will now consider the arguments made by all parties. Eventually (which might be after 1–2 years), all parties will be invited to discuss the issues orally at a formal hearing at the EPO. At this hearing, each of the issues will be addressed one by one (e.g. novelty, inventive step, etc.) and the EPO will give a decision on each of those issues. If the EP patent fails on any one issue, then it will be revoked. (On average, one-third of EP patents survive opposition proceedings intact, one-third survive, but with reduced claim scope, and one-third are revoked.) If the EP patent survives, then appeals can be filed by any of the opponents; if the patent is revoked, then Zhang's group can file an appeal. Appeals at the EPO are heard by the EPO's Technical Boards of Appeal. However, the boards are currently severely overstretched and hence the appeal process could well take a further 2–4 years, or more, to be resolved.

Since the filing of the oppositions in November 2015 against Zhang's EP 2771468 B1, six of Zhang's other EP patents have also been opposed (see endnote 9). The opponents are largely similar to those who opposed EP 2771468 B1. The procedures on each of these will be similar to those described above.

It is only once the EPO's appeal processes have been fully resolved that we will have final decisions on the validity of these EP patents and hence on who owns rights to the basic CRISPR gene editing technology in Europe.

28.3.5 Interferences at the USPTO

In the US, Zhang's group already have 13 CRISPR patents (see endnote 8), whilst all of Doudna's US patent applications are still under examination.

The key issue that is under dispute, however, is who actually first conceived the CRISPR gene editing invention and who reduced it to practice, and hence who should own rights to it in the US.

Although the US currently has a "first to file" patent system that is similar to the rest of the world, Doudna's US provisional applications were filed under the old US "first to invent" (pre-America Invents Act, 2013) system. Under this old system, it is not only the contents of the US provisional applications that need to be considered (see Box 28.3), but also the contents of laboratory notebooks and other relevant documents. Such issues are resolved in what are known as "interference" proceedings; these are judged by the USPTO's Patent Trial and Appeals Board (PTAB).

In January 2016, the USPTO (prompted by Doudna's attorneys) declared an "interference" between Doudna's US patent applications and Zhang's US patents.[12] The PTAB subsequently invited Doudna's and Zhang's attorneys to file written comments in support of their cases. A hearing took place in November 2016 and the decision of the PTAB judges issued in February 2017. The decision centered around whether – on the basis of the prokaryotic CRISPR data that Doudna's group had in their patent application – it was obvious that CRISPR would work in eukaryotes. The judges said no. Consequently, the judges ruled that Zhang's group were allowed to keep patent claims to methods of using CRISPR in eukaryotic cells; Doudna's group were only allowed to claim methods of using CRISPR in prokaryotic cells. Since they decided that there was no overlap between the patent claims of the two groups, the judges ruled that there was in fact no interference.

Not surprisingly, Doudna's group have filed an appeal to the US Court of Appeals for the Federal Circuit, and hence a final decision on this matter in the US may still be a few years away.

It should be noted that these "interference" proceedings will only seek to establish who was first to invent the CRISPR patent under US law; it will not consider issues of novelty or inventive step; these may be raised later in any US court proceedings.

28.4 Other CRISPR Patent Applications and Freedom to Operate

It will be a number of years before final decisions are given by the European and US patent offices/courts as to who actually has valid patent rights to the CRISPR gene editing technology.

In the meantime, over 700 patent applications have been filed by other inventors with claims to CRISPR-based inventions.[13] Most of these relate to new methods of using CRISPR, ways to reduce off-target effects, or variants of Cas9 or other CRISPR enzymes. Provided that the claimed inventions satisfy the general criteria for patentability (as discussed above), many of these patent applications will be granted.

However, given that Zhang's group already have granted patents which cover the CRISPR gene editing technology in broad terms, the owners of these other inventions and anyone else who is using CRISPR-based technology will need to consider whether they need to obtain licenses under the Zhang patents if they wish to exploit those inventions. Such licenses are currently being offered by the Broad Institute in many technological areas.[14]

The plethora of other CRISPR patents that are likely to be granted in the coming years will be welcomed as providing a potential source of research funding for some, but will be considered by others as being an unwelcome hindrance to the exploitation of this remarkable technology.

NOTES

1. Decision T 0272/95 (Relaxin/HOWARD FLOREY INSTITUTE) of the EPO Technical Boards of Appeal dated October 23, 2002.
2. Decision G1/98 (Transgenic plants/NOVARTIS II) of the Enlarged Board of Appeal of the EPO dated December 20, 1999.
3. Decision T19/90 (Onco-Mouse) of the Technical Boards of Appeal of the EPO dated October 3, 1990.
4. Decision G2/06 (WARF/Thomson) of the Enlarged Board of Appeal of the EPO, dated November 25, 2008; Decision C-34/10 (Brüstle) of the Court of Justice of the EU dated October 18, 2011.
5. Directive 98/44/EC of the European Parliament and of the Council of July 6, 1998. www.europa.eu.int/eur-lex/pri/en/oj/dat/1998/l_213/l_21319980730en00130021.pdf.
6. USPTO Utility Examination Guidelines, January 5, 2001 www.uspto.gov/web/offices/com/sol/og/2001/week05/patutil.htm.
7. Supreme Court of the United States No. 12–398, Association for Molecular Pathology, et al. Petitioners v. Myriad Genetics, Inc. et al. on writ of certiorari to the United States Court of Appeals for the Federal Circuit, June 13, 2013. www.supremecourt.gov/opinions/12pdf/12–398_1b7d.pdf.
8. US 8945839 B2; US 8771945 B1; US 8795965 B2; US 8865406 B2; US 8871445 B2; US 8889356 B2; US 8889418 B2; US 8895308 B1; US 8906616 B2; US 8932814 B2; US 8999641 B2; US 8697359 B1; US 8993233 B2.
9. EP 2771468 B1; EP 2784162 B1; EP 2764103 B1; EP 2896697 B1; EP 2921557 B1; EP 2898075 B1; EP 2931898 B1. The status of the oppositions may be checked at: https://register.epo.org/regviewer?lng=en. Note added to proof: EP 2825654 B1 and EP 2840140 have also been granted. All of these patents have now been opposed.
10. The published patent applications and patents may be found on Espacenet: https://worldwide.espacenet.com/?locale=en_EP.
11. PCT/US2013/032589, which was published as WO2013/176772.
12. The Broad Institute's summary of the US interference proceedings may be found at: www.broadinstitute.org/what-broad/areas-focus/project-spotlight/crispr-patent-interference-updates. The PTAB documents may be found at: https://acts.uspto.gov/ifiling/PublicView.jsp?identifier=106048&identifier2=null&tabSel=4&action=filecontent&replyTo=PublicView.jsp.
13. Patent applications in WIPO's PatentScope database (https://patentscope.wipo.int/search/en/structuredSearch.jsf) with claims that have the word "CRISPR" or "Cas9" in them.
14. www.broadinstitute.org/partnerships/office-strategic-alliances-and-partnering/information-about-licensing-crispr-genome-edi.

29 Patent, Ownership and Licensing Issues of CRISPR-based Genome Editing: Impact on Universities and their Licensees

Rodney L. Sparks

29.1 Introduction

These are exciting times in molecular biology research and recent discoveries have great potential for use in prevention and treatment of diseases. Advances in genome editing technology, particularly meganucleases, transcription activator-like effector nucleases (TALENS), zinc-finger nucleases (ZFNs), and clustered regularly interspaced short palindromicrepeats (CRISPR) have been extraordinary to the point that they have begun to revolutionize research and hold great promise for treating diseases associated with genetic mutations. The treatment could include direct editing (correction) of mutant genes or treatment with drugs or other agents found to be useful after genome editing technology. Aside from medical treatment, these technologies will also be quite valuable for research and several are already being used commercially. It is obvious that these technologies have the potential to be worth billions of dollars.

CRISPR, one of the most recent iterations of genome editing, seems to be particularly useful in some situations. This cannot be more evident than the title of a recent article entitled "Genome editing with CRISPR-Cas9: can it get any better?" (Haeussler and Concordet, 2016). As stated in their abstract. "The CRISPR-Cas revolution is taking place in virtually all fields of life sciences. Harnessing DNA cleavage with the CRISPR-Cas9 system of *Streptococcus pyogenes* has proven to be extraordinarily simple and efficient, relying only on the design of a synthetic single guide RNA (sgRNA) and its co-expression with Cas9." The three primary scientists credited with the discovery of the technology, Jennifer Doudna, Emmanuelle Charpentier and Feng Zheng, have already garnered various awards and many companies have made major investments into developing and commercializing the technology. CRISPR is also of special interest to the academic and business communities because of issues arising as to ownership and inventorship of the technology (Ledford, 2016; Noonan, 2016; Sherkow, 2015).

Most chapters in this book discuss the state-of-the-art of genome editing technology. This chapter will deviate from that and instead highlight issues in patenting, licensing and defending patented technology such as genome editing technology, using the much publicized CRISPR patent battle as the primary example (for more detailed reviews of patenting and licensing see Sibley and Sparks, 2009; Sparks, 2008, 2009; Sparks *et al.*, 2015). This chapter describes why patent inventorship and ownership are important, how they are determined, how and why a patent battle has erupted between giants in the academic community, and some of the nuances of how these disputes are settled.

29.2 Overview of Patents, Patent Ownership and Licensing

Patents are important because they give the patent owner certain exclusive rights. Under United States (US) law, there are four major requirements for patentability of an invention: novelty, usefulness, non-obviousness and statutory subject matter. There are additional requirements, but their description is beyond the scope of this chapter (Sparks, 2008, 2009). Most countries have similar requirements, although the terminology may be slightly different. Patents can be granted on things such as chemical compounds, some biological molecules, methods of making and using chemical compounds and biological molecules, devices and processes. Quite a few patents have now issued for various genome editing technologies, including CRISPR, so presumptively they all met the proper patentability requirements. As with anything else that is regulated by the government, there are many ways to challenge the validity of a patent, its ownership and its inventorship. The rules and laws governing inventorship are quite complex. Because different inventors made similar discoveries at about the same time, determining who is the inventor(s) on key CRISPR patents is critically important, but is not a simple process.

Ownership, by way of patent rights, is crucial when technology has value and when it requires great expense to develop and bring to market. Patent laws were established to promote creativity and to grant property rights to inventors to allow them to recoup their investment and earn a profit.

Ownership is also important because the owner of a patent has the power to exclude others from making, using, selling or importing their patented invention during the term of the patent, which in most countries is 20 years from the time of filing the patent application. If there is no patent protection, the developer of the technology will have to compete in the open market and might not be able to recoup the investment made to commercialize the invention. Furthermore, if licensing the technology, a potential licensee may be less likely to take a license or may offer a much lower royalty if that potential licensee needs exclusive rights to the technology to ensure that the development costs will be worth the investment.

License agreements can be exclusive or non-exclusive and allow the licensee to develop the technology and then commercialize it, while the owner (licensor) receives some small percentage of the profits in the form of royalty payments. These licenses and payments can be structured in many different ways. For example, an exclusive license can be for all rights in a technology to a single company, or specific rights to one company and a different set of rights to another company as so-called "field-of-use" licenses. An example of a non-exclusive license would be one where the same thing is licensed to many companies. Generally, the most valuable licenses are of the "exclusive" type for biotechnology and pharmaceutical inventions.

Inventors are owners of patent rights, unless or until they assign those rights to someone else. Most companies and universities, particularly in the US, require that all inventions by employees be assigned to the company or university through an employment agreement, thus making the company or university owner of the technology. US universities are required by federal law to try and commercialize all inventions made that are supported by federal grants, even if that support is minimal. Because universities do not have the capacity to fully develop and commercialize all inventions, they rely on license agreements with companies that have development and commercialization capacity.

When multiple inventors are involved, complications can arise if they are employees of different companies or universities (Sparks, 2009). For example, without an agreement in place one co-owner (or co-inventor) can sell their entire rights in the invention without having to pay anything to other co-owners (or co-inventors) because each co-owner owns an undivided interest in the technology. Similarly, a co-owner, whether an entity or an individual inventor, can grant a license to a company without the consent of the other co-owners and without having to give them a share of the license royalties. That type of license is non-exclusive and in technologies where exclusive rights are sought or needed, potential licensees may walk away from such a license agreement. Therefore, it is important for an institution that has outside collaborations to consider ownership issues carefully, particularly when the institution will be seeking to license the technology arising from such collaborations. Additional precautions may need to be taken when a current licensee or potential licensee is a co-owner because if the licensee/co-owner terminates the license agreement or does not take a license agreement, they are free to use or license the product on their own. Therefore, when a technology has the potential for great value and a large royalty stream for a university, it is imperative to establish inventorship, ownership and, concomitantly, licensing rights. This can be done by formalizing the relationship contractually prior to the collaboration to address these scenarios.

29.2.1 The Business of CRISPR and Genome Editing: High Stakes Investment and Potential High Return or Catastrophic Losses

Several companies have formed around CRISPR technology, notably startups CRISPR Therapeutics, Editas Medicine and Intellia Therapeutics. All claim some ownership or licensing rights to key CRISPR/Cas9 discoveries, either patented or in pending patent applications. Moreover, many products and services are now available using or providing genome editing technologies. For example, various CRISPR cell lines are available, including CRISPR/Cas9-modified induced pluripotent stem cells, as are CRISPR knockout and knockin mice and rats. This is true for other gene editing technologies as well.

The amount of investment into CRISPR technology and the three startup companies has been eye popping for such an early-stage technology. Editas Medicine, found by Zhang, Doudna and others, raised at least $120 million from private investors, one of them being Bill Gates. In Editas' Initial Public Offering (IPO) in early 2016, it raised another $94 million. Editas has licensed rights to Zhang's patents from the Broad Institute. Doudna later helped start Intellia Therapeutics, along with Caribou Biosciences, a company she previously co-founded. It has licensed rights from the Doudna patent application. Intellia had already raised at least $84 million when its recent IPO yielded another $108 million. Additionally, Novartis has entered into a deal with Intellia in order to couple its immunotherapy program using CART technology with Intellia's CRISPR technology. Charpentier co-founded CRISPR Therapeutics, which has raised close to $100 million, including investments from Celgene Corporation. CRISPR has recently formed alliances with Bayer and Vertex Pharmaceuticals.

Genome editing technologies other than CRISPR have also been the subject of developing patent portfolios and companies such as Cellectis and Sangamo Biosciences are heavily invested in meganuclease and engineered zinc-finger DNA-binding proteins, respectively. Other notable companies and universities with patented genome editing technologies include Dow, DuPont, Sigma-Aldrich, Chinese Academy of Sciences and Iowa State University.

Unfortunately, it is doubtful that all of the licenses and collaboration agreements in place for CRISPR patents and patent applications will remain valid because the inventorship and ownership of filed patent applications and issued patents on CRISPR technology are in dispute and have yet to be resolved. Recently, challenges have been filed against patents and pending patent applications contesting things from inventorship to patentability of the patent or patent application. Depending on the outcome, some of the investments may turn out to be monumentally bad if some of the licensees and licensors lose any or all of the rights they thought they had to the technology once the inventorship and patent disputes are resolved.

29.2.2 Inventorship in the US

Every patent application submitted to a patent office must list the inventors, and the validity and enforceability of a patent requires that the correct inventors be named on the patent (Sibley and Sparks, 2009). Fortunately, the patent rules and laws of most countries allow for correcting an error of inventorship if discovered, provided that there was no deceptive intent on the part of the inventors or those closely associated with the application when the error was made. If there was deceptive intent, the patent rights may be irrevocably lost.

Inventorship is a legal issue requiring a legal determination. It is not a moral decision and is not the same as authorship. Inventorship is a question of who actually invented the subject matter claimed in a patent. Inventorship problems often arise when applicants must determine if the invention was a joint effort and who actually contributed to the invention. Conception is the key determining factor of inventorship. Conception is the formation in the mind of the inventor, or the minds of the inventors, of a definite and permanent idea of the complete and operative invention, as it is to be applied in practice. An inventor is the person or persons who conceived the patented invention, but a co-inventor need only contribute to the conception of at least one aspect of the invention described to be named as an inventor. Interestingly, inventorship does not require that the inventor do any work to reduce the invention to practice, it can be done by someone else at the direction of the inventor. Thus, a person doing all of the physical work in the laboratory may not be an inventor, and often is not.

Most countries have "first-to-file" patent systems, meaning that if competitors invent the same thing, the first to file a patent application will be deemed the true inventor and the second to file will not be able to get a patent. Of course, that is assuming that the first-filed application still meets all of the patentability requirements. However, the US only recently switched to this system from a "first-to-invent" system. Described below are some highlights of the US patent battle, which is based on the old "first-to-invent" system, and the inventions in question fall under those rules and laws in the US.

29.3 Patent Battle Over Who Invented CRISPR and Who Owns CRISPR

Scientific credit for who discovered or invented CRISPR genome editing technology seems to be between Jennifer Doudna of the University of California at Berkeley and her colleagues, including Emmanuelle Charpentier, and Feng Zhang and his colleagues at the Broad Institute, Massachusetts Institute of Technology and Harvard College. However, for ownership and patent purposes, these determinations are a legal question to be determined by the United States Patent and Trademark Office (USPTO) or federal court system in the US. Similar determinations will be made in other countries and opposition proceedings have begun in Europe against Zhang's issued patents there.

As mentioned above, when there is joint ownership of a patent, in the absence of an agreement to the contrary, each owner has the right to independently license a patent. This creates several problems, and if the licensing rights remain separate the technology value can be greatly diminished because a license(s) in this scenario can only be a non-exclusive license. Non-exclusive licenses can be of little value to many biotechnology and pharmaceutical companies when it is important to own or have full rights to a molecule or process or to a method of treatment using a molecule or a process. Non-exclusive licenses tend to encompass such things as assays and the use of reagents and do not typically generate the type of income that a biologic or pharmaceutical can when exclusively licensed.

29.3.1 CRISPR Patent Battle in the US

An odd scenario has been playing out between Doudna and colleagues, and Zhang and colleagues, highlighting some of the quirks of patent law. Although Doudna's group published the first paper and did file a patent application, Zhang and the

Table 29.1 Summary of issued patents and applications in dispute

Patent owner[a]	US patent number	European patent number[b]
Broad Institute/ MIT/Harvard	8,697,359	EP2771468
	8,771,945	EP2784162
	8,795,965	EP2896697
	8,865,406	
	8,871,445	
	8,889,356	
	8,895,308	
	8,906,616	
	8,932,814	
	8,945,839	
	8,993,233	
	8,999,641	
University of California/ University of Vienna/ E. Charpentier	None (utility application no. 13/ 842,859, filed March 15, 2013 is still pending)	None (application pending)

[a] Broad Institute also has several more pending US applications.
[b] The European patents do not necessarily correspond to the adjacent US patent.

Broad Institute, MIT and Harvard now have at least 12 issued US patents, as well as three issued European patents (Table 29.1). Zhang is the sole inventor on the first patent (issued April 15, 2014), but has co-inventors on other patents. Additionally, Harvard has at least six CRISPR patents from George Church and his laboratory, and David Liu and his laboratory, both of whom are also affiliated with the Broad Institute. Other entities with patents issued on various aspects of CRISPR include DuPont, Agilent Technologies, University of Georgia Research Foundation, Institut Pasteur and Caribou Biosciences.

Doudna's group has yet to receive a patent, but taking advantage of patent laws in effect when their application was filed they challenged Zhang's patents by asserting that the invention was really made first by Doudna and colleagues. On January 11, 2016, following the request by Doudna and her colleagues, the USPTO declared Interference Number 106,048 between Zhang and colleagues and Doudna and colleagues. The interference battle surrounds broad patent claims to the CRISPR genome editing technology. Multiple applications and patents have been included in the proceedings by the USPTO. It is unusual to find two universities in such a fight.

Officially, the USPTO has identified the junior party to be The Broad Institute, Inc., Massachusetts Institute of Technology, and President and Fellows of Harvard College (co-owners of the patents and applications being contested) and the senior party to be The Regents of the University of California, University of Vienna and Emmanuelle Charpentier (co-owners of the Doudna and colleagues application). Under the interference rules (see in detail below), the first party to file a patent application is considered to be the senior party and the burden for proving earlier invention is placed on the junior party (for a review on determining inventorship see Sibley and Sparks, 2009).

Doudna and Charpentier filed the earliest application, filing a US provisional patent application on March 25, 2012. Zhang filed a US provisional patent application on December 12, 2012. The issues that will be considered include: Was the first filed application by Doudna a complete invention relative to what Zhang invented and filed?, as well as Who invented first and was diligent in developing the invention up until the day the application was filed? The latter requirement is a nuance of US interference proceedings. If a lack of diligence in working on an invention is found by the USPTO it will result in a loss of patent rights. Additionally, if a provisional patent application does not provide support for the later filed regular application, it can impact whether or not the regular application is even entitled to the priority date of its claimed provisional application. That can have a major impact on what is considered prior art for patentability purposes in both the US and Europe. It does appear that Zhang's application was more detailed and specifically encompassed the use of CRISPR in eukaryotic cells, while Doudna's did not. Based on information available, it appears that the European patent examiner has raised questions about Doudna's application, asserting that the filed provisional applications upon which it claims priority do not support the regular application as filed. To complicate the unusual circumstance of similar inventions filed close in time, Doudna's application hit several procedural stumbling blocks, including some issues with co-inventor Charpentier and prosecution of the application, while Zhang took advantage of a procedure that fast-tracks patent applications (Sherkow, 2015). Regarding the article published by Doudna in *Science* (Jinek *et al.*, 2012), it, in fact, published before Zhang's provisional application was filed. Interestingly, that publication and Doudna's earlier filed application were both cited against Zhang's patent application and they were able to overcome the rejection based on differences in his application relative to Doudna's technology.

Not only were Zhang's regular US applications prosecuted under special fast-track procedures that they requested, the Broad Institute took other steps to speed up the process, such as filing the first US application ahead of the required deadline for filing and by filing the regular US application in conjunction with an international Patent Cooperation Treaty (PCT) application. Usually, a regular US application is not filed until 18 months after a PCT application is filed.

29.3.2 Interference Proceedings at the USPTO

The US Congress passed the Leahy Smith America Invents Act (AIA) in 2011, enacting the broadest changes in US patent law in over six decades, and the law put an end to "interferences," but not for patent applications that were filed before that part of the law went into effect on March 16, 2013 (Leahy-Smith America Invents Act, 2011). The new law established a modified "first-to-file" patent application system, referred to as "first-inventor-to-file." Applications filed after March 16, 2013 are governed by the new rules, where the first to file an application is the inventor, although there are a few provisos in the law, such that it is not quite a true first-to-file system. Because of the filing dates of the patent applications in dispute they will be examined under the interference rules by the USPTO. It has been suggested that this may be the last great interference in the US because of the high stakes and because the interference rules were eliminated. Interferences determine who invented first.

Interference between two patent applications or between a patent and a patent application is provoked by the USPTO when there is a question of who invented first (Sparks *et al.*, 2015). There are a series of issues related to the invention that have to be addressed regarding who, what, when, where and why. When all data, documents and evidence have been compiled, the USPTO will decide who the inventor is, based on a series of rules and court cases that developed over the years. Basically, whoever invented first and remained diligent in their work on the invention until they filed their patent application is deemed to be the inventor. Things considered in the evaluation include dated lab notebooks, emails, testimony and even the quality of the first-filed patent application to ensure that it truly encompasses what is being claimed and/or what has been patented. However, there are limitations relative to regular court proceedings, such as no live testimony and limited discovery.

As described in Section 2301 of the USPTO's Manual of Patent Examining Procedure, "An interference is a contest under 35 U.S.C. 135(a) between an application and either another application or a patent. An interference is declared to assist the Director of the United States Patent and Trademark Office in determining priority, that is, which party first invented the commonly claimed invention within the meaning of 35 U.S.C. 102(g)(1)." According to US law, "An interference exists if the subject matter of a claim of one party would, if prior art, have anticipated or rendered obvious the subject matter of a claim of the opposing party and vice versa" (37 CFR 41.203). Under this system, it is possible to win a patent interference if the inventor was the first to conceive, but the last to reduce to practice.

During an interference proceeding to determine inventorship, patentability issues can also be raised. Under US patent law, the primary statutory requirements of patentability are encompassed by 35 U.S.C. §§ 101, 102, 103 and 112, and various US Supreme Court and Federal Court cases have ruled on the meanings of these sections. These sections of the code encompass, *inter alia*, statutory subject matter, novelty, obviousness, written description and enablement.

Novelty and obviousness are determined relative to what was known at the time of the invention (prior art) and written description and enablement encompass more or less how well the description and data, as filed, support the claims and breadth of the claims in light of what was known in the art and whether one of ordinary skill in the art would be able to understand and reproduce the invention upon reading the patent. The proceeding is bifurcated into a preliminary motion phase and a priority phase, and each has separate hearings and decisions. This means that in addition to the battle over priority (first-to-invent), the other party's claims can be attacked as unpatentable over other prior art as being anticipated, obvious or both, or it can be attacked as not meeting the enablement or written description requirements of US patent law.

Interference proceedings are not fast, and average about three years to completion. There are several scenarios that can play out, including something other than winner takes all. Opposing parties in an interference are allowed to settle the dispute contractually, such as becoming co-owners of the invention or coming to some sort of an agreement as to how to share potential licensing revenues, etc. Many summaries are available, providing a more in-depth review of the patents and parties involved in the interference (Ledford, 2016; Noonan, 2016; Sherkow, 2015).

Interestingly, the Broad Institute started a page at their website where they generally describe the interference and provide updates. In a July 6, 2016 update of their website, it is stated that settlement discussions have begun between the parties in the US interference proceeding. The terms of a settlement can take many forms and often include a license agreement, but they cannot be anticompetitive, must be in writing and approved by the USPTO, which can then terminate the interference proceeding.

The USPTO will be evaluating what is claimed in each patent and application in the interference. For example, a representative patent claim in the Zhang portfolio of method of use claims and systems claims is claim 1 of Zhang's first US Patent (8,697,359), which is quite a robust claim for genome editing:

1. A method of altering expression of at least one gene product comprising introducing into a eukaryotic cell containing and expressing a DNA molecule having a target sequence and encoding the gene product an engineered, nonnaturally occurring Clustered Regularly Interspaced ShortPalindromic Repeats (CRISPR)–CRISPR associated (Cas)(CRISPR-Cas) system comprising one or more vectors comprising:

    a) a first regulatory element operable in a eukaryotic cell operably linked to at least one nucleotide sequence encoding a CRISPR-Cas system guide RNA that hybridizes with the target sequence, and
    b) a second regulatory element operable in a eukaryotic cell operably linked to a nucleotide sequence encoding aType-II Cas9 protein,

    wherein components (a) and (b) are located on same or different vectors of the system, whereby the guide RNA targets the target sequence and the Cas9 protein cleaves the DNA molecule, whereby expression of the at least one gene product is altered; and, wherein the Cas9 protein and the guide RNA do not naturally occur together.

Of course, other claims are under consideration, but this is an example of what is now patented and being subjected to the interference proceeding.

29.3.3 CRISPR Patent Battle in Europe

Zhang's group has three issued patents in Europe, while Doudna's group has none (see Table 29.1). In Europe, a patent can be attacked by a process called "opposition." Multiple groups have opposed Zhang's issued European patents (EP), three of which have the December 12, 2013 filing date, claiming priority to an earlier provisional application filing date of December 12, 2012. For example, patent EP 2771468, issued to Zhang has been opposed by numerous parties: George Schlich,

Dr. Martin Grund, Regimbeau, CRISPR Therapeutics AG, Dr. Ulrich Storz, Novozymes A/S, Boxall Intellectual Property Management Ltd., Sagittarius Intellectual Property LLP and Harvey Adams. It is unusual for so many parties to oppose a European patent. Like interference proceedings in the US, opposition proceedings in the EPO can take at least a couple of years and appeals can drag on for several more years.

The EP patents to Zhang were granted despite numerous anonymous third parties filing "observations" during prosecution, which suggests that there is at least some strength to the patents if they survived the information being submitted. An "observation" is something submitted to the European Patent Office (EPO) during prosecution of the application, where a third party highlights potential patentability issues for that application. The Doudna group has one patent application pending in Europe with a priority claim to May 25, 2012. Numerous third-party observations have been filed against it, including anonymous filers, but the Broad Institute filed an observation under its own name. Quite a few entities have now filed CRISPR patent applications in Europe in addition to the University of California and the Broad Institute. These include Vilnius University, ToolGen, Sigma-Aldrich and Harvard College.

Doudna and Charpentier's application has had problems during examination, similar to the problems it was having in the US. As mentioned above, the EPO examiner asserts that the provisional patent applications from which it claims priority do not fully support the application under examination. The provisional applications were filed on May 25, 2012, October 19, 2012, January 28, 2013 and February 15, 2013. The pending application may have to assume a new priority date, that date being the one that the examiner thinks fully supports the pending application. The worst-case scenario for the application would be if the date turns out to be the day the regular application was filed. The regular application was not filed until March 15, 2013 and if the examiner cannot be convinced to accept the May 25, 2012 priority date, then their own publication in *Science* on June 28, 2012 could be used against them as prior art, as well as any other publication that occurred before the priority date ultimately established by the EPO.

29.4 Impact of the CRISPR Dispute on Future Development and Research

Previous breakthrough technologies in biotechnology have also been patented, including forms of recombinant DNA and polymerase chain reaction (PCR). For the most part, these patents did not stymie research and licenses were available or were included as part of purchased kits and reagents. Perhaps the winner here will follow suit. Currently, there are also certain reagents being made freely available by the scientists in the patent dispute.

Further complicating ownership issues and the future of licensing and business is the fact that one of Doudna's co-inventors, as listed on her patent application, Charpentier, is currently an owner under patent law because she has not assigned her rights in the invention to any entity. When the Doudna/Charpentier *et al.* paper published in *Science*, Dr. Charpentier's work affiliation was listed as Umeå University, Sweden, but she does not appear to have had an obligation to assign her rights to Umeå University. The rights to their single patent application appear to have been licensed to two separate companies (Intellia and CRISPR), the Berkeley group having licensed to Intellia and Charpentier separately forming CRISPR Therapeutics. With Zhang's primary technology licensed to Editas, once the patent and ownership battles are complete, it could be a winner-take-all scenario, but it is possible that at least some of Zhang's patents and applications may survive if the interference battle over Doudna's first application goes to Doudna's group, because Zhang continued to modify, diversify and add to the base patent that was issued.

As with all inventions, if something better than CRISPR comes along its use could be displaced. In fact, Zhang's group has already published a paper on a new protein, Cpf1, which could replace Cas9 in the CRISPR gene editing platform. Cpf1 is not in any of the early patent applications or patents using CRISPR; however, Zhang now has at least one US patent application (Zhang *et al.*, 2016) that published July 21, 2016, describing Cpf1 and it use.

Cpf1 may have certain advantages over Cas9, which could decrease the value of CRISPR/Cas9 in gene editing and a CRISPR/Cpf1 system could replace it or provide even greater value. Cas9 requires two RNA molecules to cut DNA, but Cpf1 needs only one. The two types of proteins also cut DNA at different places, which can be useful when trying to select a site to edit. In addition, Cpf1 cuts DNA in a different manner from Cas9, which cuts both strands in a DNA molecule at the same position in the form of blunt end cuts. Cuts by Cpf1 leave "sticky" ends. Because blunt ends are not as easy to work with, the generation of sticky ends by Cpf1 seems to provide another advantage over Cas9.

Of course, a newer, better, easier to use and cheaper technology than the CRISPR system may be discovered, rendering the current patent battles moot and possibly diminishing or eliminating the tremendous potential value of CRISPR.

Disclaimer

This chapter reflects the personal views and thoughts of Rodney L. Sparks, J.D., Ph.D. and is not to be construed as representing in any way the corporate views or business advice of Rodney Sparks or of the University of Virginia Licensing & Ventures Group and their Affiliates, Subsidiaries or Divisions, nor the views or advice of Cambridge University Press. The content is solely for purposes of education, discussion and illustration, and is not to be considered legal advice. Additionally, the author has no financial ownership in any companies referred to in the chapter nor any role as an attorney or consultant to any of the parties involved in the inventorship and ownership disputes described in the chapter.

REFERENCES

Haeussler M, Concordet J. 2016. Genome editing with CRISPR-Cas9: can it get any better? *J Genet Genom* **43**(5): 239–250.

Jinek M, Chylinski K, Fonfara I, *et al.* 2012. A programmable dual-RNA-guided DNA endonuclease in adaptive bacterial immunity. *Science* **337**: 816–821.

Leahy-Smith America Invents Act. 2011. One Hundred Twelfth Congress of the United States of America, Public Law 112–29, September 16, 2011, 125 Stat 284.

Ledford H. 2016. Bitter fight over CRISPR patent heats up. *Nature* **529**: 265.

Noonan KE. 2016. CRISPR interference declared. *Patent Docs*. www.patentdocs.org/2016/01/crispr-interference-declared.html (accessed November 2017).

Sherkow J. 2015. The CRISPR patent interference showdown is on: how did we get here and what comes next? *Stanford Law School, Law and Biosciences Blog*. https://law.stanford.edu/2015/12/29/the-crispr-patent-interference-showdown-is-on-how-did-we-get-here-and-what-comes-next/ (accessed November 2017).

Sibley A, Sparks R. 2009. The difficulty of determining joint inventorship, especially with regard to novel chemical compounds and their applications. *Loyola Law Technol Annu* **8**(1): 44–62.

Sparks R. 2008. Performing a preliminary assessment of patentability for a new invention: guidelines for non-patent lawyers. *Int In-House Counsel J* **2**(5): 816–827.

Sparks R. 2009. Unintended negative consequences of joint ownership of a patent. *Int In-House Counsel J* **3**(9): 1411–1420.

Sparks R, Park S, Paschall C. 2015. Recent patent legislation and court decisions in the United States: impact on validity of patents and on obtaining, licensing, and enforcing patents. *Int In-House Counsel J* **8**(31):, 3369–3380.

Zhang F, Zetsche B, Slaymaker I, Gootenberg J, Abudayyeh O. 2016. Novel CRISPR Enzymes and Systems. US Patent Application Number 14/975,085.

30 The Ethics of Human Genome Editing

Tetsuya Ishii

30.1 Introduction

By repurposing bacterial nucleases, genome editing techniques, such as zinc-finger nucleases (ZFNs; Klug, 2010), transcription activator-like effector nucleases (TALENs; Joung and Sander, 2013) and the clustered regularly interspaced short palindromic repeat (CRISPR)/(Cas9), (Hsu *et al.*, 2014), have facilitated inserting an exogenous gene, correcting a gene mutation (or copying of a variant) and disrupting an endogenous gene in mammalian cells. ZFNs and TALENs have already been used in clinical settings. A Phase I clinical trial demonstrated that the infusion of ZFN-CCR5-disrupted T cells was safe in HIV-positive patients (Table 30.1: NCT00842634; Tebas *et al.*, 2014). Four relevant Phase 1 and/or 2 trials for treating HIV infection are ongoing in the USA (Table 30.1). A compassionate use of allologous chimeric antigen receptor (CAR) 19 T cells for an infant with terminal leukemia resulted in remission (Qasim *et al.*, 2015). Of note, in the CAR T cells, both T-cell receptor (TCR) and CD52 genes were disrupted by TALEN treatment. Currently, at least seven relevant trials are ongoing in the USA and China, including gene disruption for treating HIV human papillomavirus (HPV) infection, as well as gene addition for treating hemophilia B (HB) and mucopolysaccharidosis I (MPS I; Table 30.1). With regard to germline genome editing, two reports demonstrated that human embryos can be genetically modified by CRISPR/Cas9 treatment, despite some technical problems (Kang *et al.*, 2016; Liang *et al.*, 2015).

However, genome editing utilizes artificial nucleases that can unintentionally break DNA double strands at off-target sites in the genome. Although off-target mutations in ZFN-CCR5-dirupted T cells were not investigated in the world-first genome editing trials (Tebas *et al.*, 2014), the off-target effects could create small indels that might affect subjects in clinical trials (Araki and Ishii, 2016; Cox *et al.*, 2015; Hsu *et al.*, 2014; Joung, 2015; Joung and Sander, 2013; Klug, 2010). In addition to the tragedies and subsequent setbacks in gene transfer (Barbour, 2000; Hacein-Bey-Abina *et al.*, 2008; Kimmelman, 2008), the previous attempts at germline modification (Barritt *et al.*, 2001a; 2001b; Zhang, 2003) urge us to meticulously consider when, where and how to carry

Table 30.1 Clinical trials involving somatic genome editing

Phase (Year)	Sponsor	Condition	Intervention	Study design	Trial ID
Ex vivo somatic genome editing					
Phase 1 (2009–2013)	University of Pennsylvania (USA)	HIV	Infusion of autologous CD4+ T Cells modified at CCR5 by ZFN, with interruption of antiretroviral treatment	Non-randomized. Open label. 12 patients enrolled. ZFN delivered by adenoviral vector	NCT00842634
Phase 1 (2009–2014)	Sangamo Biosciences (USA)	HIV	Dose escalation of autologous CD4+ T Cells modified at CCR5 by ZFN	Non-randomized. Open label. 19 patients enrolled. ZFN delivered by adenoviral vector	NCT01044654
Phase 1 and 2 (2011–)	Sangamo Biosciences (USA)	HIV	Dose escalation of cyclophosphamide in patients receiving autologous CD4+ T cells modified at CCR5 by ZFN	Non-randomized. Open label. Patient enrollment plan: 26. ZFN delivered by adenoviral vector	NCT01543152
Phase 1 and 2 (2014–)	Sangamo Biosciences (USA)	HIV	Repeat doses of autologous CD4+ T cells modified at CCR5 by ZFN following cyclophosphamide conditioning	Non-randomized. Open label. Patient enrollment plan: 12. Cells electroporated with ZFN mRNA	NCT02225665
Phase 1 (2015–)	University of Pennsylvania (USA)	HIV	Infusion of autologous CD4+ T Cells modified at CCR5 by ZFN with or without prior administration of cyclophosphamide	Non-randomized. Open label. Patient enrollment plan: 15. ZFN mRNA electroporated	NCT02388594

Phase 1 (2015–)	City of Hope Medical Center (USA)	HIV	Infusion of autologous CD34+ hematopoietic stem/ progenitor cells modified at CCR5 by ZFN following busulfan conditioning	Non-randomized. Open label. Patient enrollment plan: 12. Cells electroporated with ZFN mRNA	NCT02500849
In vivo somatic genome editing					
Phase 1 (2016–)	Sangamo Biosciences (USA)	HB	Intravenously delivery of ZFN-adeno-associated virus2/6 to insert a correct copy of *FIX* into the albumin locus in hepatocytes	Non-randomized. Open label. Patient enrollment plan: 9	NCT02695160
Phase 1 (2016–)	Sangamo Biosciences (USA)	MPS I	Intravenously delivery of ZFN-adeno-associated virus2/6 to insert a correct copy of *IDUA* into the albumin locus in hepatocytes	Non-randomized. Open label. Patient enrollment plan: 9	NCT02702115
Phase 1 (2016–)	Huazhong University of Science and Technology (China)	CC without invasion	Intratumorally transfection of ZFN into HPV-infected cervical epithelials to eliminate HPV infection by disrupting HPV16 and HPV18 E7 DNA.	Non-randomized. Open label. Patient enrollment plan: 20	NCT02800369

Relevant clinical trials were surveyed using ClinicalTrials.gov.
The status is on July 12, 2016. HIV: Human immunodeficiency virus, HB: Hemophilia B, MPS I: Mucopolysaccharidosis I, CC: cervical cancer, HPV: Human papillomavirus.

out human genome editing that can affect the subjects or future generations. In this chapter, the risks and benefits are weighed with regard to somatic and germline genome editing in humans. The social implications of germline genome editing are also considered.

30.2 The History of Human Gene Transfer

Conventional gene transfer attempts to achieve longer efficacy than drug treatment, by introducing a copy of a functional gene into human cells *ex vivo* or *in vivo*. To date, at least 2356 clinical trials of somatic gene transfer have been conducted worldwide since the first-in-human trial for treating adenosine deaminase-severe combined immune deficiency (ADA-SCID) in 1990 (www.abedia.com/wiley/continents.php). However, severe side effects and tragedies occurred in the subjects, leading to setbacks in the clinical development (Barbour, 2000; Hacein-Bey-Abina *et al.*, 2008; Kimmelman, 2008). Of particular note, one infant among four X-linked SCID patients died after developing leukemia following the infusion of $CD34^+$ cells in which the IL2RG gene was retrovirally transferred (Hacein-Bey-Abina *et al.*, 2008). Nonetheless, consistent efforts, such as the improvement of the viral vector, have led to the approval of some products, such as Gendicine (*in vivo* cancer gene therapy) in China (Li *et al.*, 2015), Neovasculgen (*in vivo* gene therapy for peripheral arterial disease) in Russia (Chervyakov *et al.*, 2016) and Glybera (*in vivo* gene therapy for lipoprotein lipase deficiency; Watanabe *et al.*, 2015) and Strimvelis (*ex vivo* ADA-SCID gene therapy; Schimmer and Breazzano, 2016) in the EU.

With regard to germline cells (oocytes, spermatozoa, embryos), some clinics have practiced allologous mitochondrial (DNA) transfer to oocytes (ooplasmic transfer; Barritt *et al.*, 2001a; 2001b) or embryos (pronuclear transfer; Zhang, 2003) to treat intractable infertility with which mitochondrial dysfunction might be associated. However, the practices of these two protocols raised profound ethical and social issues, thus calling for intervention by regulators in the USA and China, respectively (Hawes *et al.*, 2002; Ishii, 2015b). Currently, at least 29 countries prohibit the reproductive use of germline gene modification for safety and/or ethical reasons (Ishii, 2015a). However, mitochondrial donation in oocytes or embryos recently became permissible in order to prevent maternal transmission of severe mitochondrial diseases to progeny in the UK (Ishii, 2014, 2015b).

It is important to understand that there are currently only a few approved gene transfer products in the world. The clinical use of germline gene modification is allowed in part in the UK if the genetic intervention focuses on mitochondrial DNA and aims at disease prevention in offspring.

30.3 Subjects and Informed Consent

A small number of healthy volunteers are frequently recruited in Phase 1 trials that aim to assess the safety of a small dose of drug that is absorbed initially and ultimately excreted from the body (Li, 2001). They are compensated financially. Likewise, some researchers might contemplate enrolling healthy or medically stable volunteers in their first-in-human trials involving genome editing (Kimmelman, 2007). However, such people should not be enrolled. First, genome editing is a relatively new genetic engineering tool compared with gene transfer, and unknown risks are likely to remain. Indeed, only patients have been and will be enrolled in clinical trials involving genome editing (Table 30.1). However, the author is somewhat skeptical about the gene editing trial for eliminating HPV *in vivo* using ZFNs because it plans to recruit females with cervical cancer neoplasia "without invasion" (Table 30.1). Several verified treatment options are available at this early stage of cervical cancer (Colombo *et al.*, 2012), so using such subjects for an unproven virus elimination treatment seems to raise ethical concerns. Second, healthy volunteers are seldom recruited in gene transfer trials (www.abedia.com/wiley/indications.php). Third, genome editing is technically similar to gene transfer, which can change the DNA sequence in the genome and leave such a change *in vivo* for a relatively long period of time. If the genetic intervention is associated with some risks, it can impose long-term, possibly adverse health effects on the subjects, which outweighs the compensation for participating in the study.

In somatic genome editing trials, both the risks and benefits must be carefully explained to patients in advance (Figure 30.1; Araki and Ishii, 2016). However, genome editing has more complicated pathways of double-strand break (DSB) and non-homologous end joining (NHEJ) or homology-directed repair (HDR) than gene transfer (Hsu *et al.*, 2014; Joung and Sander, 2013; Klug, 2010). For this reason, the subjects are ideally adult patients who can sufficiently understand the explanation of the potential health risks and give legitimate informed consent. In germline genome editing, the direct subjects are unborn children. Informed consent must be obtained from prospective parents in clinics providing assisted reproductive technology (ART) such as *in vitro* fertilization (IVF), intracytoplasmic sperm injection (ICSI) and preimplantation genetic diagnosis (PGD; Ishii, 2015a). Likewise, the provision of germline genome editing in ART clinics requires the informed consent of any parents, who must fully understand the potential risks to embryos, fetuses and progeny.

30.4 Somatic Genome Editing

Since genome editing can attain efficient and precision genetic modification in human cells, it appears that the benefits will exceed the risks in advanced treatment of refractory diseases or cases where no treatment is available. However, according to the Declaration of Helsinki, the predictable risks must be carefully assessed in advance (www.wma.net/en/30publications/10policies/b3/). Furthermore, the risks must be minimized and monitored in clinical research. Somatic genome editing in humans is scrutinized for both deliberate and unintended genetic modifications. With regard to HDR, it might result in just NHEJ if a simultaneously introduced DNA template is not utilized for repairing DSB.

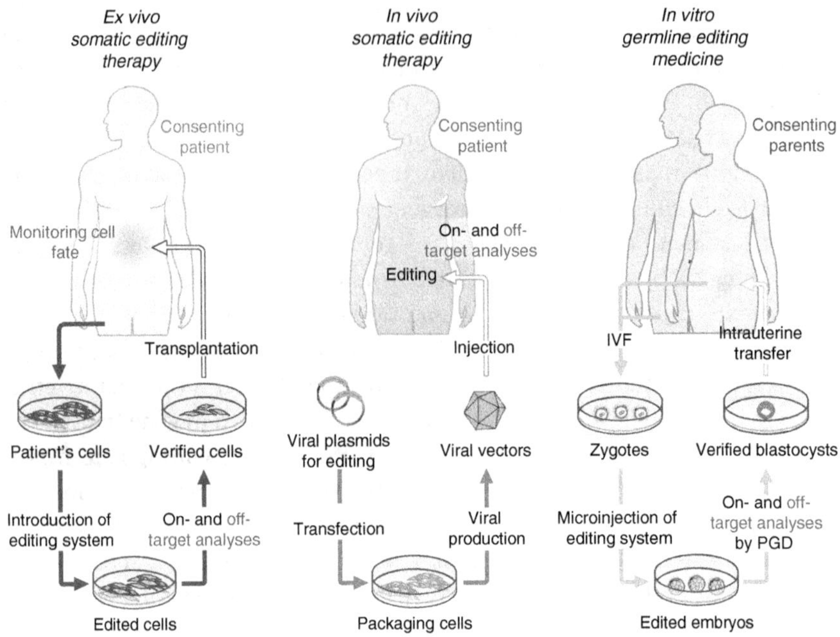

Figure 30.1 Three concepts of genome editing medicine. Autologous *ex vivo* somatic editing therapy begins by procuring somatic cells from a consenting patient. After genome editing, somatic cells are subject to on-target gene modification and off-target mutation analyses. The behavior of transplanted cells will be monitored in the patient. With regard to *in vivo* somatic editing therapy, the direct subject of genome editing is a consenting patient. A viral vector harboring the genome editing system is produced by the culturing of packaging cells. After the viral vector is delivered to the body, the patient will undergo various tests, which include the confirmation of on-target gene modification and the occurrence of off-target mutations in some tissues. *In vitro* germline editing medicine is primarily initiated by creating zygotes *in vitro*. Similarly to an assisted reproductive technology (ART) procedure that is used for male infertility, genome editing nucleases are microinjected into zygotes and are subsequently subject to preimplantation genetic diagnosis (PGD), by which no off-target mutations as well as an on-target gene modification are verified, prior to embryo transfer. IVF: *in vitro* fertilization. (Araki and Ishii, 2016. Reproduced with permission from Elsevier.) (A black-and-white version of this figure will appear in some formats. For the color version, please refer to the plate section.)

Therefore, the efficiency of HDR should be sufficiently assessed in advance. In gene addition via HDR, it is important to consider a genomic safe harbor (GSH) where a copy of a functional gene can reside without affecting the phenotypes, when introduced into the genome (Sadelain *et al.*, 2012). *ROSA26* is a well-characterized GSH in mice. Recent clinical trials for treating HB and MPS I utilize the albumin locus as a GSH (Table 30.1). However, it is vital to carefully select a locus as a GSH in each therapeutic strategy through animal experiments. Disrupting an endogenous gene via NHEJ is permissible only if clinical evidence supports that the genetic modification can truly lead to a therapeutic benefit. For instance, the case reports of "Berlin patient" (Hutter *et al.*, 2009) justified the world-first genome editing trial in which *CCR5* in T cells was intentionally disrupted *ex vivo* in order to provide the resistance to HIV infection in patients

(Table 30.1; Tebas *et al.*, 2014). However, of note, NHEJ introduces various-length indels at the target site in somatic cells (Araki and Ishii, 2016). Most of the indels are likely to result in loss-of-function mutations at a target gene by a frameshift. In contrast, some indels at a target site can lead to gain-of-function mutations via mechanisms such as amino acid substitution. Therefore, it is crucial to carefully assess the efficiency of a certain genetic modification and its resulting phenotypic changes.

Regardless of the genetic modification pathways, one major concern in human genome editing is the off-target effect (Araki and Ishii, 2016; Cox *et al.*, 2015; Joung, 2015). If the custom-made nucleases break DNA double strands at unintentional sites in the genome, the off-target DSBs could create indels that may cause cancer via mechanisms such as the disruption of the tumor-suppressor gene, *TP53*. Even if more sophisticated nucleases are used under optimized conditions, the off-target mutation risk remains substantial because artificial nucleases are introduced in human cells *ex vivo* or *in vivo*. However, in *ex vivo* somatic editing trials, this risk can be reduced if off-target mutations are investigated prior to transplantation (Figure 30.1; Araki and Ishii, 2016). If harmful off-target mutations are found, the cell transplantation can be suspended. In so doing, analyzing only potential off-target sites may be insufficient, as genome editing is designed based on a reference genome, not the patient's genome, which varies in such respects as SNPs and CNV. In retrospect, some retrovirus vectors were preferentially inserted near proto-oncogenes, such as *LMO2*, causing the onset of leukemia within several years, although the retrovirus vector had been believed to randomly integrate into the genome (Hacein-Bey-Abina *et al.*, 2008). In contrast, although whole-genome sequencing is considered useful by some for the comprehensive identification of off-target mutations, the efficacy of this approach is doubtful due to the ambiguity between indels and SNPs. Instead, whole-exome sequencing or investigation of cancer-related genes may be more effective and practical (Araki and Ishii, 2016). Of particular note, prior investigation of off-target mutations is impossible in *in vivo* somatic editing. However, the risk of an off-target effect can be reduced through preclinical research using human cells or disease model animals.

Such risk–benefit analyses view *ex vivo* somatic editing more favorably than *in vivo* somatic editing (Figure 30.1; Araki and Ishii, 2016). Nonetheless, the benefit may exceed the risk even in *in vivo* genome editing, if patients with severe or life-threatening diseases are recruited. In each genome editing trial, the risk–benefit balance must be cautiously reviewed. It is also important to follow up patients who undergo somatic editing therapy for at least five to ten years, as suggested by the onset of T-cell leukemia in infants as late as 31–68 months after the infusion of $CD34^{+}$ cells (Hacein-Bey-Abina *et al.*, 2008). Long-term follow-up, which may impose a burden on volunteers, is mandatory to protect them and assess the safety and efficacy for future genome editing therapies.

30.5 Germline Genome Editing

As mentioned above, germline genome editing, if ever implemented, will require parental consent because the direct subject is an unborn child who cannot consent before birth (Figure 30.1). Under what conditions does parental consent justify germline intervention from the point of view of a child's welfare? Consider previous germline modifications for intractably infertile patients. Although ooplasmic transfer resulted in about 30 live births, some cases led to a miscarriage and an elective abortion because the fetuses were affected with Turner syndrome (Barritt *et al.*, 2001b). Furthermore, one child born of ooplasmic transfer was diagnosed with pervasive developmental disorder. Pronuclear transfer led to a triplet pregnancy, but no live births occurred: two fetuses died of respiratory distress and cord prolapse after the selective reduction of a fetus (Zhang, 2003). Such germline modifications raised profound ethical concerns about overzealous clinical practices (not clinical trials) for infertility treatment (Hawes *et al.*, 2002). Although genome editing differs from mitochondrial transfer, germline genome editing for treating difficult infertilities is unthinkable, because the risky germline modification is conducted primarily based on the parental wish, not the prospective child's welfare.

A potential purpose, based on the child's welfare point of view, may be the prevention of genetic disease in offspring because the first human embryo editing study, despite no mention of available options such as PGD, aimed to correct a mutation responsible for the onset of an autosomal-recessive β-thalassemia (Liang *et al.*, 2015). In addition, some speakers at the International Human Gene-editing Summit suggested that the most likely purpose of germline editing is preventing the inheritance of genetic disease in the offspring of prospective parents, who wish to have a healthy child without using donor gametes (www.nationalacademies.org/gene-editing/Gene-Edit-Summit/index.htm). Disease prevention in offspring seems compelling in the cases of definite inheritance of a severe genetic condition, where PGD is inapplicable, such as in mitochondrial disease carriers with a high mutation load, autosomal dominant diseases in which one or both parents is/are homozygous (e.g. Huntington's disease) and autosomal recessive diseases where both parents are homozygous (e.g. cystic fibrosis; Ishii, 2015b, 2015a). Such uses of germline editing can be considered to be primarily based on the welfare of a prospective child. In such cases, the benefits of corrective genome editing will be likely to exceed the risks (embryo arrest, implantation failure, miscarriage, stillbirth and the failure of disease prevention) when genome editing advances further. One might assert that such cases are too rare to initiate such risky germline editing (Lander, 2015). However, the rarity of such cases is not a sufficient reason for ethical rejection. Present-day medicine is expected to establish effective procedures for such rare conditions.

However, clinical and social issues may emerge. If a resultant child who underwent germline editing to prevent one disease is born with another disease, can we determine whether it occurred due to an off-target effect or spontaneous mutation? Indeed, previous discussions did not clarify how Turner syndrome and pervasive developmental disorder are derived from ooplasmic transfer (Hawes *et al.*, 2002). Although PGD may confirm gene correction prior to embryo transfer, it appears to be difficult to comprehensively investigate off-target mutations due to chromosomal mosaicism in human embryos (Taylor *et al.*, 2014). Additionally, can corrective genome editing be employed at ART clinics without inviting its misuses? ART has raised many ethical issues in society thus far. For instance, PGD has been used for non-medical sex selection (sex selection for social reasons, such as family balancing) in countries with lax regulations (Whittaker, 2011), although mouse experiments suggest that offspring who undergo the biopsy for PGD are at high risk of late onset neurodevelopmental and metabolic diseases (Sampino *et al.*, 2014; Yu *et al.*, 2009).Genome editing can also be used to change visible traits in offspring, such as eye and hair colors (Ishii, 2015a). Such misuses are likely to occur in some countries, since many parents have certain desires for their unborn children. Such potential clinical and social issues discourage the implementation of germline genome editing in clinics.

30.6 A Reflection from Japan

In Japan, the gene transfer trial requiresa two-step review by an institutional review board and the Ministry of Health, Labour and Welfare (MHLW). Although approximately 50 gene transfer trials have been carried out, there are currently no approved products (www.nihs.go.jp/cgtp/cgtp/sec1/gt_prtcl/prtcl-j3.html). Some have complained about the strict regulations on human gene transfer. Remarkably, the double layers of review have prevented severe side effects in previous trials. In 2015, the MHLW enacted new guidelines, the Guidelines for Clinical Research such as Gene Therapy (2015), which add disease prevention alongside disease treatment and are relevant in *in vivo* somatic editing therapy. Clinical trials of *ex vivo* somatic editing therapy would be regulated under the Act on Securement of Safety of Regenerative Medicine, etc.

Clinical trials involving germline genome editing are still prohibited under the Guidelines (Ishii, 2015a). However, the rough-and-ready use of germline genome editing could occur in ART centers, as Japan is one of the most ART-active countries in the world and has no ART-relevant legislation, except for the prohibition of human cloning (Ishii, 2015a).

30.7 Conclusions and Path Forward

Genome editing has been rapidly applied in clinical settings. However, despite the great benefits of robust genetic modification, clinical uncertainties remain on account of off-target effects. In the first-in-human trials for somatic genome editing, the risks and benefits should be cautiously contemplated to ensure the safety of patients and social trust in this field. To obtain legitimate informed consent, researchers should form a consensus on how to assess off-target effects and investigate the risks in both preclinical and clinical studies as much as possible (Araki and Ishii, 2016; Joung, 2015). Germline genome editing appears to be appropriate in cases where there is a high probability of transmitting a severe genetic condition to progeny. Nonetheless, germline editing is, at present, inadvisable due to the difficulties in risk assessment and potential misuse in countries without explicit regulations concerning ART.

It is vital to inform the public of the current status of genome editing medicine, as well as gene transfer, and conduct ongoing discussions in order to shape the social norm regarding appropriate uses of genome editing in humans.

REFERENCES

Araki M, Ishii T. 2016. Providing appropriate risk information on genome editing for patients. *Trends Biotechnol* **34**: 86–90.

Barbour V. 2000. The balance of risk and benefit in gene-therapy trials. *Lancet* **355**: 384.

Barritt JA, Brenner CA, Malter HE, Cohen J. 2001a. Mitochondria in human offspring derived from ooplasmic transplantation. *Hum Reprod* **16**: 513–516.

Barritt JA, Brenner CA, Malter HE, Cohen J. 2001b. Rebuttal: interooplasmic transfers in humans. *Reprod Biomed Online* **3**: 47–48.

Chervyakov YV, Staroverov IN, Vlasenko ON, *et al.* 2016. [Remote results of treatment of patients with chronic lower-limb ischaemia by means of indirect revascularization and gene therapy]. *Angiol Sosud Khir* **22**: 29–34, 36–37.

Colombo N, Carinelli S, Colombo A, *et al.* 2012. Cervical cancer: ESMO Clinical Practice Guidelines for diagnosis, treatment and follow-up. *Ann Oncol* **23**(Suppl. 7): vii27–vii32.

Cox DB, Platt RJ, Zhang F. 2015. Therapeutic genome editing: prospects and challenges. *Nat Med* **21**: 121–131.

Hacein-Bey-Abina S, Garrigue A, Wang GP, *et al.* 2008. Insertional oncogenesis in 4 patients after retrovirus-mediated gene therapy of SCID-X1. *J Clin Invest* **118**: 3132–3142.

Hawes SM, Sapienza C, Latham KE. 2002. Ooplasmic donation in humans: the potential for epigenic modifications. *Hum Reprod* **17**: 850–852.

Hsu PD, Lander ES, Zhang F. 2014. Development and applications of CRISPR-Cas9 for genome engineering. *Cell* **157**: 1262–1278.

Hutter G, Nowak D, Mossner M, *et al.* 2009. Long-term control of HIV by CCR5 Delta32/Delta32 stem-cell transplantation. *N Engl J Med* **360**: 692–698.

Ishii T. 2014. Potential impact of human mitochondrial replacement on global policy regarding germline gene modification. *Reprod Biomed Online* **29**: 150–155.

Ishii T. 2015a. Germ line genome editing in clinics: the approaches, objectives and global society. *Brief Funct Genom* **16**: 46–56.

Ishii, T. 2015b. Germline genome-editing research and its socioethical implications. *Trends Mol Med* **21**: 473–481.

Joung JK. 2015. Unwanted mutations: standards needed for gene-editing errors. *Nature* **523**: 158.

Joung JK, Sander JD. 2013. TALENs: a widely applicable technology for targeted genome editing. *Nat Rev Mol Cell Biol* **14**: 49–55.

Kang X, He W, Huang Y, *et al.* 2016. Introducing precise genetic modifications into human 3 PN embryos by CRISPR/Cas-mediated genome editing. *J Assist Reprod Genet* **33**: 581–588.

Kimmelman J. 2007. Stable ethics: enrolling non-treatment-refractory volunteers in novel gene transfer trials. *Mol Ther* **15**: 1904–1906.

Kimmelman J. 2008. The ethics of human gene transfer. *Nat Rev Genet* **9**: 239–244.

Klug A. 2010. The discovery of zinc fingers and their applications in gene regulation and genome manipulation. *Annu Rev Biochem* **79**: 213–231.

Lander ES. 2015. Brave new genome. *N Engl J Med* **373**: 5–8.

Li AP. 2001. Screening for human ADME/Tox drug properties in drug discovery. *Drug Discov Today* **6**: 357–366.

Li Y, Li B, Li CJ, Li LJ. 2015. Key points of basic theories and clinical practice in rAd-p53 (Gendicine) gene therapy for solid malignant tumors. *Expert Opin Biol Ther* **15**: 437–454.

Liang P, Xu Y, Zhang X, *et al.* 2015. CRISPR/Cas9-mediated gene editing in human tripronuclear zygotes. *Protein Cell* **6**: 363–372.

Qasim W, Jal Amrolia S, Samarasinghe S, *et al.* 2015. First clinical application of talen engineered universal CAR19 T cells in B-ALL. *Blood* **126**: 2046.

Sadelain M, Papapetrou EP, Bushman FD. 2012. Safe harbours for the integration of new DNA in the human genome. *Nat Rev Cancer* **12**: 51–58.

Sampino S, Zacchini F, Swiergiel AH, *et al.* 2014. Effects of blastomere biopsy on post-natal growth and behavior in mice. *Hum Reprod* **29**: 1875–1883.

Schimmer J, Breazzano S. 2016. Investor outlook: rising from the ashes; GSK's European approval of Strimvelis for ADA-SCID. *Hum Gene Ther Clin Dev* **27**: 57–61.

Taylor TH, Gitlin SA, Patrick JL, *et al.* 2014. The origin, mechanisms, incidence and clinical consequences of chromosomal mosaicism in humans. *Hum Reprod Update* **20**: 571–581.

Tebas P, Stein D, Tang WW, *et al.* 2014. Gene editing of CCR5 in autologous CD4 T cells of persons infected with HIV. *N Engl J Med* **370**: 901–910.

Watanabe N, Yano K, Tsuyuki K, Okano T, Yamato M. 2015. Re-examination of regulatory opinions in Europe: possible contribution for the approval of the first gene therapy product Glybera. *Mol Ther Methods Clin Dev* **2**: 14066.

Whittaker AM. 2011. Reproduction opportunists in the new global sex trade: PGD and non-medical sex selection. *Reprod Biomed Online* **23**: 609–617.

Yu Y, Wu J, Fan Y, *et al.* 2009. Evaluation of blastomere biopsy using a mouse model indicates the potential high risk of neurodegenerative disorders in the offspring. *Mol Cell Proteomics* **8**: 1490–1500.

Zhang JEA. 2003. Pregnancy derived from human nuclear transfer. *Fertil Steril* **80**: 56.

31 Regulating Human Genome Editing: Negotiating Ethical Concerns Through Science and Policy

Sarah Chan

31.1 Introduction

Since the early days of recombinant DNA technology, human genetic modification has been the subject of much discussion in bioethics, with the prospect of human germline (or heritable) genetic modification attracting particular concerns. Until recently, however, the technologies available for genetic modification were simply not practical for use in this way. Now, with the advent of genome editing techniques such as CRISPR/Cas9 that are vastly more precise and efficient than previous methods, human germline genetic modification has suddenly become a much more real possibility, sparking renewed ethical debate and calls for regulation from scientists, the public and policy-makers alike. But what form should this regulation take, what should be permitted or prohibited, and how?

My aim in this chapter is to examine ethical concerns related to human genome editing in order to identify the issues that present key sites for regulation. I argue that while research involving human embryos and germline genetic modification may continue to pose ethical questions, there are ways in which regulation can help to negotiate these issues by providing, not moral answers, but practical solutions.

31.2 Objections to Human Genome Editing

A widely acknowledged concern over human gene editing is the risk of unforeseen adverse side effects. Previous attempts at gene therapy using earlier techniques were marred by a number of high-profile failures, such as the death of Jesse Gelsinger, one of the participants in a 1999 gene therapy trial, who suffered an immune response to the viral vector used (Marshall, 1999; Savulescu, 2001) and a series of gene therapy trials in children for X-linked severe combined immunodeficiency disease, which, although effective in treating the target disorder, also resulted in a number of patients developing leukemia as the result of insertional mutagenesis (Ginn *et al.*, 2013; Hacein-Bey-Abina *et al.*, 2008). Hindsight provides opportunities to criticize such cases on scientific and ethical grounds, and gene editing technologies, with their increased targeting accuracy and precision, may reduce some of the uncertainty associated with previous methods of genetic modification. Other aspects of risk and uncertainty in the use of gene editing, however, such as off-target effects, remain to be characterized.

Of course, nothing is entirely without risk; this is not a principled objection to the use of genome editing as such. It is important, however, that the development and eventual provision of clinical gene editing applications be carefully managed in order to balance safety and efficacy with patient needs and demands; this will be a key task for regulators in the near future (Chan and Medina Arellano, 2016).

Another line of objection rests on the idea that genetic modification itself is somehow wrong or harmful, regardless of the actual effect of the modification: that, perhaps, it infringes the sanctity of the human genome or is "contrary to human dignity" (UNESCO, 1998). Such objections have tended to focus on germ line, rather than somatic modifications. While claims based on human dignity and the supposed inviolability of the genome in its natural state tend to be vague at best (Kimmelman, 2008; Macklin, 2003), they also point towards a deeper concern regarding germline genetic modification, invoking issues such as choice and consent, or the right to an unmodified genome. These will be considered further below. Thus far, however, no scientists have come forward seriously to propose immediate germline reproductive uses of gene editing, though it remains the subject of ethical and regulatory concern.

The most controversial uses of genome editing to date have undoubtedly been those concerning human embryos. Even aside from the prospect of genetic modification, the use of human embryos in research itself remains far from a settled moral issue, as the debate over human embryonic stem cell (hESC) research continues to demonstrate. At the heart of the problem is whether or not human embryos possess some special status that makes it morally impermissible to use them in research, especially research that would lead to their destruction.

31.3 Regulating the Human Embryo

The moral status of the human embryo has been considered at great length, in numerous discussions that I shall not recap here; suffice to say that considerable disagreement remains. Nevertheless, despite the persisting lack of consensus on the moral front, a number of countries have succeeded in developing a compromise regulatory position regarding embryo and stem cell research that is acceptable, or at least accepted. What this position is varies among countries: the UK, for example, allows a range of uses, including the creation of embryos for research purposes, but all human embryo research is strictly controlled and monitored under a licensing scheme; some countries permit the use of embryos that are left over from IVF procedures ("spare" or supernumerary embryos), but not their creation for research; in other countries embryo research is prohibited, although research on hESC lines derived elsewhere may be allowed. Some countries yet lack specific regulation on embryo research altogether (Isasi, 2009; Isasi and Knoppers, 2006; see endnote pg. 462).

The key point for the regulation of genome editing is this: in jurisdictions where an adequate regulatory solution already exists to the moral problem of the embryo and where embryo research is permitted by the regulations in force, we should not allow the debate over genome editing to stall at the question of whether embryo research is permissible *at all*, especially if the effect of this is the persistence of a regulatory vacuum with respect to embryo genome editing. Regulatory stalemate as a response to moral disagreement is undesirable in most circumstances, as it serves the goals of nobody: not those who oppose the practice and wish to see it banned, nor those who wish to pursue the practice in question with the confidence of state-endorsed legitimacy and certainty over what is permitted, nor the often less polarized majority who are not entirely opposed, but may have concerns that an ongoing lack of regulation fails to address. It is even less desirable when the stalemate arises as the result of a moral objection to which a regulatory solution has already been found. Hence, while review of existing embryo research legislation may be needed to ensure that genome editing is adequately covered (Chan *et al.*, 2015), this ought not to be an excuse to reopen the embryo debate and leave genome editing unaddressed.

31.4 The Harm of Heritability?

The major worry over germline gene editing is the heritability of the modifications, which will thus potentially have an effect on future generations. This is what we might call an "escalator problem"; that is to say, by and large it does not create new ethical problems, but rather exacerbates existing ones. In the case of heritable genome editing, the risks are magnified because adverse effects have the potential to affect more people, not just our children but our children's children and so forth. Likewise, any harm that is done to children by being born with genetic modifications over which they had no choice, or to which they did not consent, is supposedly multiplied because these modifications will also be passed on to their descendants, who also had no choice and did not consent.

We should bear in mind, however, that if the risks of harm are magnified, so are the benefits. Should a heritable adverse effect occur, the affected individual may choose to avoid passing it on by using reproductive selection, or indeed electing not to reproduce at all; potentially a hard choice, but one that many people with genetic disease already face. If on the other hand the modification is successful, the burden of potentially passing on genetic disease is also lifted from their descendants.

As to whether children are harmed by their lack of choice or consent to genome editing procedures, we must remember that nobody is given a choice about the genes they are born with, or indeed about being born at all! We might say that if we decide to use genome editing in reproduction, to alter a gene that would otherwise lead to some genetic disease, the children born as a result are denied a choice over whether they wanted to be genetically modified. Yet if we choose *not* to use genome editing and allow them to be born with the disease, they have equally been denied the choice of whether to be cured (Chan and Harris, 2006).

When it comes to gene editing our children, then, given that the choice is not theirs either way, we are not depriving them of a choice by deciding for them one way or another. It might be relevant, however, to ask ourselves what they would probably prefer if given a choice, or what the best choice for them would be. Embryos, like infants and young children, are able neither to refuse nor to consent, but they nevertheless can and indeed in some cases should be given medical treatment. The choice falls on those responsible to make decisions in their best interests, which may well include receiving treatment such as gene editing.

Perhaps, then, the concern is not harm to individuals caused by gene editing, but the potential to alter the future genome of humanity in the collective sense? Even somatic modifications or non-genetic medical interventions may do this, however, by allowing individuals to survive and reproduce who would not otherwise have done so. Beyond this, we make choices that will affect the future human gene pool when we decide whether, when and with whom to reproduce naturally. While we might say that these individual decisions are hardly likely to have an effect on the human race as a whole, it should be noted that on the scale at which genome editing is likely to be used, at least initially, species-level effects are equally improbable. Further, if we are worried about species-level effects on the collective genome, we ought to be far more concerned about existing environmental and social conditions that are having population-level effects on the survival, wellbeing and reproduction of large numbers of humans: these are also affecting the "genome of humanity," and with unknown long-term consequences.

31.5 Research vs. Reproduction

These arguments will not convince everyone, and human reproductive germline gene editing will undoubtedly remain controversial for some time. For regulatory purposes, however, it is crucial to note that all of these issues about harm to children and future generations become moot if the embryo is not implanted. A key distinction for regulation, therefore, should be between research uses and reproductive uses of gene editing: if what we are concerned to prevent is the production of genetically modified children, a ban on *all* embryo gene editing is neither necessary nor justified.

In the case of the UK, the research–reproduction distinction is given effect by legislative provisions that specify the types of embryos "permitted" to be implanted (Human Fertilisation and Embryology Act Human, 1990, as amended 2008, s3ZA(4)); an embryo that has undergone genome editing would not be considered a "permitted embryo" and its implantation would be illegal. In whatever way it may be recognized, this distinction will also be an important one for other jurisdictions in the regulation of gene editing; as we noted in the Hinxton Group statement on this issue: "concerns about human genome editing for clinical reproductive purposes should not halt or hamper application to scientifically defensible basic research" (Chan *et al.*, 2015).

31.6 Basic vs. Application-directed Research

Can we or ought we to regulate research use of human embryos differently depending on the aims of the research? The second published study on human embryo gene editing (Kang *et al.*, 2016) targeted a gene for HIV resistance; this was criticized because, scientists argued, it was aimed at an eventual reproductive application (Callaway, 2016b). In contrasting this with the work being carried out in the UK under license from the Human Fertilisation and Embryology Authority (the UK regulatory body that oversees uses of human embryos), some scientists sought to draw a distinction between basic research aimed at "answering questions intrinsic to embryology," and research that might serve as "proof-of-principle" for creating a genetically modified human.

Is it reasonable, though, to base our ethical evaluation and regulation of research on supposed end-goals that are very far removed from the actual experiments in question? While specific clinical and preclinical research may be clearly directed at the development of applications, it would be impractical and counterproductive to attribute intention to basic research and to attempt to regulate it on that basis: science is by its nature unpredictable, and any discovery may have a variety of possible uses.

31.7 The Fear of "Slippery Slopes" and the Limitations of Regulatory Fences

A common objection in relation to many technologies is that a seemingly innocuous use will be the first step on a "slippery slope" to (one presumes, although it is often not specified) some disastrous or thoroughly undesirable "pit of despair." Gene editing is no exception: in this case, the allegation is usually that allowing somatic genome editing will inevitably lead to reproductive germline genome editing, or that therapeutic genome editing will inevitably lead to genetic enhancement (Lanphier *et al.*, 2015). It is far from universally agreed that genetic enhancement would necessarily be a bad or undesirable place to end up, but let us grant for the sake of argument that it is. We might still say that the possibility of a slippery slope is not a definitive objection: if we can guard adequately against "slippage,", we need not fear that allowing some uses of gene editing will necessarily open the door to all.

The question then is: how can or should we regulate against the "slippery slope?" Can we guarantee to prevent reproductive germline applications? Or, having decided that clinical reproductive genome editing to cure serious disease were permissible, could we then guarantee to prevent "enhancement" uses? Imagining a hypothetical worst-case scenario, if one "rogue scientist" did decide to take the forbidden step and attempt to engineer a genetically enhanced super-athlete, could we make absolutely sure this would not happen?

Clearly, regulation can provide a strong deterrent by way of penalizing unauthorized uses of genome editing. By creating a framework of operation for permitted research, it can also make it more difficult for research to take place outside of this framework and hence discourage forms of activity that are not permitted. Laws, however, can only do so much; someone willing to go against the consensus of the scientific community and social views in taking the step to reproductive germline genome editing might well be willing to break laws and risk punishment in order to do so. To use an analogy, appropriate regulation can build a fence across the slippery slope, to strongly discourage and to penalize attempts to cross the line, but it cannot stop people from climbing over the fence. Should we then put a stop to genome editing research, for fear of possible fence-climbers?

It is unlikely that halting any further research and arresting the development of genome editing technology at the stage we have currently reached would be effective in preventing such a scenario. Any sufficiently motivated and moderately resourced scientist would likely be able to pursue the necessary research on their own, based on the tools and materials available for genome editing research in general. To this extent, then, the "slippery slope" argument contains a kernel of truth: a certain degree of technological development is necessary for this work to be possible. Such an operation would be much more difficult to carry out in a world where nobody was producing the necessary reagents and materials, without access to DNA synthesis technologies and so on.

In this sense, therefore, any progress in genome editing research can be seen as a step towards the "slippery slope," in that it could *enable* potential work towards germline genetic enhancement. However, this does not mean germline enhancement is an inevitable consequence of genome editing research, nor is it the only consequence that should be taken into account. The progression of science and technology is not a unidirectional slope but a "garden of forking paths"; the point that could lead to the possible future uses we wish to avoid is also the same point that must be reached before we can achieve the benefits of genome editing that we wish to attain.[1]

The "slippery slope," then, even if we cannot absolutely guard against slippage, must be considered in terms of all its possible end-points, not just the worst one. If we focus our evaluation of genome editing solely on one potential undesirable use and ignore all the other, possibly beneficial, uses, we cannot hope to develop a fair or balanced view as to whether we ought to pursue or prohibit this technology. In order to prevent *any* possibility of "slipping," we would have to give up a lot of current and future worthwhile research to ensure that we were not incidentally enabling a designer baby. We must ask ourselves whether giving up the benefits would be worth it to avert this harm, if it even is a harm; I strongly suspect it would not.

Furthermore, the fears implied in invoking the slippery slope argument often relate not to misuse by one or two renegade scientists, but to the possibility of widespread societal use and acceptance of a practice which some currently believe to be unacceptable. There is, then, a way in which regulation can be effective against the slippery slope, though not in forever closing off a possibility that we currently view as undesirable – indeed, regulation ought not to do this: imagine if our more conservative forebears had somehow managed to inscribe the permissibility of slavery, or a ban on homosexuality or miscegenation, or permanently into law. The purpose of regulation is not to foreclose any possibility of moral evolution and bind our future societies to the (possibly erroneous) beliefs of the present. What it should do, however, is to provide confidence in negotiating "slippery" terrain, and ensure that we do not "slip" before we are ready. This role of the "regulatory handrail" in providing guidance and surety through morally contentious ground is an important one.

31.8 Scientific Solutions to Ethical Problems?: The Role of Regulation

An occasional response to ethical conflict has been to attempt to "hardwire" regulatory or ethical solutions into the science itself. While this may seem a useful additional safeguard to supplement laws and guidelines, a closer examination reveals that it is not always appropriate or effective.

For example, in an attempt to defuse ethical concerns about germline editing, the early Chinese embryo genome editing studies (Kang *et al.*, 2016; Liang *et al.*, 2015) used tri-pronuclear (3PN) embryos, which would not develop past a certain stage and certainly not into genetically modified children. But, as noted above, the best way to ensure that genome-edited embryos do not develop into children is simply not to implant them. Why, then, use non-viable embryos for this study, where implantation was never intended? Indeed, the study was criticized scientifically for this: the differences of 3PN embryos make them a poor model for a proof of principle intended to apply to normal diploid embryos, and they do not solve any moral problems. Of course embryo research is controversial – but if the embryo itself is what is of moral value, the use of "spare" (supernumerary) embryos or embryos that will not develop past a certain point is not a real solution to the ethical problem of embryo genome editing, any more than to the problem of embryonic stem cell research (Holm, 1993).

Another attempt at "ethical hardwiring" is seen in the US National Academy of Sciences report on mitochondrial replacement therapy, which recommends that the technique should be allowed, but be limited to the creation of male children only "as a safety precaution" (Reardon, 2016). Because mitochondria are maternally inherited, this is intended to ensure that the "modification" (not a sequence change as such, but a novel combination of mitochondrial and nuclear DNA) is not passed on. But this apparently ingenious use of biological properties to limit the transmission of donated mitochondria masks what appears to me a puzzling ethical inconsistency. We may be worried about limiting the spread of any chance and as-yet-unknown adverse effects of MRT that might be caused by the combination of inherited nuclear genes and donated mitochondria. But what about women who choose not to undertake MRT – perhaps, in the face of this proposed restriction, because they wish to have daughters as well as sons? Is it ethically neutral for their children to pass on the very likely and well-characterized adverse effects of having a mitochondrial disease? Are we more afraid of the unknown than of a known serious harm, and is this fear sufficient to impose such a strong constraint on reproductive liberty for those women who do seek to use MRT for the sake of their children's health?

A final example: in June 2016 the "HGP-Write" project (Boeke *et al.*, 2016), with the ultimate stated aim of synthesizing a human genome, was launched amidst a flurry of ethical attention (Callaway, 2016a; Endy and Zoloth, 2016; Pollack, 2016). Critics jumped (perhaps too rapidly) from the idea of creating a synthetic genome to creating synthetic humans, dredging up fears about human cloning, parentless children, genetically engineered bioweapons and other familiar scare-tropes along the way (Endy and Zoloth, 2016). Responding to these concerns, project leader Jef Boeke sought to allay concerns by emphasizing that the aim was to synthesize cells, not entire organisms, and proposing the introduction of a "germline firewall" by which such cells would be "engineered to make reproduction impossible"(Callaway, 2016a), for example by altering their karyotype so that germ cells and gametes could not be produced.

With HGP-Write, as with genome editing, however, the worry is not about "the germline" in a scientific sense, but about the prospect of genetically modified children who will pass those modification on to their descendants. The possibility of creating germ cells, gametes and even embryos with edited or synthetic genomes should be no cause for concern, so long as the end-goal is not reproduction: the scientific and ethical concepts at stake in these debates are not always the same. Further, although the ethical objections were dressed in the flamboyant guise of "synthesis[ing] Einstein's genome," the real and far more immediate concern was the apparent lack of ethical consideration and wider engagement in relation to the project, something that a scientific solution does nothing to address.

Using science in place of ethical deliberation, engagement and appropriate regulation is thus the wrong tool for the job: we can perhaps engineer biosafety, but not safety from "ethical dangers." Attempting to use science for regulatory hardwiring assumes that the ethical problem is susceptible to a purely scientific solution, when often this is not the case.

Furthermore, to attempt to address the ethical issues via scientific design-based regulation often fails because it mistakes the role of regulation itself. Regulation does not solve the ethical problem; indeed in many cases it leaves the controversial moral issues unsolved. Writing the 14-day limit for the use of human embryos into law did not mean that embryos have no moral value before 14 days of development and suddenly acquire inviolable moral status after this point. What regulation does, or ideally should do, is provide a practical means of negotiating ethically contested territory, promoting mutual trust and enabling further discourse between stakeholders with different interests. This has been the case for regulation of embryo research in the UK (Brazier, 1999; Wilson, 2011) and has proven to be relatively successful, to date at least. We should demand the same of regulation in relation to gene editing.

Acknowledgement

The author would like to acknowledge the Arts and Humanities Research Council for supporting a travelling fellowship that contributed to this work (AH/N000579 "Technology, ethics and human nature").

NOTE

A summary of various regulatory positions on stem cell research in European countries can also be found at http://www.eurostemcell.org/stem-cell-regulations.

1. I have previously made this point in a slightly different way (Chan and Medina Arellano, 2016).

REFERENCES

Boeke JD, Church G, Hessel A, *et al.* 2016. Genome engineering: the Genome Project-Write. *Science* **353**: 126–127.

Brazier M. 1999. Regulating the reproduction business? *Med Law Rev* **7**: 166–193.

Callaway E. 2016a. Plan to synthesize human genome triggers mixed response. *Nature* **534**: 163.

Callaway E. 2016b. Second Chinese team reports gene editing in human embryos. *Nature* **530**: 18.

Chan S, Donovan PJ, Douglas T, *et al.* 2015. Genome editing technologies and human germline genetic modification: the Hinxton Group Consensus Statement. *Am J Bioeth* **15**: 42–47.

Chan S, Harris J. 2006. The ethics of gene therapy. *Curr Opin Mol Ther* **8**: 377–383.

Chan S, Medina Arellano M-D-J. 2016. Genome editing and international regulatory challenges: lessons from Mexico. *Ethics Med Publ Health* **2**(3): 426–434.

Endy D, Zoloth L. 2016. Should we synthesise a human genome? *Cosmos*, 12 May, https://cosmosmagazine.com/society/should-we-synthesise-a-human-genome (accessed August 2016).

Ginn SL, Alexander IE, Edelstein ML, Abedi MR, Wixon J. 2013. Gene therapy clinical trials worldwide to 2012: an update. *J Gene Med* **5**: 65–77.

Hacein-Bey-Abina S, Garrigue A, Wang GP, *et al.* 2008. Insertional oncogenesis in 4 patients after retrovirus-mediated gene therapy of SCID-X1. *J Clin Invest* **118**: 3132–3142.

Holm S. 1993. The spare embryo: a red herring in the embryo experimentation debate. *Health Care Anal* **1**: 63–66; discussion 67–68.

Human Fertilisation and Embryology Act 1990. www.legislation.gov.uk/ukpga/1990/37/contents (accessed November 2017).

Isasi RM. 2009. Policy interoperability in stem cell research: demystifying harmonization. *Stem Cell Rev* **5**: 108–115.

Isasi RM, Knoppers BM. 2006. Mind the gap: policy approaches to embryonic stem cell and cloning research in 50 countries. *Eur J Health Law* **13**: 9–25.

Kang X, He W, Huang Y, *et al.* 2016. Introducing precise genetic modifications into human 3 PN embryos by CRISPR/Cas-mediated genome editing. *J Assist Reprod Genet* **33**: 581–588.

Kimmelman J. 2008. The ethics of human gene transfer. *Nat Rev Genet* **9**: 239–244.

Lanphier E, Urnov F, Haecker SE, Werner M, Smolenski J. 2015. Don't edit the human germ line. *Nature* **519**: 410–411.

Liang P, Xu Y, Zhang X, *et al.* 2015. CRISPR/Cas9-mediated gene editing in human tripronuclear zygotes. *Protein Cell* **6**: 363–372.

Macklin R. 2003. Dignity is a useless concept. *BMJ* **327**: 1419–1420.

Marshall E. 1999. Gene therapy death prompts review of adenovirus vector. *Science* **286**: 2244–2245.

Pollack A. 2016. Scientists talk privately about creating a synthetic human genome. *New York Times*, 13 May, www.nytimes.com/2016/05/14/science/synthetic-human-genome.html (accessed August 2016).

Reardon S. 2016. US panel greenlights creation of male "three-person" embryos. *Nature* **530**: 142.

Savulescu J. 2001. Harm, ethics committees and the gene therapy death. *J Med Ethics* **27**: 148–150.

UNESCO. 1998. Universal Declaration on the Human Genome and Human Rights. http://unesdoc.unesco.org/images/0012/001229/122990eo.pdf (accessed November 2017).

Wilson D. 2011. Creating the "ethics industry": Mary Warnock, *in vitro* fertilization and the history of bioethics in Britain. *Biosocieties* **6**: 121–141.

Index

CPSIA information can be obtained
at www.ICGtesting.com
Printed in the USA
LVHW061935090821
694911LV00005B/518